The Science of Botany

The Science of Botany

Edited by **Kiara Woods**

Callisto Reference
New York

Published by Callisto Reference,
106 Park Avenue, Suite 200,
New York, NY 10016, USA
www.callistoreference.com

The Science of Botany
Edited by Kiara Woods

International Standard Book Number: 978-1-63239-720-1 (Hardback)

Printed in China.

Contents

Preface

I am honored to present to you this unique book which encompasses the most up-to-date data in the field. I was extremely pleased to get this opportunity of editing the work of experts from across the globe. I have also written papers in this field and researched the various aspects revolving around the progress of the discipline. I have tried to unify my knowledge along with that of stalwarts from every corner of the world, to produce a text which not only benefits the readers but also facilitates the growth of the field.

Botany refers to the branch of science that deals with the study of plants. It is an extensive research based study of plant biology. Some of the techniques included in this vast field are live cell imaging, optical microscopy, electron microscopy, etc. This book will provide detailed analysis and data about this subject. It attempts to understand the multiple branches that fall under the discipline of botany and gives detailed information about them. This book is a valuable compilation of topics, ranging from the basic to the most complex advancements in this field. Students and scientists actively engaged in this field will find this book full of crucial and unexplored concepts. It will also serve as a beneficial source of reference for botanists, agriculturists and professionals associated with this field.

Finally, I would like to thank all the contributing authors for their valuable time and contributions. This book would not have been possible without their efforts. I would also like to thank my friends and family for their constant support.

Editor

1

Arabidopsis Qc-SNARE genes *BET11* and *BET12* are required for fertility and pollen tube elongation

Pablo Bolaños-Villegas[1*], Cian-Ling Guo[2] and Guang-Yuh Jauh[2*]

Abstract

Key message: *BET11* and *12* are required for pollen tube elongation.

Abstract: Pollen tubes are rapidly growing specialized structures that elongate in a polar manner. They play a crucial role in the delivery of sperm cells through the stylar tissues of the flower and into the embryo sac, where the sperm cells are released to fuse with the egg cell and the central cell to give rise to the embryo and the endosperm. Polar growth at the pollen tube tip is believed to result from secretion of materials by membrane trafficking mechanisms. In this study, we report the functional characterization of Arabidopsis *BET11* and *BET12,* two genes that may code for Qc-SNAREs (soluble *N*-ethylmaleimide-sensitive factor attachment protein receptors). Double mutants (*bet11/bet12*) in a homozygous/heterozygous background showed reduced transmission of the mutant alleles, reduced fertilization of seeds, defective embryo development, reduced pollen tube lengths and formation of secondary pollen tubes. Both *BET11* and *BET12* are required for fertility and development of pollen tubes in *Arabidopsis*. More experiments are required to dissect the mechanisms involved.

Keywords: *Arabidopsis thaliana*, SNAREs, Fertility, Pollen tube elongation

Background

Unlike in animals, sperm cells from flowering plants are non-motile and need to be delivered to the ovule by means of a pollen tube (Kawashima and Berger 2011). Once delivered, one sperm fuses with the egg cell to result in the diploid zygote and the other fuses with the secondary nucleus to generate the triploid endosperm (Berger et al. 2008; Kawashima and Berger 2011).

For the pollen tube to perform this task, it needs to elongate in a polar way. This growth mechanism requires highly coordinated interactions between the secretory machinery, the cell wall biosynthetic machinery and the cytoskeleton (Domozych et al. 2013). For instance, newly synthesized proteins first associate with the endoplasmic reticulum (ER), then are transported to the appropriate subcellular compartment (El-Kasmi et al. 2011). This process involves the fusion of ER-derived vesicles with the cis-Golgi cisternae, followed by protein sorting to the plasma membrane or lysosomes at the trans-Golgi network (TGN), a compartment also involved in endocytosis (El-Kasmi et al. 2011; Morita and Shimada 2014). The TGN performs two main functions: it sorts cargo destined for the plasma membrane and endosomes and receives cargo from endosomal compartments (Uemura and Nakano 2013).

Several proteins take part in membrane fusion events that occur in tip-growing plant cells. Examples are Rab-GTPases, which regulate the membrane recruitment and activity of tethering factors; Rab effectors such as phosphatidylinositol kinases; ATP-driven chaperones from the Sec1/Munc18 family, also called SM proteins; and soluble *N*-ethylmaleimide-sensitive factor attachment protein receptors (SNAREs) (Alpadi et al. 2012; Guo and McCubbin 2012; Karnik et al. 2013; Ohya et al. 2009). SNAREs are believed to be the principal determinants of membrane fusion specificity (De Benedictis et al.

*Correspondence: pablo.bolanosvillegas@ucr.ac.cr; jauh@gate.sinica.edu.tw
[1] Fabio Baudrit Agricultural Experimental Station, University of Costa Rica, La Garita de Alajuela, P.O. Box 183-4050, Alajuela, Costa Rica
[2] Institute of Plant and Microbial Biology, Academia Sinica, 128 Sec. 2, Academia Rd, Nankang, Taipei 11529, Taiwan
Full list of author information is available at the end of the article

2013). These proteins, 150-300 amino acids long, feature a membrane-anchored C-terminus, α-helical heptad repeats, and a cytosolic SNARE motif that functions during membrane fusion (Guo and McCubbin 2012; Uemura et al. 2012). A fusogenic SNARE complex comprises four SNAREs, including three with a central glutamine (Gln, Q) residue within the SNARE motif (the Q SNAREs), classified as Qa-, Qb-, and Qc-, and the R-SNARE, which features a central arginine residue (Arg, R) (Fujiwara et al. 2014; Uemura et al. 2012). In these complexes, the Q-SNAREs usually reside on the target membrane, whereas the R-SNARE resides on the vesicle membrane (Karnik et al. 2013). *trans*-QabcR–SNARE complexes are believed to function in individual intracellular fusion steps (Furukawa and Mima 2014), their topology varies in stringency (Alpadi et al. 2012), and formation of a fusion-competent membrane microdomain requires the presence of specific lipids such as sterols, diacylglycerol, and phosphoinositides (Furukawa and Mima 2014).

Arabidopsis has about 60 different genes in six groups that encode SNARE proteins (Lipka et al. 2007); most are members of protein families with more than one gene (Sanderfoot 2007). This large number of genes may be associated with the need in higher plants for polarized secretion (Sanderfoot 2007). Nonetheless, despite the sheer number of genes, most Q-SNAREs, including the Qa-SNAREs, show partial functional redundancy (Fujiwara et al. 2014; Morita and Shimada 2014). Examples are the *Arabidopsis* Qa-SNARE genes *SYP41, SYP42,* and *SYP43,* which are involved in secretory and vacuolar transport and are needed to maintain the morphology of the Golgi apparatus and the TGN (Uemura et al. 2014). Several other SNARE genes involved in post-Golgi membrane trafficking include the Qb-SNAREs *VTI11* and *12* and the Qc-SNAREs *SYP51* and *52* (De Benedictis et al. 2013). Transient expression of SYP51 and SYP52 in protoplasts revealed functional redundancy in vesicle sorting to the vacuole, localization to the TGN and endocytic compartments, and an inhibitory effect on fusion when accumulated on the tonoplast (De Benedictis et al. 2013).

In this study, we functionally characterized two *Arabidopsis* genes, *BET11* and *BET12* (At3g58170 and At4g14455), believed to encode Qc-SNAREs (Uemura et al. 2004). These genes are Bet1/Sft1-like SNAREs that share 60 % aminoacid sequence identity with yeast *Sft1* gene, and when overexpressed are able to suppress the temperature-sensitive growth defect in *sft1-1* (Tai and Banfield 2001). A previous study determined that these two Arabidopsis genes (previously known as AtBS14a/b) are expressed in all plant tissues, including flowers, leaves, stem, roots and suspension cells (Uemura et al. 2004). We found that both genes were required for fertility, and that the N-terminal GFP fusion proteins localized to the Golgi apparatus, as observed in WT protoplasts. These results are in agreement with results by Uemura et al. (2004), who found that treatment of protoplasts with BFA, an inhibitor of Arf-GTPase guanine nucleotide exchange factors, caused accumulation of BET11-GFP and BET12-GFP at the Golgi. In that report the Golgi apparatus was labeled with marker Venus-SYP31, and colocalized well with BET11 and BET12 (Uemura et al. 2004). Our characterization of phenotypes in stable T-DNA double mutants suggested that the combined activity of *BET11* and *BET12* may be required for embryo development and proper pollen tube extrusion. Nonetheless, more work may be required to elucidate mechanistic and regulatory details related to the role the genes play at the Golgi apparatus in pollen tubes.

Methods

Molecular analysis of *BET11/12*

From a screen for potential gametophytic mutants, we chose the *Arabidopsis At3g58170* and *At4g14455* loci (*BET11* and *BET12*) for study (Additional file 1: Table S1). Then from a subset of 18 mutant lines corresponding to 12 SNARE genes, we chose the Columbia T-DNA lines SAIL_509_C09 (*bet11-1*) and SALK_124063 (*bet12-1*) for study. Genes were selected by high expression in *Arabidopsis* pollen as reported by GENEVESTIGATOR (https://www.genevestigator.com). The respective amino acid sequences were analyzed by T-Coffee, a consistency-based multiple sequence alignment program (Di Tommaso et al. 2011) and the on-line repositories UNIPROT (http://www.uniprot.org) and SWISS-MODEL (swiss-model.expasy.org). Plants from each line were genotyped with specific primer pairs for their corresponding T-DNA inserts and WT locus. Single homozygous lines did not show obvious vegetative or reproductive phenotypes, but homozygous/heterozygous (HM/HZ) double mutants for *bet11-1* and *bet12-1* (and vice versa) showed reduced seed set. No double HM mutants were recovered.

Plant materials and growth conditions

The WT (ecotype Columbia) and T-DNA insertional lines, SAIL_509_C09 (*bet11-1*), and SALK_124063 (*bet12-1*) were obtained from the Arabidopsis Biological Resource Center (Columbus, OH, USA; http://abrc.osu.edu). Seeds were surface-sterilized in 30 % sodium hypochlorite and germinated on half-strength Murashige and Skoog medium without sucrose, then stratified at 4 °C for 96 h in the dark. Seedlings were grown at 21 °C under a 16-h photoperiod and 60 % relative humidity for approximately 5 days after the emergence of the radicle. Seedlings were then transferred to soil and genotyped with specific primer pairs for their corresponding T-DNA inserts and the

WT locus. The primer sequences were for BET11-LP, 5′-GAGTAAGCCTGCCTCTGGTTC-3′, BET11-RP: 5′-TAGTACCCTGCCACGGTACAG-3′; BET12-LP: 5′-TCAAGCAAGCGGTTATGATTC-3′, BET12-RP: 5′-CACGAAAACTTACGCTTCTGG-3′, LB1:5′-GCCT TTTCAGAAATGGATAAATAGCCTTGCTTCC-3′, and LBb1.3: 5′-ATTTTGCCGATTTCGGA AC-3′. Complementation lines on the double mutant backgrounds were developed by transformation with the vector pZP221 carrying the native promoters and the full genomic sequence for either *BET11* or *BET12*, with a *GFP* tag on the N-terminus. Seedlings from the T_1 generation were selected on agar based on resistance to antibiotic G418 and genotyped for the corresponding T-DNA insert and GFP sequence. A modified LP primer was used for complemented *bet12-1* lines. The sequence was as follows: BET12-modLP: 5′-CGGTTGGTTCACTAGTCTCT-3′.

Quantitative PCR (qPCR)

Total RNA was extracted with the RNAeasy Plant Minikit (Qiagen, http://www.qiagen.com/) from WT, *bet11-1* and *bet12-1* seedlings. First-stranded cDNA was prepared from total RNA with the murine leukemia virus reverse transcriptase system (Promega, http://www.promega.com/). For quantitative PCR, a Power SYBR Green I Master Mix (Applied Biosystems, http://www.appliedbiosystems.com) was used with 150–200 nM primers, 20 ng/μL cDNA, and 50 μL reverse transcriptase reaction product. Reactions were run and analyzed on the AB 7500 Real Time PCR System (Applied Biosystems). Melting curve analyses and negative controls were used to exclude artifacts and low specificity. Reactions were performed in triplicate and averaged. Primers specific for the 3′ end of transcripts were designed by use of PRIMER EXPRESS 3.0 (Applied Biosystems). The primer sequences were for BET11, forward, 5′-CCATAGATCCAGGTGAATTCTGG-3′, and reverse, 5′-GCG GTTATGAGTATCGACCTCTTC-3′; BET12, forward, 5′-AGAGACTAGTGAACCAACCGA-3′, and reverse, 5′-ACAAAACTGCTAACATGAACCCA-3′; and ACT2, forward, 5′-GGCTCCTCTTAACCCAAAGGC-3′, and reverse, 5′-CACACCATCACCAGAATCCAGC-3′ as a normalization control.

Analysis of subcellular localization of BET11 and BET12

Protoplasts prepared from leaves of 4-week-old Arabidopsis plants were co-transformed with *proBET11:BET11:GFP*, *proBET12:BET12:GFP*, Golgi marker construct *35S:Man1:GFP* and nuclear marker construct *35S:ERF4:mRFP* (not shown). Transformed protoplasts were observed by two-photon laser confocal microscopy and analyzed with use of Zeiss LSM Image Browser 3.5. *In planta* observations were not carried out.

Genetic analyses

BET11/12 double HM/HZ or HZ/HM mutants (*bet11/−bet12/+*, *bet11/+bet12/−*) were cross-pollinated with the WT to determine transmission efficiency of *bet11-1* and *bet12-1* alleles. The mutants were used as pollen receptors and donors. Transmission of the T-DNA alleles was analyzed by PCR with sampling of 100 progenies per cross with the WT and negative controls. Results were used to calculate the expected Mendelian segregation rate (E), the actual observed segregation rate (O) and the transmission efficiency rate (O/E × 100).

Morphological characterization of *bet11-1* and *bet12-1* single and double mutants

Images of seed sets were recorded after dissection of at least 30 siliques from 7-week-old plants under a Lumar V12 fluorescence stereomicroscope (Zeiss, http://microscopy.zeiss.com/microscopy) connected to an AxioCam MRc5 CCD unit. In vitro pollen germination was performed according to Lin et al. (2014). Pollen was incubated in the dark for 4 h at 25 °C, then observed under a light microscope. Pollen was collected from at least three different plants, and 100 pollen grains were used to estimate pollen viability, with three replicates. Analysis of in vivo pollen-tube growth was as described (Szumlanski and Nielsen 2009; Lin et al. 2014). Emasculated pistils from (a) WT, (b) single homozygous *bet11-1* and *bet12-1*, (c) *bet11/− bet12/+* and (d) *bet11/+ bet12/−* were cross-pollinated and collected after 48 h. Mature pollen morphology was analyzed by staining with 1.5 mg/mL 4′, 6-diamino-2-phenylindole (DAPI) (Vector Laboratories, http://www.vectorlabs.com) solution for 15 min in the dark, then observed under an Olympus BX51 epifluorescence microscope coupled to an Olympus DP70 CCD unit (Olympus, http://www.olympus-global.com/en/corc/company/lifescience).

Results

The mutant alleles *bet11-1* and *bet12-1* show distorted segregation rates

A preliminary in silico analysis of the expression of several Arabidopsis fusogenic genes by Genevestigator (http://www.genevestigator.org) revealed that some were differentially expressed in reproductive tissues. *BET11* (At3g58170) showed expression in pollen, which suggested a role during plant reproduction, while no data was available for *BET12* (At4g14455). Both genes encode SNARE proteins of the Qc family, whose members feature a conserved glutamine residue within the coiled-coil SNARE motif (Uemura et al. 2004). T-DNA insertional mutants were available for each gene: line *bet11-1*, containing an insertion in exon 5 of *BET11* (Fig. 1a) and *bet12-1*, containing an insertion in exon 2 of *BET12* (Fig. 1a).

Our analyses of the phenotype of *bet11-1* and *bet12-1* single homozygous plants did not reveal any obvious defects in vegetative or reproductive growth. Moreover, PCR genotyping indicated that the mutants segregated in a 3:1 Mendelian ratio, which suggests that transmission of the T-DNA allele was more or less efficient. We could not isolate double homozygous *bet11/bet12* mutants. In fact, the corresponding homozygous/heterozygous mutants segregated in highly distorted, non-Mendelian ratios. In the *bet11/−bet12/+* mutant (HM/HZ), the *bet11-1*

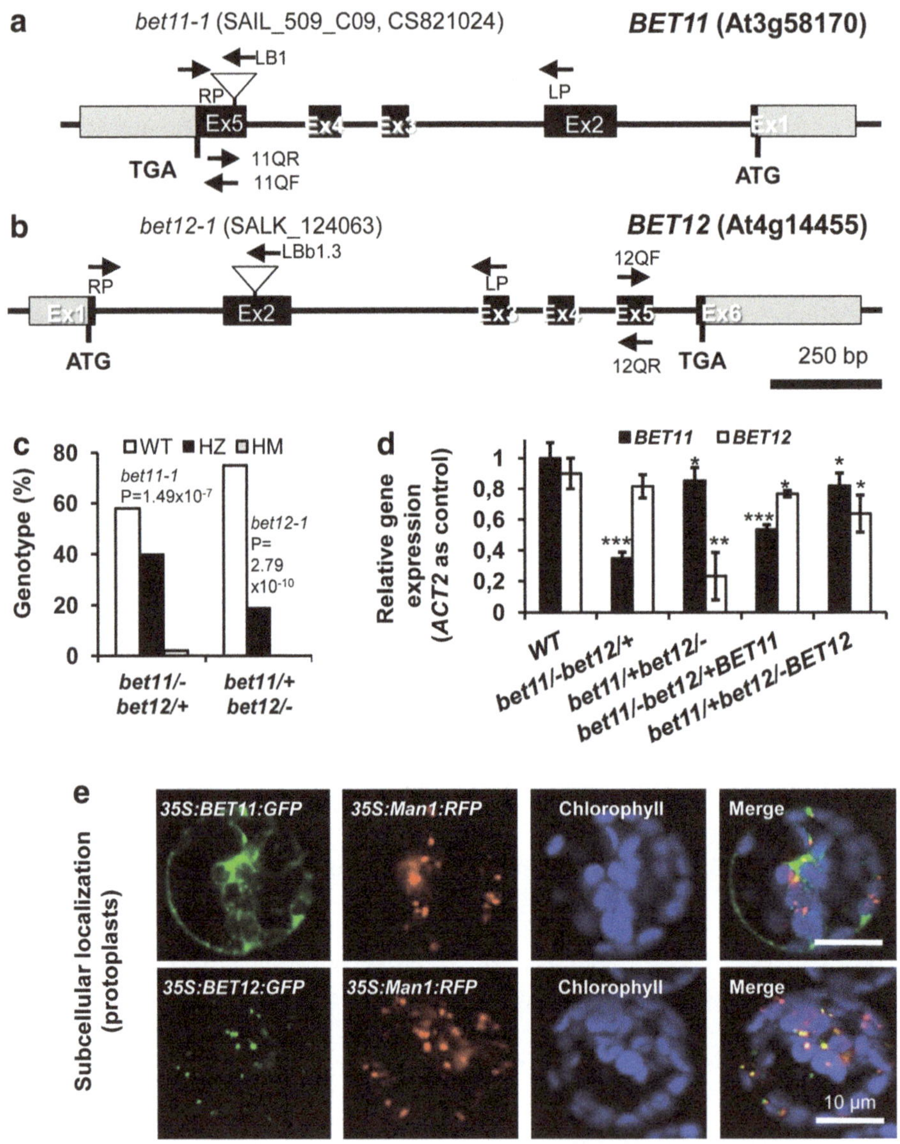

Fig. 1 Molecular and cellular characterization of *Arabidopsis BET11* and *BET12* genes. **a** Diagram of *BET11* gene (At3g58170); exons appear as *black boxes*. The T-DNA insert in mutant line *bet11-1* (SAIL_509_C09) is on exon 5, the ATG start codon appears to the *right*, and the TGA stop codon is to the *left*. UTRs appear in *grey*. The location of the complementary primer sequences for genotyping and quantitative PCR appear as *arrowheads*. **b** Diagram of *BET12* gene (At4g14455). The T-DNA insert in mutant line *bet12-1* (SALK_124063) is on exon 2. **c** Segregation of *bet11-1* and *bet12-1* alleles in double mutants. Non-Mendelian segregation of the homozygous alleles is statistically significant, which suggests severe defects in transmission. **d** Quantification of relative expression of the 3′ end of *BET11* and *BET12* in the WT, double mutants and complementation lines. **e** Subcellular localization of BET11 and BET12 N-terminal, GFP-tagged proteins in *Arabidopsis* protoplasts. Both proteins showed partial colocalization with the Golgi marker Man1 (Man1:RFP). *Scale bars* 10 μm. Data are mean ± SD ($n = 3$) from three biological samples. *$P < 0.1$, **$P < 0.01$, ***$P < 0.001$ (Student *t* test) compared to the WT

T-DNA allele segregated with $P = 1.49 \times 10^{-7}$, and in the *bet11/+ bet12/−* mutant (HZ/HM), the *bet12-1* T-DNA allele segregated with $P = 2.79 \times 10^{-10}$ (Fig. 1c). These extremely significant values suggested reduced efficiency in the transmission of the T-DNA alleles caused by gametophyte or embryo lethality.

To better understand the reason for these defects, we analyzed the expression of the *BET11* and *BET12* genes in the WT and corresponding complementation lines, which carried the full genomic sequence driven by the native promoter. The expression of both genes was lower in the mutants than the WT (Fig. 1d). For instance, for homozygous *bet11-1*, the expression of *BET11* was reduced to 0.35 ± 0.04 (vs. 1.00 ± 0.10 in the WT) and 0.82 ± 0.09 in the heterozygous background, whereas for homozygous *bet12-1*, the expression of *BET12* was reduced to 0.23 ± 0.15 (vs. 0.90 ± 0.10 in the WT), and 0.77 ± 0.07 in the heterozygous background. For the complementation lines, the expression of both genes was increased to 0.54 ± 0.06 and 0.64 ± 0.12, respectively (Fig. 1d). Therefore, the *bet11/−* and *bet12/−* alleles appeared to be knockdown alleles that show reduced expression at the 3′ end.

BET11 and BET12 proteins may reside at the Golgi apparatus

To better understand the function of the putative BET11 and BET12 proteins, we examined the localization patterns of their respective N-terminal, GFP-fusion proteins in *Arabidopsis* leaf protoplasts isolated from 4-week old *Arabidopsis* plants. Both GFP-tagged proteins showed some level of colocalization with the Golgi marker Man1 (Man1:RFP), especially BET12, which suggests that both proteins may reside at the Golgi apparatus (Fig. 1e). These results are in line with previous work (Uemura et al. 2012) and are in agreement with bioinformatic modeling of the function of both genes. For instance, alignment of amino acid sequences by use of T-Coffee suggested conservation between the putative Arabidopsis BET11 and BET12 proteins compared to the yeast homolog Sft1 (Additional file 2: Figure S1a). Results from the UNIPROT database indicated that both loci appear to encode canonical Qc-SNARE proteins that feature a SNARE coiled-coil domain, a transmembrane domain, a vesicular topological domain, and for BET12, a predicted phospho-serine site at residue 56 (Additional file 2: Figure S1b–c) that may regulate posttranslational dynamics. Moreover, SWISS-MODEL results suggest that each locus encodes a single chain able to assemble as a heterotetramer (Additional file 2: Figure S1b–c). This high level of similarity between both predicted proteins may indicate functional redundancy.

Transmission of the homozygous mutant alleles is reduced in pollen

To better determine the extent of *BET11* and *BET12* function, we cross-pollinated *bet11/bet12* double mutants and the WT. Seeds were collected and progenies were genotyped by PCR to calculate the transmission efficiency of the *bet11-1* and *bet12-1* alleles (Table 1). For the *bet11* HM/*bet12* HZ line, maternal transmission of the heterozygous *bet12-1* allele showed distorted segregation of 21 % and a transmission efficiency of 42 %, which suggests defects in the maternal transmission of the allele. Even lower values in transmission of *bet12-1* occurred when the allele was transmitted paternally through pollen (22 %). For the *bet11* HZ/*bet12* HM line maternal transmission of the heterozygous *bet11-1* allele led to a segregation rate of 12 % and a transmission efficiency of 24 %. Paternal transmission of the heterozygous *bet11-1* allele led to a further reduction in the segregation rate, to 7 %, and a transmission efficiency of 14 %. Therefore,

Table 1 Transmission of the heterozygous *bet11-1* and *bet12-1* alleles is impaired in *bet11/12* double mutants

Cross	Segregating T-DNA allele	Expected segregation rate (E)	Observed segregation rate (O)	Transmission efficiency (O/E) × 100 (%)
bet11 HM/*bet12* HZ (female) × WT (male)	*bet12*	50	21	42
WT (female) × *bet11* HM/*bet12* HZ	*bet12*	50	11	22
bet12 HM/*bet11* HZ (female) × WT (male)	*bet11*	50	12	24
WT (female) × *bet12* HM/*bet11* HZ	*bet11*	50	7	14

Results show the segregation rates of the *bet11-1* and *bet12-1* alleles after reciprocal crosses involving the wild type (WT) and *bet11/bet12* double mutants in either the homozygous/heterozygous (HM/HZ) or the HZ/HM condition. From left to right, the table indicates the expected Mendelian segregation rate (E), the actual observed segregation rate (O) and the transmission efficiency (O/E × 100). For the *bet11* HM/*bet12* HZ line, maternal transmission efficiency of the *bet12-1* heterozygous (HZ) allele was low (42 %). Even lower values in the transmission efficiency of *bet12-1* allele were observed when the T-DNA allele was transmitted paternally through pollen (22 %). For the *bet11* HZ/*bet12* HM line, maternal and paternal transmission efficiencies of the heterozygous *bet11-1* allele were low (24 and 14 %). Data was collected from 100 segregating seedlings per cross simultaneous reduction in activity of *BET11* and *BET12* genes may lead to impaired function of the male gametophyte and defects in embryo development.

BET11 and *BET12* genes are required for fertilization

Analysis of seed development in mutant siliques from double mutants indicated that the *bet11/−bet12/+* (HM/HZ) mutant showed significant failure to fertilize (white arrowheads, Fig. 2a), whereas the *bet11/+ bet12/−* (HZ/HM) mutant showed both fertilization failure and embryo developmental defects (white and orange arrowheads, Fig. 2a). The respective complementation lines (HM/HZ backgrounds) showed normal seed development. On analysis of seeds per silique, the WT and single mutants *bet11/−* and *bet12/−* had a mean of 47–43 seeds per silique, whereas the *bet11/−bet12/+* and *bet11/+ bet12/−* double mutants had only about half the amount of seeds, 20–21 ($P < 0.01$ by Student's *t* test). Moreover the *bet11/+ bet12/−* double mutant had a mean of 1–3 defective seeds per silique, so the *bet11/−* and *bet12/−* alleles may affect fertility differently, for instance *bet11-1/− bet12/+* showed fertilization failure in 36 % of all samples (grey column segment, Fig. 1b), while *bet11-1/+ bet12-1/−* (HM) showed both embryo developmental defects and fertilization failure in 11 and 17 % of all samples, respectively (black column segment, Fig. 2b). These results strongly suggest that both *BET11* and *BET12* are required for fertility and embryo development.

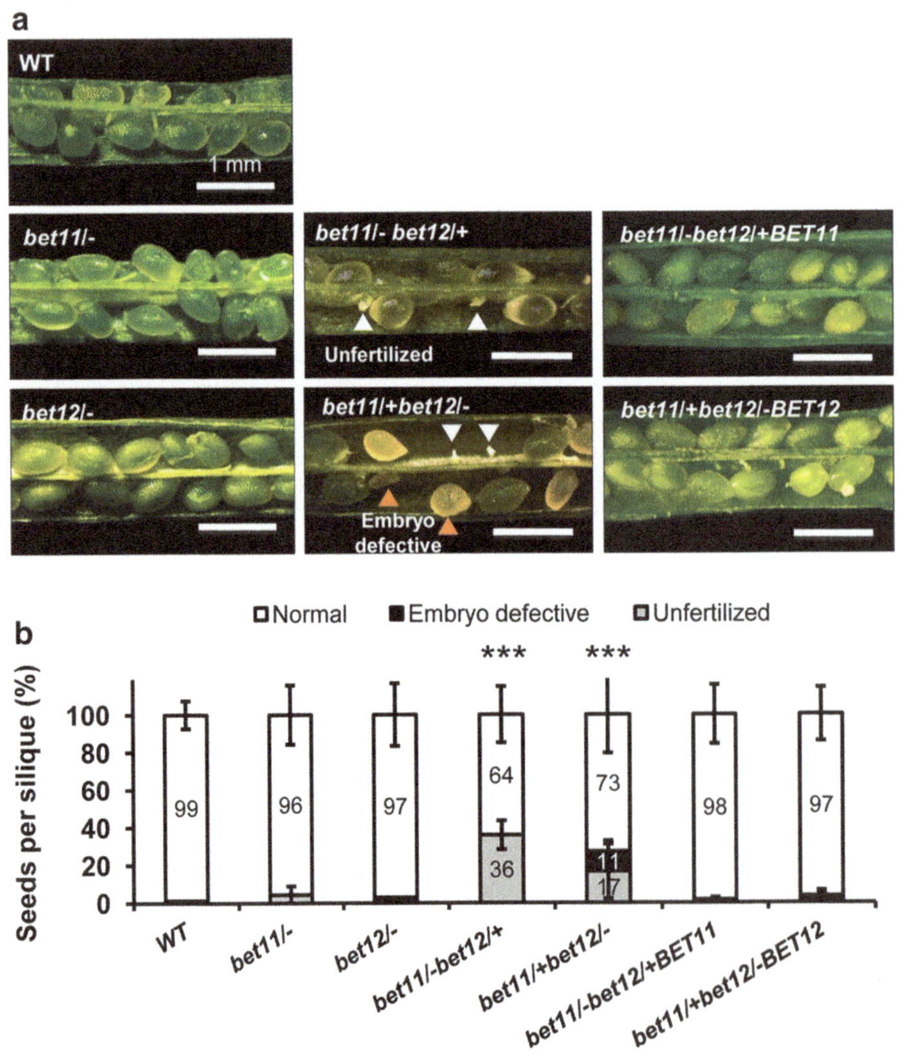

Fig. 2 *Arabidopsis BET11* and *BET12* genes are required for fertility. **a** Seed development in the WT (*left*), single homozygous T-DNA lines (*left*), double mutants (homozygous/heterozygous) (*middle*) and complementation lines (*right*). *Scale bar* 1 mm. Data are mean ± SD ($n = 3$) from three biological samples. *$P < 0.1$, **$P < 0.01$, ***$P < 0.001$ (Student's *t* test) compared with the WT

BET11 and *BET12* genes are required for pollen tube elongation

In order to study pollen tube growth dynamics, pollen samples were first germinated in vitro and then were analyzed under the light microscope. In general, results from in vitro germination experiments showed that the single mutants exhibited reduced pollen tube length, and that this reduction became more severe in *bet11/bet12* double mutants (Fig. 3Aa–Ac); moreover, about 4 % of double mutant pollen grains extruded a secondary pollen tube (Fig. 3Ad–Ae). A tube of 100–200 μm long developed in about 60 % of WT pollen grains, and about 33 % grew tubes longer than 200 μm (Fig. 3B). However, in *bet11-1* and *bet12-1* single homozygous mutants, many of the pollen tubes (33–46 %) were 50–100 μm, followed by 100–200 μm (33–21 %), and <50 μm (7–25 %). The remainder were 100–200 μm. In the *bet11/bet12* double mutants (HM/HZ, and HZ/HM) no pollen tubes were >200 μm. The germination rate of the mutants was also different. For instance, in the WT about 77 % of pollen grains were able to germinate (Fig. 3C), but in *bet11-1* and *bet12-1* single homozygous mutants, this rate decreased to 63 and 43 %, respectively. In double mutants, the germination rate decreased significantly to 25 % in *bet11/- bet12/+* and only 20 % in *bet11/+ bet12/−* (Fig. 3C). Thus, *BET11* and *BET12* activity may be crucial for proper pollen tube extrusion and elongation, and

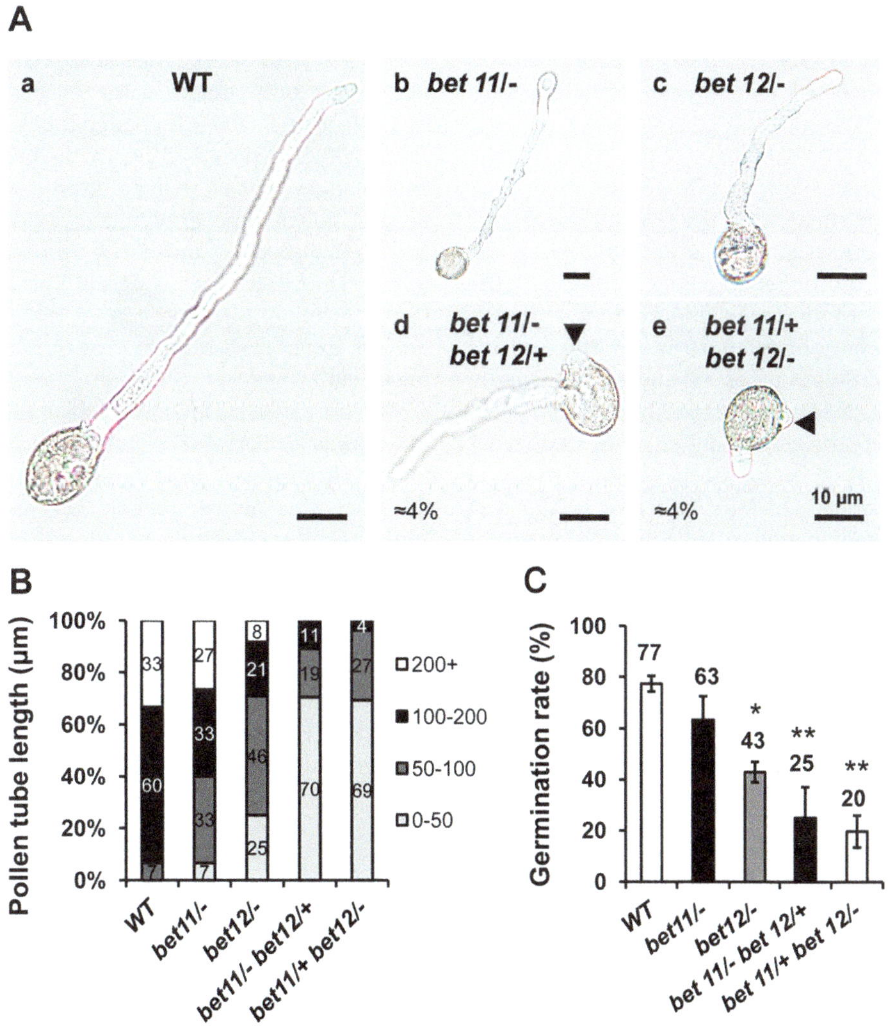

Fig. 3 *BET11/12* mutants show extra pollen tubes and reduced pollen tube length in vitro. **A** Images from pollen tubes after germination. Shown are the WT *a*, single mutants *bet11/−* *b*, *bet12/−* *c*, and double mutants (HM/HZ) *bet11/− bet12/+* *d*, and *bet11/+ bet12/−* *e*. **B** Quantification of pollen tube length (in μm). **C** Germination rate. Data are mean ± SD ($n = 100$) from three different biological samples. *P < 0.1, **P < 0.01, ***P < 0.001 (Student's *t* test) compared to the WT. *Scale bars* 10 μm

reduced activity of both genes may lead to reduced pollen germination and pollen tube length and formation of secondary pollen tubes.

Moreover, in vivo pollen germination experiments with WT pollen in single and double mutant *bet11/bet12* flowers did not reveal any apparent defects in the germination and elongation of pollen tubes within female tissues (Additional file 3: Figure S2a, upper panel). Use of the single mutants *bet11/−* and *bet12/−* as pollen donors and WT flowers as the female did not reveal any defects (Additional file 3: Figure S2a, lower panel). Nonetheless pollen from the *bet11/−bet12/+* and *bet11/+ bet12/−* double mutants appeared to show poor germination and elongation; in fact, few pollen tubes were observed within female tissues after 48 h (Additional file 3: Figure S2a, lower panel, right side). Moreover, staining with DAPI indicated no significant differences in the morphology of mutant pollen grains; most grains (98–99 %) were tricellular and contained an identifiable vegetative cell and two sperm nuclei (Additional file 3: Figure S2b–c).

Discussion

Plant reproduction starts with the deposition of pollen grains onto the stigma of the pistil. After the grains germinate, the pollen tubes grow through the transmitting tract of the style and septum to reach the embryo sac, where double fertilization takes place (Kaya et al. 2014). Pollen tubes expand by tip growth, whereby the apex of the cell grows much faster than the other sides, thereby generating an elongated structure (Kaya et al. 2014). In this study we show that two putative, highly similar Qc-SNARE genes, *BET11* and *BET12*, are involved in pollen tube growth and that double mutants in a homozygous/heterozygous condition show reduced transmission of the heterozygous alleles (Table 1), defects in fertilization and embryo development (Fig. 2), reduced pollen tube elongation and the extrusion of secondary pollen tubes (Fig. 3), which suggests possible loss of polarity. The respective N-terminal GFP-tagged proteins mostly localized to the Golgi in protoplasts, so they may play a role during endomembrane trafficking (Fig. 1).

The SNAREs and other fusogenic determinants have important roles during pollen tube development. For instance, development of secondary pollen tubes (as in *bet11/bet12* mutants) has been linked to reduced expression of the *PIP5K4* gene, which encodes the phosphatidylinositol-4-monophosphate 5-kinase 4, an enzyme in charge of the conversion of phosphatidylinositol (4,5)-bisphosphate (or PtdIns(4,5)P_2). This molecule plays key roles in actin dynamics, vesicle trafficking, and ion transport (Sousa et al. 2008). Lily pollen tubes overexpressing the actin-binding protein (ABP) LILIM1 showed protrusion of multiple tubes from one pollen grain and aggregation of FM4-64-labeled compartments, which suggests possible cross-talk between endomembrane trafficking and cytoskeletal organization (Wang et al. 2008). In fact, Rho GTPases such as ROP1 and interacting protein RIC3 are believed to function as spatial regulators of ABPs including F-actin, profilin and formin and thus help determine growth polarity (Chebli et al. 2013; Cheung et al. 2010).

Within the Ras superfamily of proteins, Rab GTPases are thought to be required for selective vesicle attachment to target vesicles (Carr and Rizo 2010). When bound to GDP, RABs are inactive, but when bound to GTP, they become activated and interact with several effector proteins (Ung et al. 2013), including SNAREs, during the tethering of cargo at post-Golgi compartments (Cai et al. 2007). In pollen tubes, loss of Arabidopsis *RABA4D* expression led to disrupted polar growth and changes in cell wall patterning and compromised in vivo pollen tube growth and guidance (Szumlanski and Nielsen 2009). Conversely, overexpression of C-terminal–truncated versions of the Rho guanine exchange factor AtRopGEF12 also caused disturbed pollen tube growth, including isotropic growth (Zhang and McCormick 2007).

Moreover systems biology model based on the behavior of *Arabidopsis* RABA4D required the activity of SNAREs for vesicle fusion and GTPases for budding (Kato et al. 2010). In this model, specific GTPases favor budding from specific compartments at specific rates, whereas the rate of vesicle fusion depends on the rate of SNARE pairing between the vesicle and the target (Kato et al. 2010). However it is not known whether this model applies to *bet11/bet12* mutants.

Recent work has shown that the *Arabidopsis* Qc-SNARE SYP61 and Qa-SNARE SYP121 are required for coordinated targeting to the plasma membrane of aquaporin PIP2;7, a process that requires anteroretrogade passage through the TGN during secretion or during retrograde uptake from the plasma membrane (Hachez et al. 2014). This process also appears to mediate expansion of root hairs (Hachez et al. 2014), which are cells well known for showing SNARE-dependent polarized tip growth (Larson et al. 2014).

In *Arabidopsis*, defects in the expression of SNARE genes induce collapse of Golgi compartments into the endoplasmic reticulum (ER) (Chatre et al. 2005). A similar phenomenon was reported for the R-SNARE gene *SEC22* in *Arabidopsis* mutants (El-Kasmi et al. 2011), which show fragmentation of Golgi stacks, presumably because of blocked membrane fusion during anterograde or retrograde ER to Golgi trafficking (El-Kasmi et al. 2011). A similar process might occur in *bet11/bet12* double mutant pollen tubes. Nonetheless, examination of endomembrane compartments by transmission electron microscopy is required to address this question and help determine whether *bet11/bet12* mutants are required for maintenance of proper structure of the Golgi or TGN in pollen tubes.

Protein-driven membrane fusion events are essential in all organisms (Pérez-Vargas et al. 2014). They are crucial to achieve intracellular trafficking, neurotransmitter secretion, cell mating and fertilization. They are also key to the development of tissues and organs in multicellular organisms (Pérez-Vargas et al. 2014).

In this study, we characterized the role of two putative *Arabidopsis* Qc-SNARE genes, *BET11* and *BET12*, in plant growth. Results from the characterization of T-DNA double mutants strongly suggest that these two genes perform largely overlapping functions during fertilization. Both are required during polarized extrusion of pollen tubes, as evidenced by reduced pollen tube length and the development of secondary pollen tubes (Table 1). Genetic data also suggests a role for both genes during embryo development, especially for *BET12* (Table 1). These roles seem to be a departure from the role that bet1p plays in yeast, in which the *bet1-1* allele leads only to minor growth defects as compared with the WT (Rogers et al. 2013). This difference in behavior could be interpreted as an evolutionary gain of function for BET Qc-SNAREs in multicellular eukaryotes. If so, *BET11/BET12* homologs in plants and metazoans may have a specialized role during polarized growth and embryo development.

In conclusion, *Arabidopsis BET11* and *BET12* genes are required for polarized growth in pollen and for embryo development. Nonetheless, more work is required to identify the role of the BET11 and BET12 in anterograde and retrograde traffic, whether fragmentation of the Golgi apparatus occurs in *bet11/bet12* mutants, and to determine the type of cargo specified by BET11 and BET12 in vesicles. We believe that this study may pave the way for further characterization of the function played by these two genes during reproduction and uncover interactions unknown at the present time.

Additional files

Additional file 1: Table S1. Genetic screen of suspected fusogenic factors. Alleles were selected based on expression levels in silico, possible ER/Golgi localization and the availability of T-DNA lines. Double mutants were synthesized only for alleles corresponding to *BET11/12* and *SFT11/12*. The *SFT11/12* double mutant remains uncharacterized.

Additional file 2: Figure S1. The Arabidopsis Qc-SNARE genes *BET11* and *BET12* encode membrane integral proteins. **a** Alignment of amino acid sequences with the consistency-based T-Coffee program (http://www.tcoffee.org) revealed structural conservation (67.1 %) between the Arabidopsis BET11 and BET12 proteins and the yeast homolog Sft1. The residue color scheme shows the primary library support for the alignment of the considered residue on a scale from 0 (blue, poorly supported) to 9 (dark red, strongly supported). **b** and **c** left, diagram of the structural domains of BET11 and BET12 proteins adapted from UNIPROT (http://www.uniprot.org). Both proteins feature an N-terminal SNARE coiled-coil domain (red), a C-terminal transmembrane domain (red), and a C-terminal vesicular topological domain (green); however, only BET11 features a phospho-serine at residue 56. Right, 3D structures according to the SWISS-MODEL repository (swissmodel.expasy.org). Both BET11 and BET12 are believed to assemble as single chains that may associate as heterotetramers. Symbols: * = identity match,: = high structural homology,. = high similarity.

Additional file 3: Figure S2. Pollen grains from *BET11/12* mutants show reduced pollen tube growth in vivo but no developmental defects before germination. **a** Top row, confocal microscopy of cross of *bet11/12* single and double mutants used as pollen donors (♂) with the WT as the receptor (♀). Bottom row, the WT was used as a pollen receptor (♀), with *bet11/12* single and double mutants as pollen donors. At least 3 flowers from 3 different plants were analyzed per line. **b** DAPI staining of pollen grains by confocal microscopy. **c** Quantification of tricellular pollen abundance in the WT, single *bet11/12* mutants and *bet11/bet12* double mutants. No statistically significant differences were observed. Data are mean ± SD ($n = 100$) from three different biological samples.

Authors' contributions

PBV and GYJ designed the research. PBV and AG performed the experiments. PBV and GYJ analyzed data. PBV and GYJ wrote the article. All authors read and approved the final manuscript.

Author details

[1] Fabio Baudrit Agricultural Experimental Station, University of Costa Rica, La Garita de Alajuela, P.O. Box 183-4050, Alajuela, Costa Rica. [2] Institute of Plant and Microbial Biology, Academia Sinica, 128 Sec. 2, Academia Rd, Nankang, Taipei 11529, Taiwan.

Acknowledgements

The authors thank Ms. Mei-Jane Fang for assistance with DNA sequencing and capillary PCR (DNA Analysis Core Laboratory, Academia Sinica, Taiwan). The authors also thank Dr. Frédéric Berger for valuable input at the initial stages of this project. This work was supported by research grants from Academia Sinica (Taiwan), the National Science and Technology Program for Agricultural Biotechnology (NSTP/AB, 098S0030055-AA), Taiwan, and the National Science Council (NSF; 99-2321-B-001-036-MY3), Taiwan, to G.Y. Jauh; and a study-abroad contract from the University of Costa Rica to P. Bolaños-Villegas.

Compliance with ethical guidelines

Competing interests

The authors declare that they have no competing interests.

References

Alpadi K, Kulkarni A, Comte V, Reinhardt M, Schmidt A, Namjoshi S, Mayer A, Peters C (2012) Sequential analysis of Trans-SNARE formation in intracellular fusion. PLoS Biol 10:e1001243

Berger F, Hamamura Y, Ingouff M, Higashiyama T (2008) Double fertilization: caught in the act. Trends Plant Sci 13:437–443

Cai H, Reinisch K, Ferro-Novick S (2007) Coats, tethers, Rabs, and SNAREs work together to mediate the intracellular destination of a transport vesicle. Dev Cell 12:671–682

Carr CM, Rizo J (2010) At the junction of SNARE and SM protein function. Curr Opin Cell Biol 22:488–495

Chatre L, Brandizzi F, Hocquellet A, Hawes C, Moreau P (2005) Sec22 and Memb11 are v-SNAREs of the anterograde endoplasmic reticulum-Golgi pathway in tobacco leaf epidermal cells. Plant Physiol 139:1244–1254

Chebli Y, Kroeger J, Geitmann A (2013) Transport logistics in pollen tubes. Mol Plant 6:1037–1052

Cheung AY, Niroomand S, Zou YJ, Wu HM (2010) A transmembrane formin nucleates subapical actin assembly and controls tip-focused growth in pollen tubes. Proc Natl Acad Sci USA 107:16390–16395

De Benedictis M, Bleve G, Faraco M, Stigliano E, Grieco F, Piro G, Dalessandro G, Di Sansebastiano GP (2013) AtSYP51/52 functions diverge in the post-Golgi traffic and differently affect vacuolar sorting. Mol Plant 6:916–930

Di Tommaso P, Moretti S, Xenarios I, Orobitg M, Montanyola A, Chang JM, Taly JF, Notredame C (2011) T-Coffee: a web server for the multiple sequence alignment of protein and RNA sequences using structural information and homology extension. Nucleic Acids Res 39:W13–W17

Domozych DS, Fujimoto C, LaRue T (2013) Polar expansion dynamics in the plant kingdom: a diverse and multifunctional journey on the path to pollen tubes. Plants 2:148–173

El-Kasmi F, Pacher T, Strompen G, Stierhof YD, Muller LM, Koncz C, Mayer U, Jurgens G (2011) Arabidopsis SNARE protein SEC22 is essential for gametophyte development and maintenance of Golgi-stack integrity. Plant J 66:268–279

Fujiwara M, Uemura T, Ebine K, Nishimori Y, Ueda T, Nakano A, Sato MH, Fukao Y (2014) Interactomics of Qa-SNARE in *Arabidopsis thaliana*. Plant Cell Physiol 55:781–789

Furukawa N, Mima J (2014) Multiple and distinct strategies of yeast SNAREs to confer the specificity of membrane fusion. Sci Rep 4:4277. doi:10.1038/srep04277

Guo F, McCubbin AG (2012) The pollen-specific R-SNARE/longin PiVAMP726 mediates fusion of endo- and exocytic compartments in pollen tube tip growth. J Exp Bot 63:3083–3095

Hachez C, Laloux T, Reinhardt H, Cavez D, Degand H, Grefen C, De Rycke R, Inzé D, Blatt MR, Russinova E, Chaumont F (2014) *Arabidopsis* SNAREs SYP61 and SYP121 coordinate the trafficking of plasma membrane aquaporin PIP2;7 to modulate the cell membrane water permeability. Plant Cell. doi:10.1105/tpc.114.127159

Karnik R, Grefen C, Bayne R, Honsbein A, Kohler T, Kioumourtzoglou D, Williams M, Bryant NJ, Blatt MR (2013) *Arabidopsis* Sec1/Munc18 protein SEC11 is a competitive and dynamic modulator of SNARE binding and SYP121-dependent vesicle traffic. Plant Cell 25:1368–1382

Kato N, He HY, Steger AP (2010) A systems model of vesicle trafficking in *Arabidopsis* pollen tubes. Plant Physiol 152:590–601

Kawashima T, Berger F (2011) Green love talks; cell-cell communication during double fertilization in flowering plants. AoB Plants plr015. doi: 10.1093/aobpla/plr015

Kaya H, Nakajima R, Iwano M, Kanaoka MM, Kimura S, Takeda S, Kawarazaki T, Senzaki E, Hamamura Y, Higashiyama T, Takayama S, Abe M, Kuchitsu K (2014) Ca^{2+}-activated reactive oxygen species production by *Arabidopsis* RbohH and RbohJ is essential for proper pollen tube tip growth. Plant Cell 26:1069–1080

Larson ER, Domozych DS, Tierney ML (2014) SNARE VTI13 plays a unique role in endosomal trafficking pathways associated with the vacuole and is essential for cell wall organization and root hair growth in arabidopsis. Ann Bot 114:1147–1159

Lin SY, Chen PW, Chuang MH, Juntawong P, Bailey-Serres J, Jauh GY (2014) Profiling of translatomes of in vivo-grown pollen tubes reveals genes with roles in micropylar guidance during pollination in *Arabidopsis*. Plant Cell 26:602–618

Lipka V, Kwon C, Panstruga R (2007) SNARE-ware: the role of SNARE-domain proteins in plant biology. Ann Rev Cell Dev Biol 23:147–174

Morita M, Shimada T (2014) The plant endomembrane system—a complex network supporting plant development and physiology. Plant Cell Physiol 55:667–671

Ohya T, Miaczynska M, Coskun U, Lommer B, Runge A, Drechsel D, Kalaidzidis Y, Zerial M (2009) Reconstitution of Rab- and SNARE-dependent membrane fusion by synthetic endosomes. Nature 459:1091–1097

Pérez-Vargas J, Krey T, Valansi C, Avinoam O, Haouz A, Jamin M, Raveh-Barak H, Podbilewicz B, Rey FA (2014) Structural basis of eukaryotic cell-cell fusion. Cell 157:407–419

Rogers JV, Arlow T, Inkellis ER, Koo TS, Rose MD (2013) ER-associated SNAREs and Sey1p mediate nuclear fusion at two distinct steps during yeast mating. Mol Biol Cell 24:3896–3908

Sanderfoot A (2007) Increases in the number of SNARE genes parallels the rise of multicellularity among the green plants. Plant Physiol 144:6–17

Sousa E, Kost B, Malhó R (2008) *Arabidopsis* phosphatidylinositol-4-monophosphate 5-kinase 4 regulates pollen tube growth and polarity by modulating membrane recycling. Plant Cell 20:3050–3064

Szumlanski AL, Nielsen E (2009) The Rab GTPase RabA4d regulates pollen tube tip growth in Arabidopsis thaliana. Plant Cell 21:526–544

Tai WCS, Banfield DK (2001) AtBS14a and AtBS14b, two *Bet1/Sft1*-like SNAREs from *Arabidopsis thaliana* that complement mutations in the yeast *SFT1* gene. FEBS Lett 500:177–182

Uemura T, Nakano A (2013) Plant TGNs: dynamics and physiological functions. Histochem Cell Biol 140:341–345

Uemura T, Ueda T, Ohniwa RL, Nakano A, Takeyasu K, Sato MH (2004) Systematic analysis of SNARE molecules in *Arabidopsis*: dissection of the post-Golgi network in plant cells. Cell Struct Funct 29:49–65

Uemura T, Kim H, Saito C, Ebine K, Ueda T, Schulze-Lefert P, Nakano A (2012) Qa-SNAREs localized to the trans-Golgi network regulate multiple transport pathways and extracellular disease resistance in plants. Proc Natl Acad Sci USA 109:1784–1789

Uemura T, Suda Y, Ueda T, Nakano A (2014) Dynamic behavior of the trans-Golgi network in root tissues of *Arabidopsis* revealed by super-resolution live imaging. Plant Cell Physiol 55:694–703

Ung N, Brown MQ, Hicks GR, Raikhel NV (2013) An approach to quantify endomembrane dynamics in pollen utilizing bioactive chemicals. Mol Plant 6:1202–1213

Wang HJ, Wan AR, Jauh GY (2008) An actin-binding protein, LlLIM1, mediates calcium and hydrogen regulation of actin dynamics in pollen tubes. Plant Physiol 147:1619–1636

Zhang Y, McCormick S (2007) A distinct mechanism regulating a pollen-specific guanine nucleotide exchange factor for the small GTPase Rop in *Arabidopsis thaliana*. Proc Natl Acad Sci USA 104:18830–18835

2

Phenolic constituents and antioxidant properties of five wild species of *Physalis* (Solanaceae)

José Roberto Medina-Medrano, Norma Almaraz-Abarca*, M. Socorro González-Elizondo, José Natividad Uribe-Soto, Laura Silvia González-Valdez and Yolanda Herrera-Arrieta

Abstract

Background: Fruits of wild species of the genus *Physalis* are consumed as food and calyces and leaves are used in traditional medicine. The phenolic composition of the species of this genus have been scarcely studied. To contribute to a better knowledge for the use of all the potential of these wild species of plants, leaves, fruits, and calyces of five wild species of the genus were analyzed for their phenolic composition and antioxidant properties.

Results: Important tissue- and species-dependent variations were found. Calyces of *Physalis subulata* showed the highest contents of phenolics (176.58 mg of gallic acid equivalents/g dry tissue), flavonoids (39.63 mg/g dry tissue), and phenolic acids (50.57 mg of quercitrin equivalents/g dry tissue), and its leaves displayed the highest total antioxidant capacity (3.59 mg of ascorbic acid equivalents/mL) and one of the highest reduction powers (0.54 μg of ascorbic acid equivalents/mL). A high performance liquid chromatography with photodiode array detection analysis revealed a total of 28 phenolic compounds in foliar tissues (mainly kaempferol-3-*O*-glycosides), 16 in fruits (mainly phenolic acids), and 16 in calyces (mainly kaempferol-3-*O*-glycosides); the profiles of these compounds in the three types of tissue were species-specific.

Conclusions: The studied species of *Physalis* are important sources of phenolics with relevant antioxidant activity. The current results indicate that phenolic profiles are valuable specific chemical markers and can be relevant as food tracing and authenticity indicators for plant-based preparations involving species of *Physalis*.

Keywords: *Physalis*, Antioxidant capacity, Phenolic profiles, Phenolic content

Background

Plants are a promising source of beneficial compounds for human health and represent a starting point for the discovery and development of drugs for the treatment of human diseases (Kouloura et al. 2014). There is an impressive plant diversity worldwide, in which the chemical variation is represented by a great deal of bioactive compounds. The interest in wild plants as a source of natural antioxidants is valid as a healthier and cheaper alternative to the synthetic ones, which have been perceived as toxic and carcinogenic (Krishnaiah et al. 2011).

*Correspondence: nalmaraz@ipn.mx; noralab@yahoo.com
Centro Interdisciplinario de Investigación para el Desarrollo Integral Regional, Unidad Durango, Instituto Politécnico Nacional, Sigma 119, Fraccionamiento 20 de Noviembre II, 34220 Durango, Dgo., Mexico

The family Solanaceae is constituted by around 2300 species (D'Arcy 1991), which are significant sources of phytochemicals (Eick 2008). This family includes important cultivated species, such as chili (*Capsicum annuum* L.), tomato (*Lycopersicon esculentum* Mill.), and potato (*Solanum tuberosum* L.). The family also includes numerous wild species, like most belonging to the genus *Physalis*, which is the fifth largest genus of the family, including 70–90 species (Whitson and Manos 2005). Mexico, with about 50 species of *Physalis* growing in its territory, is considered the center of origin, diversity (D'Arcy 1991), and domestication of this genus (Santiaguillo et al. 1994). The fruits of many species of *Physalis* have been important elements in the culinary traditions of the people of Mesoamerica, consumed fresh or cooked, and calyces and leaves have been used in folk medicine since pre-Columbian times. Calyces are also consumed as seasoning and leaves as food (Hernández and Yáñez 2009). However, at present, only a few species are cultivated, *P. ixocarpa*, *P. peruviana*, and *P. alkekengi* are among them.

Phenolic compounds are synthesized and accumulated in practically all plant tissues. Many of these compounds, mainly flavonoids and phenolic acids, have biological properties with medical implications (Zhang and Cui 2005). The antioxidant capacity is one of the most important of these properties, because oxidation damages nucleic acids, proteins, lipids, and other macromolecules, producing cardiovascular disorders, neurodegenerative diseases, and cancer (Ames et al. 1993). Flavonoids and phenolic acids have relevant antioxidant properties (Barriada-Bernal et al. 2014). The antioxidant capacity of plant phenolic extracts depends on their concentration (Dobre et al. 2010), but also on their accumulation profiles inside plant tissues (Barriada-Bernal et al. 2014). The concentration is affected by environmental conditions, age, and phenological stage (Almaraz-Abarca et al. 2013), while the qualitative phenolic profiles are more stable and vary among different groups of plants with a species-specific tendency (Emerenciano et al. 2001).

Few species of *Physalis* have been analyzed for their phenolic composition and antioxidant properties, among them are *P. ixocarpa* (González-Mendoza et al. 2011), *P. peruviana* (Rockenbach et al. 2008; Wu et al. 2009), *P. angulata* (Ismail and Alam 2001), and *P. alkekengi* var. *franchetii* (Diaz et al. 2012). The aim of the present study was to evaluate the abundance and diversity of phenolics and the antioxidant properties of leaves, edible fruits, and calyces of five wild species of *Physalis* [*P. angulata* L., *P. patula* Mill., *P. hederifolia* A. Gray var. *hederifolia*, *P. solanacea* (Schltdl.) Axelius, and *P. subulata* Rydb.] from Durango, Mexico to determine their potential as sources of natural antioxidants. The significance of the phenolic profiles as specific chemomarkers and their potential as tool for food tracing and authenticity was also evaluated.

Methods

Plant material

Leaves, fruits, and calyces of mature individuals (blooming and bearing fruits) of six natural populations of *Physalis* were collected in different locations of Durango, Mexico (Table 1). Tissues of four individuals per population were randomly sampled, combined, and three pools of samples were formed and separately analyzed. Voucher specimens were deposited at the Herbarium CIIDIR. Samples were individually dried in a ventilated oven at 40 °C, and then ground in a domestic blender. The dry, ground tissues were kept in paper bags at room temperature, in a desiccator with silica, in darkness until analysis.

Preparation of extracts

Phenolic compounds were extracted from dry ground tissues (1 g) by maceration in 20 mL of 80 % methanol (v/v), for 24 h, by shaking at 100 rpm, in darkness, at room temperature. The extracts were centrifuged (4800*g*) for 5 min at room temperature and the supernatants formed the total extracts. Different aliquots were taken to be used in the determination of phenolic content, antioxidant potential, and HPLC–DAD analysis.

Determination of total phenolics

The concentrations of total phenolics of each sample were determined using the Folin-Ciocalteu reagent, according to Nurmi et al. (1996). The phenolic contents were calculated by using a calibration curve of gallic acid ($A_{760} = 0.003[\text{gallic acid}] - 0.0241$, $r = 0.9985$) constructed with five concentrations of this compound (0–92 µg/mL). Phenolic contents were expressed as milligrams of gallic acid equivalents per gram of dry tissue (mg GAE/g dt).

HPLC–DAD analysis

The HPLC–DAD analysis was carried out according to Campos and Markham (2007). Aliquots (20 µL) were analyzed in a Perkin Elmer Series 200 HPLC system and a Perkin Elmer Brownlee Analytical C18 column (4.6 × 250 mm, 5 µm), using an acidified

Table 1 Collection data for the Mexican studied species of *Physalis*

Sample	Curatorial number	Species	Location	Latitude N	Longitude W	Altitude (m)	Date
401	42854	*P. angulata*	Durango, Durango	24°08′15.4″	104°31′58.5″	1864	8/08/2012
407	42861	*P. hederifolia* var. *hederifolia*	Durango, Durango	23°43′24.9″	104°24′8.9″	1975	21/08/2012
412	42866	*P. solanacea*	Durango, Durango	24°14′14.6″	104°27′34″	1872	4/09/2012
413	42867	*P. patula*	Durango, Durango	24°14′14.6″	104°27′34″	1872	4/09/2012
426	42882	*P. solanacea*	Nombre de Dios, Durango	23°58′19.3″	104°19′28.2″	1841	2/10/2012
427	42885	*P. subulata*	Nombre de Dios, Durango	23°47′50.6″	103°51′9.44″	2234	5/10/2012

acetonitrile–water gradient. Water adjusted to pH 2.5 with orthophosphoric acid was solvent A, and acetonitrile was solvent B, mixed following the gradient: starting with 100 % A, decreasing to 91 % over the next 12 min, to 87 % over the next 8 min, to 67 % over the next 12 min, to 57 % over the next 10 min, and held at this level until the end of the 60 min analysis. Chromatograms were plotted at 260 and 340 nm. Spectral data for all peaks were accumulated in the range of 200–400 nm by using a diode array detection (Perkin Elmer Series 200). Structural information of compounds was obtained by direct comparisons of retention time (RT) and UV spectra of resolved compounds with those of standards caffeic acid (RT: 53.13, λ_{max}: 239sh, 295sh, 318), *p*-coumaric acid (RT: 37.2, λ_{max}: 293sh, 308), quercetin (RT: 47.05, λ_{max}: 255, 268sh, 299sh, 370), quercitrin (quercetin-3-*O*-rhamnoside, RT: 38.54, λ_{max}: 255, 264sh, 295sh, 348), morin (RT: 45.4, λ_{max}: 254, 264sh, 298sh, 354), hesperidin (RT: 39.34, λ_{max}: 284, 335sh), and naringenin (RT: 52.25, λ_{max}: 289, 335sh). Structural information was also obtained from the compilations of Mabry et al. (1970) and Campos and Markham (2007). The phenolic profile of each extract was formed by all compounds resolved in their respective chromatograms. Quantitative determinations were made by an external standard method, with the commercial references quercitrin (for flavonols), naringenin (for dihydroflavonoids), and *p*-coumaric acid (for phenolic acids), by area measurements, using standard curves (area $= 3 \times 10^6 + 5 \times 10^7$ [quercitrin], correlation coefficient $r = 0.992$; area $= 2 \times 10^6 + 5 \times 10^7$ [naringenin], correlation coefficient $r = 0.996$; area $= 3 \times 10^6 + 4 \times 10^7$ [*p*-coumaric acid], correlation coefficient $r = 0.999$). The contents were expressed as micrograms of quercitrin equivalents, naringenin equivalents, or *p*-coumaric acid equivalents/g dry tissue (QiE/g dt, NE/g dt, and CE/g dt, respectively). The sum of individual flavonoid concentrations in one sample represented the total flavonoid concentration, expressed as milligram per gram of dry tissue (mg/g dt), and the sum of individual phenolic acids corresponded to the total phenolic acid concentration, expressed as milligram per gram of dry tissue (mg/g dt).

Free radical scavenging activity

Determination of the free radical scavenging activity was carried out following the DPPH* method described by Barriada-Bernal et al. (2014). Decreases in absorbance (at 523 nm) of an initial DPPH* solution (62 μg/mL methanol) against increasing flavonoid concentrations of samples were plotted to determine, by linear regression, the efficient concentration at 50 %, defined as the amount of antioxidant needed to decrease by 50 % the initial DPPH* concentration (EC_{50}). The following curve made with DPPH* between 2 and 60 μg/mL was used to estimate the DPPH* concentration (μg/mL) in the reaction medium: $A_{523nm} = 0.0019 + 0.0309$ [DPPH*], correlation coefficient $r = 0.9996$. Antiradical activities were expressed in terms of EC_{50} in microgram per milliliter (μg/mL). Quercetin and epicatechin were analyzed in the same manner as references.

Total antioxidant capacity (TAC)

Total antioxidant capacity was evaluated according to Prieto et al. (1999). In this method, the reduction of Mo (VI) to Mo (V) by an antioxidant, forming a green phosphate/Mo (V) complex at an acidic pH is measured. TAC values were expressed as milligrams of ascorbic acid equivalents/mL (AAE/mL), which were calculated using the following curve: $A_{695} = -0.2365 + 4.2133$ [ascorbic acid], correlation coefficient $r = 0.9987$, constructed with six concentrations of ascorbic acid between 0.1 and 1 mg/mL. Quercetin (0.1 mg/mL) and epicatechin (0.1 mg/mL) were analyzed in the same manner as references.

Iron reducing power

The iron reducing power method is based on the measurement of the formation of Fe^{2+} from Fe^{3+} in the presence of antioxidants; Fe^{2+} was estimated according to Siddhuraju and Becker (2003). Results were expressed in terms of IC_{50} in micrograms of ascorbic acid equivalents per milliliter (μg AAE/mL), calculated from the following curve of ascorbic acid: $A_{700} = 25.372$ [ascorbic acid] $- 0.0242$, correlation coefficient $r = 0.9956$, constructed with five concentrations of ascorbic acid between 0.03 and 0.1 mg/mL. The highest absorbance values indicated the greatest reducing capacity. Quercetin and epicatechin were analyzed in the same manner as references.

Data analysis

All the assays were carried out for three independent pools of each sample. Data were subjected to an analysis of variance ($p \leq 0.05$), and means were separated by Tukey test. Correlations between different parameters were carried out with Pearson test, by using the SPSS statistics 17.0 software. A principal component analysis (PCA), considering all the quantitative parameters, was carried out by using Past 3.0; the contribution of each parameter for the differentiation of samples was evaluated. For the different tissues (leaves, fruits, and calyces), the phenolic profiles were made up of all compounds present in the respective HPLC–DAD chromatogram. Each compound was treated as a single chemical character. A binary matrix coded by 1 (presence) or 0 (absence) formed by all individual samples vs. all resolved compounds for each type of tissue was subjected to a cluster analysis (paired group algorithm and Jaccard similarity measure) by using Past 3.0.

Results

Phenolic, flavonoid and phenolic acid contents

The phenolic contents are displayed in Table 2. Significant tissue and species-dependent variations were found. Among the foliar tissues, the leaves of *P. patula* accumulated the highest level of phenolics (129 mg/g dt) and among the fruits, those of *P. hederifolia* var. *hederifolia* were the richest ones in total phenolics (86.51 mg/g dt). The calyx phenolic content of *P. subulata* was the highest found in the current study (176.58 mg/g dt).

Table 2 shows the flavonoid contents found in the different tissues of the analyzed species of *Physalis*. Significant tissue and species-dependent variations were found. The foliar flavonoid contents explained between 7.55 and 56.86 % of total phenolics, the highest level was found for *P. angulata* (23.036 mg/g dt). Flavonoid contents in fruits represented between 0.0 and 16.91 % of total phenolics, the fruits of *P. subulata* showed the highest level (5.462 mg/g dt). Flavonoid levels in calyces (representing between 3.08 and 30.43 % of total phenolics) ranged from 3.67 mg/g dt in *P. solanacea* (sample 412) to 39.63 mg/g dt in *P. subulata.*

Species- and tissue-dependent variations were also found in the phenolic acid contents (Table 2). In leaves, phenolic acids corresponded to between 2.11 and 16.55 % of total phenolics and *P. angulata* leaves showed the highest level (6.7 mg/g dry tissue). In fruits, phenolic acids represented between 0.0 and 21.25 % of total phenolics; the fruits of *Physalis angulata* were the richest in these type of phenolics (7.84 mg/g dt). Excepting the calyces of *P. hederifolia* var. *hederifolia*, which practically did not accumulate phenolic acids, these compounds explained from between 2.86 and 28.63 % of total phenolics. The content of phenolic acids in the calyces of *P. subulata* (50.57 mg/g dt) was outstanding.

Phenolic profiles

Compounds **22** (λ_{max}: 255, 264sh, 294sh, 355), **F7** (λ_{max}: 255, 264sh, 294sh, 355), and **C11** (λ_{max}: 255, 264sh, 294sh, 355) were proposed as rutin because of its spectral data corresponded to those reported by Campos and Markham (2007) for that flavonol (λ_{max}: 255, 264sh, 294sh, 355).

The HPLC–DAD analysis revealed a total of 28 phenolic compounds in the leaves of the five species of *Physalis* analyzed (the respective chromatograms are shown in Fig. 1). Twenty of those compounds were flavonols (among them, 14 were kaempferol-3-*O*-glycosides and 5 were quercetin-3-*O*-glycosides) and 8 were phenolic acids. The respective retention times and λ_{max} are displayed in Table 3. The HPLC chromatograms of fruits (Fig. 2) revealed a total of 16 compounds: 11 phenolic acids, 3 flavonols (2 quercetin-3-*O*-glycosides and 1 kaempferol-3-*O*-glycoside), and 2 dihydroflavonols (retention times and λ_{max} are displayed in Table 3). A total of 16 phenolic compounds were unveiled by the chromatograms of calyces (Fig. 3), among which 12 were flavonols (10 kaempferol-3-*O*-glycosides and 2 quercetin-3-*O*-glycosides) and 4 were phenolic acids (Table 3). The concentration of each phenolic compound found in the different analyzed species is shown in Table 3.

Cluster analysis

The results of the cluster analysis based on the matrices constructed with each foliar, fruit, and calyx phenolic profile of each sample of *Physalis* are displayed in Fig. 4. All species were discriminated one from each other at levels of similarity from around 0.06 (Jaccard similarity measure, according to the foliar phenolic profiles) to around 0.42 (according to the calyx phenolic profiles).

Antioxidant assays

The results of the free radical scavenging activity, total antioxidant capacity, and iron reducing power are shown in Table 2. Specific and tissue dependent variations were found in the three antioxidant assays.

Correlation analysis

While the kinetic evaluation of all the antioxidant assays was highly related to flavonoid contents ($0.9731 < r < 0.999$), the correlation analysis revealed lower associations between the antioxidant properties and the phenolic, flavonoid, and phenolic acid contents in the samples (0.224 between calyx total flavonoids and calyx IC_{50} <Pearson correlation value <0.856 between calyx total phenolics and calyx IC_{50}).

Principal component analysis

The results of a PCA, based on the quantitative determinations (phenolic, flavonoid, and phenolic acid contents and the estimation of antioxidant capacity by three methods) are showed in the Fig. 5. Three principal components accounted for 93.77 % of total variance, being the calyx phenolic content (PC1) the main one, taking 73.467 %, with the highest relative discriminating power (eigenvalue 4860.76). The calyx antiradical activity (PC2) explained the 12.425 % of total variance (eigenvalue 822.079), and the foliar phenolic content, the 7.88 % (eigenvalue 521.505).

Discussion

Phenolic, flavonoid, and phenolic acid contents

Phenolic compounds contribute to the organoleptic and antioxidant properties of foods and they are among

Table 2 Phenolics, flavonoids, phenolic acid contents and antioxidant evaluations of five wild species of *Physalis*

Species and reference compounds	Phenolic content (mg AGE/g dt)	Flavonoid content (mg QE/g dt)	Phenolic acid content (mg CE/g dt)	DPPH scavenging capacity (EC_{50}) (μg/mL)	Total antioxidant capacity (TAC) (mg AAE/mL)	Iron reducing power (IC_{50}) (μg AAE/mL)
Leaves						
P. patula (413)	129.06 ± 0.66 a	13.363 ± 0.01 d	4.053 ± 0.049 cd	22.96 ± 0.62 d	2.45 ± 0.08 b	0.43 ± 0.04 b
P. subulata (427)	118.01 ± 3.34 b	20.152 ± 0.033 c	5.566 ± 0.040 b	8.83 ± 0.28 f	3.59 ± 0.26 a	0.54 ± 0.01 a
P. solanacea (426)	89.83 ± 3.3 c	9.365 ± 0.050 e	4.671 ± 0.046 d	8.18 ± 0.21 f	1.80 ± 0.04 c	0.49 ± 0.04 ab
P. solanacea (412)	86.66 ± 6.32 c	6.543 ± 0.014 f	2.388 ± 0.034 e	28.21 ± 0.96 c	2.10 ± 0.10 bc	0.44 ± 0.001 b
P. hederifolia var. *hederifolia* (407)	58.75 ± 5.18 d	21.265 ± 0.006 bc	1.242 ± 0.075 f	59.17 ± 1.42 a	2.35 ± 0.06 b	0.11 ± 0.001 d
P. angulata (401)	40.51 ± 0.71 e	23.036 ± 0.100 a	6.706 ± 0.007 a	38.74 ± 1.04 b	2.44 ± 0.18 b	0.20 ± 0.004 c
Quercetin				5.29 ± 0.39 g	0.08 ± 0.002 d	0.09 ± 0.003 d
Epicatechin				14.33 ± 0.29 e	0.11 ± 0.005 d	0.07 ± 0.004 d
Fruits						
P. hederifolia var. hederifolia (407)	86.51 ± 4.03 a	4.80 ± 0.031 b	Not found	0.87 ± 0.03 f	0.59 ± 0.006 e	0.06 ± 0.002 e
P. patula (413)	53.14 ± 3.95 b	1.392 ± 0.007 d	7.417 ± 0.067 a	9.14 ± 0.17 b	1.07 ± 0.04 a	0.10 ± 0.001 b
P. solanacea (426)	46.29 ± 1.93 bc	2.499 ± 0.023 c	1.253 ± 0.020 d	3.11 ± 0.05 d	0.82 ± 0.005 c	0.06 ± 0.002 e
P. solanacea (412)	40.39 ± 5.32 cd	4.642 ± 0.020 b	2.773 ± 0.032 c	2.51 ± 0.18 e	0.72 ± 0.010 d	0.03 ± 0.001 f
P. angulata (401)	36.92 ± 2.57 cd	Not found	7.849 ± 0.050 a	3.05 ± 0.19 c	0.92 ± 0.006 b	0.02 ± 0.00 g
P. subulata (427)	32.29 ± 0.93 d	5.462 ± 0.033 a	4.258 ± 0.03 b	2.15 ± 0.18 e	0.58 ± 0.15 e	0.15 ± 0.001 a
Quercetin				5.29 ± 0.39 c	0.08 ± 0.002 f	0.09 ± 0.003 c
Epicatechin				14.33 ± 0.29 a	0.11 ± 0.005 f	0.07 ± 0.004 d
Calyces						
P. subulata (427)	176.58 ± 12.55 a	39.633 ± 0.079 a	50.570 ± 0.125 a	60.29 ± 0.30 a	2.07 ± 0.014 a	0.34 ± 0.02 b
P. solanacea (426)	170.78 ± 7.06 a	5.269 ± 0.059 d	5.805 ± 0.042 b	12.20 ± 0.17 d	1.56 ± 0.040 b	0.55 ± 0.03 a
P. solanacea (412)	76.68 ± 4.99 b	3.678 ± 0.049 e	4.054 ± 0.038 c	27.82 ± 0.34 c	0.63 ± 0.013 d	0.02 ± 0.001 e
P. patula (413)	91.90 ± 11.90 b	5.339 ± 0.215 d	5.719 ± 0.024 b	0.94 ± 0.05 f	1.41 ± 0.017 b	0.24 ± 0.01 c
P. hederifolia var. hederifolia (407)	54.15 ± 1.64 c	16.48 ± 0.153 b	Traces	7.68 ± 0.08 e	1.03 ± 0.002 c	0.06 ± 0.000 d
P. angulata (401)	33.18 ± 1.43 c	8.829 ± 0.032 c	1.389 ± 0.009 d	31.94 ± 3.03 b	2.10 ± 0.16 a	0.09 ± 0.01 d
Quercetin				5.29 ± 0.39 e	0.08 ± 0.002 e	0.09 ± 0.003 d
Epicatechin				14.33 ± 0.29 d	0.11 ± 0.005 e	0.07 ± 0.004 d

The values represent the mean and standard deviation for three independent samples. Different letters in the same column for each type of tissue mean significant differences ($p < 0.05$)

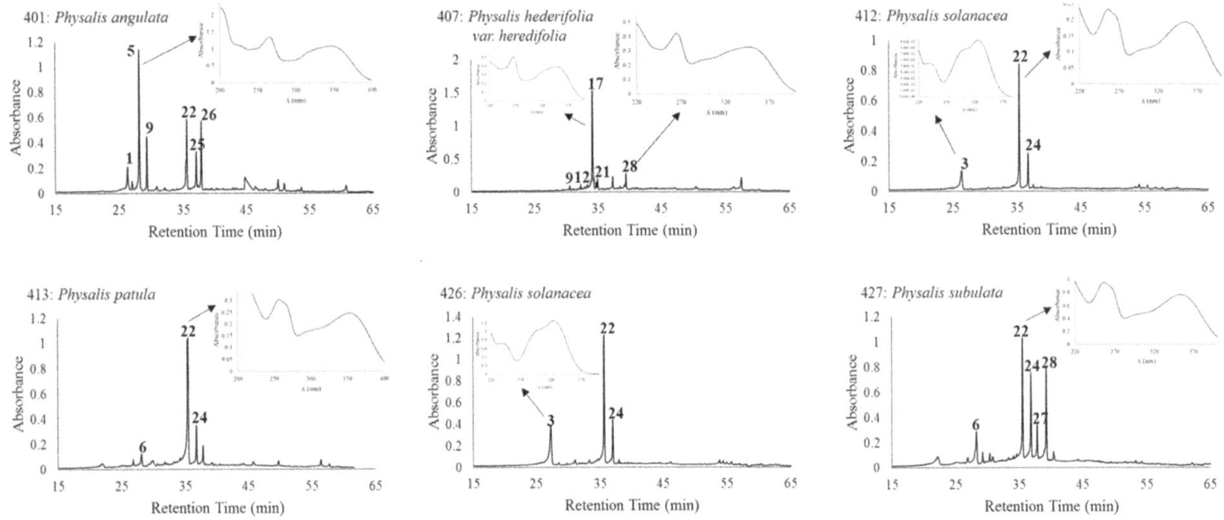

Fig. 1 HPLC chromatograms and UV spectra of the major foliar phenolics of five species of *Physalis*. The HPLC chromatograms were registered at 260 nm and the UV spectra of the resolved compounds were obtained from 200 to 400 nm. The number of compounds corresponds to those of Table 3

the most important nutraceuticals of interest in food industry (Tapas et al. 2008). Some, like phenolic acids, play important roles in fruit maturation, prevention of enzymatic browning, and food preservation (Robbins 2003). Phenolics also represent important tools for food tracing and to prove authenticity of foods (Zimmermann and Galensa 2007). The understanding of their abundance, diversity, and distribution in the plant kingdom and in the different plant tissues offers the opportunity of developing new drugs to improve human health, and to develop new varieties with enhanced accumulation of these compounds through breeding programs.

The potential of different species to synthesize and accumulate flavonoids, and phenolics in general, is the result of the interaction of genetic and environmental factors (Veit et al. 1995) and different species of plants can synthesize and accumulate widely variable levels (Diaz et al. 2012). Aside from the differences in the flavonoid levels due to diverse environmental conditions of growing and to the specific variations, differences between tissues can be expected as the result of the differential regulation of the genic expression in well differentiated cells.

The total phenolic concentrations observed in the leaves and fruits of the analyzed species of *Physalis* were comparable to those reported by Wang and Lin (2000) in the leaves (30.9–129.2 mg/g dry matter) and fruits (9.16–23.10 mg/g dry matter basis) of several cultivars of black raspberry, red raspberry, and strawberries, all considered as important source of phenolics. To our knowledge, this is the first report of phenolic contents in calyx for the species here analyzed of *Physalis* and reveals that these structures can accumulate relevant phenolic levels. The present results indicate that the species analyzed of *Physalis* are important sources of total phenolics, mainly accumulated in leaves and calyces.

Comparatively, the found foliar flavonoid contents were lower than those reported by Wu et al. (2009) for the leaves of *P. peruviana* (37.39–226.19 mg/g dry extract, depending on the type of solvent used for the extraction). Excepting *P. angulata*, in whose fruits no flavonoids were found, the fruit flavonoid contents found in the present study (Table 2) were higher than that reported for ripe fruits of *Lycium barbarum* (Fructus Lycii) (0.72 mg/g dry fruit), whose importance as source of antioxidants has been reported by Le et al. (2007). The flavonoid levels found in the calyces of *P. subulata* (Table 2) can be comparable to that reported for the aqueous extract of *P. peruviana* leaves (37.39 mg/g) by Wu et al. (2009). Flavonoids were mainly accumulated in leaves and calyces.

Studies on determinations of phenolic acids in species of *Physalis* have been mainly focused on fruits (Rockenbach et al. 2008); however, the current results reveal that leaves and calyces of the species of *Physalis* can accumulate these phenolics at similar or even higher levels than fruits (Table 2).

Phenolic profiles

Quercetin derivatives, kaempferol derivatives, and phenolic acids were the major phenolic compounds

Table 3 Phenolic compounds found in the leaves, fruits, and calyces of five wild species of *Physalis*

Number of compound	RT (min)	λ_{max}	Type of phenolic compound	Species
Leaves				
1	26.50 ± 0.11	256, 267sh, 319, 360sh	Flavonol	*P. angulata* (2.476)
2	22.57 ± 0.36	240sh, 296sh, 326	Phenolic acid	*P. subulata* (2.184)
3	27.37 ± 0.07	240sh, 296sh, 325	Phenolic acid	*P. solanacea* (412: 2.288; 426: 4.671)
4	27.23 ± 0.11	256, 266sh, 347	Quercetin-3-*O*-glycoside	*P. angulata* (1.544)
5	28.41 ± 0.02	243sh, 265, 317sh, 345	Kaempferol-3-*O*-glycoside	*P. angulata* (5.094)
6	28.36 ± 0.32	233sh, 291sh, 324	Phenolic acid	*P. angulata* (1.281) *P. patula* (2.467) *P. hederifolia* var *hederifolia* (1.242) *P. subulata* (3.381)
7	28.73 ± 0.05	227sh, 243sh, 265, 344	Kaempferol-3-*O*-glycoside	*P. subulata* (1.522)
8	29.63 ± 0.24	243sh, 290sh, 322	Phenolic acid	*P. patula* (1.586)
9	29.65 ± 0.09	227sh, 265, 343	Kaempferol-3-*O*-glycoside	*P.angulata* (2.329) *P. hederifolia* var. *hederifolia* (1.412) *P. subulata* (1.524)
10	31.17 ± 0.07	229sh, 268sh, 308	Phenolic acid	*P. angulata* (1.417)
11	31.87 ± 0.00	231sh, 267sh, 312	Phenolic acid	*P. angulata* (1.268)
12	32.21 ± 0.05	256, 266sh, 353	Quercetin-3-*O*-glycoside	*P. hederifolia* var. *hederifolia* (1.403)
13	32.43 ± 0.03	226sh, 295sh, 312	Phenolic acid (probably *p*-coumaric acid)	*P. angulata* (1.341)
14	32.93 ± 0.00	265, 318sh, 354	Kaempferol-3-*O*-glycoside	*P. hederifolia* var. *hederifolia* (1.331)
15	33.29 ± 0.00	228sh, 266, 320sh, 343	Kaempferol-3-*O*-glycoside	*P. hederifolia* var. *hederifolia* (1.515)
16	33.82 ± 0.00	228sh, 265, 317sh, 345	Kaempferol-3-*O*-glycoside	*P. hederifolia* var. *hederifolia* (1.897)
17	34.16 ± 0.00	227sh, 244sh, 265, 318sh, 345	Kaempferol-3-*O*-glycoside	*P. hederifolia* var. *hederifolia* (6.868)
18	34.29 ± 0.00	228sh, 265, 317sh, 346	Kaempferol-3-*O*-glycoside	*P. hederifolia* var. *hederifolia* (1.538)
19	34.58 ± 0.00	236sh, 270sh, 290sh, 324	Phenolic acid	*P. angulata* (1.397)
20	34.75 ± 0.10	265, 294sh, 344	Kaempferol-3-*O*-glycoside	*P. hederifolia* var. *hederifolia* (1.583)
21	34.90 ± 0.03	228sh, 245sh, 266, 345	Kaempferol-3-*O*-glycoside	*P. hederifolia* var. *hederifolia* (1.631)
22	35.69 ± 0.20	255, 264sh, 294sh, 355	Quercetin-3-*O*-glycoside (rutine)	*P. angulata* (5.175) *P. patula* (5.431) *P. solanacea* (412: 4.388; 426: 6.882) *P. subulata* (5.892)
23	34.85 ± 0.60	255,266sh, 296sh, 355	Quercetin-3-*O*-glycoside	*P. patula* (3.627)
24	36.24 ± 0.34	256, 266sh, 296sh, 353	Quercetin-3-*O*-glycoside	*P. patula* (2.421) *P. solanaceous* (412: 2.154, 426: 4.682) *P. subulata* (4.599)
25	36.47 ± 0.16	227sh, 265, 294sh, 346	Kaempferol-3-*O*-glycoside	*P. angulata* (2.837)
26	37.40 ± 0.22	227sh, 265, 294sh, 346	Kaempferol-3-*O*-glycoside	*P. angulata* (3.577) *P. patula* (1.879)
27	38.20 ± 0.08	265, 293sh, 343	Kaempferol-3-*O*-glycoside	*P. subulata* (2.597)
28	39.44 ± 0.12	264, 293sh, 346	Kaempferol-3-O-glycoside	*P. hederifolia var. hederifolia* (2.032) *P. subulata* (3.977)
Fruits				
F1	26.648 ± 0.489	293sh, 325	Phenolic acid	*Physalis solanacea* (412: 2.773; 426: 1.253) *P. subulata* (2.470)
F2	27.389 ± 0.187	296sh, 322	Phenolic acid	*P. angulata* (1.534) *P. patula* (1.144) *P. subulata* (1.787)
F3	29.821 ± 0.000	292sh, 326	Phenolic acid	*P. angulata* (1.536)
F4	30.716 ± 0.000	283sh, 311	Phenolic acid	*P. patula* (1.163)
F5	33.488 ± 0.000	226sh, 265, 346	Kaempferol-3-*O*-glycoside	*P. hederefolia var hederifolia* (4.807)
F6	34.47 ± 0.000	291sh, 311	Phenolic acid	*P. patula* (1.232)

Table 3 continued

Number of compound	RT (min)	λ_{max}	Type of phenolic compound	Species
F7	35.138 ± 0.323	255, 264sh, 294sh, 355	Quercetin-3-*O*-glycoside	*P. patula* (1.392) *P. solanacea* (412: 2.320, 426: 1.282) *P. subulata* (1.765)
F8	37.55 ± 0.000	281, 324sh	Dihydroflavonol	*P. subulata* (1.765)
F9	39.193 ± 0.000	277, 303sh	Dihydroflavonol	*P. subulata* (1.932)
F10	45.69 ± 0.000	286sh, 312	Phenolic acid	*P. patula* (1.134)
F11	45.954 ± 0.000	283sh, 317	Phenolic acid	*P. angulata* (1.593)
F12	46.937 ± 0.000	291sh, 311	Phenolic acid	*P. patula* (1.282)
F13	50.091 ± 0.397	255, 267sh, 296sh, 355	Quercetin-3-*O*-glycoside	*P. solanaceous* (412: 2.322; 426: 1.217)
F14	52.628 ± 0.000	292sh, 321	Phenolic acid	*P. angulata* (1.610)
F15	55.928 ± 0.000	290sh, 312	Phenolic acid	*P. patula* (1.429)
F16	57.497 ± 0.000	281sh, 318	Phenolic acid	*P. angulata* (1.573)
Calyces				
C1	23.055 ± 0.054	246sh, 290sh, 322	Phenolic acid	*Physalis angulata* (1.389) *Physalis subulata* (25.085)
C2	27.855 ± 0.287	246sh, 292, 326	Phenolic acid	*Physalis solanacea* (412: 2.034; 426: 3.100) *P. patula* (3.085) *Physalis subulata* (12.132)
C3	28.574 ± 0.000	226sh, 266, 287sh, 347	Kaempferol-3-*O*-glycoside	*Physalis angulata* (1.549)
C4	29.180 ± 0.040	242sh, 286sh, 322	Phenolic acid	*Physalis patula* (2.634) *Physalis subulata* (13.352)
C5	29.641 ± 0.000	228sh, 265, 345	Kaempferol-3-*O*-glycoside	*Physalis angulata* (1.558)
C6	31.400 ± 0.317	246sh, 316	Phenolic acid	*Physalis solanacea* (412: 2.020; 426: 2.704)
C7	33.172 ± 0.036	229sh, 266, 318sh, 344	Kaempferol-3-*O*-glycoside	*Physalis hederifolia var. hederifolia* (3.084)
C8	33.572 ± 0.146	227sh, 2656, 316sh, 346	Kaempferol-3-*O*-glycoside	*Physalis hederifolia* var. *hederifolia* (4.214)
C9	33.605 ± 0.000	229sh, 266, 319sh, 344	Kaempferol-3-*O*-glycoside	*Physalis hederifolia var. hederifolia* (3.023)
C10	34.016 ± 0.000	225sh, 265, 340	Kaempferol-3-O-glycoside	*Physalis hederifolia var. hederifolia* (3.017)
C11	35.990 ± 0.064	255, 264sh, 294sh, 355	Quercetin-3-*O*-glycoside, (Rutin)	*Physalis angulata* (2.087) *Physalis solanacea* (412: 2.032; 426: 2.671) *Physalis patula* (3.080) *Physalis subulata* (13.885)
C12	37.407 ± 0.135	267sh, 265, 293sh, 340	Kaempferol-3-*O*-glycoside	*Physalis angulata* (1.320) *Physalis patula* (2.124)
C13	37.55 ± 0.000	256, 266sh, 295, 352	Quercetin-3-*O*-glycoside	*Physalis subulata* (13.278)
C14	38.606 ± 0.000	265, 293sh, 346	Kaempferol-3-*O*-glycoside	*Physalis solanaceous* (412: 1.646; 426: 2.631)
C15	38.437 ± 0.175	225sh, 265, 291sh, 346	Kaempferol-3-*O*-glycoside	*Physalis angulata* (2.335) *Physalis hederifolia var. hederifolia* (3.146)
C16	39.897 ± 0.000	226sh, 265, 345	Kaempferol-3-*O*-glycoside	*Physalis subulata* (12.469)

Figures in brackets mean concentration (mg/g dry tissue)

synthesized by the five species analyzed of *Physalis*, and were accumulated with a species-specific tendency.

The leaves of *P. angulata* accumulated the highest number of phenolics (12 compounds) and those of *P. solanacea* the lowest (3 compounds) (Table 3). Rutin (compound **22**) was found in all species (as one of the dominant compounds), except in *P. hederifolia* var. *hederifolia*, whose leaves were rich in

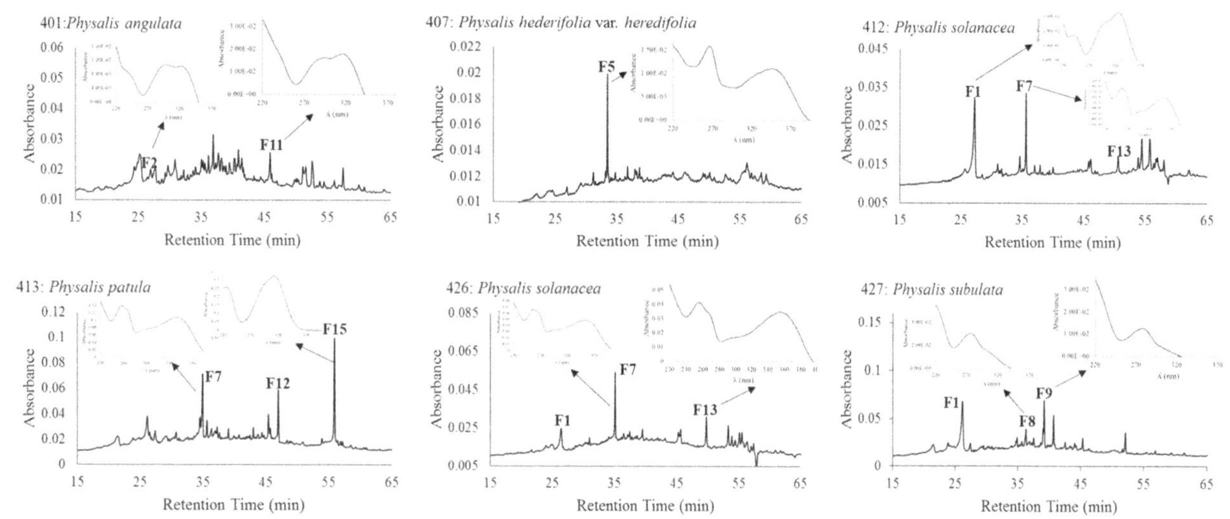

Fig. 2 HPLC chromatograms and UV spectra of major fruit phenolics of five species of *Physalis*. The HPLC chromatograms were registered at 260 nm and the UV spectra of the resolved compounds were obtained from 200 to 400 nm. The number of compounds corresponds to those of Table 3

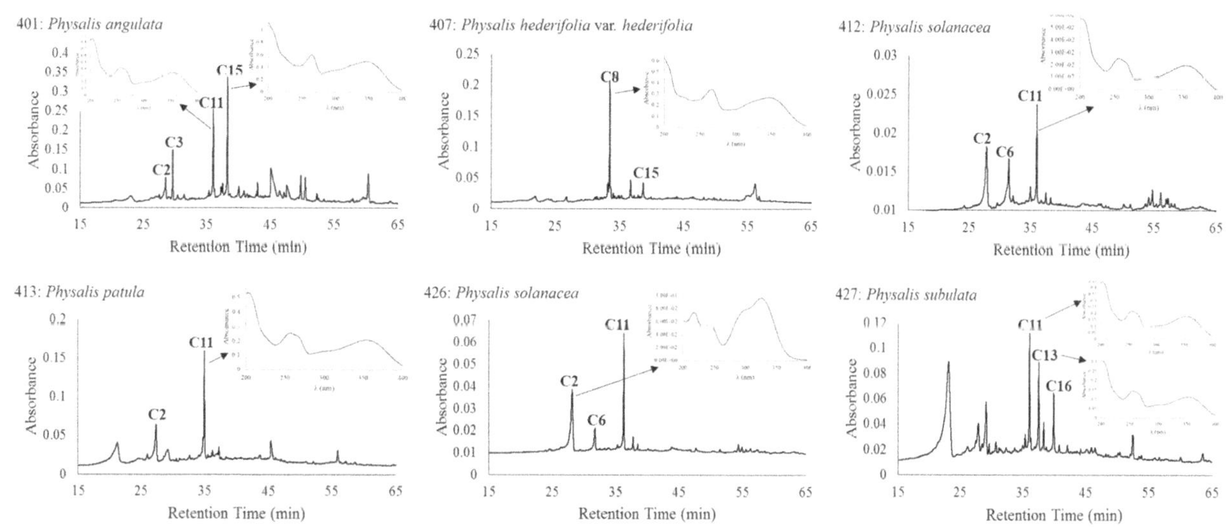

Fig. 3 HPLC chromatograms and UV spectra of major calyx phenolics of five species of *Physalis*. The HPLC chromatograms were registered at 260 nm and the UV spectra of the resolved compounds were obtained from 200 to 400 nm. The number of compounds corresponds to those of Table 3

kaempferol-3-*O*-glycosides (9 among 11 compounds were kaempferol derivatives). Rutin has also been reported in aerial parts of *Physalis orizabae* (Maldonado et al. (2012). Myricetin-3-*O*-neohesperidoside (λ_{max}: 266, 299sh, 356) was reported in methanol extracts of leaves of *P. angulata* by Ismail and Alam (2001); however, the current results revealed no myricetin derivatives for any of the analyzed species.

In fruits, phenolic acids were the dominant compounds (Table 3), except for *P. hederifolia* var. *hederifolia*. A similar dominance has been reported for the fruits of *P. peruviana* (Rockenbach et al. 2008). Relevant interspecific differences were observed in the phenolic composition, for example, for the fruits of *P. hederifolia* var. *hederifolia* the only phenolic compound found was one kaempferol-3-*O*-glycoside (**F5**) (Fig. 2), this contrasts with the composition found for the other four species and with the results reported for *P. peruviana* (Rockenbach et al. 2008). Rutin and myricetin have been found in the fruits of *P. peruviana*, rutin being present as a

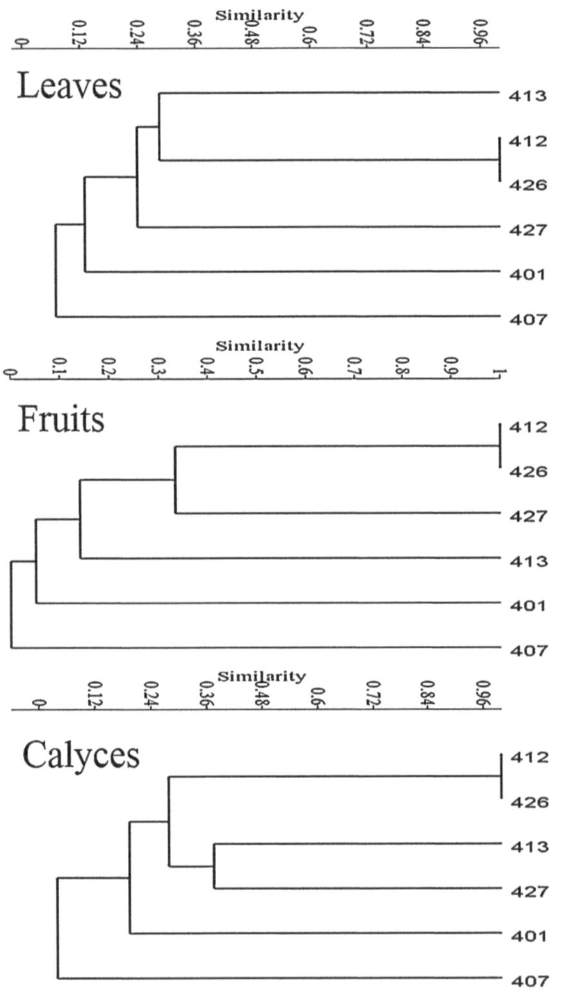

Fig. 4 Results of the cluster analysis based on the phenolic profiles of five species of *Physalis*. The dendrogram was generated with the Paired Group Algorithm and the Jaccard's Similarity Measure, for comparing foliar, fruit and calyx phenolic profiles for five wild species of *Physalis*. The reference number of samples corresponds to those of Table 1

predominant compound (Licodiedoff et al. 2013). Rutin (compound **F7**) was also present in the fruits of *P. patula*, *P. solanacea*, and *P. subulata*, and in the two first species it was one of the dominant compounds (Fig. 2). However, none of the spectral data found in the present study corresponded to those of myricetin reported by Campos and Markham (2007) (λ_{max}: 252, 263sh, 298sh, 371) or by Mabry et al. (1970) (λ_{max}: 254, 272sh, 301sh, 374). The fruits of *P. subulata* contained dihydroflavonoids (**F8** and **F9**), which are less common phenolics in plant kingdom than flavonoids. The concentrations of **F8** and **F9** were similar to those of most phenolic acids (Table 3). Dihydroflavonoids can be important inhibitors of tumor development, antioxidant, and analgesic (Wong and Rabie 2009). The fruits of *P. patula* accumulated the highest number of phenolic compounds (seven).

In calyces of *P. hederifolia* var. *hederifolia*, as in its leaves and fruits, the dominant compounds were kaempferol-3-*O*-glycosides, and this was the only species in whose calyces, rutin (compound **C11**) was missing. Rutin has also been reported in the calyx of *P. solanacea* by Pérez-Castorena et al. (2013). The number of compounds accumulated in the calyces varied from 4 in *P. solanacea* and *P. patula* to 7 in *P. subulata*.

The concentrations of the individual phenolics were variable between species and between tissues. The highest level was found for compound **C1,** a phenolic acid accumulated in the calyces of *P. subulata* (Table 3). Comparatively, the individual phenolic acid concentrations of fruits and calyces were higher than the concentrations reported by Rockenbach et al. (2008) for the fruits of *Physalis peruviana*.

Cluster analysis

The results of the cluster analysis suggests that the phenolic profiles obtained by HPLC–DAD (Fig. 4) were species-specific, independent from location and time of collection, as indicated for the two samples of *P. solanacea* (Jaccard similarity value of 1), in such a way that each species can be typified by its phenolic profile. These phenol profiles can be important specific chemotaxonomic markers for *Physalis* and may contribute to solve taxonomic controversies around the delimitation of species mentioned by several authors (Whitson and Manos 2005). The species-specific feature of the profiles can be used also as a tool to typify, trace, and define the authenticity of fruits and preparations of these species. The chemical diversity revealed for the five wild analyzed species of *Physalis* can be a starting point to improve the quality of fruit of the cultivated species of the genus, and could represent relevant allelic forms. Our results corroborate previous reports about the specificity of phenolic profiles (Almaraz-Abarca et al. 2013).

Antioxidant properties and correlation analysis

All samples displayed important antioxidant properties. As scavengers of the free radical DPPH highlighted the fruits of *P. hederifolia* var. *hederifolia* (Table 2). All extracts showed higher scavenging activity than that reported by Chang et al. (2008) for the fruits of *P. peruviana* (EC_{50} value around 300 µg/mL). All extracts showed higher power to reduce Mo (VI) to Mo (V) than quercetin and epicatechin (Table 2), both considered as powerful antioxidants (Tapas et al. 2008). Leaves of *P. subulata* displayed until 44-fold higher values of TAC than any of two standards. All the foliar extracts showed higher iron reduction potential than the standards quercetin and

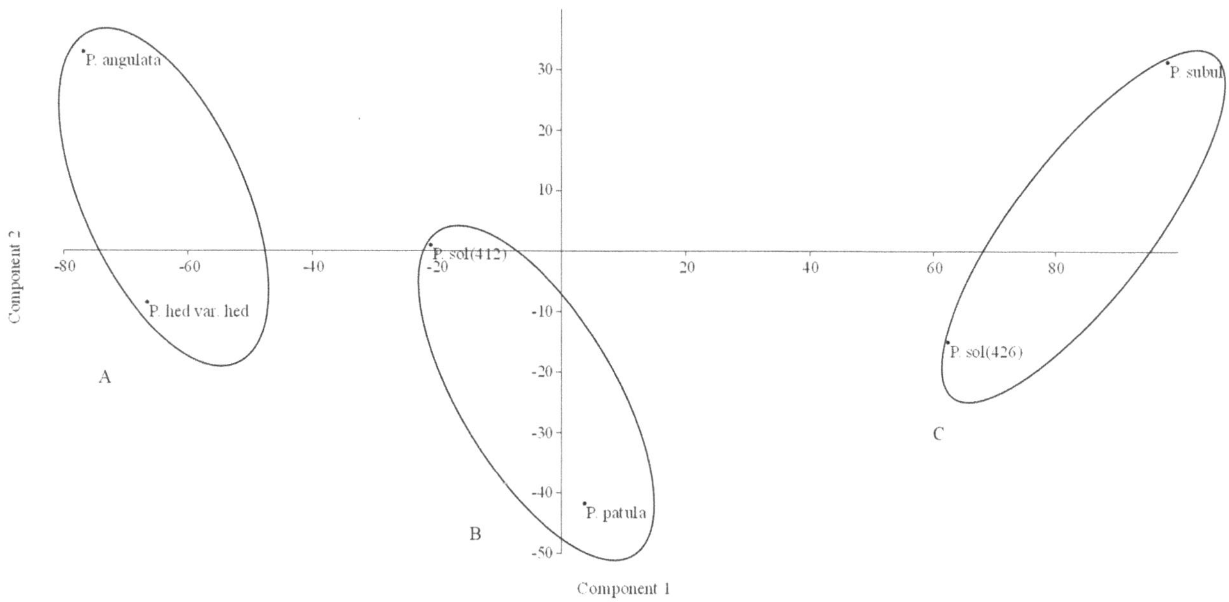

Fig. 5 Results of a principal component analysis (PCA). The contents of total phenolics, flavonoids and phenolic acids, along with the estimations of antioxidant capacities of five wild species of *Physalis* were analyzed using PCA. *Group A*: samples collected in August; *group B*: samples collected in September; *C*: samples collected in October

epicatechin (Table 2). However, the highest value, which was reached by the extract of calyces of *P. solanacea* (426) was lower than the value reported by Diaz et al. (2012) for the fruits of *P. alkekengi* ($IC_{50} = 36.58$ mg/g dry weight).

The variations found among the two samples of *P. solanacea* suggest that as the phenolic profiles were the same for both samples (Figs. 1, 2, 3), plant age caused variations in the concentrations of phenolics (Tables 2, 3), all which affected the antioxidant properties of tissues. These kind of effects have been reported for other plant species (Del Baño et al. 2003). In fruits, the accumulation of other antioxidants, like ascorbic acid, could affect the antioxidant properties. Synthesis of ascorbic acid has been reported in the fruits of *P. peruviana* (Rop et al. 2012) and may occur also in the species of *Physalis* here analyzed.

The low association between the antioxidant properties and the phenolic, flavonoid, and phenolic acid contents found in the present study agree with the reports of other authors (Morais et al. 2011; Barriada-Bernal et al. 2014), who suggested that the antioxidant potential of a given total extract depends not only on the amount of phenolic compounds present in it but also on the types of phenolics accumulated, their proportions, as well as the presence of compounds of different chemical nature.

Principal component analysis

The results of the PCA revealed a tendency of grouping based on the developmental stage (Fig. 5), group A included samples collected in August (*P. hederifolia* var. *hederifolia* and *P. angulata*), group B was formed by samples collected in September (*P. patula* and *P. solanacea* 412), and group C was formed by samples collected in October (*P. solanacea* 426 and *P. subulata*). These results reveal the effect of time of collection and the associated developmental stage, on the phenolic, flavonoid, and phenolic acids contents and the antioxidant properties of different tissues of five wild species of *Physalis*. Systematic studies concerning the seasonal variation of phenolics of the species of *Physalis* are needed, as it has been done for some species of *Betula* (Raal et al. 2015), and concerning the developmental variation, as it has been carried out for *Rosmarinus officinalis* (Del Baño et al. 2003). Those studies would contribute to identify the best collecting time and developmental stage to obtain the highest concentration of some phenolic compounds and the environmental conditions in which they are produced.

Conclusions

The wild analyzed species of *Physalis* synthesize important amounts of diverse phenolic compounds. The phenolic profiles of leaves, fruits, and calyces were species-specific; this reveals phenolic patterns as significant specific chemomarkers in the genus. The dominant phenolic compounds accumulated in leaves and calyces were 3-*O*-glycoside derivatives of kaempferol, and the dominant ones in fruits were phenolic acids; all these compounds are recognized for their important activities as antioxidants. Because of their phenolic compositions, *P. angulata, P. patula, P. hederifolia var. hederifolia, P. solanacea, and P. subulata* may be considered as relevant potential sources of natural antioxidants to be used in the food and pharmaceutical industries. The present phenolic fingerprinting contributes to generate a better knowledge for using all the potential of these wild species of plants. The chemical variability of wild species of *Physalis* offers the opportunity for domesticating and developing new cultivars with enhanced accumulation of phenolics having positive effects on human health. The results inform about the importance of consuming these plant species and support the use of calyces and leaves in traditional medicine. The not always high correlation between phenolic or flavonoid contents and antioxidant activities suggests the relevance of the phenolic profiles, besides the contents of phenolics, in determining these activities.

Abbreviations

AAE: ascorbic acid equivalents; CE: *p*-coumaric acid equivalents; DAD: diode array detector; DPPH: 2,2-diphenyl-1-picrylhydrazyl; dt: dry tissue; EC_{50}: efficient concentration at 50 %; GAE: gallic acid equivalents; HPLC: high performance liquid chromatography; IC_{50}: half maximal inhibitory concentration; NE: naringenin equivalents; PCA: principal component analysis; PC1: principal component 1; PC2: principal component 2; QiE: quercitrin equivalents; TAC: total antioxidant capacity; UV: ultraviolet.

Authors' contributions

NAA conceived and designed the study. JRMM made collections of plant material and carried out the antioxidant capacity evaluations. LSGV and JNUS performed the HPLC-DAD analysis. NAA, LSGV, and JNUS analyzed and interpreted the data. MSGE and YHA conducted the botanical identification of the plants and made a critical revision, providing important intellectual content. All authors read and approved the final manuscript.

Authors' information

JRMM: Master in Science at National Polytechnic Institute (CIIDIR Durango IPN), his area of expertise is in Food Biotechnology. NAA: Dr., researcher at CIIDIR Durango IPN, her area of expertise is in Secondary Metabolism. MSGE: Dr., researcher at CIIDIR Durango IPN, she is Associate Editor of Acta Botanica Mexicana, her area of expertise is in Plant Systematics and Ecology. JNUS: Master in Science, researcher at CIIDIR Durango IPN, his research has been focused on Plant Production under Protected Environment. LSGV: Dr., researcher at CIIDIR Durango IPN, her area of expertise is on Environmental Analytical Chemistry. YHA: Dr., researcher at CIIDIR Durango IPN, her area of expertise is in Plant Systematics and Ecology.

Acknowledgements

The authors thank the Comisión de Operación y Fomento a las Actividades Académicas, Instituto Politécnico Nacional for the stimuli for research, and to Consejo Nacional de Ciencia y Tecnología for the grant (265393) to one of the authors.

Compliance with ethical guidelines

Competing interests

The authors declare that they have no competing interests.

References

Almaraz-Abarca N, González-Elizondo MS, Campos MG, Ávila-Sevilla ZE, Delgado-Alvarado EA, Ávila-Reyes JA (2013) Variability of the foliar phenol profiles of the *Agave victoriae-reginae* complex (Agavaceae). Bot Sci 91:295–306

Ames BM, Shigena MK, Hagen TM (1993) Oxidants, antioxidants and the degenerative diseases of aging. Proc Natl Acad Sci USA 90:7915–7922

Barriada-Bernal LG, Almaraz-Abarca N, Delgado-Alvarado EA, Gallardo-Velázquez T, Ávila-Reyes JA, Torrres-Morán MI, González-Elizondo MS, Herrera-Arrieta Y (2014) Flavonoid composition and antioxidant capacity of the edible flowers of *Agave durangensis* (Agavaceae). CyTA J Food 12:105–114

Campos MG, Markham KR (2007) Structure information from HPLC and on-line measured absorption spectra-flavone, flavonols and phenolic acids. Coimbra University Press, Coimbra

Chang JC, Lin CC, Wu SJ, Lin DL, Wang SS, Miaw CL, Ng LT (2008) Antioxidative and hepatoprotective effects of *Physalis peruviana* extract against acetaminophen-induced liver injury in rats. Pharm Biol 46:724–731

D'Arcy WG (1991) The Solanaceae since 1976, with a review of its biogeography. In: Hawkes JG, Lester M, Nee RN, Estrada N (eds) Solanaceae III: taxonomy, chemistry, and evolution. Royal Botanic Gardens, London

Del Baño MJ, Lorente J, Castillo J, Benavente-García O, del Río JA, Ortuño A, Quirin KW, Gerard D (2003) Phenolic diterpenes, flavones, and rosmarinic acid distribution during the development of leaves, flowers, stems and roots of *Rosmarinus officinalis*. Antioxidant activity. J Agric Food Chem 51:4247–4253

Diaz P, Jeong SC, Lee S, Khoo C, Koyyalamudi SR (2012) Antioxidant and anti-inflammatory activities of selected medicinal plants and fungi containing phenolic and flavonoid compounds. Chin Med J 7:26

Dobre I, Dâdei G, Patrascu L, Elisei AM, Segal R (2010) The antioxidant activity of selected Romanian honeys. AUDJG Food Technol 34:67–73

Eick E (2008) Solanaceae and Convolvulaceae: Secondary metabolites. Biosynthesis, chemotaxonomy, biological and economic significance. Springer, Heidelberg

Emerenciano VP, Militão JS, Campos CC, Romoff P, Kaplan MA, Zambon M, Brant AJ (2001) Flavonoids as chemotaxonomic markers for Asteraceae. Biochem Syst Ecol 29:947–957

González-Mendoza D, Martínez-Ascencio D, Hau-Poox A, Mendez-Trujillo V, Grimaldo-Juárez O, Santiaguillo-Hernández J, Cervantes-Díaz L, Avilés-Marin S (2011) Phenolic compounds and physiochemical analysis of *Physalis ixocarpa* genotypes. Sci Res Essays 6:3808–3814

Hernández J, Yáñez S (2009) Aprovechamiento tradicional de las especies de *Physalis* en México. R Geo Agric 43:81–86

Ismail N, Alam M (2001) A novel cytotoxic flavonoid glycoside from *Physalis angulata*. Fitoterapia 72:676–679

Kouloura E, Genta-Jouve G, Pergola C, Krauth V, Litaudon M, Benaki D, Wertz O, Michel S, Mikros E, Skaltsounis L, Halabalaki M (2014) Dereplication and metabolomics strategies for the discovery of bioactive natural products: the Acronychia example. Planta Med 80:SL3

Krishnaiah D, Sarbatly R, Nithyyanandam R (2011) A review of the antioxidant potential of medicinal plant species. Food Bioprod Process 89:217–233

Le K, Chiu F, Ng K (2007) Identification and quantification of antioxidants in *Fructus lycii*. Food Chem 105:353–363

Licodiedoff S, Koslowski LAD, Ribani RH (2013) Flavonols and antioxidant activity of *Physalis peruviana* L. fruits at two maturity stages. Acta Sci Technol 35:393–399

Mabry TJ, Markham KR, Thomas MB (1970) The systematic identification of flavonoids. Springer, New York

Maldonado E, Gutiérrez R, Pérez-Castorena AL, Martínez M (2012) Orizabolide, a new withanolide from *Physalis orizabae*. J Mex Chem Soc 56:128–130

Morais M, Moreira L, Feás X, Estevinho LM (2011) Honeybee-collected pollen from five Portuguese natural parks: palynological origin, phenolic content, antioxidant properties and antimicrobial activity. Food Chem Toxicol 49:1096–1101

Nurmi K, Ossipov V, Haukioja E, Pihlaja K (1996) Variation of total phenolic content and low-molecular-weight phenolics in foliage of the mountain birch trees (*Betula pubescens* ssp. tortuosa). J Chem Ecol 22:2033–2050

Pérez-Castorena AL, Hernández IZ, Martínez M, Maldonado E (2013) Chemical study of calyxes and roots of *Physalis solanaceus*. Rec Nat Prod 7:230–233

Prieto P, Pineda M, Aguilar M (1999) Spectrophotometric quantitation of antioxidant capacity through the formation of a phosphomolybdenum complex: specific application to the determination of Vitamin E. Anal Biochem 269:337–341

Raal A, Boikova T, Püssa T (2015) Content and dynamics of polyphenols in *Betula* spp. leaves naturally growing in Estonia. Rec Nat Prod 9:41–48

Robbins RJ (2003) Phenolic acids in foods: an overview of analytical methodology. J Agric Food Chem 51:2866–2887

Rockenbach II, Rodrigues E, Cataneo C, Gonzaga LV, Lima A, Mancini-Filho J, Fett R (2008) Phenolic acids and antioxidant activity of *Physalis peruviana* L. Alimentos e Nutrição 19:271–276

Rop O, Mlcek J, Jurikova T, Valsikova M (2012) Bioactive content and antioxidant capacity of cape gooseberry fruit. Cent Eur J Biol 7:672–679

Santiaguillo HJF, López RMA, Peña LA, Cuevas SC, Sahagún CJ (1994) Distribución, colecta y conservación de germoplasma de tomate de cáscara (*Physalis ixocarpa* Brot.). Rev Chapingo Ser Hortic 2:125–129

Siddhuraju P, Becker K (2003) Antioxidant properties of various solvent extracts of total phenolic constituents from three different agroclimatic origins of drumstick tree (*Moringa oleifera* Lam.) leaves. J Agric Food Chem 51:2144–2155

Tapas AR, Sakakar DM, Kakde RB (2008) Flavonoids as nutraceuticals: a review Trop J Pharm Res 7:1089–1099

Veit M, Beckert C, Höhne C, Bauer K, Geiger H (1995) Iterspecific and intraspecific variation of phenolics in the genus *Equisetum* subgenus *Equisetum*. Phytochemistry 38:881–891

Wang SY, Lin HS (2000) Antioxidant activity in fruits and leaves of blackberry, raspberry, and strawberry varies with cultivar and developmental stage. J Agric Food Chem 48:140–146

Whitson M, Manos PS (2005) Untangling *Physalis* (Solanaceae) from the physaloids: a two-gene phylogeny of the Physalinae. Syst Bot 30:216–230

Wong RWK, Rabie ABM (2009) Grapefruit flavonoids: naringenin and naringinin. In: Keller RB (ed) Flavonoids. Biosynthesis, biological effects and dietary sources. Nova Science Publishers, New York, pp 141–180

Wu SJ, Chang SP, Lin DL, Wang SS, Hou FF, Ng LT (2009) Supercritical carbon dioxide extract of *Physalis peruviana* induced cell cycle arrest and apoptosis in human lung cancer H661 cells. Food Chem Toxicol 47:1132–1138

Zhang Q, Cui H (2005) Simultaneous determination of quercetin, kaempferol, and isorhamnetin in phytopharmaceuticals of *Hippophae rhamnoides* L. by high-performance liquid chromatography with chemiluminescence detection. J Sep Sci 28:1171–1178

Zimmermann BF, Galensa R (2007) One for all-all for one: proof of authenticity and tracing of foods with flavonoids. Analysis of proanthocyanidins in barley and malt. Eur Food Res Technol 224:385–393

3

Cambial activity and xylogenesis in stems of *Cedrus libani* A. Rich at different altitudes

Aylin Güney*, Danielle Kerr, Ayça Sökücü, Reiner Zimmermann and Manfred Küppers

Abstract

Background: The dynamics of cambial activity and xylogenesis provide information on how and to what extent wood formation respond to climatic variability. The Lebanon Cedar (*Cedrus libani* A.Rich) is a montane tree species which is distributed along a wide altitudinal range in the northeastern Mediterranean region, currently considered as a potential forest species for Central Europe with respect to climate change. This study provides first data on intra-annual growth dynamics at cellular level using the microcore technique for a montane Mediterranean tree species at different altitudes within and outside its natural range.

Results: Microcores were collected fortnightly in the growing season of 2013 in order to study temporal dynamics of cambial activity and xylogenesis in stems of *C. libani* at different altitudes in the Taurus Mountains (1000 – 2000 m a.s.l.) and at a plantation at Bayreuth (330 m a.s.l.; Germany). The dormant cambium consisted of about 5 cells at the Turkish sites and 7 cells at Bayreuth. Cambial activity set in, when daily minimum temperatures exceeded 0 °C and daily means of air and stem temperature exceeded 5 °C. Xylogenesis started between April and May, ended approximately the end of September to the beginning of October and lasted 134 (at tree line) to 174 days (at the lowest Turkish site). Mean ring widths varied from 0.55 to 3.35 mm, with highest values observed at Bayreuth very likely resulting from a steady water supply during growing season. Means of daily cell production rates varied from 0.73 to 0.12. Samples containing traumatic resin ducts occurred only rarely and where not used for analysis.

Conclusions: In *C. libani*, onset and dynamics of cambial activity and xylogenesis are triggered by daily means of stem and air temperatures whereas water availability has a higher influence on growth rates and cessation of wood formation. Within sites, duration of xylogenesis does not significantly differ with respect to age and tree size. *C. libani* grows well outside its natural range and thus may be a promising species for forestation in Central Europe with respect to climate change. We suggest further studies on if/how traumatic resin ducts influence tree ring width.

Keywords: *Cedrus libani*; Cambium; Xylogenesis; Microcoring; Cell differentiation; Wood formation; Stem temperature; Traumatic resin ducts

Background

Wood formation is a highly dynamic process regulated by physiological and environmental factors. It can be subdivided in onset, rate and duration of cell differentiation (Plomion et al. 2001; Deslauriers et al. 2003; Deslauriers and Morin 2005). Xylem cells are produced by the cambium and go through the phases of enlargement, secondary wall thickening, lignification and cell death (Plomion et al. 2001; Rossi et al. 2006c). Information on wood formation dynamics are necessary (a) to understand how and to what extent cambial activity and xylogenesis (with the different stages of cell development) respond to climatic variability and (b) to predict how forecasted climate change may influence tree growth and on the long hand forest productivity (Moser et al. 2009; Camarero et al. 2010; Vieira et al. 2014). Dendrochronological studies in general provide information on tree growth on an inter- annual to multi- centennial time scale and lack information as to when main events of tree ring formation take place within a single growing season. Therefore, intra- annual studies provide more detailed information on patterns and processes of xylogenesis within one growing season (Deslauriers et al.

* Correspondence: aysokucu@uni-hohenheim.de
University of Hohenheim, Institute of Botany, Garbenstrasse 30, 70599 Stuttgart, Germany

2003, Rossi et al. 2007). The microcoring method (Rossi et al. 2006b) is easy to apply and allows to monitor the dynamic processes of cambial activity and xylogenesis. Some studies used the microcore method to investigate wood formation of Mediterranean tree species with the aim to determine the influence of climate parameters on cambial activity and xylogenesis. For example, Vieira et al. (2014) found that maritime pine cambial activity appears to be controlled by temperature (which triggers the onset of growth) and reduced water availability (which leads to cessation of growth). Camarero et al. (2010) analyzed intra-annual patterns of three conifer species from continental Mediterranean climates and showed that plasticity in xylogenesis patterns enables species to occupy sites with more variable climatic conditions. However, to our knowledge there are no studies present which investigate cambial activity and xylogenesis of a montane (oreophilous) Mediterranean tree species along an altitudinal gradient which could contribute to a better understanding on how the dynamics of wood formation change with altitude and varying site-specific conditions. Lebanon cedar (*Cedrus libani* A. Rich) is a drought tolerant conifer which is distributed along a wide altitudinal range (600 – 2300 m a.s.l) in Turkey, the Lebanon and Syria. Because of its stress tolerance and high timber quality many efforts in reforestation and afforestation have been undertaken within and outside its natural range (Boydak 2007; Boydak and Caliskan 2014; Fady et al. 2003; Khuri et al. 2000; Schütt et al. 2008; Uyar et al. 1990). *C. libani* appears to be a promising species to substitute tree species in Central Europe which might suffer from the expected climatic changes in their recent natural distribution areas. However, little is known about its growth dynamics. The objective of the study was to determine with the use of the microcore method (a) how cambial activity and xylogenesis in stems of *C. libani* take place along an altitudinal gradient (critical timings of the onset of cell production, termination of wood formation, cell number and ring width) (b) how dynamics of cambial activity and xylogenesis are linked with contemporary microclimatic conditions (air temperature, rain, drought) and (c) how xylogenesis of *C. libani* occur under central European climate conditions where a uniform water supply is available throughout the year (no severe summer drought). Based on the information of xylogenesis we present an initial insight on the effect of spring temperature on the onset of growth and the relative role of vegetation period on seasonal growth when drought periods are present or missing.

Methods

Study species

Cedrus libani A. Rich is an evergreen conifer of the *Pinaceae* family (Boydak and Calikoglu 2008).Today, *C. libani* has its largest distribution in Turkey with the purest stands located mainly between 800 and 2200 meters in the western Taurus Mountains where it reaches ages up to 1000 years (Dirik 2000; Akkemik 2003; Avci and Carus 2005; Carus and Avci 2005; Ducrey et al. 2008; Kurt et al. 2008). Stand sites are characterized by a summer dry season primarily between July and September. Annual mean temperatures range from 7.5 to 15 °C, with possible air temperature extremes as low as −35 °C and up to +40 °C (Aussenac 1984; Atalay 1987; Senitza 1989). Cedar forests contain a high amount of endemic species which make them biologically and ecologically valuable ecosystems (Kavgaci et al. 2010). Being highly adaptive to temperature extremes and summer drought, *C. libani* is one of the major commercial tree species in Turkey besides *Pinus brutia* and *Abies cilicica* (Brooks et al. 2008), and it consists of valuable wood which is very durable and versatile in application and use (Boydak 2003).

Study sites

The study was conducted simultaneously in Turkey in the Southwestern Taurus Mountains at the Cedar Research Forest (CRF) near Elmali and in Germany at the Ecological-Botanical Gardens (EBG) at Bayreuth. Four study sites were chosen in the CRF along an altitudinal transect ranging from 1055 m to 1960 m a.s.l. and one at the EBG at 335 m a.s.l. (Table 1). Parts of the CRF that are forested mainly consist of pure *Cedrus libani* stands which range from 1000 m up to about 2200 m a.s.l., mixed occasionally with some *Juniperus* individuals. The transect stretches from the outer border of the forest near the Avlan lake to the upper timberline at 1990 m a.s.l. on a northwest-facing slope. The CRF is characterized by the Mediterranean mountain climate with relatively cold winters and a drought period during summer (from June to September) with a mean annual temperature of 7.3 °C (measured temperature extremes are −31 °C and +34 °C) and a mean annual precipitation of 644 mm (1970–2003, Turkish State Meteorological Service). The local climate in Bayreuth is subcontinental with a mean annual temperature of 7.9 °C (minimum: −25 °C, maximum: 36.1 °C) and mean annual precipitation of 724 mm (1971–2000; Foken et al. 2004). At the CRF clay and clay loam soils were dominant with pH varying from 7.2 to 8.1 (Basaran et al. 2008). At the EBG soil material was a mix of clay and sandy loam and pH varies from 4.4 to 5.3. At each site three individuals of *C. libani* with upright stems and no visible damages or parasitic infestations were selected (Table 1). *C. libani* individuals at the EBG and the CRF are of the same origin. All trees selected were adult and formation of juvenile wood was completed when sampling occurred (Bao et al. 2001; Plomion et al. 2001). Meteorological data such as stem temperature of the sampled trees and air temperature were recorded with thermocouples throughout the whole study period at each site

Table 1 Location of the study sites and characteristics of the sampled trees

ID	Site	CC	Latitude Longitude	Altitude (m a.s.l.)	Forest type	Population density (trees ha^{-1})	DBH (m)	Height (m)	Age at DBH (year)	LAI (m^2 m^{-2})
T1	CRF Elmali	TR	36.58422°N 30.03077°E	1960	Natural forest (Tree line)	478	0.6 ± 0.73	9.2 ± 3.7	42- 360 (152 ± 180)	0.95 ± 0.75
T2	CRF Elmali	TR	36.58520°N 30.02035°E	1665	Natural forest	501	0.39 ± 0.12	20 ± 2.1	84- 146 (112 ± 31)	4.7 ± 0.69
T3	CRF Elmali	TR	36.57771°N 29.98542°E	1355	Natural forest	533	0.4 ± 0.09	22.8 ± 1	90- 119 (101 ± 15)	1.49 ± 0.38
T4	CRF Elmali	TR	36.57800°N 29.96881°E	1055	Natural forest	334	0.44 ± 0.08	18.1 ± 3.5	65- 85 (75 ± 10)	2.0 ± 0.82
O1	EBG Bayreuth	D	49.925915°N 11.582851°E	335	Plantation	-	0.34 ± 0.03	12.3 ± 1.6	35	-

The diameter at breast height (DBH), height of the sampled trees and the leaf area index (LAI) are presented as means with standard deviations (SD). The age of the sampled trees is shown as age range at DBH and as means with SD (in parentheses). At the site O1, all sampled trees were of same age, population density and LAI were not determined. ID: Site identification. CC: Country code. TR: Turkey. D: Germany

respectively. Measurements were taken every 10 min and stored as hourly averages in data loggers (DL2e Data Logger, Delta-T Devices Ltd. Cambridge). For analysis, daily minimum, mean and maximum air temperatures as well as mean stem temperatures were calculated. Values of leaf area index (LAI) were obtained using the LAI-2000 plant canopy analyzer (Li-Cor, Inc., Lincoln, NE) at the Turkish sites (Table 1). Individuals at the EBG were planted in a single line from North to South with no significant shading, so LAI was not determined for this site. Precipitation data (monthly sums) of Bayreuth were derived from the meteorological station at the Ecological-Botanical Gardens of Bayreuth. For the Turkish site monthly sums of precipitation were derived from the meteorological station at Elmali (Turkish State Meteorological Service) which is situated circa 15 km afar of the CRF. At the sites T1 and T3 WatchDog weather stations (WatchDog 200 Series, Spectrum Technologies, Inc., IL, USA) were installed to measure precipitation with a 15 min frequency. Values were than computed as monthly sums. Due to data gaps at the measurement sites (T1 and T3) we plotted respective monthly sums of precipitation data available from the CRF with continuous precipitation data from Elmali to check if a linear relationship was present. Thereafter we performed nonlinear regressions (dynamic curve fit, polynomial, quadratic) using SigmaPlot (SigmaPlot software for Windows Version 10.0, Systat Software Inc., 2006) to compute and fill missing precipitation data at T1 and T3 (Fig. 3). An additional file shows the regression analysis we used to model missing precipitation data [see Additional file 1].

Microcoring and histological analyses

Every 2 weeks during the growing season from March to October 2013, wood microcores were collected in each tree using a "Trephor" (Rossi et al. 2006b). The Trephor is a hand- sized tool with a cutting tube which is inserted into the wood using a rubber mallet. Samples were taken from stems at, as well as 50 cm below and above breast height (1.3 m) from the stem side parallel to the slope. Using a chisel, the outer bark was removed carefully before sampling at the sampling point was carried out. Microcore samples were 2 mm in diameter and between 5–20 mm in length, containing parts of the phloem, the cambium and the latest year rings of the xylem. To avoid tissue damage and formation of traumatic resin ducts (Fig. 1f) or other wood reactions, samples were taken following a spiral pattern along the stem surface leaving a minimum space of 5 cm between extraction points (Deslauriers et al. 2003). Immediately after sampling, microcores were placed in Eppendorf tubes (2 ml) containing a FAA fixative solution (formalin - acetic acid - alcohol, in a ratio of 5: 5: 90). To avoid tissue damage they were stored as soon as possible in a refrigerator until further use. After dehydration, using an ethanol series (70 % - 70 % - 90 % - 96 % - 100 % - 100 %; two hours steps), samples were embedded in glycolmethacrylate (Technovit 7100; Heraeus Kulzer, Wehrheim, Germany) with the transversal side facing upwards. Transverse sections were cut to a thickness of 3–5 µm using a rotary microtome (Microm HM 340 E), stained with fuchsine, safranin and astrablue (Etzold 1983) for about 45 min and fixed onto microscope slides with Roti (R) - Histokitt. In order to investigate cambial activity and wood formation dynamics, to distinguish cells in the cambial zone and to differentiate between cells according to their stage of xylogenesis (Fig. 1), transverse sections were examined and photographed on a Zeiss Axioplan microscope. The cambial zone consists of cambial initials and undifferentiated derivates with thin walls and small radial diameters (Fig. 1a). The onset of growth was determined by the first appearance of cells undergoing radial enlargement, resulting in cells larger than in the cambial zone and

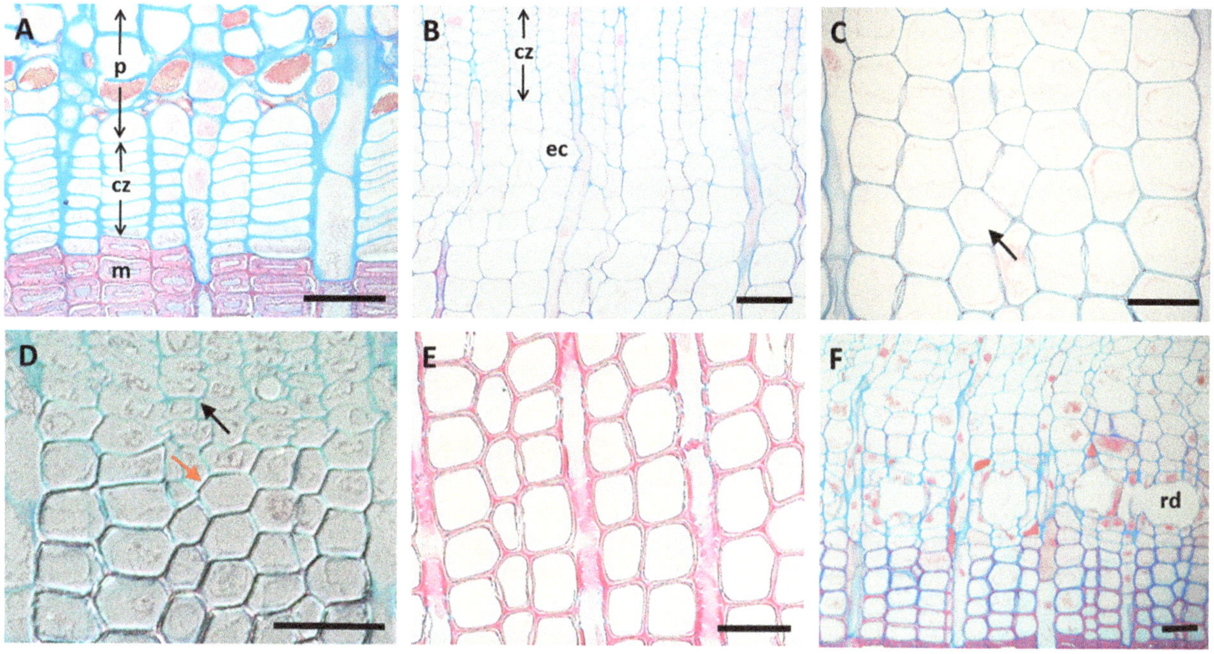

Fig. 1 Cross sections of *Cedrus libani* stem wood. **a** Cambial zone (cz) during dormancy at the beginning of March with cambial initials and derivates, mature tracheids (m) of the previous year with thick secondary walls, and phloem (p) (**b**) Cambial zone during the growing season in June, enlarging cells (ec) (**c**) Wall thickening cells. The black arrow points at the intact protoplast of the cell (**d**) Enlarging cells which are not birefringent (black arrow) and wall thickening cells which are birefringent (red arrow) under polarized light (**e**) Mature cells with empty cell bodies and completely lignified secondary walls which appear intensely pink. **f** Resin ducts (rd) in the early wood during June. The bottom of the graphs are towards center of the stem. Scale bars: 50μm

characterized by thin primary cell walls (Rossi et al. 2006a, 2007) (Fig. 1b). When radial expansion of tracheids ceased, secondary walls began to develop. In this so-called wall thickening stage cells still contained the protoplast (Fig. 1c) and started to appear birefringent under polarized light because of the special arrangement of cellulose micro fibrils in the secondary wall (Abe et al. 1997; Rossi et al. 2007) (Fig. 1d). Mature cells had empty cell bodies due to protoplast autolysis and completely lignified secondary walls (Fig. 1e). The end of growth was considered when all cells reached maturity (Rossi et al. 2006a, 2008b). In each sample the number of cell types were counted and widths of the actual performed tree ring were measured along three radial files and then averaged for subsequent analyses (Deslauriers et al. 2003; Rossi et al. 2006a, 2007; Thibeault-Martel et al. 2008). For each tree cell production rates were computed dividing the total number of cells formed within 2013 by the number of days spent in xylogenesis. Though it is not well known if and how much the presence of traumatic resin ducts influences wood formation and tree ring width we avoided samples with resin ducts in our study. In some sections where one part of the developing tree ring contained resin ducts and another part was free of them, it became clear that resin ducts generally did not affect the width of tree rings. However, at few occasions tree rings containing resin ducts appeared to be wider in radial direction as compared to parts without such ducts but studies with respect to this also show no clear pattern (Fahn et al. 1979; Wimmer and Grabner 1997).

Statistical analysis

Univariate comparisons were used for means of cell number in dormant cambium, final number of xylem cells produced, daily cell production rate and annual tree ring width on the five sites. Homogeneity of variances among sites was assessed by Levene's Test. In order to unveil whether xylogenesis (e.g. cell numbers, ring widths) was affected by site (altitude) we used Analysis of Variance (ANOVA). ANOVA and Tukey tests were performed for number of cells in dormant cambium when the requirements for homogeneity of variances were met, otherwise Welch's ANOVA and Games-Howell tests were performed (for total number of xylem cells, cell production rate and ring width). Statistical analyses were performed using SPSS (IBM SPSS Statistics for Windows, Release 20.0.0, 2011)

Results

From March to October 2013, calculated means of air temperature and stem temperature differed between all sites. Highest values were found at T4 with 16.9 °C for mean air temperature and 16.2 °C for mean stem temperature and lowest values at O1 with 11.7 °C and 12.3 °C, respectively. Mean air temperatures and mean stem temperatures were 15.3 °C and 15.1 °C for T3, 13.1 °C and 13.2 °C for T2 and 12.9 °C and 13.1 °C for T1. Precipitation was evenly distributed throughout the growing season at O1 whereas summer drought was present at the Turkish sites (Fig. 3).

During dormancy, when no cells have formed newly, the number of cells in the cambial zone (Fig. 1A) varied between four and nine (Table 2) showing significant differences among sites (ANOVA, $F = 38.19$, $P < 0.001$) with highest cell number at O1 (7.7 ± 1 cells; Tukey Test, $P < 0.001$). Among the Turkish sites (T4 to T1) cell numbers were similar and not significantly different (Tukey Test, $P > 0.05$). At all sites, the dormant period of the cambium lasted approximately from the beginning of August until April the following year. Cambial activity was first observed on April 3rd (day of the year, DOY: 94) at T4, followed by O1 (DOY 110) and the sites T3 to T1 (Fig. 2). Highest cambial activity was observed shortly after its initiation in spring and lasted 50 to 60 days. During this period the maximum number of cambial cells was observed at O1 (15 ± 2 cells).

Once cells started to enlarge and the number of enlarging cells peaked in May and June, the number of cambial cells consequently decreased. Wall thickening was initiated approximately twenty days after the first cells reached the enlargement stage and it lasted for about 126 ± 17 days. Cambial activity, cell enlargement and wall thickening occurred sequentially forming a bell-shaped curve when the number of cells in a particular stage is plotted against time. Wall thickening was the stage that lasted the longest during xylogenesis (Fig. 2). The number of mature cells and the total number of cells formed at each sampling time point resulted in a sigmoidal curve (Fig. 2). When all cells had reached maturity and when no cells in the wall thickening stage were any longer present, xylem cell differentiation and thus growth of the year ring was considered to be complete. Growth ended first at T3 on 29th of July (DOY 211) and latest at O1 on October 10th (DOY 284). About four to five months were required to form a tree ring with the longest duration at T4 (174 days) and shortest at the highest site T1, with 134 days of xylogenesis (Table 2). For the means in total number of cells at the end of xylogenesis a high standard variation is obvious at T4 and T1, whereas it is low at O1, T3 and T2. Since within- group variances were not equal (Levene's test, $P < 0.01$), a Welch's ANOVA was performed, showing significant differences among sites ($P < 0.001$). The total number of xylem cells was significantly higher at O1 compared to T3, T2 and T1 (Games Howell Test, $P < 0.001$) and significantly higher at T4 compared to T3 and T2 (Games Howell Test, $P < 0.05$), while there were no significant differences comparing T3 and T2 (Games Howell Test, $P > 0.05$).

Mean values of daily cell production rates varied from 0.1 and 0.7 cells per day (Table 2, unequal within- group variances (per site); Welch's ANOVA, $P < 0.001$). Cell production rates at O1 were significantly higher than at T3, T2 and T1 (Games Howell Test, $P < 0.001$) and significantly higher at T4 compared to T3 and T2 (Games Howell Test, $P < 0.05$), but did not differ significantly between sites T2 and T3 (Games Howell Test, $P > 0.05$). When measuring tree ring width from microcores, cells in the cambial zone were not included, so that tree ring formation started with the first presence of enlarging cells which is equivalent to the onset of growth (Table 2). When comparing it with weather conditions, it becomes evident that prior to the initiation of growth a pre-

Table 2 Timing of main events and characteristics of tree ring formation in stems of *Cedrus libani*

Site	O1	T4	T3	T2	T1
Observation	(335 m)	(1055 m)	(1355 m)	(1665 m)	(1960 m)
Number of cells in dormant cambial zone	7.7 ± 1.0	5.5 ± 0.5	4.9 ± 0.6	5.0 ± 0.5	5.0 ± 0.6
Onset of cambial activity	Apr 19 (110)	Apr 3 (94)	May 3 (124)	May 3 (124)	May 3 (124)
Onset of growth	Apr 30 (121)	Apr 10 (101)	May 3 (124)	May 3 (124)	May 15 (136)
Start of cell wall thickening	May 19 (140)	May 7 (128)	June 7 (155)	June 7 (155)	June 7 (155)
Dormant cambium	Sep 6 (250)	Aug 26 (239)	July 29 (211)	Aug 12 (225)	Aug 12 (225)
End of growth	Oct 10 (284)	Oct 1 (275)	Sep 24 (268)	Sep 26 (270)	Sep 26 (270)
Duration of cambial activity (days)	140	145	87	101	101
Duration of xylogenesis (days)	163	174	144	146	134
Total number of xylem cells	120.1 ± 8.5	96.6 ± 34.0	18.0 ± 6.4	31.6 ± 8.3	45.3 ± 13.4
Cell production rate (cells/day)	0.73 ± 0.05	0.55 ± 0.19	0.12 ± 0.04	0.21 ± 0.05	0.33 ± 0.1
Ring width (mm)	3.35 ± 0.65	2.76 ± 0.99	0.55 ± 0.19	1.01 ± 0.38	1.36 ± 0.39

For each site, number of cells in dormant cambial zone, radial growth, total number of xylem cells and cell production rate are presented as means with standard deviations. Numbers in parentheses indicate day of year (DOY) of first observation of the specified characteristics during the growing season 2013.

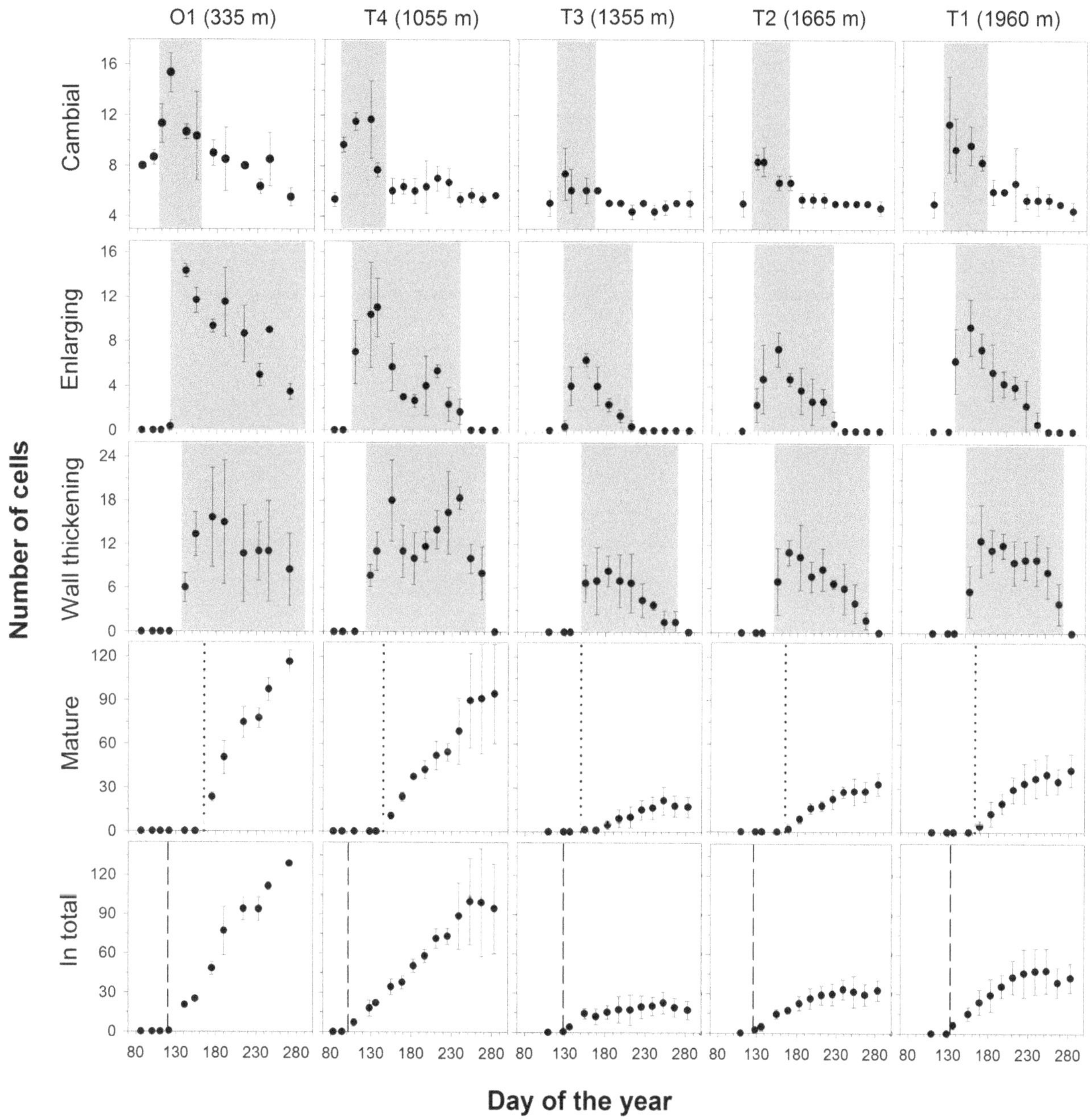

Fig. 2 Timing and dynamics of cambial activity and xylogenesis in stems of *Cedrus libani* during 2013. Number of cells in the cambial zone, in radial enlargement phase, in wall thickening phase, mature cells and total number of xylem cells are shown for the German (O1) and the four Turkish sites (T1 –T4). Error bars represent standard deviations among trees per site. Periods of cambial, enlargement and wall thickening activity are highlighted in gray. Dotted vertical lines indicate the time when first mature cells could be observed, dashed vertical lines the time when enlarging cells emerged thus wood formation was initiated

running fortnight free of frost is required (Fig. 3). From there onwards, daily means of air and stem temperature were above 5 °C and stayed above this level during the whole growing season.

Maximum radial growth occurred between May and July. Tree ring widths at the end of the growing season varied significantly among sites (Welch's ANOVA, $P < 0.001$). Lowest ring widths at the end of growth occurred at T3 with 0.55 ± 0.19 mm and highest at O1 with 3.35 ± 0.65 mm (Table 2). Ring widths were significantly higher at O1 compared to T3, T2 and T1 (Games Howell Test, $P < 0.001$) and significantly higher at T4 compared to T3 and T2 (Games Howell Test, $P < 0.05$). The highest within-site variances regarding ring widths were observed at T4 with 2.76 ± 0.99 mm.

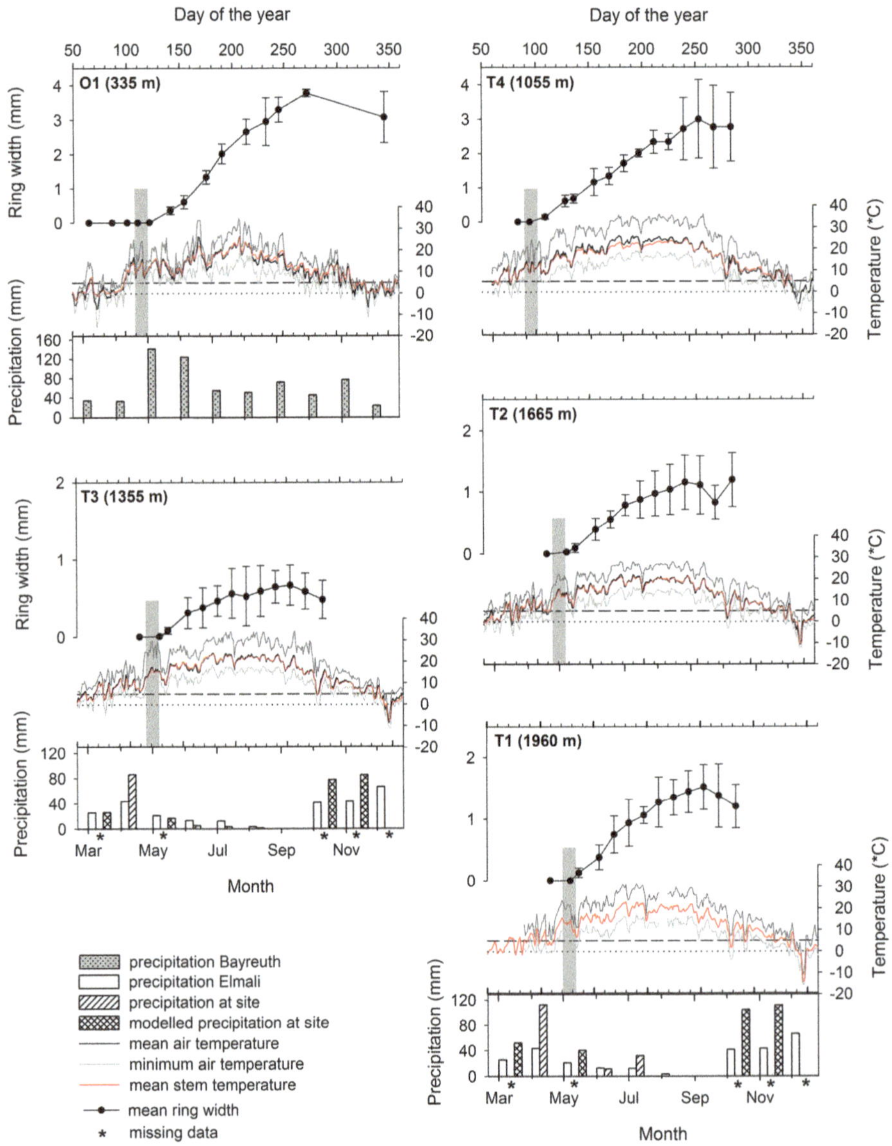

Fig. 3 Tree ring formation in *Cedrus libani* with site specific microclimatic data. Shown are tree ring width resulting from radial growth with error bars indicating standard deviation among trees, daily maximum (dark gray line), mean (black line) and minimum (gray line) air temperature and daily mean stem temperature of sampled trees (red line) during the growing season of 2013. Vertical gray bands highlight the specific weather conditions over 14 days pre-running the first observation of enlarging cells and thus the beginning of radial increment. Precipitation data (monthly sum) are shown for O1 (grey, dotted bars), for T3 and T1. White bars show precipitation data from the Elmali Meteorological Station, hatched bars indicate precipitation data measured at the respective site, double-hatched bars show modelled precipitation data where site-specific precipitation data were missing (indicated with a star)

Discussion

Several studies have been carried out regarding intra-annual wood formation in conifer and broadleaf trees (e.g. Camarero et al. 2010; Vavrcik et al. 2013), but information on how dynamics of wood formation occur in trees along an altitudinal gradient in the northeastern Mediterranean Mountains is completely missing. This study investigated cambial activity and xylogenesis in stems of *C. libani* along an altitudinal gradient in the Taurus Mts during the growing season of 2013. At the same time *C. libani* individuals were investigated at a plantation (EBG) under central European climate conditions where summer drought is largely missing. A clear relationship between temperature and cambial activity was obvious at all studied sites. Dynamics of xylogenesis differed between sites but showed no significant trend from low to high altitudes regarding duration of xylogenesis and the number of xylem cells formed during one vegetation period. Growth performance seemed to depend on water availability which was best at the German site.

Number of cambial cells in the dormant cambium was significantly higher at the German site. Studies on other coniferous and broadleaf tree species (Gryc et al. 2011; Thibeault-Martel et al. 2008; Vavrcik et al. 2013) give similar number of dormant cells as observed at the Turkish sites (mean of 5 cells). Gričar et al. (2009)) report 3–11 cells in silver firs and a positive linear relationships between the number of dormant cambium cells and the total number of newly formed xylem cells. This leads to the conclusion that trees under good growth conditions are able to invest in a higher number of cambial cells which in return results in higher growth rates comparable with the results obtained in our study. This is especially noticeable with regard to the results from T4 (1055 m); here, duration of ring formation lasted longer than at O1 but annual ring width and the number of xylem cells formed at the end of the growing season were lower. In total, O1 showed best growth performance. The likely most important factor favoring growth at Bayreuth appears to be the absence of a drought period during summer and the uniform distribution of rainfall during the whole growing season. *C. libani* is a drought tolerant species (Ladjal et al. 2007), trees at the Turkish sites showed continuous growth during the growing season. However with respect to the results of other studies (Akkemik 2003; Touchan et al. 2005; Ducrey et al. 2008), tree ring width is positively correlated with the number of days with rainfall during the growing season and optimal growth performance is limited by lack of water during the growing season. This appears to cause significantly lower growth rates at the three upper Turkish sites, whereas because of its position close to the Avlan Lake, climate conditions at T4 may be more humid and temperate providing more favorable growth conditions within the Turkish sites. Many studies report that temperature is the main trigger for the onset of cambial activity and xylogenesis. It was observed in the Mediterranean species *Pinus pinaster* (Vieira et al. 2014) and especially in cold climates like boreal forests or tree line ecotones (Schmitt et al. 2004; Deslauriers and Morin 2005; Rossi et al. 2007). Threshold stem temperatures for the onset of growth are reported to be 7–9 °C (Rossi et al. 2007, 2008a; James et al. 1994). Investigations on Chinese pine (*Pinus tabulaeformis* Carr.) showed that cell division corresponded to the time when daily minimum temperature was above 0 °C (Liang et al. 2009). This is consistent with our observations in *C. libani*, where threshold temperatures for the onset of growth appeared to be 0 °C for daily minimum air temperatures and at least 5 °C for daily means of air and stem temperatures. When growth ceases in *C. libani*, lack of water seems to be the limiting factor while temperature would still favor growth. Vieira et al. (2014) reported that low water availability triggers pine wood formation in the Mediterranean area. Due to low water availability at the end of the drought period, growth ceases in *C. libani* at the end of September at the Turkish sites. For there is sufficient water supply at the German site, wood formation appears to cease when temperatures fall below a required minimum. When cambial activity finished, xylogenesis still continues with enlargement, wall thickening and lignification (Rossi et al. 2007). Our results clearly show that a lower number of cells in the dormant cambium as well as a low and short cambial activity result in a short growth duration and a small tree ring with only a small number of newly formed cells. According to the investigation of Rossi et al. (2007) at high altitude in the eastern Italian Alps on larch, stone pine and Norway spruce, duration of overall wood formation varies between 90 to 140 days. With time spans varying from 134 to 174 days, wood formation in *C. libani* lasted obviously longer. Observed differences in the duration of wood formation resulted mainly from varying timings in the onset of growth.

Sampled trees at our sites varied largely in age, but duration of cambial activity and xylogenesis did not significantly differ regardless of tree age and tree size within sites. When comparing the youngest (42 years) and the oldest tree (360 years) at the highest site T1 (1960 m), the latter showed a lower number of total cells formed during the growing season, but the onset, duration and end of growth appeared to be almost the same. This was observed at all sites, and this observation is also supported by hourly dendrometer measurements taken at the same sampled trees over the growing season (data not shown). These observations differ from results of Rossi et al. (2008a) who observed that timings and durations of xylogenesis differed between adult (50–80 years) and old conifers (200–350 years) at the alpine timberline, with a 2–3 weeks shorter cambial activity for the old conifers. It was also stated by various authors that the period of tree ring formation becomes longer the more cells are produced along a radial row (Gričar et al. 2005; Rossi et al. 2006b). This trend is in agreement with our Turkish sites. However, when comparing O1 and T4, it is fully evident that O1 produces significantly more xylem cells in an at least 10 day shorter growth period. The higher number of dormant cells therefore seems to host an important advantage for growth and allows a differentiation of more xylem cells in a shorter period. It can therefore not generally be concluded, that a longer period of ring formation leads to a higher number of cells. The highest number of cells in an active cambium was observed to be 17 at the German site and 15 at the Turkish site T4, which are much higher than observed by Thibeault-Martel et al. (2008) with 9 cambial cells in *Picea mariana* and 10–11 cells in *Abies balsamea*.

Along the altitudinal gradient, we expected a decreasing duration of cambial activity and xylogenesis from low to high sites because of decreasing temperatures. In fact, means of air and stem temperature during the growing season decreased significantly with altitude. But, while T4 (1055m) showed best growth performance as expected, the next site T3 (1355 m) showed lowest growth performance. Growth performance was significantly higher at T2 and much higher at T1 (tree line). Besides temperature and water availability this result could be affected by other parameters, such as stand density and light. T3 has the highest tree density, this combined with dryer conditions could result in the very low growth even though LAI was low. In comparison, tree density in T2 is lower than in T3 but the LAI showed the highest value for the Turkish sites. *C. libani* is a light demanding species (Senitza 1989). Among the upper three stands, the highest site becomes open towards the timberline, so light competition is the lowest here. Although the slope is north-west facing, due to the fact that it is located at the top of a mountain, it is exposed to the sun almost all day round. The same applies for O1, where trees are aligned in a single row orientated from North to South, and no significant shading occurs. The better light supply implies higher rates of photosynthesis which may in return positively influence growth performance. Sites T3 and T2 were more north- facing, denser in stand cover implying higher competition for light. Lower photosynthesis rates and stronger competition may be a reason for significantly lower growth performance at T3 and T2, but other reason that were not apparent (e.g. varying soil conditions within sites) may also be relevant, thus need further investigation.

Conclusions

The xylem in stems of *C. libani* showed differences in onset, duration and end of cambial activity and xylogenesis as well as growth rates with respect to climate and site conditions. Temperature, especially daily means of air and stem temperature, appeared to be the main trigger when it comes to the onset of cambial activity and xylogenesis, whereas water availability and other factors, e.g. stand structure, light exposure etc., affect the total annual increment. Under the same site conditions (or within the respective sites) trees showed no significant differences in timing and duration of cambial activity and xylogenesis regardless of plant age and size. Cessation of growth is affected either by temperature or- as typical for the natural range habitats- by restricted water availability. Obviously, a temperate climate with an evenly distributed rainfall during the growing season is more favorable for *C. libani* than a Mediterranean climate. Our results therefore underline the potential of *C. libani* for afforestation outside its natural range with respect to climate change. Further studies on the relationship between traumatic resin ducts and tree ring width are necessary for a better understanding of the wood formation processes.

Abbreviations

CRF: Cedar Research Forest; EBG: Ecological-Botanical Gardens;
DBH: Diameter at breast height; DOY: Day of the year; LAI: Leaf area index.

Competing interests

The authors declare that they have no competing interests.

Authors' contributions

AG designed the study, conducted sampling at the Turkish sites and statistical analyses. DK designed the protocol for histological analyses and conducted sampling at the German site. AG, DK and AS did sample preparation in the laboratories and histological analysis with the microscope. AG wrote the manuscript with contributions from DK, AS, RZ and MK. All authors have revised the manuscript and approved the final manuscript.

Authors' information

AG conducted this work as part of her PhD thesis. DK conducted the sampling portion at the EBG Bayreuth as part of her Master's thesis.

Acknowledgments

The preparation and analysis of the microcores which this publication is based on, Aylin Güney and Ayça Sökücü were funded by the Federal Ministry of Education and Research (01PL11003).The responsibility for the content lies with the authors. Field work, like sampling and meteorological monitoring was carried out within the project funded by the Federal Ministry of Education and Research (01DL12041) and in cooperation with the Southwest Anatolian Forest Research Institute in Antalya ant the Ecological Botanical Gardens in Bayreuth. We would like to thank all who have supported our research, with special thanks to Melahat Şahin, Neşat Erkan, Gregor Aas, Viviana Horna, Catuscia Simonato and Anne Heller.

References

Abe H, Funada R, Ohtani J, Fukazawa K (1997) Changes in the arrangement of cellulose microfibrils associated with the cessation of cell expansion in tracheids. Trees 11(6):328–332

Akkemik Ü (2003) Tree rings of *Cedrus libani* at the northern boundary of its natural distribution. IAWA Journal 24(1):63–73

Atalay I (1987) Sedir ormanlarinin yayilis gösterdigi alanlar ve yakin cevresinin genel ekolojik özellikleri ile sedir tohum transfer rejyonlamasi (General ecological properties of natural occurrence areas of cedar (*Cedrus libani* A. Rich) forests and regioning of seed transfer of cedar in Turkey). Orman Genel Müdürlügü yayini (Ankara) 663 (in Turkish with English Summary)

Aussenac G (1984) Le cèdre, essai d'interprétation bioclimatique et écophysiologique. Bull Soc Bot Fr Actual Bot 131:385–398

Avci M, Carus S (2005) The impact of cedar processionary moth [Traumatocampa ispartaensis (Doganlar & Avc.) (Lepidoptera: Notodontidae)] outbreaks on radial growth of Lebanon cedar (*Cedrus libani* A.Rich) trees in Turkey. J Pest Sci 78:91–98

Bao FC, Jiang ZH, Jiang XM, Lu XX, Luo XQ, Zhang SY (2001) Differences in wood properties between juvenile wood and mature wood in 10 species grown in China. Wood Science and Technology 35(4):363–375

Basaran MA, Basaran S, Bas N, Kaçar S, Tolunay D, Makineci E, Kavgaci A, Deniz G (2008) Determining the actual state of Cedar Research Forest Elmali by GIS based digital maps. South-West Anatolia Forest Research Institute, Antalya, Turkey, p 331

Boydak M (2003) Regeneration of Lebanon cedar (*Cedrus libani* A. Rich.) on karstic lands in Turkey. Forest Ecology and Management 178:231–243

Boydak M (2007) Reforestation of Lebanon Cedar (*Cedrus libani* A. Rich.) in Bare Karstic Lands by Broadcast Seeding in Turkey. In Options méditerranéennes, Series A: Mediterranean Seminars (No. 75, pp. 33–43)

Boydak M, Calikoglu M (2008) Biology and silviculture of Lebanon Cedar (*Cedrus libani* A.Rich.). OGEM - VAK, Ankara, Turkey, p 228

Boydak M, Caliskan S (2014) Agaclandirma. Ormanciligi Gelistirme ve Orman Yanginlari ile Mücadele Hizmetlerini Destekleme Vakfi Yayini, 693s, Ankara (in press)

Brooks JR, Jiang L, Ozcelik R (2008) Compatible stem volume and taper equations for Brutian pine, Cedar of Lebanon, and Cilicica fir in Turkey. Forest Ecology and Management 256(1):147–151

Camarero JJ, Olano JM, Parras A (2010) Plastic bimodal xylogenesis in conifers from continental Mediterranean climates. New Phytologist 185(2):471–480

Carus S, Avci M (2005) Growth Loss of Lebanon Cedar (*Cedrus libani*) stands as Related to Periodic Outbreaks of the Cedar Shoot Moth (Dichelia cedricola). Phytoparasitica 33(1):33–48

Deslauriers A, Morin H (2005) Intra-annual tracheid production in balsam fir stems and the effect of meteorological variables. Trees 19:104–408. doi:10.1007/s00468-004-0398-87

Deslauriers A, Morin H, Begin Y (2003) Cellular phenology of annual ring formation of *Abies balsamea* in the Quebec boreal forest (Canada). Canadian Journal of Forest Research 33:190–200

Dirik H (2000) Effet du stress hydrique osmotique sur la germination des grainez chez les provenances de Cedre du Liban (*Cedrus libani* A.Rich.) d'origine Turque. Ann. For. Sci 57:361–367

Ducrey M, Huc R, Ladjal M, Guehl JM (2008) Variability in growth, carbon isotope composition, leaf gas exchange and hydraulic traits in the eastern Mediterranean cedars *Cedrus libani* and C. brevifolia. Tree Physiology 28:689–701

Etzold H (1983) Eine kontrastreiche, simultane Mehrfachfärbung für pflanzenanatomische Präparate: Fuchsin-Safranin-Astrablau. Mikrokosmos, Jg 72:213–219

Fady B, Lefèvre F, Reynaud M, Vendramin GG, Dagher-Kharrat MB, Anzidei M, Bariteau M (2003) Gene flow among different taxonomic units: evidence from nuclear and cytoplasmic markers in Cedrus plantation forests. Theoretical and Applied Genetics 107(6):1132–1138

Fahn A, Werker E, Ben-Tzur P (1979) Seasonal effects of wounding and growth substances on development of traumatic resin ducts in *Cedrus libani*. New Phytol 82:537–544

Foken T, Lüers J, Lauerer M, Aas G (2004) Im Ökologisch-Botanischen Garten: Dem Klima auf der Spur, Freundeskreis des Ökologisch-Botanischen Gartens der Universität Bayreuth e.V., Bayreuth

Gričar J, Čufar K, Oven P, Schmitt U (2005) Differentiation of terminal latewood tracheids in silver fir trees during autumn. Annals of botany 95(6):959–965

Gričar J, Krže L, Čufar K (2009) Number of cells in xylem, phloem and dormant cambium in silver fir (Abies alba), in trees of different vitality. Iawa Journal 30(2):121–133

Gryc V, Vavrčík H, Vichrová G (2011) Monitoring of xylem formation in Norway spruce in the Czech republic 2009. Wood research 56(4):467–478

James JC, Grace J, Hoad SP (1994) Growth and photosynthesis of Pinus sylvestris at its altitudinal limit in Scotland. Journal of Ecology, 297–306

Kavgacı A, Başaran S, Başaran MA (2010) Cedar forest communities in Western Antalya (Taurus Mountains, Turkey). Plant Biosystems 144(2):271–287

Khuri S, Shmoury MR, Baalbaki R, Maunder M, Talhouk SN (2000) Conservation of the *Cedrus libani* populations in Lebanon: history, current status and experimental application of somatic embryogenesis. Biodiversity and Conservation 9:1261–1273

Kurt Y, Kaya N, Isik K (2008) Isozyme Variation in Four Natural Populations of *Cedrus libani* A.Rich. in Turkey. Turk J Agric For 32:137–145

Ladjal M, Deloche N, Huc R, Ducrey M (2007) Effects of soil and air drought on growth, plant water status and leaf gas exchange in three Mediterranean cedar species. Cedrus atlantica, C. brevifolia and C. libani. Trees 21:201–213

Liang E, Eckstein D, Shao X (2009) Seasonal cambial activity of relict Chinese pine at the northern limit of its natural distribution in North China–Exploratory results. IAWA journal 30(4):371–378

Moser L, Fonti P, Büntgen U, Esper J, Luterbacher J, Franzen J, Frank D (2009) Timing and duration of European larch growing season along altitudinal gradients in the Swiss Alps. Tree Physiology 30:225–233. doi:10.1093/treephys/tpp108

Plomion C, Leprovost G, Stokes A (2001) Wood formation in trees. Plant physiology 127(4):1513–1523

Rossi S, Deslauriers A, Andofillo T (2006a) Assessment of cambial activity and xylogenesis by microsampling tree species: an example at the alpine timberline. IAWA J 27:383–394

Rossi S, Anfodillo T, Menardi R (2006b) Trephor: a new tool for sampling microcores from tree stems. IAWA J 27:89–97

Rossi S, Deslauriers A, Anfodillo T, Morin H, Saracino A, Motta R, Borghetti M (2006c) Conifers in cold environments synchronize maximum growth rate of tree-ring formation with day length. New Phytologist 170(2):301–310

Rossi S, Deslauriers A, Anfodillo T, Carraro V (2007) Evidence of threshold temperatures for xylogenesis in conifers at high altitudes. Oecologia 152(1):1–12

Rossi S, Deslauriers A, Anfodillo T, Carrer M (2008a) Age-dependent xylogenesis in timberline conifers. New Phytologist 177(1):199–208

Rossi S, Deslauriers A, Griçar J, Seo JW, Rathgeber CB, Anfodillo T, Morin H, Levanic T, Oven P, Jalkanen R (2008b) Critical temperatures for xylogenesis in conifers of cold climates. Global Ecology and Biogeography 17(6):696–707

Schmitt U, Jalkanen R, Eckstein D (2004) Cambium dynamics of Pinus sylvestris and Betula spp. in the northern boreal forest in Finland. Silva Fennica 38:167–178

Schütt P, Weisberger H, Schuck HJ, Lang UM, Stimm B, Roloff A (2008) Lexikon der Nadelbäume. Nicol Verlagsgesellschaft Hamburg, S. 639

Senitza E (1989) Waldbauliche Grundlagen der Libanonzeder (*Cedrus libani* A.Rich.) im Westtaurus/Türkei. Dissertation. Universität für Bodenkultur in Wien, p. 225

Thibeault-Martel M, Krause C, Morin H, Rossi S (2008) Cambial activity and intra-annual xylem formation in roots and stems of Abies balsamea and Picea mariana. Annals of Botany 102:667–674

Touchan R, Xoplaki E, Funkhouser G, Luterbacher J, Hughes MK, Erkan N, Akkemik Ü, Stephan J (2005) Reconstructions of spring/summer precipitation for the Eastern Mediterranean from tree-ring widths and its connection to large-scale atmospheric circulation. Climate Dynamics 25:75–98

Uyar N, Argimak Z, Topak M (1990) Lübnan sedirinde (*Cedrus libani* A. Richard) tohum temini ve islah calismalari (Studies on the production and improvement of Lebanon cedar seeds). In: Proceedings of the International Cedar Symposium, October 22–27, 1990, Antalya, Ormancılık Araş,. Ens. Muhtelif Yayınlar (Ankara) 59, 248–259 (in Turkish with English abstract)

Vavrcik H, Gryc V, Mensik, L, Baar J (2013) Xylem formation in Fagus sylvatica during one growing season. Dendrobiology, 69

Vieira J, Rossi S, Campelo F, Freitas H, Nabais C (2014) Xylogenesis of Pinus pinaster under a Mediterranean climate. Annals of Forest Science 71:71–80. doi:10.1007/s13595-013-0341-5

Wimmer R, Grabner M (1997) Effects of climate on vertical resin duct density and radial growth of Norway spruce [Picea abies (L.) Karst.]. Trees 11(5):271–276

4

Three new species of *Begonia* endemic to the Puerto Princesa Subterranean River National Park, Palawan

Rosario Rivera Rubite[1], Mark Hughes[2], Patrick Blanc[3], Kuo-Fang Chung[4], Hsun-An Yang[5], Yoshiko Kono[5], Grecebio J D Alejandro[6], Llogene B De Layola[1], Arthur Gregory N Virata[1] and Ching-I Peng[5*]

Abstract

Background: *Begonia* is a mega-diverse genus of flowering plants prone to generating micro-endemic species, especially on limestone habitats. During fieldwork in the Puerto Princesa Subterranean River National Park, Palawan (Philippines), three species were encountered which did not match any previously described from the region.

Results: Following morphological, anatomical, molecular phylogenetic and cytological investigation a hypothesis of three new species is supported. The three new species belong to a clade endemic to Palawan and Borneo.

Conclusions: The limestone habitats in the Puerto Princesa Subterranean River National Park environs support a unique flora. The description of three new species from a small area within the park demonstrates how much remains to be discovered there, and the importance of its continued protection.

Keywords: Limestone, Endemic, New species, Conservation

Background

The Philippines is a biodiversity hotspot, with an exceptional concentration of endemic species undergoing exceptional loss of habitat (Myers et al. 2000). Some of the most extensive areas of intact forest in the country are found in Palawan, where there are reasons for optimism due to the implementation of conservation through devolved local government and civil society groups (Posa et al. 2008). Puerto Princesa Subterranean River National Park (PPSRNP) is managed by the City Government of Puerto Princesa in partnership with the Department of Environment and Natural Resources (DENR), and is the first national park in the Philippines to be managed at this level. It was designated a World Heritage Site in 1999 (World Heritage Commitee 2000), and voted as one of the *New 7 Wonders of Nature*, boosting tourism as much as 300% (Fitzgerald 2011, 2012). Palawan as a whole is designated by UNESCO as a Biosphere Reserve, and by Birdlife International as an Important Bird and Biodiversity Area (BirdLife International 2015).

The PPSRNP contains a spectacular karst limestone landscape and eight different forest formations; forest on ultramafic soil, forest on limestone soil, montane forest, freshwater swamp forest, lowland evergreen tropical rainforest, riverine forest, beach forest, and mangrove forest. This mix of forest types and substrates means the PPSRNP harbours a very rich flora, but much remains to be discovered. During recent fieldwork in the PPSRNP we found three new species of *Begonia* within a distance of only a few hundred meters, adding these distinct species to the 14 known *Begonia* species already recorded from Palawan (Hughes and Coyle 2009; Hughes et al. 2010, 2011). Based on morphology, chromosome cytology and molecular phylogenetic analysis we place the three new species in *Begonia* sect. *Baryandra*. Although the species occur very close to one another at altitudes of less than 50 m, they grow in mutually exclusive habitats. The first species *Begonia taraw* C.-I Peng, R. Rubite & M. Hughes is located around the mouth of the underground river, growing tenaciously on the vertical limestone

*Correspondence: bopeng@sinica.edu.tw
[5] Biodiversity Research Center, Academia Sinica, Taipei 115, Taiwan
Full list of author information is available at the end of the article

cliffs. Nearby in karst forest growing in rock crevices is the second species *Begonia hughesii* R. Rubite and C.-I Peng. The third species, *Begonia tagbanua* M. Hughes, C.-I Peng & R. Rubite, grows on clay soil banks inside the forest.

Methods

Chromosome cytology

Somatic chromosomes of *B. taraw* (*Ching-I Peng 23463, 23464*), *B. hughesii* (*Ching-I Peng 23466, 23475*), *B. tagbanua* (*Ching-I Peng 23471*) were examined using root tips. The procedures of pre-treatment, fixation and staining for chromosome observations followed Peng et al. (2014). Voucher specimens are deposited in HAST.

Cryo scanning electron microscopy

Fresh leaves of *B. taraw*, *B. hughesii* and *B. tagbanua* were dissected and attached to a stub. The samples were frozen with liquid nitrogen slush, then transferred to a sample preparation chamber at −160°C and etched for 15 min at −85°C. After etching, the temperature reached −130°C for sample fracturing and coating. After coating, the samples were transferred to the SEM chamber and observed at −190°C with a cryo scanning electron microscope (FEI Quanta 200 SEM/Quorum Cryo System PP2000TR FEI).

Phylogenetic analyses

A total of 33 species from *Begonia* sect. *Baryandra* was sampled, including the 3 new species from Palawan (each represented by two individuals) and a further 5 from Palawan (*Begonia blancii, B. cleopatrae, Begonia suborbiculata, B. wadei, B. woodii*). In addition three species of *Begonia* sect. *Reichenheimia* were sampled, with two species of *Begonia* sect. *Coelocentrum* as an outgroup based on the tree topology in Thomas et al. (2012). Vouchers and genbank accession numbers are in Additional file 1: Table S1. *Begonia* sect. *Baryandra* has recently been recircumscribed and confirmed as monophyletic (Rubite et al. 2013). Total genomic DNA was extracted from young leaves and buds using DNeasy Plant Mini Kits (Qiagen, USA). Four chloroplast non-coding regions (*ndh*A intron, *ndh*F–*rpl*32 spacer, *rpl*32–*trn*L spacer, *trn*C–*trn*D spacer) were amplified as in Hughes et al. (2015) with the PCR primer sequences from Demesure et al. (1995) and Thomas et al. (2011). Forward and reverse reads for all regions were assembled and aligned using Geneious Pro (Biomatters, New Zealand), with the alignments being subsequently manually edited in BioEdit 7.1.3 (Hall 1999); inversions were offset and the final alignment was 6905 bases long, with 39 bases excluded due to alignment uncertainty. The optimal model of DNA sequence evolution was assessed with jModeltest 2.1.3 (Posada et al. 2012) using the Aikake Information Criterion (AIC) to estimate the model with the closest fit to the data. The GRT + G + I model was the most probable (AIC weight of 0.98). Bayesian phylogenetic analysis was carried out using MrBayes 3.2.1 (Ronquist et al. 2012), treating the dataset as a single partition. The analysis consisted of two runs with four chains each, run for 10,000,000 generations with a sample tree taken every 1,000. The effective sample size was >5,600 for all model parameters, and the average standard deviation of split frequencies was 0.0016. The proportion of successful state exchanges between adjacent chains ranged between 0.62 and 0.68 for both runs. Convergence between runs and attainment of stationarity was runs was further comfirmed by examining parameter distrubutions in Tracer v1.6 (Rambaut et al. 2014). The first 25% of sampled trees were discarded as burn-in, and the remainder summarised as a maximum clade credibility tree visualised using FigTree v.1.4.0 (Rambaut 2009).

Results

Chromosome cytology

Recently, Philippine *Begonia* species belonging to section *Diploclinium* were transferred to section *Baryandra*, based on molecular analysis (Rubite et al. 2013). Among 47 species in sect. *Baryandra*, chromosome numbers were previously reported for five species, namely *B. blancii* ($2n = 30$, Hughes et al. 2011); *B. fenicis* ($2n = 26$, Oginuma and Peng 2002; $2n = 56$, Kokubugata and Madulid 2000); *B. parva* ($2n = 36 + 2f$, Legro and Doorenbos 1969); *B. nigritarum* ($2n = 44$, Doorenbos et al. 1998); and *B. suborbiculata* ($2n = 30$, Hughes et al. 2011). Somatic chromosomes at metaphase of the three new species are reported here for the first time: $2n = 28$ for *Begonia taraw* (Figure 1a); $2n = 30$ for *B. hughesii* (Figure 1b) and *B. tagbanua* (Figure 1c).

Leaf anatomy and vestiture

Begonia taraw

Adaxial surface nearly glabrous, epidermis cells slightly bullate (Figure 2a, b); cross section ca. 820 μm thick; epidermis single-layered on both surfaces, hypodermis 2-layered; palisade parenchyma cells 1-layered; spongy parenchyma cells ca. 3-layered (Figure 2b); abaxial surface with glandular hairs and multiseriate trichomes, stomata in clusters of 3–7, helicocytic, flat, subsidiary cells 3 (Figure 2c).

Begonia hughesii

Adaxial surface with glandular trichomes (Figure 2d); cross section ca. 870 μm thick; epidermis single-layered on both surfaces, hypodermis 2-layered; palisade parenchyma cells 1-layered; spongy parenchyma cells ca. 4-layered (Figure 2e); abaxial surface with glandular

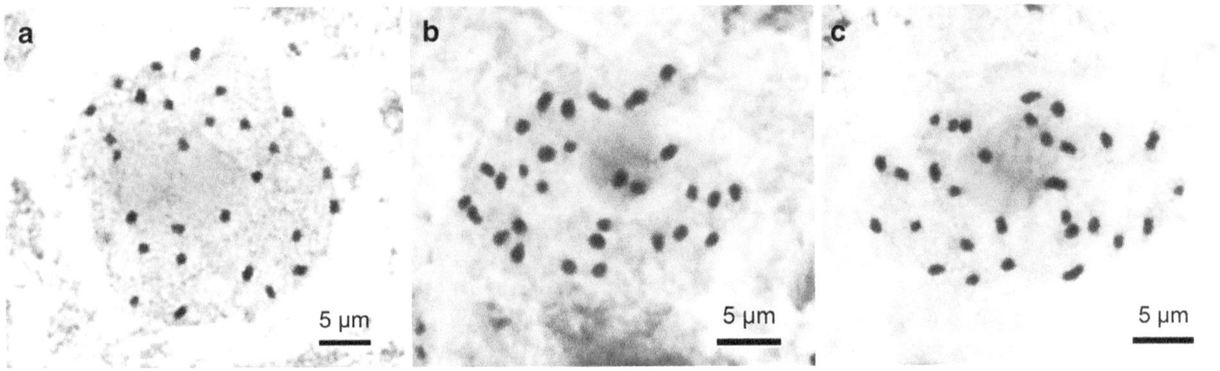

Figure 1 Somatic chromosomes at metaphase of *Begonia*. **a** *B. taraw* (2n = 28 [0.9–1.6 μm], Peng 23464); **b** *B.hughesii* (2n = 30 [1.0–2.1 μm], Peng 23466); **c** *B. tagbanua* (2n = 30 [0.9–2.0 μm], Peng 23471).

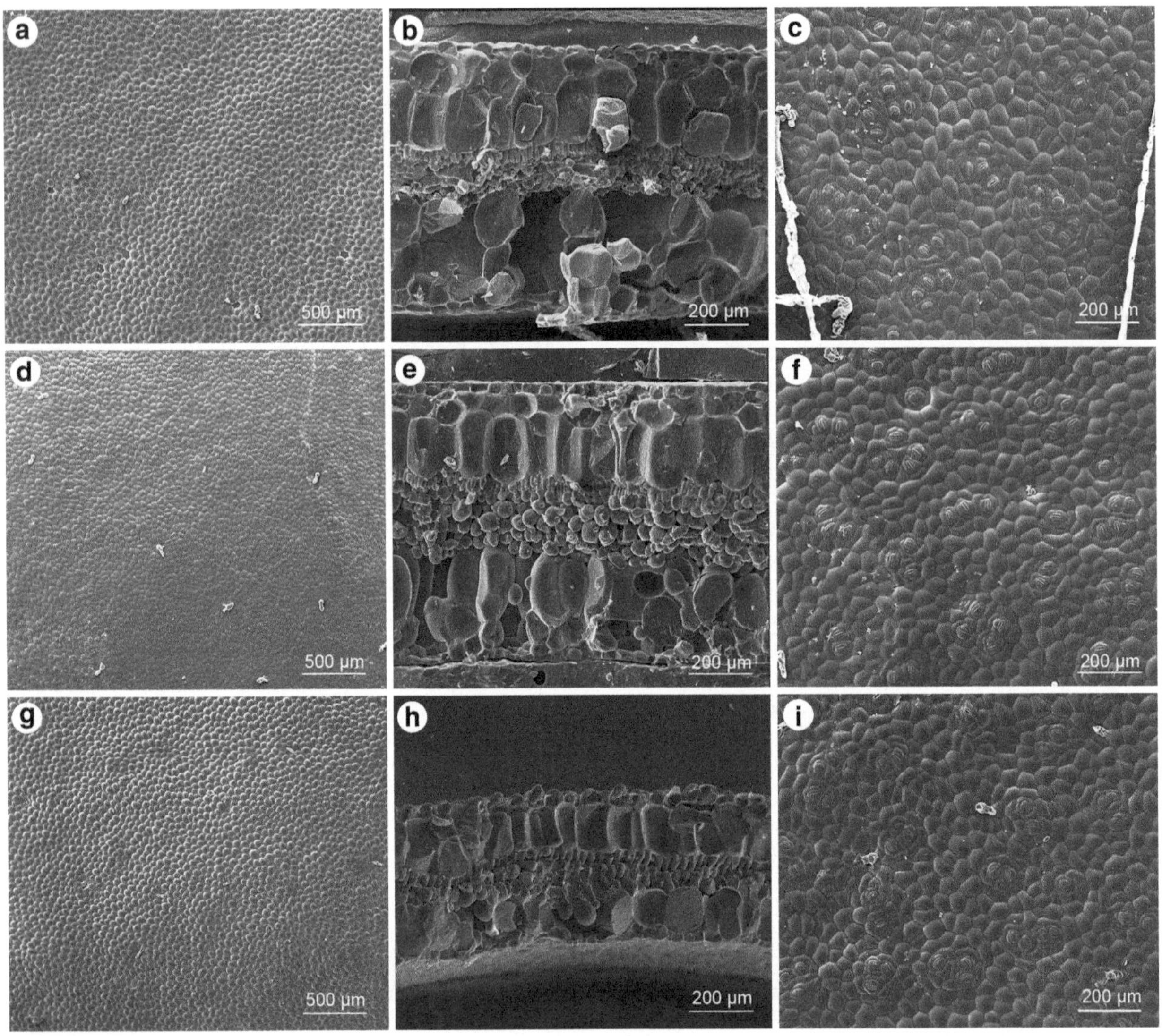

Figure 2 Leaf SEM microphotographs of *Begonia*. **a–c** *Begonia taraw*; **d–f** *B. hughesii*; **g–i** *B. tagbanua*; **a**, **d**, **g**, adaxial surface; **b**, **e**, **h**, cross section; **c**, **f**, **i**, abaxial surface.

trichomes, stomata single or 2–5 clustered, helicocytic, flat, subsidiary cells 3 (Figure 2f).

Begonia tagbanua

Adaxial surface with conoidal cells and glandular trichomes (Figure 2g, h); Cross section ca. 490 μm thick; epidermis single-layered on both surfaces, hypodermis 1-layered; palisade parenchyma cells 1-layered; spongy parenchyma cells ca. 2-layered (Figure 2h); abaxial surface with glandular trichomes, stomata 2–4 clustered, helicocytic, flat, subsidiary cells 3 (Figure 2i).

Phylogenetic relationships

The three new species *B. hughesii*, *B. tagbanua* and *B. taraw* are strongly supported as being members of *Begonia* sect. *Baryandra*, and belong to a clade mostly endemic to Palawan (Figure 3). *Begonia hughesii* is not monophyletic according to its chloroplast genotype, whereas *B. tagbanua* and *B. taraw* are supported as monophyletic whilst having some within-species polymorphism as shown by the non-zero branch lengths. The two samples of *Begonia hughesii* were taken from different localities; sample 1 (*Peng 23466*) was collected in littoral forest near Sabang, and sample 2 (*Peng 23475*) was collected at Ugong Rock.

Taxonomic treatment

1. *Begonia taraw* C.-I Peng, R. Rubite & M. Hughes, sp. nov. § *Baryandra* (Figures 4, 5)

TYPE: PHILIPPINES. Palawan, Puerto Princesa, Puerto Princesa Subterranean River National Park, elev. ca. 5 m, 10°12′1″N, 118°55′32″E, 2 Nov. 2011, *Ching-I Peng 23464*, with Kuo-Fang Chung, Chien-I Huang, Rosario Rubite (holotype PNH, isotype E, HAST).

Begonia taraw can be differentiated from the only other peltate species in Palawan, *Begonia gutierrezii*, in having longer petioles which are sparsely lanate with appressed hairs (not erect 4 mm long hairs), short internodes on the rhizome (not stoloniferous and elongate), larger leaves, glabrous stipules, and a conspicuous ring of fused fleshy hairs at the base of most petioles.

Lithophytic rhizomatous herb 20–35 cm tall. *Stem* 11–16 mm diameter, glabrous, internodes 5–10 mm long. Stipules broadly triangular, asymmetric, 13–15 × 15–17 mm, keeled, acuminate, entire, glabrous, persistent, maroon becoming dark brown and papery with age. *Leaves* on erect petioles; petioles 13–20 cm long, red at the base, greener apically, terete, 5–7 mm in diameter at the base becoming more slender apically, sparsely lanate when young, the hairs appressed, subglabrous with age, the hairs rubbing off easily, broad-based 10 mm long fleshy red hairs fused into a ring at the base of the petiole; lamina peltate, petiole inserted 3–3.5 cm from midrib, distinctly coriaceous, not variegated, in various colour forms on different plants from pale to very dark matt green above, whitish to pinkish green below, broadly ovate, 6–8 × 9–10 cm, upper surface mostly glabrous, sometimes with a few scattered 2 mm fine long brown hairs between the veins, the pale eye above the petiole attachment with a few hairs; lower surface sparsely lanate on the veins and lamina; venation palmate, main veins 9, pink or green, slightly prominent beneath, flat above; apex obtuse; margin very shallowly sinuate-dentate; edges reddish becoming green when mature; stomata in clusters of 3–7. *Inflorescence* erect, 40–58 cm long, bisexual, male and female flowers open at the same time as the inflorescence matures, cymose, each unit consisting of a central male flower with a lateral female and lateral further unit, branching up to 6 times, with 40–80 flowers in total, not mass-flowering; primary peduncle 30–45 cm long, 5–7 mm diameter at the base becoming more slender apically, base is reddish becoming green at the upper third part, sparsely lanate when young, subglabrous with age, the hairs rubbing off easily, secondary peduncles 1.5–4 cm long, subglabrous; bracts broadly rounded-triangular, conduplicate, ca. 10 × 12 mm, entire, glabrous, deciduous, overlapping, translucent pale green or pinkish green. *Male flowers* pedicel 8–15 mm long, glabrous; tepals 4, uniformly white or pink, entire, glabrous; outer 2 oblong, 13–17 × 10–13 mm; inner 2 obovate, folded and retuse, 11–14 × 8–9 mm; androecium with ca. 50 stamens, sessile; filaments united at the base, unequal, 1.25–2 mm long, longer filaments found in the centre of the androecium; anthers yellow, 1.75 mm long, oblong-oblanceolate, slits lateral tending to unifacial towards the apex, running for 3/4 the length of the anther, connective extended. *Female flowers* pedicel 20–30 mm long, glabrous, with 2 min fleshy 0.5 mm long hair-like bracteoles spaced 1–2 mm below the ovary; ovary 12 × 20 mm including wings, green or pink; capsule ca. 9 × 6 mm, 3-locular, placentae bifid; wings 3, unequal, larger wing ca. 9 × 14 mm, shallowly angular to rounded, flat (not cucullate), smaller 2 wings ca. 11 × 4 mm, truncate; tepals the same as the male flowers; styles 3, bifid, stigmatic surface spirally twisted. *Fruit* drying pale brown, the same size and shape as the ovary. Somatic chromosome number, $2n = 28$.

Distribution, habitat and ecology Begonia taraw is currently known only from the type locality and a further field observation from Lions Cave ca. 500 m away. The species grows on vertical limestone cliffs semi-shaded by broadleaf forest at the entrance of the Puerto Princesa underground river. The species was flowering and fruiting when collected in early November. In cultivation in the greenhouse of Academia Sinica in Taipei, Taiwan, it flowered and fruited from July to December.

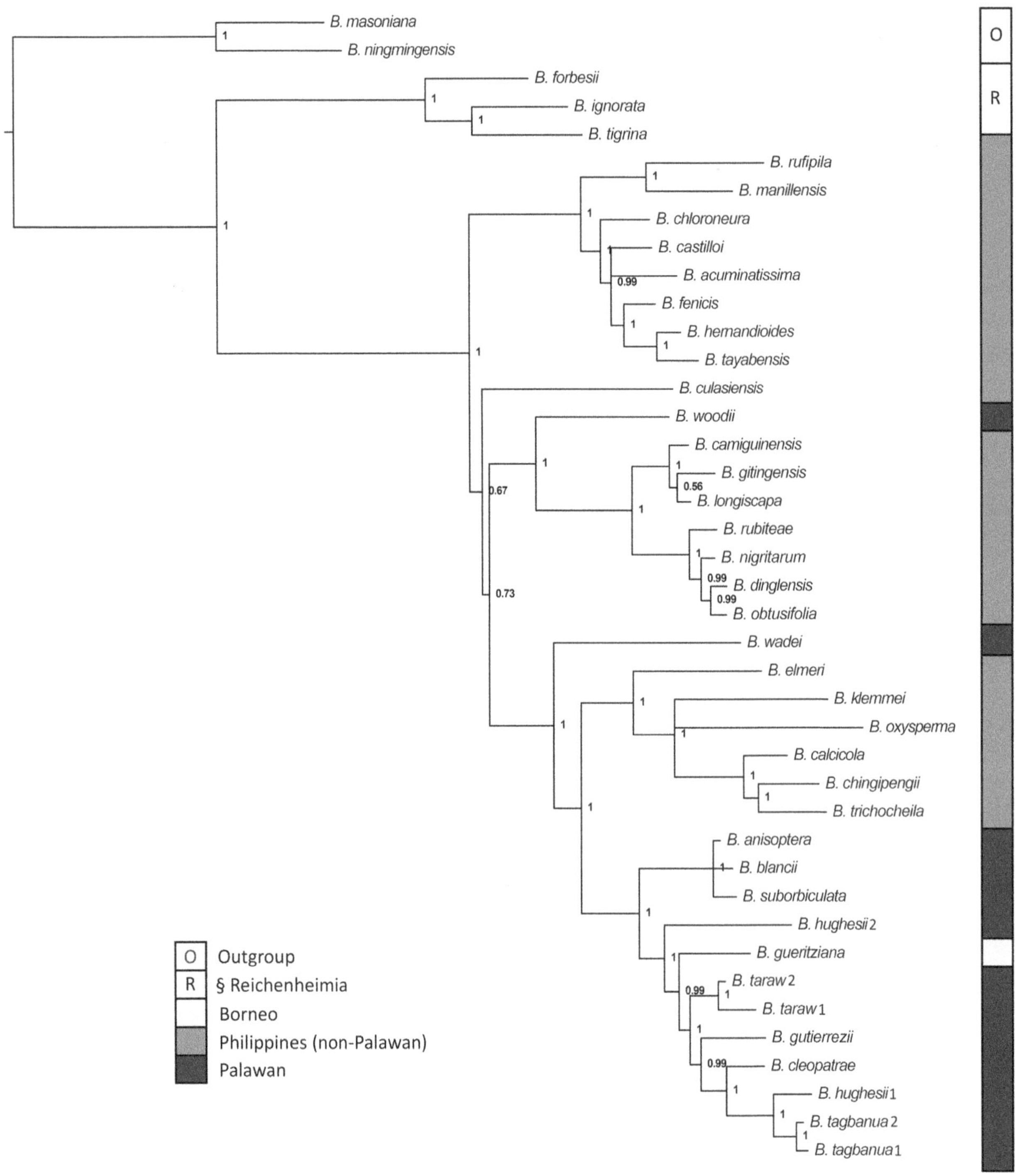

Figure 3 Phylogeny of *Begonia* sect. *Baryandra*. Generated from a Bayesian analysis of chloroplast DNA sequence data, the 50% majority-rule consensus tree is shown. The *figures* at the nodes are clade support values (posterior probabilities).

Etymology The species is named after the local word "taraw" which means karst limestone in Tagbanua language.

Additional specimen examined (paratype) PHILIPPINES. Palawan, Puerto Princesa Subterranean River National Park, elev. ca. 5 m, 10°12′1″N, 118°55′32″E, 2 Nov. 2011, *Ching-I Peng 23463*, with Kuo-Fang Chung, Chien-I Huang, Rosario Rubite (HAST).

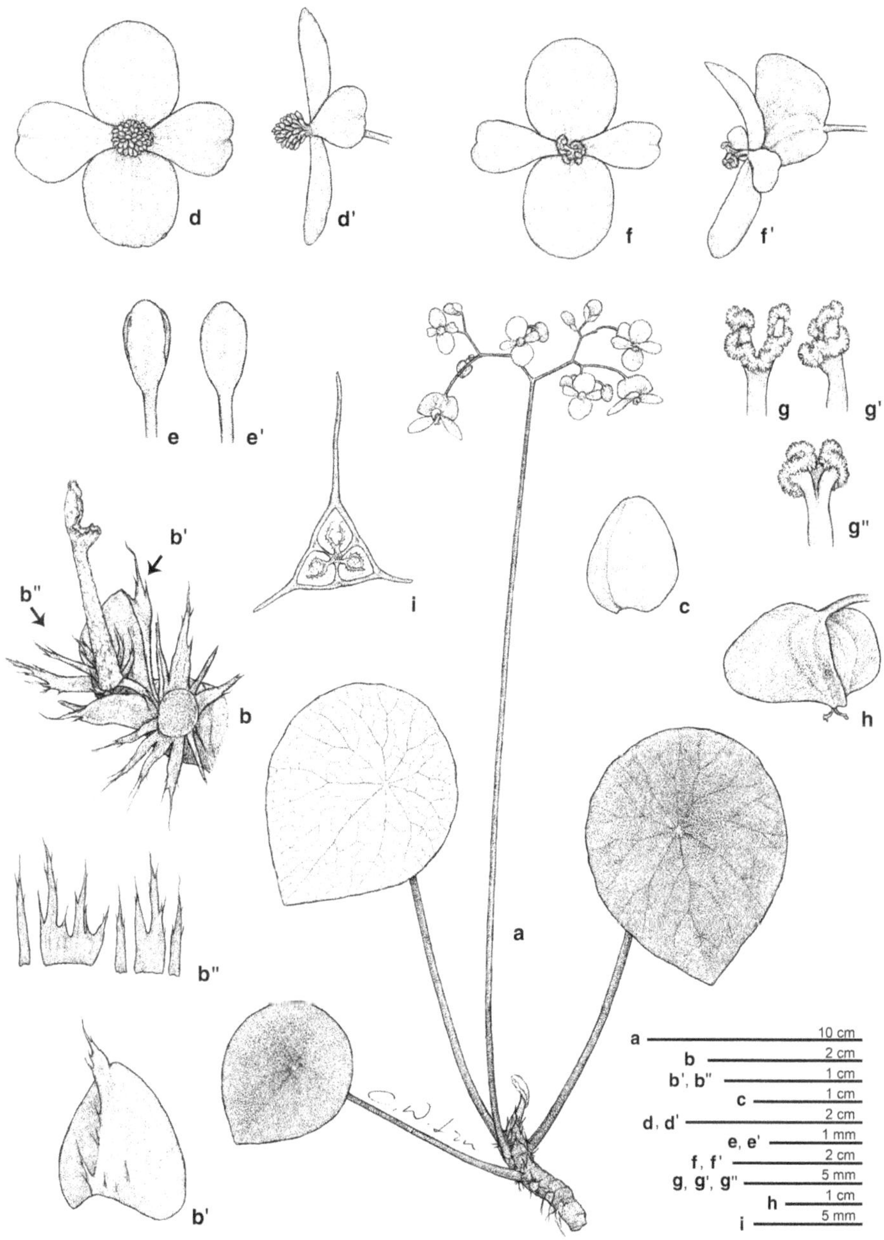

Figure 4 *Begonia taraw* C.-I Peng, R. Rubite, & M. Hughes. **a** Habit; **b** showing the base of petiole; **b'**, stipule; **b"**, fleshy hairs fused into a ring; **c** bract; **d, d'** staminate flower; **e, e'** stamen; **f, f'** carpellate flower; **g, g', g"** style and stigma; **h** capsule; **i** cross section of ovary. All from *C.-I Peng 23463* (HAST). Line drawing by Che-Wei Lin.

Proposed IUCN red list category VUD2.The species is a rare, very narrow endemic, and as such has an area of occupancy and extent of occurrence under the threshold for Critically Endangered, (IUCN 2012). However, none of the other criteria, based on number of locations and decline or fluctuation in range or population apply. *B. taraw* is best considered as belonging to the Vulnerable category, under VUD2, a category often applied to naturally rare and narrowly endemic taxa which, although not under immediate threat, have such a narrow range that unpredictable localised habitat degradation puts them at risk of rapidly fulfilling the criteria for Critically Endangered or Extinct.

2. *Begonia hughesii* R. Rubite & C.-I Peng, sp. nov. § *Baryandra* (Figures 6, 7)

TYPE: PHILIPPINES. Palawan Island, Palawan Province, Puerto Princesa City, Puerto Princesa Subterranean River

Figure 5 *Begonia taraw* C.-I Peng, R. Rubite & M. Hughes. **a** Habit and habitat; **b** inflorescence; **c** petiole; **d** leaf adaxial surface; **e** leaf abaxial surface, red type; **f** young leaf, stipules and fleshy hairs fused into a ring at the base of the petiole; **g** bracts and young inflorescence; **h** staminate flower, face view; **i** 4-tepaled carpellate flower, face view; **j** 5-tepaled carpellate flower, face view; **k** staminate flower, side view; **l** carpellate flower, side view; **m** cross section of ovary. All from *C.-I Peng 23463* (HAST) except E & J from *C.-I Peng 23464* (HAST).

National Park, elev. ca. 5 m, 10°12′1″N, 118°55′32″E, 2 Nov. 2011, *Ching-I Peng 23466*, with Kuo-Fang Chung, Chien-I Huang, Rosario Rubite (holotype PNH, isotype HAST).

Begonia hughesii is most similar to *B. acclivis*, but differs in having leaves which are glabrous above (not with scattered 2 mm long hairs) and uniform bright green (not markedly variegated), and dimorphic stipules which are either hairy on the keel only or hairy all over the abaxial surface.

Lithophytic rhizomatous herb ca. 20 cm tall. *Stem* 5–10 mm thick, tomentose (2–3 mm brown hairs) becoming glabrous in parts with age, internodes 3–8 mm long. Stipules triangular, asymmetric, 10–12 × 9–11 mm,

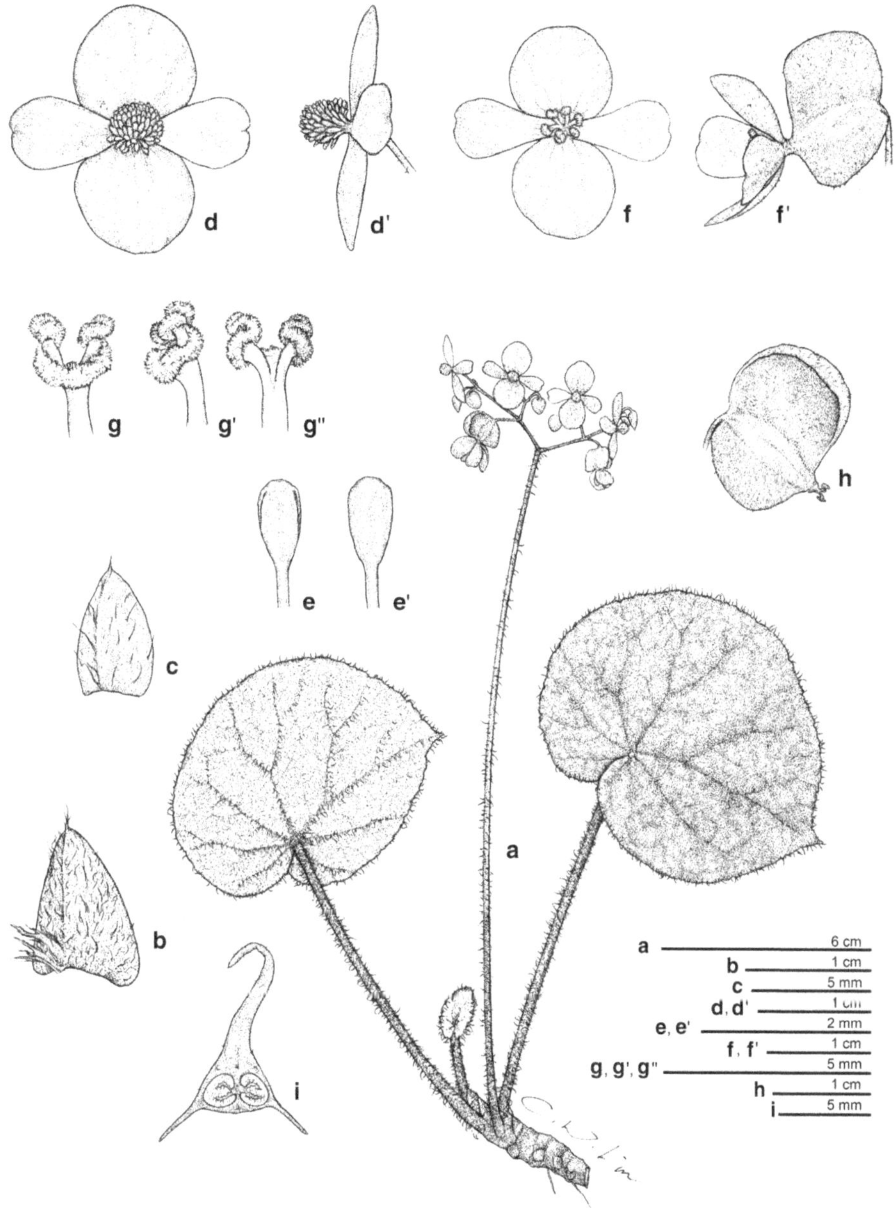

Figure 6 *Begonia hughesii* R. Rubite & C.-I Peng. **a** Habit; **b** stipule; **c** bract; **d, d'** staminate flower; **e, e'** stamen; **f, f'** carpellate flower; **g, g', g''** style and stigma; **h** capsule; **i** cross section of ovary. All from *C.-I Peng 23466* (HAST). Line drawing by Che-Wei Lin.

heteromorphic, alternately either with brown hairs on the keel only, or tomentose all over abaxially, hairs brown, 2–3 mm long, keeled, keel extending at the tip for ca. 2 mm, ring of hairs at base of petiole absent. *Leaves* on erect petioles; petioles 6–15 cm long, 4–6 mm in diameter, terete, red, tomentose even when mature and with 2 hair types, one linear, the other thicker towards the base; lamina basifixed, thick and fleshy in life, drying opaque and stiff, ovate, 7–10 × 5–7.5 cm, apple-green above, pale green below, glabrous above except for minute pubescence near the petiole attachment, below hairy on the veins only with ca. 2 mm long hairs, venation palmate, main veins 7; margin maroon, broadly and shallowly undulate, dentate-sinuate, ciliate with pink hairs (1–1.5 mm); tip shortly acuminate; stomata in clusters of 2–5. *Inflorescence* erect, ca. 20 cm long, bisexual, male and female flowers open at the same time as the inflorescence matures, cymose, branching 4 times; primary peduncle ca. 15 cm long, with scattered hairs (much less dense than on the petioles), bracts small, ovate, 1.5–2 × 1.5–2 mm, with minute scattered glandular

Figure 7 *Begonia hughesii* R. Rubite & C.-I Peng. **a** Habit and habitat; **b** cultivated plant at anthesis; **c** inflorescence; **d** rhizome, stipules and petioles; **e** bracts and young inflorescence; **f** carpellate flower, face view; **g** carpellate flower, side view; **h** leaf adaxial surface; **i** leaf abaxial surface; **j** capsule; **k** cross section of ovary. All from *C.-I Peng 23466* (HAST).

hairs. *Male flowers*: pedicel 7–10 mm long, with sparse minute glandular hairs; tepals 4, uniformly very pale pink or white, entire, with a few minute glandular hairs adaxially; outer 2 sub-orbicular 8–9 × 6–7 mm; inner 2 obovate, slightly retuse, 6 × 3 mm; androecium with ca. 40 stamens, sessile; filaments free, unequal, 1–1.75 mm long, longer filaments found in the centre of the androecium; anthers yellow, 1.25 mm long, oblong-oblanceolate, slits lateral becoming unifacial at the apex; connective extended. *Female flowers*: pedicel 7–10 mm long, with sparse glandular hairs; ovary 10 × 14 mm including wings, pink; capsule 9 × 5 mm, two-locular, placentae bifid; wings 3, unequal, larger wing cucullate, smaller wings rounded; tepals the same as in the male flowers; styles 3, stigmatic surface spirally twisted. *Fruit* pedicel 7–12 mm, drying pale brown, 9–14 × 14–18 mm, recurved with the 2 lateral wings (9–14 × 3–4 mm, round) forming a splash cup, the larger wing markedly cucullate, 9–14 × 7–9 mm, apex truncate. Somatic chromosome number, $2n = 30$.

Distribution, habitat and ecology *Begonia hughesii* herbarium material has been collected from the type locality near the entrance to the Underground River and on Ugong Rock near Sabang and on limestone outcrops at the base of Mt. St. Paul. Further photographic evidence reveal records from the other side of the Babuyan River at PPURNP and within the forest on Cleopatra's Needle (Jonah Van Beijnen, pers. comm.). The species grows on eroded limestone in semi-shade under broadleaf forest, and was flowering and fruiting when collected in early November. In cultivation in the greenhouse of Academia Sinica in Taipei, Taiwan, it flowered and fruited from July to December.

Etymology The species epithet is named after Dr Mark Hughes of the Royal Botanical Garden Edinburgh, who studies extensively Southeast Asian begonias including Philippine *Begonia*.

Additional specimens examined (paratypes) PHILIPPINES. Palawan, Sabang, Ugong Rock, 10°5′13″N, 118°51′16″E, elev. ca. 60 m, in limestone outcrops, semishaded, locally abundant, 2 Nov. 2011, *Ching-I Peng 23475*, with Kuo-Fang Chung, Chien-I Huang, Rosario Rubite (HAST); Palawan, Mt. St. Paul, elev. 300–600 m, slope forest on karst limestone, 03-05 May 2012, *Hughes & Tandang MH1726* (HAST); Palawan, Puerto Princesa Subterranean National Park, Sabang, jungle trail between CPS and underground river, elev. 50 m, 29 May 2012, M. Hughes, C. Puglisi, D. Tandang & Julius *CP305* (E, PNH).

Proposed IUCN Red list category LC. The species has a fairly wide distribution within the PPURNP and surrounding forests on Mt St Paul and Cleopatra's Needle, and providing these areas remain well managed *B. hughesii* can be considered as Least Concern.

3. *Begonia tagbanua* M. Hughes, C.-I Peng & R. Rubite, sp. nov. § *Baryandra* (Figures 8, 9)

TYPE: PHILIPPINES. Palawan, Puerto Princesa, Puerto Princesa Subterranean River National Park, elev. ca. 35 m, 10°12′1″N, 118°55′32″E, 2 Nov. 2011, *Ching-I Peng 23471*, with Kuo-Fang Chung, Chien-I Huang, Rosario Rubite (holotype PNH, isotype A, E, HAST, MO).

Begonia tagbanua is most similar to *B. suborbiculata*, but differs in having less succulent leaves which are rugolose (not smooth), sparsely puberulous above (not glabrous), and fruits which are 3-winged (not 5-winged).

Terrestrial or lithophytic rhizomatous herb ca. 10 cm tall. *Stem* 6–8 mm in diameter, with sparse 1–2 mm brown hairs becoming glabrous with age, internodes 4–5 mm long. Stipules triangular, 14–16 × 4–5 mm, greenish to pink, with sparse hairs, keel present, apex fimbriate, margin entire. *Leaves* on petioles relaxed against the substrate; petiole 11–13 cm long and 4–6 mm in diameter, terete, succulent, maroon densely covered with brown straight hairs (1–2 mm), simple 2 mm long red hairs in a ring at the base of the petiole; blade ovate to suborbicular; 2 colour forms: one is adaxial surface dark green with brown patches at interveins and abaxial surface maroon, while another is adaxial surface plain apple green and abaxial surface light green; adaxial surface has white straight hairs (2–3 mm) arising from a green or brown dot while abaxial surface has white straight hairs (2–3 mm) at veins and shorter hairs (1–1.5 mm) at interveins, base cordate, lobes rounded, sinus overlap 1 cm, apex obtuse, margin shallowly undulate, denticulate, ciliate with brown straight hairs (1 mm), primary veins 7–8, stomata in clusters of 2–4. *Inflorescence* erect, 22–31 cm long, axillary, arising directly from the rhizome, bisexual, male and female flowers open at the same time as the inflorescence matures, 4 times dichotomously branching; peduncle pink 16–23 cm long, 3–4 mm in diameter, erect, with sparse light brown hairs (1–1.5 mm); bracts rounded-triangular, 4 × 2 mm, glabrous; pedicels pink 1.6–1.8 cm erect to ascending in staminate flowers, 1.5 cm ascending to horizontal in carpellate flowers. Male flowers: pedicel ca. 10 mm, glabrous; tepals 4, pink, glabrous, outer pair orbicular with 10–12 veins; 11–12 × 10–11 mm, inner pair obovate folded and retuse, 8–13 × 7–8 mm; stamens 30–40; filaments 1 mm long, united at the base; anthers yellow, rounded, 0.5 mm long. Female flowers: pedicel ca. 15 mm, ovary 10 × 14 mm including wings, pink; capsule 9 × 4 mm, two-locular, placentae bifid; wings 3, unequal, larger wing cucullate, smaller wings rounded; tepals 4, pink, glabrous, outer pair suborbicular 13–16 × 10–12 mm; inner pair obovate, folded, retuse 11–13 × 4–6 mm; styles 3, stigmatic surface spirally twisted. *Fruit* pedicel 1–1.3 cm;

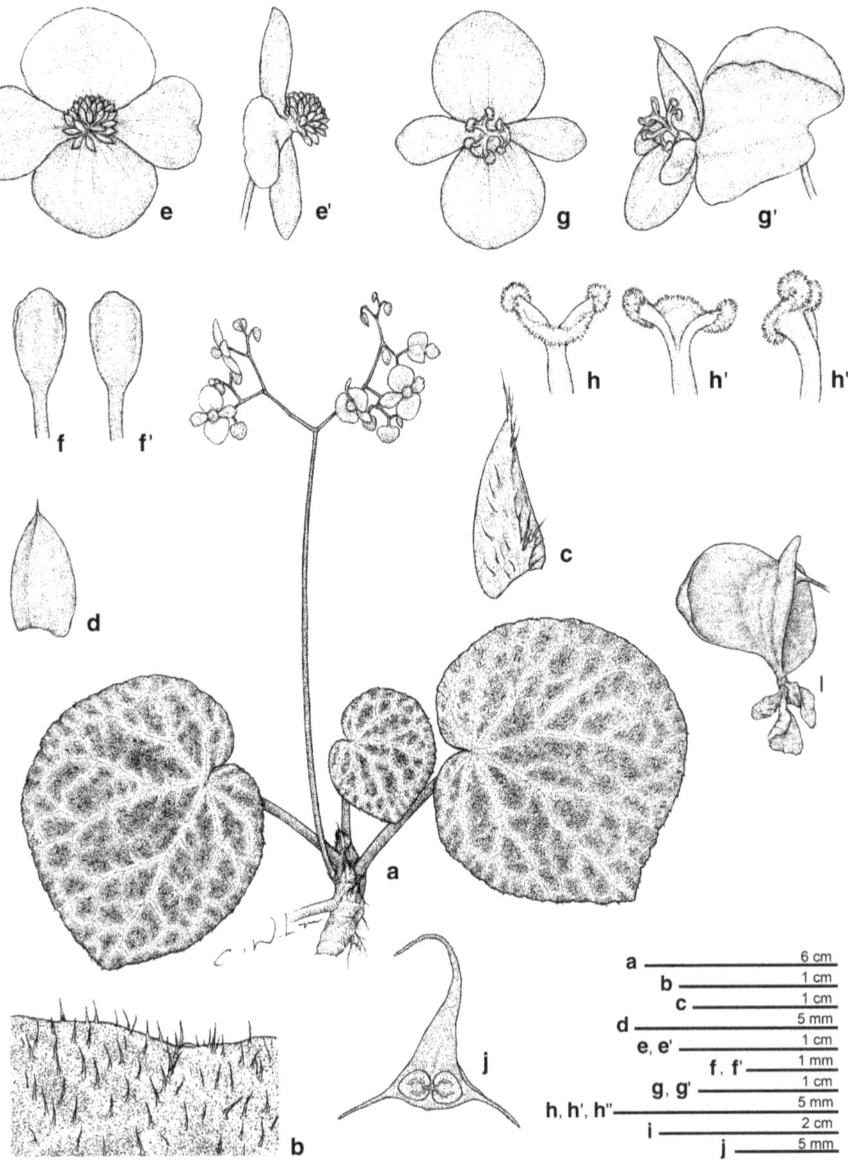

Figure 8 *Begonia tagbanua* M. Hughes, C.-I Peng & R. Rubite. **a** Habit; **b** indumentum on leaf adaxial surface; **c** stipule; **d** Bract; **e**, **e'**, staminate flower; **f**, **f'** stamen; **G**,G', carpellate flower; **g**, **h'**, **h''**, style and stigma; **i** Capsule; **j** cross section of ovary. All from *C.-I Peng 23471* (HAST). Line drawing by Che-Wei Lin.

capsule rounded in outline, glabrous, 8–12 × 18–20 mm unequally 3-winged, adaxial wing strongly cucullate shorter than the capsule 7–10 × 8–11 mm; lateral wings 8–12 × 5–6 mm forming a splash cup. Somatic chromosome number, $2n = 30$.

Distribution, habitat and ecology Begonia tagbanua is currently known from the type locality in PPSRNP, and from the foothills at the base of Mt. St. Paul at ca. 200 m altitude (M. Hughes field observation). In PPSRNP it grows quite abundantly along trail sides in bare soil, under semi-shaded broadleaf forest. Away from trails it can be found growing on steep banks of bare soil such as can be found at the base of tree roots. The species was flowering and fruiting when collected in early November. In cultivation in the greenhouse of Academia Sinica in Taipei, Taiwan, it flowered and fruited from July to December.

Etymology The species is named after the Tagbanuas, one of the indigenous peoples of the type locality.

Additional specimens examined (paratypes) PHILIPPINES. Palawan, Puerto Princesa, Puerto Princesa Subterranean River National Park, elev. ca. 35 m,

Figure 9 *Begonia tagbanua* M. Hughes, C.-I Peng & R. Rubite. **a** Habit and habitat; **b** cultivated plant at anthesis; **c** carpellate flower, side view; **d** carpellate flower, face view; **e** staminate flower, face view; **f** staminate flower, side view; **g** stipules; **h** leaf adaxial surface, showing indumentum; **i** capsule; **j** cross section of ovary. All from *C.-I Peng 23471* (HAST).

10°12′1″N, 118°55′32″E, 2 Nov. 2011, *Ching-I Peng P23472*, with Kuo-Fang Chung, Chien-I Huang, Rosario Rubite (HAST); Palawan, Puerto Princesa Subterranean National Park, Sabang, jungle trail between CPS and underground river, elev. 50 m, 10°11′53″N, 118°55′5″E, 29 May 2012, M. Hughes, C. Puglisi, D. Tandang & Julius *CP301* (E, PNH).

Proposed IUCN Red list category LC. In common with *B. hughesii* and *B. taraw*, *B. tagbanua* is a narrow endemic restricted to the PPSRNP environs. However it is not an obligate lithophyte, and can colonise recently dug clay soil banks by trail sides, and indeed the largest population observed was along the coastal trail toward the subterranean river cave entrance. Hence the appropriate conservation category for the species is Least Concern.

Discussion

The variety of habitats in terms of forest types and substrates, and in particular the variety of limestone habitats, may play a part in promoting species richness at this relatively low altitude for *Begonia* (Chung et al. 2014). Further exploration towards the karst summit of the nearby Mount St. Paul did not yield any more *Begonia* species, contrary to what we would expect for this predominantly montane genus. The phylogenetic results show that all three species harbour chloroplast polymorphism, which is congruent with *Begonia* species existing as populations with a high degree of genetic isolation (Hughes and Hollingsworth 2008; Twyford et al. 2014). Both *B. tagbanua* and *B. taraw* resolved as monophyletic in the chloroplast phylogeny, whereas *B. hughesii* appears as polyphyletic. Natural hybrids have been reported a number of times in *Begonia* (Peng and Chiang 2000; Peng and Ku 2009; Peng et al. 2010) and it is possible that *B. hughesii* and *B. tagbanua* share a common chloroplast lineage due to a past hybridisation event.

Conclusions

The study further highlights the importance of the Puerto Princesa Subterranean River National Park as a remarkable World Heritage Site, with its combination of unique geology and endemic flora. The presence of three narrowly endemic and rare species in the most heavily visited part of the reserve means continued care must be taken to ensure trails are well maintained and that they are respected by visitors, and that visitor numbers are managed appropriately. Banks by trail sides can provide an ideal habitat for *Begonia*, particularly if the understory vegetation along the trail is left as intact as possible to provide shade and preserve the micro-climate. Ex-situ conservation collections also have a role to play in ensuring a secure future for the three species described here, giving the potential for reintroduction if a chance event such as drought or typhoon should destroy or drastically reduce the wild populations.

The fact that three new species were found in such close proximity demonstrates how much remains to be discovered in the PPSRNP, and how even small scale forest disturbance could result in extinctions. The study supports the formation of a larger forest reserve encompassing both PPSRNP and the adjacent Cleopatra's Needle Mountain Range, which also has two recently discovered endemic *Begonia* (Hughes et al. 2010), and doubtless many other endemic plant species awaiting description.

Authors' contributions
RR, MH, PB, GA, KFC and CIP carried out field observations of the new taxa and conceived of the study. RR, HAY, LL and AV drafted the species descriptions. YK carried out the Chromosome cytology. HAY carried out the cryo scanning electron microscopy studies. MH carried out the molecular phylogenetic analysis. All authors read and approved the final manuscript.

Author details
[1] Department of Biology, College of Arts and Sciences, University of the Philippines Manila, Padre Faura, Manila 1000, Philippines. [2] Royal Botanic Garden Edinburgh, 20a Inverleith Row, Edinburgh EH3 5LR, UK. [3] CNRS, 3 rue Michel-Ange, 75794 Paris, France. [4] School of Forestry and Resource Conservation, National Taiwan University, Taipei 106, Taiwan. [5] Biodiversity Research Center, Academia Sinica, Taipei 115, Taiwan. [6] College of Science and Research Centre for the Natural and Applied Sciences, University of Santo Tomas, España, Manila 1015, Philippines.

Acknowledgements
We thank the Palawan Council for Sustainable Development, the Palawan Protected Areas Management Board, Palawan City Council and the Department of Environment and Natural Resources for granting access and research permits; Chien-I Huang (HAST) for assistance in the field trip; Guan-Bin Chen for cryo-SEM studies, Che-Wei Lin (TAIF) for the line drawings, and Jonah Van Beijnen for comments on the manuscript. The curators of BM, E, HAST, K, PNH herbaria are thanked for providing access to specimens. The Royal Botanic Garden Edinburgh is supported by the Scottish Governments Rural and Environment Science and Analytical Services Division. This research was supported in part by grants from the Royal Society of Edinburgh Bilateral Programme, University of the Philippines System Enhanced Creative Work and Research Grant (ECWRG 2014-09), Ministry of Science and Technology, Taiwan (NSC 101-2621-B-001-003-MY3; NSC 98-2621-B-001-002-MY3) and Academia Sinica, Taiwan.

Compliance with ethical guidelines

Competing interests
The authors declare that they have no competing interests.

References

BirdLife International (2015) Endemic Bird Area Factsheet, Palawan. http://www.birdlife.org. Accessed 16 Mar 2015

Chung K-F, Leong W-C, Rubite R, Repin R, Kiew R, Liu Y et al (2014) Phylogenetic analyses of *Begonia* sect. *Coelocentrum* and allied limestone species of China shed light on the evolution of Sino-Vietnamese karst flora. Bot Stud 55:e1. doi:10.1186/1999-3110-55-1

Demesure B, Sodzi N, Petit RJ (1995) A set of universal primers for amplification of polymorphic non-coding regions of mitochondrial and chloroplast DNA in plants. Mol Ecol 4:129–131. doi:10.1111/j.1365-294X.1995.tb00201.x

Doorenbos J, Sosef MSM, de Wilde JJFE (1998) The sections of *Begonia* including descriptions, keys and species lists. Wageningen Agric Univ Pap 98:1–266

Fitzgerald E (2011) Tourist arrivals triple thanks to N7W PPUR promotion. http://news.n7w.com/2011/07/26/tourist-arrivals-triple-thanks-to-n7w-ppur-promotion/?utm_source=rss&utm_medium=rss&utm_campaign=tourist-arrivals-triple-thanks-to-n7w-ppur-promotion. Accessed 21 Apr 2015

Fitzgerald E (2012) PPUR officially confirmed as one of the New7Wonders of Nature. http://news.n7w.com/2012/01/28/3743/?utm_source=rss&utm_medium=rss&utm_campaign=3743. Accessed 21 Apr 2015

Hall TA (1999) BioEdit: a user-friendly biological sequence alignment editor and analysis program for Windows 95/98/NT. Nucleic Acids Symp Ser 41:95–98 **(citeulike-article-id:691774)**

Hughes M, Coyle C (2009) *Begonia* Section *Petermannia* (Begoniaceae) on Palawan (Philippines), including two new species. Edinb J Bot 66:205–211. doi:10.1017/S0960428609005307

Hughes M, Hollingsworth PM (2008) Population genetic divergence corresponds with species-level biodiversity patterns in the large genus *Begonia*. Mol Ecol 17:2643–2651. doi:10.1111/j.1365-294X.2008.03788.x

Hughes M, Coyle C, Rubite RR (2010) A revision of *Begonia* section *Diploclinium* on the Philippine island of Palawan, including five new species. Edinb J Bot 67:123–140

Hughes M, Rubite RR, Kono Y, Peng C (2011) *Begonia blancii* (sect. *Diploclinium*, Begoniaceae), a newspecies endemic to the Philippine island of Palawan. Bot Stud 52:203–209

Hughes M, Rubite RR, Blanc P, Chung K-F, Peng C-I (2015) The Miocene to Pleistocene colonization of the Philippine archipelago by *Begonia* sect. *Baryandra* (Begoniaceae). Am J Bot 102:695–706. doi:10.3732/ajb.1400428

IUCN (2012) IUCN Red List Categories and Criteria: Version 3.1. 2nd edn. IUCN, Gland, Cambridge

Kokubugata G, Madulid DA (2000) Chromosomal study of four plant-taxa in Batan Island, the Philippines and the Yaeyama Group, Ryukyu Islands, Japan. Natl Sci Museum Monogr 18:139–144

Legro RAH, Doorenbos J (1969) Chromosome numbers in *Begonia* 1. Neth J Agric Sci 17:189–202

Myers N, Mittermeier RA, Mittermeier CG, da Fonseca GA, Kent J (2000) Biodiversity hotspots for conservation priorities. Nature 403:853–858. doi:10.1038/35002501

Oginuma K, Peng C-I (2002) Karyomorphology of Taiwanese *Begonia* (Begoniaceae): taxonomic implications. J Plant Res 115:225–235. doi:10.1007/s102650200028

Peng C-I, Chiang T-Y (2000) Molecular confirmation of unidirectional hybridization in *Begonia* × *Taipeiensis* Peng (Begoniaceae) from Taiwan. Ann Missouri Bot Gard 87:273–285

Peng C, Ku S (2009) *Begonia* × *chungii* (Begoniaceae), a new natural hybrid in Taiwan. Bot Stud 50:241–250

Peng CI, Liu Y, Ku SM, Kono Y, Chung KF (2010) *Begonia* × *breviscapa* (Begoniaceae), a new intersectional natural hybrid from limestone areas in Guangxi, China. Bot Stud 51:107–117

Peng C-I, Wang H, Kono Y, Yang H-A (2014) *Begonia wui-senioris* (sect. *Platycentrum*, Begoniaceae), a new species from Myanmar. Bot Stud 55:e13. doi:10.1186/1999-3110-55-13

Posa MRC, Diesmos AC, Sodhi NS, Brooks TM (2008) Hope for threatened tropical biodiversity: lessons from the Philippines. Bioscience 58:231–240. doi:10.1641/B580309

Posada D, Taboada GL, Doallo R, Darriba D (2012) jModelTest 2: more models, new heuristics and parallel computing. Nat Methods 9:772

Rambaut A (2009) FigTree v1.4.2. Software package. http://tree.bio.ed.ac.uk. Accessed 4 Feb 2015

Rambaut A, Suchard MA, Xie W, Drummond AJ (2014) Tracer v1.6. Software package. http://beast.bio.ed.ac.uk/Tracer. Accessed 2 Jul 2014

Ronquist F, Teslenko M, van der Mark P, Ayres DL, Darling A, Höhna S et al (2012) MrBayes 3.2: efficient Bayesian phylogenetic inference and model choice across a large model space. Syst Biol 61:539–542. doi:10.1093/sysbio/sys029

Rubite RR, Hughes M, Alejandro GJ, Peng C-I (2013) Recircumscription of *Begonia* sect. *Baryandra* (Begoniaceae): evidence from molecular data. Bot Stud 54:e38. doi:10.1186/1999-3110-54-38

Thomas DC, Hughes M, Phutthai T, Rajbhandary S, Rubite R, Ardi WH et al (2011) A non-coding plastid DNA phylogeny of Asian *Begonia* (Begoniaceae): evidence for morphological homoplasy and sectional polyphyly. Mol Phylogenet Evol 60:428–444. doi:10.1016/j.ympev.2011.05.006

Thomas DC, Hughes M, Phutthai T, Ardi WH, Rajbhandary S, Rubite R, Twyford AD, Richardson JE (2012) West to east dispersal and subsequent rapid diversification of the mega-diverse genus Begonia (Begoniaceae) in the Malesian archipelago. J Biogeogr 39:98–113. doi:10.1111/j.1365-2699.2011.02596.x

Twyford A, Kidner C, Ennos R (2014) Genetic differentiation and species cohesion in two widespread Central American *Begonia* species. Heredity (Edinb) 112:382–390. doi:10.1038/hdy.2013.116

World Heritage Commitee (2000) Twenty-third session, Marrakesh, Morocco

5

Proteomic analysis of broccoli (*Brassica oleracea*) under high temperature and waterlogging stresses

Hsin-Hung Lin[1†], Kuan-Hung Lin[2†], Su-Ching Chen[2], Yu-Hsing Shen[3] and Hsiao-Feng Lo[1*]

Abstract

Background: The production of broccoli (*Brassica oleracea*) is largely reduced by waterlogging and high temperature stresses. Heat-tolerant and heat-susceptible broccoli cultivars TSS-AVRDC-2 and B-75, respectively, were used for physiological and proteomic analyses. The objective of this study was to identify TSS-AVRDC-2 and B-75 proteins differentially regulated at different time periods in response to waterlogging at 40 °C for three days.

Results: TSS-AVRDC-2 exhibited significantly higher chlorophyll content, lower stomatal conductance, and better H_2O_2 scavenging under stress in comparison to B-75. Two-dimensional liquid phase fractionation analyses revealed that Rubisco proteins in both varieties were regulated under stressing treatments, and that TSS-AVRDC-2 had higher levels of both Rubisco large and small subunit transcripts than B-75 when subjected to high temperature and/or waterlogging.

Conclusions: This report utilizes physiological and proteomic approaches to discover changes in the protein expression profiles of broccoli in response to heat and waterlogging stresses. Higher levels of Rubisco proteins in TSS-AVRDC-2 could lead to increased carbon fixation efficiency to provide sufficient energy to enable stress tolerance under waterlogging at 40 °C.

Keywords: Ribulose bisphosphate carboxylase; Broccoli; Heat and waterlogging stresses; Differentially expressed proteins

Background

Broccoli (*Brassica oleracea var. italica*) is a member of the Brassicaceae family. This cole crop is not only economically important but also provides nutrients that have anticancer characteristics (Aggarwal and Ichikawa 2005). Under optimal temperatures of 18 to 25 °C, this plant grows normal flower buds and forms one large central head. However, when temperature exceeds 30 °C, heads become loose and branchy, tend to develop leaf-like structures, and initial floret development is disrupted (Bjorkman and Pearson 1998). Taiwan is located in tropical and subtropical regions where daily temperatures exceed 30 °C during summer (June to September), thus reducing the quality and quantity of broccoli produced. Moreover, typhoons always bring heavy rains in summer, and flooding from typhoon rainfall is a major risk to fresh-market broccoli production when the soil is water-saturated.

Previous studies report that when plants are subjected to high temperature and waterlogging, the water and chlorophyll in leaves are decreased but the level of hydrogen peroxide in the cells is increased, resulting in reduced plant growth and development (Moller and Kristensen 2004; Kumutha et al. 2009; Pucciariello et al. 2012). Fortunately, plants have developed several mechanisms to protect themselves from stressful abiotic environments, some of which overlap. For example, pretreating *Arabidopsis* with 38 °C for 90 min results in a higher tolerance to waterlogging (Banti et al. 2010). Additionally, waterlogging can also stimulate the expression of heat shock factors and proteins as effectively as high temperature (Banti et al. 2010). These findings demonstrate that responses to different stress factors in

* Correspondence: hflo@ntu.edu.tw
†Equal contributors
[1]Department of Horticulture and Landscape Architecture, National Taiwan University, Taipei, Taiwan
Full list of author information is available at the end of the article

plants are coordinated by complex and interconnected signaling pathways. Different sets of genes have been identified that activate against various abiotic stresses, including heat, flood, drought, cold, and intense light (Kreps et al. 2002; Seki et al. 2002; Rizhsky et al. 2004). However, little is known about how the simultaneous occurrence of multiple abiotic stresses, as opposed to individual stresses, damage crop production (Rizhsky et al. 2004; Mittler 2006; Atkinson and Urwin 2012). Therefore, uncovering the physiological mechanisms whereby plants can withstand combined waterlogging and high temperature stressing is greatly desired.

Proteomic analysis is a powerful approach for revealing differentially expressed proteins under given conditions. Liu et al. discovered a large number of differentially expressed proteins from broccoli florets treated with N^6-benzylaminopurine, illuminating a complex network that provides comprehensive information on post-harvest yellowing response mechanisms (Liu et al., 2011; Liu et al., 2013). Using two-dimensional liquid phase fractionation (PF2D) and matrix-assisted laser desorption ionization time-of-flight mass spectrometry (MALDI-TOF MS), we identified 31 differentially expressed proteins from heat-tolerant and heat-susceptible broccoli cultivars TSS-AVRDC-2 and B-75, respectively, under high temperature and/or waterlogging stresses. We then further cloned the stress-responsive Rubisco genes and their transcript levels under stressing were determined. The possible mechanism whereby TSS-AVRDC-2 broccoli develops florets during summer is also discussed herein.

Methods

Plant materials, culturing, and heat- and flood-stress treatments

Seeds of broccoli (*B. oleracea*) varieties TSS-AVRDC-2 and B-75 were obtained as gifts from Mr. L.C. Chung (see Acknowledgments) at AVRDC (Vegetable Research and Development Center) - the World Vegetable Center, Tainan, Taiwan. B-75 is a heat-sensitive variety that requires optimum growing temperatures for satisfactory production. However, TSS-AVRDC-2 is a heat-tolerant variety suitable especially for warm-subtropical regions such as southern Taiwan where average day temperatures reach as high as 40 °C during summer. Seeds of TSS-AVRDC-2 and B-75 were immersed in distilled-deionized (dd) water in darkness for 24 h and germinated on wetted Whatman filter papers for three days to ensure uniform germination. Seedlings were then transplanted into 5-inch plastic pots containing a mixture of peat and moss (4:1, v:v) and placed in a growth chamber under 300 μmol m^{-2} s^{-1} light with a 16 h photoperiod provided by fluorescent and incandescent light. The temperature was maintained at 22 °C in the light and 18 °C in the dark, with a relative humidity (RH) of 80 %. Plants were watered with a half-strength Hoagland solution (Hoagland and Arnon 1950) three times a week to maintain optimal irrigation and growth for 40 days before imposing heat and waterlogging stresses.

Pots of TSS-AVRDC-2 and B-75 plants were divided into four groups and transferred to 20 °C without waterlogging (C, control), 20 °C with waterlogging (F), 40 °C without waterlogging (H), and 40 °C with waterlogging (HF) for periods of 0, 12, 24, 48, and 72 h (for chlorophyll content and H_2O_2 content experiments) or 0, 24, 36, 48, 60, 72, 96, 120 and 144 h (for stomatal conductance experiment) in four growth chambers under the above-described conditions. In the waterlogging treatments, pots were randomly placed in 28 × 14 × 14 cm plastic buckets and subjected to flooding by filling the buckets with tap water to 5 cm above the soil surface. Pots were removed from the buckets at different times following flooding, and plants were removed and leaves from each plant were clipped, frozen in liquid nitrogen, and stored at −80 °C in an ultrafreezer until use. Three replicates from each time interval for the four treatments were randomly placed in a growth chamber. The experiment was performed twice independently for a randomized design of growth environment, sampling day, and physiological analysis.

Measurement of chlorophyll content (CC), stomatal conductance (SC), and H_2O_2 content (HC)

Characteristic physiological responses of plants to different treatments were evaluated to determine if the defense mechanisms of the genotypes are related to their CC, SC, and HC values. Relative CC unit leaf area was determined according to a previous study (Davenport et al. 2005) using a SPAD (Soil Plant Analysis Development) analyzer (SPAD-502 Chlorophyll Meter, Konica Minolta, Tokyo, Japan). Stomatal conductance values were measured with a Porometer (Decagon Devices, Pullman, WA). The sensor was equilibrated by considering water vapor in the ambient atmosphere as 0 % relative humidity and wetted filter paper at 100 % relative humidity. After equilibration, each leaf was inserted into the head of the Porometer sensor and a measurement taken (Limm et al. 2009). HC measurement was previously described (Wei et al. 2012). Briefly, each leaf was homogenized in liquid nitrogen and ice-cold extraction buffer (50 mM sodium phosphate buffer, pH 6.8, and 1 mM hydroxylamine) then added. The mixture was centrifuged at 12,000 *g* for 10 min at 4 °C and an equal amount of titanium reagent (0.1 % $TiCl_2$ in 20 % H_2SO_4) added to the supernatant. The titanium reagent mixtures were centrifuged at 12,000 *g* for 10 min and supernatants measured at 410 nm absorbance. H_2O_2 content was computed from a standard curve prepared from H_2O_2 solutions of known concentrations.

Data shown in Figs. 1, 2 and 3 represent the mean of at least two independent sets of experiments with similar results. Measurements of physiological parameters were analyzed by analysis of variance (ANOVA) with completely randomized design. For significant values, means were separated by the least significant difference (LSD) test at $p \leq 0.05$ using PC SAS 8.2 (SAS Institute, Cary, NC, USA).

Protein isolation and quantification

Two grams of frozen leaf samples were ground into powder in liquid nitrogen, extracted with 7 ml extraction buffer (5 % [v/v] 2-mercaptoethanol, 2 % [w/v] polyvinylpolypyrrolidone, 0.5 % [w/v] sodium dodecyl sulfate, 10 % [v/v] glycerol, and 60 mM Tris–HCl, pH 6.8), and centrifuged for 30 min at 13,000 *g* at 4 °C. Supernatants were transferred to a new tube and ice-cold acetone was added, and placed at −20 °C for 1 h. Precipitated proteins were collected after centrifuging at 5,000 *g* for 30 min at 4 °C. Pellets were then washed with ice-cold acetone three times, vacuum-dried, and kept at −80 °C until use. Protein concentration was determined using the Bradford method (Protein Assay, Bio-Rad Laboratories, Hercules, CA), and bovine serum albumin (BSA) was used as a protein standard.

2-D liquid phase fractionation analysis and protein identification by MALDI-TOF MS and database search

A first-dimension high performance chromatofocusing column (Beckman Coulter, CA, USA) was pre-equilibrated with starting buffer (20 mM Tris–HCl, pH 8.5) until column pH reached 8.3. Samples were then injected into the column at a flow rate of 0.2 ml/min and column effluent was monitored at an absorbance of 280 nm. The pH gradient was created by an elution buffer (polybuffer 74, pH 4.0; GE Healthcare, NJ, USA) and fractions were collected every 0.3 pH unit. In the second dimension, fraction separation was performed using a 4.6 x 30 mm nonporous C18 HPRP column (Beckman Coulter) at 50 °C at a flow rate of 0.75 ml/min. Solvent A (0.1 % w/v trifluroracetic acid) and solvent B (0.08 % w/v acetonitrile) were used to create a gradient. The gradient consisted of solvent A at 100 % for 10 min and solvent B at 0 to 60 % for 30 min, and eluent was collected every 15 s, and was monitored at an absorbance of 214 nm. Protein peaks from those treated and untreated plants were compared using DeltaVue software (Beckman Coulter, Inc. Fullerton, CA, USA). DeltaVue allowed side-by-side viewing of the second-dimension runs for two samples, and was used to compare and quantify the number of differential expressed protein between them. Each band shown in this chromatogram represented a

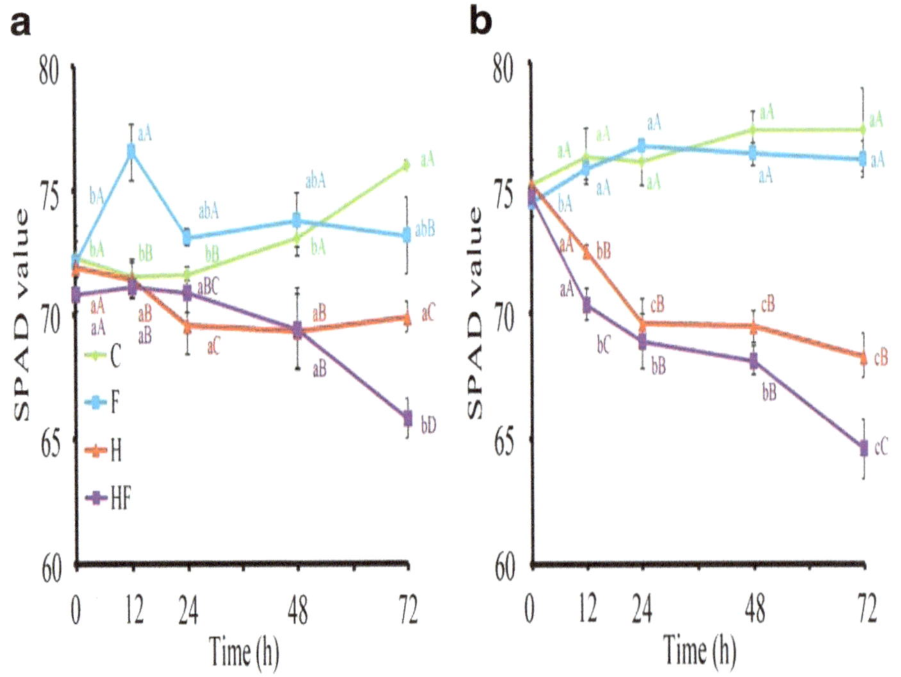

Fig. 1 Chlorophyll content in stressed plants. Leaf SPAD values in TSS-AVRDC-2 (**a**) and B-75 (**b**) plants under control (C), waterlogging (F), 40 °C (H), and waterlogging at 40 °C (HF) treatments for 0, 12, 24, 48, and 72 h. Values represent the means of five independent plants. Values represent the means of five independent plants. Means with the same small letters are not significantly different among times within the same row, based on LSD at $p \leq 0.05$ under ANOVA. Means with the same capital letters are not significantly different between treatments within the same time point, based on LSD at $p \leq 0.05$ under ANOVA

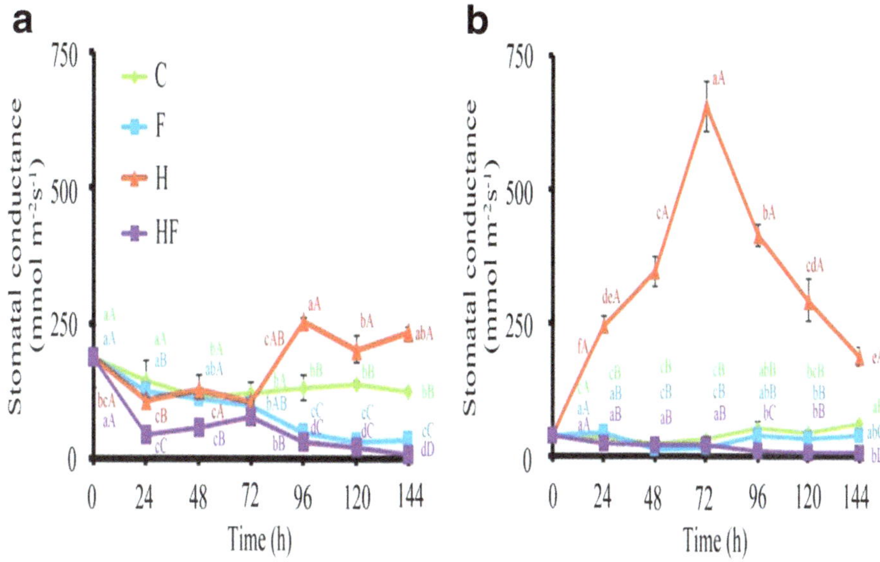

Fig. 2 Stomatal conductance in stressed plants. Stomatal conductance in TSS-AVRDC-2 (**a**) and B-75 (**b**) plants under control (C), waterlogging (F), 40 °C (H), and waterlogging at 40 °C (HF) treatments for 0, 24, 72, 48, 96, 120, and 144 h. Values represent the means of five independent plants. Values represent the means of five independent plants. Means with the same small letters are not significantly different among times within the same row, based on LSD at $p \leq 0.05$ under ANOVA. Means with the same capital letters are not significantly different between treatments within the same time point, based on LSD at $p \leq 0.05$ under ANOVA

singly separated protein and the relative intensity of the colors was directly proportional to the difference in protein concentration. The up-regulation and down-regulation of the proteins were identified (Additional file 1: Figures S1 and Additional file 2: Figure S2).

The eluents were dried under a vacuum and the pellet dissolved in a reducing solution (50 mM ammonium bicarbonate and 10 mM dithiothreitol) for 60 min at 60 °C. After reduction steps, protein solutions were digested at 37 °C for 16 h with 5 ng/μL of trypsin. Tryptic digestion

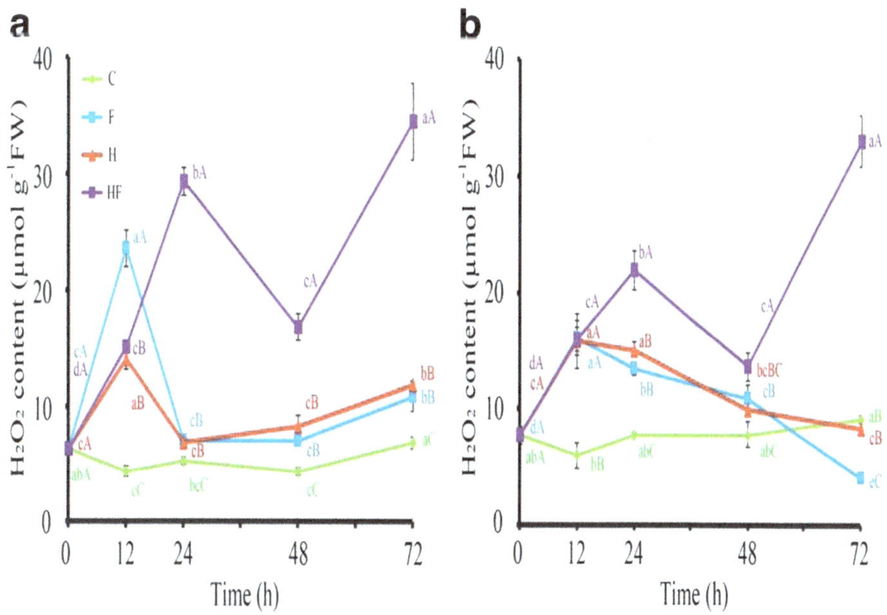

Fig. 3 H_2O_2 content in stressed plants. Leaf H_2O_2 content in TSS-AVRDC-2 (**a**) and B-75 (**b**) plants under control (C), waterlogging (F), 40 °C (H), and waterlogging at 40 °C (HF) treatments for 0, 12, 24, 48, and 72 h. Values represent the means of five independent plants. Means with the same small letters are not significantly different among times within the same row, based on LSD at $p \leq 0.05$ under ANOVA. Means with the same capital letters are not significantly different between treatments within the same time point, based on LSD at $p \leq 0.05$ under ANOVA

was stopped by adding 1 % (v/v) formic acid. Digested proteins were further analyzed by MALDI-TOF MS. Mass spectra values were searched against NCBI and Swiss-Prot protein databases using the Mascot search program (http://www.matrixscience.com). The search parameters were set as follows: 1 missed cleavage, fixed modification, peptide charge ±1, and variable modifications were carbamidomethyl of cysteine and oxidation of methionine. Trypsin was specified as the proteolytic enzyme. Peptide tolerance and MS mass tolerance were 50 ppm and 0.25 Da, respectively. Peptide mass fingerprinting match confidence was based on the MOWSE score and confirmed by accurate overlapping of matched peptides with mass spectrum major peaks. Scores greater than 67 ($p < 0.05$) were considered positive. Peaks with multiple proteins detected by MS were not considered. Only significant hits, as defined by MASCOT probability analysis ($p < 0.05$), were accepted. To identify differentially expressed proteins, the protein peaks from waterlogging (F), 40 °C (H), and 40 °C with waterlogging (HF) at 12-h and 24-h time points were compared to those from the control sample (20 °C).

Amplification and cloning of RubS and RubL from TSS-AVRDC-2 and B-75

For amplification and cloning experiments, genomic (g) DNA was isolated from mature TSS-AVRDC-2 and B-75 leaves based on the CTAB-based method from a previous study (Han et al. 2012). In brief, each leaf sample was ground into powder in liquid nitrogen, extracted with 65 °C preheated CTAB extraction buffer (2 % [v/v] CTAB, 2 % [w/v] polyvinylpolypyrrolidone, 2 % [w/v] PVP-40000, 1.4 M NaCl, 2 % 2-mercaptoethanol, 100 mM Tris HCl, pH 8.0 and 20 mM EDTA, pH 8.0), incubated at 65 °C for 1 h, and centrifuged for 20 min at 13,000 *g* at 4 °C. An equal volume of chloroform was added to the supernatant. The supernatant was transferred and precipitated with isopropanol. Total RNA was isolated from leaves of TSS-AVRDC-2 and B-75 with Trizol reagent (Invitrogen, CA, USA) according to manufacturer instructions. In brief, each leaf sample was ground into powder in liquid nitrogen and extracted with Trizol reagent. Another 0.2 volume of chloroform was added and centrifuged for 15 min at 13,000 *g* at 4 °C. A 0.5 volume of isopropanol was added to the supernatant to precipitate RNA at 4 ° C for 2 h and centrifuged for 20 min at 13,000 *g* at 4 °C. The pellet was washed with 75 % ethanol twice and re-suspended in diethyl pyrocarbonate-treated deionized water. The quality of the total RNA was resolved by formaldehyde gel. Complementary (c) DNA was synthesized by using 1 ug of total RNA, Moloney Murine Leukemia Virus Reverse Transcriptase, and polyT primer as described previously (Chen et al. 2008). The gene specific primer pairs RubS-F (5-atggcttcctctatgctctcc-3) and RubS-R (5- gcaccggtgaagcttggtgg-3), and RubL-F (5-atgtcaccacaaacagagac-3) and RubL-R (5-gatctccttccatacttcac-3), were designed according the sequences of AtRubS and AtRubL and used for gene cloning. The gDNA and cDNA of TSS-AVRDC-2 and B-75 were treated as templates to amplify RubS and RubL sequences. The PCR products were purified by Qiagen gel extraction kits (Qiagen, Valencia, CA). The PCR products were ligated with a CloneJET PCR Cloning Kit (Thermo Scientific, Rochester, USA) using T4 DNA ligase. The inserts from individual colonies were amplified by PCR using the JetF (5-cgactcactatagggagagcgg-3) and JetR (5-aagaacatcgattttccatggcag-3) primers and then sequenced. DNA sequences were aligned and compared with reference sequences using ClustalW (http://www.ebi.ac.uk/Tools/msa/clustalw2).

Northern blot analysis

Total RNA samples were extracted from plant leaves under stressed treatments to determine the expression levels of RubL and RubS in leaves of TSS-AVRDC-2 and B-75. Total RNA samples were separated by formaldehyde gel and blotted to Hybond N+ membranes (GE Healthcare, NJ, USA) using a trans-blot semidry transfer cell (Bio-Rad) for 1 h at 400 mA, followed by UV cross-linking. The radiolabeled RubL and RubS probes were produced by PCR reactions using TSS-AVRDC-2 cDNA clones as templates in the presence of α-^{32}P-dATP with RubL-F and -R and RubS-F and RubS-R primers, respectively. PCR reactions were labeled with α-^{32}P-ATP using High Fidelity *Taq* DNA Polymerase (Thermo Scientific). Blots used for the analysis were prehybridized with hybridized buffer (6X SSC, 5X Denhardt's Solution, and 0.5 % sodium dodecyl sulfate) at 55 °C for 2 h. After the radiolabeled probes were added, hybridization was performed under the same conditions for another 16 h. Blots were washed twice in half-diluted saline-sodium citrate buffer and 0.1 % sodium dodecyl sulfate for 10 min, followed by washing once in 10 % saline-sodium citrate and 0.1 % sodium dodecyl sulfate for 15 min. Radioactive blots were then displayed on a Phosphor Imager (Molecular Dynamics, CA, USA). Ribosomal (r) RNA was analyzed after ethidium bromide staining and treated as a load control. The ratios of RubL and RubS RNA to 28S rRNA in each reaction were calculated and the ratio of the control reaction was treated as a value of one for determining the relative ratios of other reactions. To quantify expression levels, the radioactive signals and amounts of 28S rRNA were measured using computing laser densitometer ImageJ software (http://imagej.nih.gov/ij/), with radioactive signals being normalized by the amount of 28S rRNA. The expression level of treatment C was treated as 1 for normalizing the relative expression levels of H, F, and HF treatments. Each experiment was replicated three times.

Results and discussions

Phenotypic and physiological characteristics of broccoli plants treated by stresses

Plants display complex phenotypic and physiological responses to heat and flood stresses. It is necessary to identify physiological characteristics that reflect their complex underlying genetic make-up. Physiological phenotypes of 40-day-old TSS-AVRDC-2 and B-75 plants under control (C) and various stress treatments (F, H, and HF) for 12, 24, 48, and 72 h were observed. Leaf chlorophyll content (CC) is an indicator of photosynthetic efficiency when plant responding to stresses (Augustine et al. 2015). Significant changes in CC (measured as SPAD) were observed between and among genotypes under stress. Figure 1 shows that the SPAD values of both cultivars under C and F conditions were higher than those under H and HF conditions after 12 h. A significant reduction in SPAD values were observed in both cultivars treated by HF over time, but the reduction rate in B75 was more rapid than in TSS-AVRDC-2, indicating that the CC of both cultivars were not influenced by waterlogging and that TSS-AVRDC-2 maintained higher CC values than B-75 under high temperature and waterlogging treatments, except for after 24 h. The SC parameter was chosen as an indirect proxy of plant physiology under stress regarding the level of leaf water loss and CO_2 uptake. SC values of B-75 plants under condition H from 24 h to120 h were significantly higher than in TSS-AVRDC-2 plants (Fig. 2), indicating that transpiration in B-75 was increased due to stomata being open at high temperatures. Moreover, the SC values of both cultivars under treatment H after 72 h were higher than in other treatments. No significant differences in SC values were observed between F and HF in both cultivars after 72 h. These results suggest that stomata were closed when plants were waterlogged and that stomata closure under waterlogging and high temperatures was faster than under either waterlogging or high temperature treatments alone. Excess accumulation of H_2O_2 can lead to oxidative stress in plants, which then triggers cell death. HC elevates under waterlogging and high temperatures stresses in plants (Simova-Stoilova et al. 2012; Xu et al. 2014). Therefore, HC was assessed over time under various treatments (Fig. 3). Compared to B-75 plants, TSS-AVRDC-2 plants had lower HC under H treatment from 24 to 48 h. Furthermore, the endogenous H_2O_2 content was elevated when both cultivars were exposed to HF stresses. Thus, high temperature plus waterlogging caused an overproduction of H_2O_2 in both genotypes, their H_2O_2 scavenging systems being unable to eliminate the extra H_2O_2. A combination of waterlogging and high temperature induced larger amounts of H_2O_2 than individual stresses in both cultivars.

Figures 1, 2 and 3 show that two cultivars displayed different physiological responses to various stress treatments. TSS-AVRDC-2 maintained significantly higher chlorophyll levels than B-75 under high temperature treatments at 24 h (Fig. 1), indicating that TSS-AVRDC-2 has a better photosynthetic efficiency than B-75, displaying its characteristically high temperature tolerance. Typically, CC is reduced in amount by stressful conditions (Xu et al. 2014). This is consistent with our observations that chlorophyll loss was more pronounced in heat-sensitive B75 plants in accordance with the more pronounced and increased visible symptoms of leaf injury. In addition, the SC values of B-75 were largely increased from 0 h to 72 h under high temperature, whereas TSS-AVRDC-2 exhibited low SC values in order to prevent water loss (Fig. 2). Although wilted leaves were observed in TSS-AVRDC-2 and B-75 under stress treatments for at least 24 h, SC values in both genotypes under F and HF treatment over time were low (Fig. 2). These results suggest that waterlogging results in stomata closure, which prevents transpiration. H_2O_2 acts as a ubiquitous marker of signaling events for the induction of adaptive stress responses and oxidative stress (Mittler et al. 2011). This dual function should be tightly regulated by the balancing of H_2O_2 production and scavenging (Mittler et al. 2004). When plants were treated with H_2O_2, their tolerances to high temperature and waterlogging were increased (Uchida et al. 2002; Banti et al. 2010). Plants often suffer from a simultaneous combination of high temperature and waterlogging. In our study, TSS-AVRDC-2 had a better H_2O_2 control system than B-75, and low levels of H_2O_2 increased tolerance to high temperature stress in TSS-AVRDC-2 (Fig. 3). However, waterlogging plus high temperature damaged broccoli plants more seriously than either high temperature or waterlogging alone, through the overproduction of H_2O_2. At the physiological level, the many effects of flooding and heat stresses indicate the importance of protecting plants from oxidative damage caused by the overproduction of H_2O_2.

Differentially expressed proteins in plant leaves among stress treatments

Changes in protein accumulation under stress likely are directly related to the physiological phenotypic response of plants to stress. To investigate the effects of single and combined stress treatments on the differential expression of proteins, leaf total protein was extracted from stressed 40-day-old TSS-AVRDC-2 and B-75 plants at 12 h and 24 h. Fifteen and 16 protein peaks were identified from TSS-AVRDC-2 (Table 1; Additional file 1: Figure S1) and B-75 (Table 2; Additional file 2: Figure S2), respectively. These identified proteins emerged as key participants in stress tolerance. Among 31 identified proteins,

Table 1 Identification of differentially expressed stress-related proteins in TSS-AVRDC-2 by MALDI-TOF-MS

Treatment	Stress Time (h)	Peak	Homologous protein	Species	PI	S[a]	Accession number
F	12	1	Putative F-box protein At5g41500	*Arabidopsis thaliana*	8.9	74	FB277_ARATH
	24	2	DNA-directed RNA polymerase subunit beta	*Geobacillus thermodenitrificans*	8.9	86	RPOC_GEOTN
		3	Conserved oligomeric Golgi complex subunit	*Magnaporthe grisea*	4.9	66	COG6_MAGO7
		4	ATP synthase subunit beta	*Brimeura amethystina*	5.2	76	ATPB_BRIAM
		5	Rubisco large chain	*Brassica oleracea*	5.8	96	RBL_BRAOL
		6	Mitochondrial 2-oxoglutarate/malate carrier protein	*Rattus norvegicus*	9.8	76	M2OM_RAT
H	24	7	ATP synthase subunit beta	*Nephroselmis olivacea*	5.2	90	ATPB-NEPOL
		8	Phosphopantetheine adenylyltransferase	*Silicibacter sp.*	5.4	76	EEW57108
		9	Ribosome-binding factor A	*Salmonella choleraesuis*	5.9	62	RBFA_SALCH
		10	Pentatricopeptide repeat-containing protein At2g01740	*Arabidopsis thaliana*	6.0	65	PP141_ARATH
HF	12	11	Putative F-box protein At5g41500	*Arabidopsis thaliana*	8.9	74	FB277_ARATH
		12	Eukaryotic translation initiation factor 3	*Saccharomyces cerevisiae*	5.4	54	EIF3I YEAST
	24	13	Rubisco small chain	*Brassica napus*	8.2	88	PO5346
		14	Ribosome-binding factor A	*Salmonella choleraesuis*	5.9	54	RBFA_SALCH
		15	Pentatricopeptide repeat containing protein At2g01740	*Arabidopsis thaliana*	6.0	65	PP141_ARATH

[a]: Probability-based molecular weight search (Mowse) score
F, waterlogging; H, 40 °C; HF, waterlogging at 40 °C

the expressed levels of RubL and RubS in TSS-AVRDC-2 and B-75 plants under stress showed distinct differences. In TSS-AVRDC-2 plants, RubL protein (peak 5) was up-regulated under F treatment for 24 h compared to controls, and RubS protein (peak 13) was up-regulated under HF treatment for 24 h compared to controls (Table 3). Furthermore, two RubS proteins (peaks 24, 27) in B-75 were up-regulated by either for 12 h HF treatment or 24 h H treatment, while RubL proteins (peak 16) were down-regulated by F, H and HF treatments for 12 h

Table 2 Identification of differentially expressed stress-related proteins in B-75 by MALDI-TOF-MS

Treatment	Stressed Time (h)	Peak	Homologous protein	Species	PI	S[a]	Accession number
C	12	16	Rubisco large chain	*Brassica oleracea*	5.8	78	RBL_BRAOL
F	12	17	ATP synthase subunit beta	*Acutodesmus obliquus*	5.2	66	ATPB ACUOB
	24	18	Inorganic pyrophosphatase	*Corynebacterium efficiens*	4.5	70	Q8FMF8
		19	tRNA pseudouridine synthase B	*Mycobacterium avium*	8.2	74	A0QIX1
H	12	20	DNA-directed RNA polymerase subunit beta	*Staphylococcus haemolyticus*	6.1	61	RPOC_STAHJ
		21	Probable O-sialoglycoprotein endopeptidase	*Leptospira borgpetersenii Serovar hardjo-bovis*	5.7	75	ABJ76496
		22	Cysteinyl-tRNA synthetase	*Lactococcus lactis subsp.*	4.8	72	SYC_LACLS
	24	23	Formin like protein 19	*Arabidopsis thaliana*	9.1	56	Q9FF14
		24	Rubisco small chain	*Brassica napus*	8.2	98	PO5346
		25	Putative uncharacterized protein yibJ	*Escherichia coli*	5	60	P32109
		26	Thiamine biosynthesis protein thiC	*Shewanella halifaxensis*	5.1	78	B0TRP7
HF	12	27	Rubisco small chain	*Brassica napus*	8.2	97	RBS1_BRANA
		28	tRNA 2-thiocytidine biosynthesis protein ttcA	*Sorangium cellulosum*	7.0	76	TTCA_SORC5
		29	Myosin-1	*Canis familiaris*	5.6	83	MYH1_CANFA
	24	30	Putative uncharacterized protein yibJ	*Escherichia coli*	5	68	P32109
		31	Haptoglobin	*Mesocricetus auratus*	5.7	81	O35086

[a]Probability-based molecular weight search (Mowse) score
F, waterlogging; H, 40 °C; HF, waterlogging at 40 °C

Table 3 Highly differentially expressed protein pattern in stressed TSS-AVRDC-2

Stressed time (h)	Homologue protein	Flooding	40 °C	Flooding at 40 °C
12	Putative F-box protein At5g41500	+	-	+
	Eukaryotic translation initiation factor 3	-	-	+
24	DNA-directed RNA polymerase subunit beta	+	-	-
	Rubisco large chain	+	-	-
	Rubisco small chain	-	-	+
	Conserved oligomeric Golgi complex subunit	+	-	-
	ATP synthase subunit beta	+	+	-
	Mitochondrial 2-oxoglutarate/malate carrier protein	+	-	-
	Phosphopantetheine adenylyltransferase	-	+	-
	Ribosome-binding factor A	-	+	+
	Pentatricopeptide repeat-containing protein	-	+	+

+, up-regulated protein; –, down-regulated protein

Table 4 Highly differentially expressed protein pattern in stressed B-75

Stressed time (h)	Homologue protein	Flooding	40 °C	Flooding at 40 °C
12	Rubisco large chain	-	-	-
	Rubisco small chain	-	-	+
	ATP synthase subunit beta	+	-	-
	DNA-directed RNA polymerase subunit beta	-	+	-
	Probable O-sialoglycoprotein endopeptidase	-	+	-
	Cysteinyl-tRNA synthetase	-	+	-
	tRNA 2-thiocytidine biosynthesis protein ttcA	-	-	+
	Myosin-1	-	-	+
24	Rubisco small chain	-	+	-
	Inorganic pyrophosphatase	+	-	-
	tRNA pseudouridine synthase B	+	-	-
	Formin like protein 19	-	+	-
	Putative uncharacterized protein yibJ	-	+	+
	Thiamine biosynthesis protein thiC	-	+	-
	Haptoglobin	-	-	+

+, up-regulated protein; –, down-regulated protein

(Table 4). Rubisco proteins are responsible for carbon fixation in plants and were regulated by stressing (Bose et al. 1999). To confirm these results, the RNA expression levels of RubL and RubS in TSS-AVRDC-2 and B-75 under stressing were analyzed.

In the present study, we analyzed physiological and proteomics data in different genotypes of broccoli under high-temperature and waterlogging stresses for different time periods. For survival, plants must respond to flood and/or heat stresses differently from the way they regulate protein expression during biochemical and physiological adaptations. Photosynthesis is one of the systems that is most sensitive to high-temperature stress (Salvucci 2008; Wang et al. 2015). Photosynthesis systems were protected from the damages caused by waterlogging and/or heat in TSS-AVRDC-2, which led to physiological phenomena such as high CC and low SC and HC in leaves. Chlorophyll functioning is not decreased in stressed thermo-tolerant plants (Li et al. 2003; Xu et al. 2014). A temperature of 40 °C reduces photosynthesis, which seems to change the enzymatic response of a CO_2-fixing enzyme, Rubisco (Markus et al. 1981). Thus, RubL and RubS were selected to do the further tests.

RubL and RubS transcripts were regulated when stressed

In our study, Rubisco subunits (small and large chain) related to photosynthesis were differentially expressed and regulated by combination treatments and between genotypes (Tables 1 and 2). RubL and RubS sequences were cloned from gDNA and cDNA of TSS-AVRDC-2 and B-75 using RubL-F and -R and RubS-F and -R in pairs. Cloned sequences were aligned and compared with *Arabidopsis* RubL and RubS genes. The sequences of RubL in gDNA were 97.4 % matched with reference sequences from *Arabidopsis* (Additional file 3: Figure S3), yet the RubS gene shows two introns in the gDNA sequences (Additional file 4: Figure S4). The cDNA identity of RubL and RubS between TSS-AVRDC-2 and B-75 was higher than 99 %. The transcript levels of RubL and RubS in TSS-AVRDC-2 and B-75 were analyzed by Northern blot. RubL and RubS transcripts in both genotypes under control treatment were observed (Fig. 4). Compared to the transcript level of RubL in TSS-AVRDC-2 controls at 72 h, RubL transcripts were increased by 18 % but decreased by 39 % and 25 % under F, H, and HF treatments, respectively (Fig. 4a). However, in B-75 plants, RubS transcripts decreased 75 %, 81 %, and 86 % at F, H and HF, respectively, compared to controls. RubS transcripts of TSS-AVRDC-2 decreased 51 % at condition F at 72 h, but increased by 10 % and 62 % under H and HF conditions, respectively, compared to controls (Fig. 4b). The transcript level of RubL in B-75 decreased 55 %, 14 %, and 5 % by F, H, and HF treatments at 72 h, respectively, compared to controls. These results suggest that the transcript abundance of RubL in TSS-AVRDC-2 was higher than that of B-75 under all stress treatments.

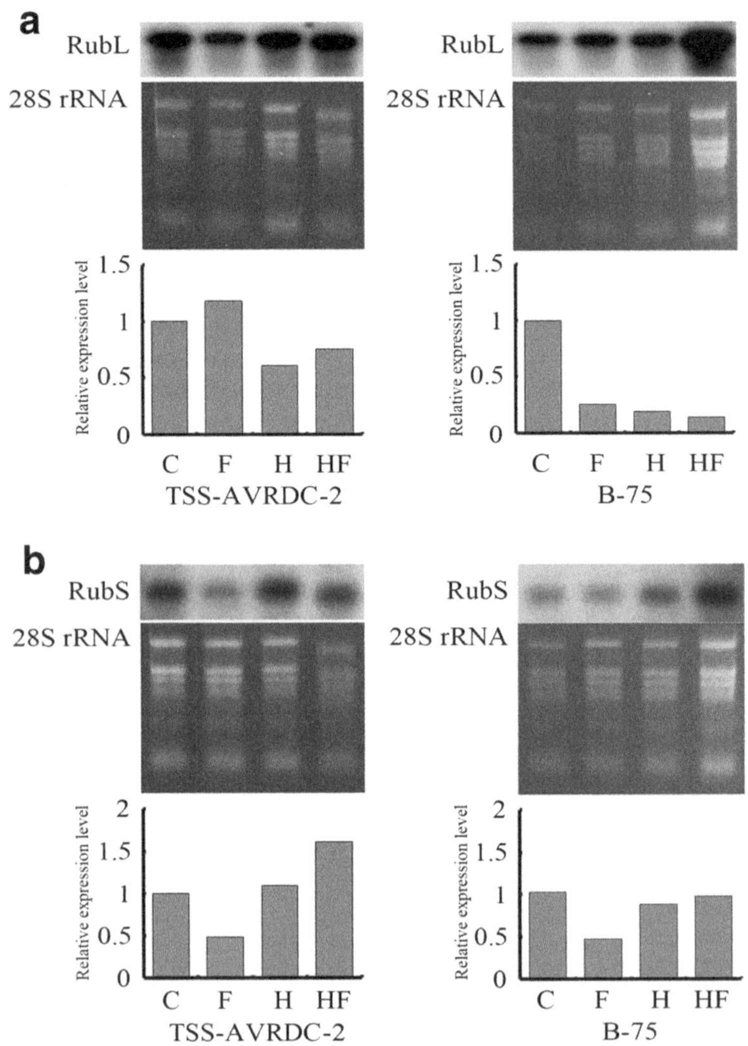

Fig. 4 Expression levels of Rubisco genes in stressed plants. Expression analysis of leaf Rubisco large (RubL, **a**) and small (RubS, **b**) RNA in TSS-AVRDC-2 and B-75 plants under control (C), waterlogging (F), 40 °C (H), and waterlogging at 40 °C (HF) treatments for 72 h. Total RNA was analyzed by RNA gel blotting with cDNA sequences of RubL and RubS subunits as specific gene probes. Ethidium bromide staining of 28S rRNA gel was shown as a loading control. The ratios of RubL and RubS RNA to 28S rRNA in each reaction were calculated, and the ratio of the control reaction was treated as a value of one for determining the relative ratios of the other reactions

To increase energy metabolism, the expressions of proteins related to redox homeostasis and response to stimuli were up-regulated, thereby maintaining physiological balance during stress. An up-regulation in Rubisco levels may also indicate an increase in the photorespiration rate. Rubisco's large and small subunits were both identified as accumulated proteins, and subsequently RubL and RubS genes were cloned from TSS-AVRDC-2 and B-75. The coding sequences of RubL and RubS between TSS-AVRDC-2 and B-75 were highly similar (Additional file 3: Figures S3 and Additional file 4: Figure S4), but their gene expression levels were different under stress treatments (Fig. 4). Compared to controls, the relative expression level of the RubL gene in TSS-AVRDC-2 plants under F stressing was 1.18, whereas the RubL transcript of the B-75 plant was 0.25 (Fig. 4a). However, the relative expression levels of the RubS gene did not change too much in either TSS-AVRDC-2 (0.49) or B-75 (0.45) under F stressing (Fig. 4b). A similar trend applied to the high temperature condition. The relative expression levels of the RubS gene did not change too much in either TSS-AVRDC-2 (1.1) or B-75 (0.86) under H stressing (Fig. 4b). However, the relative expression level of the RubL gene in B-75 under H stressing was 0.19, whereas the RubL transcript of TSS-AVRDC-2 only dropped to 0.61 compared to controls (Fig. 4a). These results indicate that TSS-AVRDC-2 is tolerant to high temperature in summer via the transcriptional control of RubL, and that the maintenance of RubL transcripts may allow for a better photosynthetic efficiency in its tolerance to high temperature. Rubisco is the main enzyme responsible for the carboxylation reaction in plants. High temperatures decrease the stability and also depress the carboxylation reaction rate of Rubisco. Miller et al. (2013) demonstrated that the thermo-tolerant *Synechococcus* lineage containing higher Rubisco stability than other lineages. The Rubisco large subunit content, Rubisco activity, and CO_2 assimilation capacity were protected from high temperature stress damage by chloroplast-targeted DnaJ (CDJ) in tomato (Wang et al. 2015). The CDJ overexpressed plants exhibited higher chlorophyll contents and fresh weights than the CJD knock-down plants. Meanwhile, heat-tolerant rice cultivar, N 22, had higher thermostability of the Rubisco protein than heat-sensitive cultivar, IR 8, under high temperature (Bose et al. 1999). These results indicate that heat-tolerant cultivar has a protective mechanism against thermal degradation of Rubisco. The maintenance of higher Rubisco subunit protein may help plants survival from the heat damages.

In this study, distinct protein profiles of TSS-AVRDC-2 and B-75 were generated and used for identifying differentially expressed protein profiles. Some of the proteins identified have been well characterized in terms of their response to stresses, while others have not been well studied with respect to their roles in the stress responses of plants. A ribosomal-binding factor A family *HrpA* was found in heat treated *Mycobacterium bovis* Bacillus Calmette-Gue'rin (Ohara et al. 1997; Tabira et al. 2000). Also, arabidopsis ribosome-binding factor A homolog *RBF1* is also essential for plastid protein biosynthesis (Fristedt et al. 2014). The increasing of ribosomal-binding factor may help in the translation of proteins, and further help plants tolerate heat stress. Our results show that the abundance of ribosome-binding factor A (peaks 9 and 14) in TSS-AVRDC-2 plants was also up-regulated after 24 h of H and HF treatment as compared to controls. Protein synthesis pathways were up-regulated in TSS-AVRDC-2 and helped TSS-AVRDC-2 tolerate high temperatures. Taken together, the genetic difference was supported by the high number of proteins differentially regulated between genotypes. Our data suggest that the early response of broccoli to flooding and high temperature might be an important stress adaptation for survival following not only hypoxia and heat, but also direct damage to cells by flooding and high temperature. All of the identified proteins likely work cooperatively to reestablish cellular homeostasis under stress. These findings are important for farming in high temperature areas and wetlands or other areas subject to short and intense rainfall events.

Conclusions

Environmental stresses represent the most limiting conditions for horticultural productivity and plant exploitation worldwide. The long term goal of our work is to help breed a competitively higher flood- and heat-tolerant broccoli to be grown in lowlands during summer. The identification of the unique stress-responsive proteins from TSS-AVRDC-2 under stressing, such as RubL, RubS, and ribosome- binding factor A, will allow further dissection of the genetic basis of this transgressive performance in its offspring. Our results not only provide information for selecting lines having better tolerance to waterlogging and heat stresses, but also provide a basis for understanding broccoli metabolic pathways and their cross-talk under stress.

Additional files

Additional file 1: Figure S1. Representatives of second-column separation in TSS-AVRDC-2 fractions. Leaf proteins from TSS-AVRDC-2 plants treated with 22 °C (C; red lines), flooding at 22 °C (F), 40 °C (H), and waterlogging at 40 °C (FH) for 12 and 24 h were separated with a two dimensional protein fractionation (PF2D) system. The Y-axis represents the optical densitometry pattern (OD at 214 nm). The X-axis represents retention time in minutes according to the concentration of acetonitrile in the mobile phase. Red lines indicate the absorbance after treatment C. Green lines indicate the absorbance from treatment F (a to d), treatment H (e to g), and treatment HF (h to l), and the detected peaks were designated by numbers.

Additional file 2: Figure S2. Representatives of the second column separation of B-75 fractions. Leaf proteins from B-75 plants treated with 22 °C (C; red lines), flooding at 22 °C (F), 38 °C (H), and waterlogging at 40 °C (FH) for 12 and 24 h were separated with PF2D. The Y-axis represents optical densitometry pattern (OD at 214 nm). The X-axis represents retention time in minutes according to the concentration of acetonitrile in the mobile phase. Red lines indicate the absorbance from treatment C. Green lines indicate the absorbance from treatment F (b and d), treatment H (a and e to j), and treatment HF (k to o), and the detected peaks were designated by numbers.

Additional file 3: Figure S3. RubL DNA sequence comparison analyses by ClustalW. cDNA sequences of RubL from TSS-AVRDC-2 and B-75 were compared with RubL in *Arabidopsis*, including AtRubL (ATCG00490). 'Star' indicates identical residues in all sequences. The line above the ATG letters indicates the translation start site.

Additional file 4: Figure S4. RubS DNA sequence comparison analyses by ClustalW. cDNA sequences of RubS from TSS-AVRDC-2 and B-75 were compared with RubS in *Arabidopsis*, including AtRubS (AT5G38420). 'Star' indicates identical residues in all sequences. The line above ATG letters indicates the translation start site.

Abbreviations

CC: Chlorophyll content; H_2O_2: Hydrogen peroxide; MALDI-TOF-MS: Matrix-assisted laser desorption ionization time-of-flight mass spectrometry; PF2D: Two-dimensional protein fractionation; RubL: Ribulose bisphosphate carboxylase large chain; RubS: Ribulose bisphosphate carboxylase small chain; SC: Stomatal conductance.

Competing interests

The authors declare that they have no competing interests.

Authors' contributions

H-H L and K-H L carried out conception of the research, and wrote the manuscript. H-H L, S-J C, and Y-H S performed the experiment. H-F L supervised the entire study. All authors have read and approved the final manuscript.

Acknowledgments

We thank Mr. Lein-Chong Chung (now retired) from the Cruciferae Breeding Unit, AVRDC-The World Vegetable Center, Tainan, Taiwan, for the generous gift of broccoli seeds. This research was supported by grants from the Council of Agriculture (96-98 AS-6421.3.2.FD-Z2), and the National Science Council (NSC 97-2313-B-002-047-MY3), Taiwan, ROC.

Author details
[1]Department of Horticulture and Landscape Architecture, National Taiwan University, Taipei, Taiwan. [2]Department of Horticulture and Biotechnology, Chinese Culture University, Taipei, Taiwan. [3]Institute of plant and microbial biology, Academia Sinica, Taipei, Taiwan.

References

Aggarwal BB, Ichikawa H (2005) Molecular targets and anticancer potential of indole-3-carbinol and its derivatives. Cell Cycle 4:1201–1215

Atkinson NJ, Urwin PE (2012) The interaction of plant biotic and abiotic stresses: from genes to the field. J Exp Bot 63:3523–3543

Augustine SM, Narayan JA, Syamaladevi DP, Appunu C, Chakravarthi M, Ravichandran V, Subramonian N (2015) *Erianthus arundinaceus* HSP70 (EaHSP70) overexpression increases drought and salinity tolerance in sugarcane (*Saccharum* spp. hybrid). Plant Sci 232:23–34

Banti V, Mafessoni F, Loreti E, Alpi A, Perata P (2010) The heat-inducible transcription factor HsfA2 enhances anoxia tolerance in Arabidopsis. Plant Physiol 152:1471–1483

Bjorkman T, Pearson KJ (1998) High temperature arrest of inflorescence development in broccoli (Brassica oleracea var. italica L.). J Exp Bot 49:101–106

Bose A, Tiwari BS, Manas KC, Gupta S, Ghosh B (1999) Thermal stress induces differential degradation of Rubisco in heat-sensitive and heat-tolerant rice. Physiol Plantarum 105:89–94

Chen YC, Lin HH, Jeng ST (2008) Calcium influxes and mitogen-activated protein kinase kinase activation mediate ethylene inducing ipomoelin gene expression in sweet potato. Plant Cell Environ 31:62–72

Davenport JR, Perry EM, Lang NS, Stevens RG (2005) Leaf spectral reflectance for nondestructive measurement of plant nutrient status. Horttechnology 15:31–35

Fristedt R, Scharff LB, Clarke CA, Wang Q, Lin C, Merchant SS, Bock R (2014) RBF1, a plant homolog of the bacterial ribosome-binding factor RbfA, acts in processing of the chloroplast 16S ribosomal RNA. Plant Physiol 164:201–215

Han B, Xu S, Xie YJ, Huang JJ, Wang LJ, Yang Z, Zhang CH, Sun Y, Shen WB, Xie GS (2012) ZmHO-1, a maize haem oxygenase-1 gene, plays a role in determining lateral root development. Plant Sci 184:63–74

Hoagland DR, Arnon DI (1950) The water-culture method for growing plants without soil. Calif Agr Expt Sta 347:1–32

Kreps JA, Wu Y, Chang HS, Zhu T, Wang X, Harper JF (2002) Transcriptome changes for Arabidopsis in response to salt, osmotic, and cold stress. Plant Physiol 130:2129–2141

Kumutha D, Ezhilmathi K, Sairam RK, Srivastava GC, Deshmukh PS, Meena RC (2009) Waterlogging induced oxidative stress and antioxidant activity in pigeonpea genotypes. Biol Plant 53:75–84

Li HY, Chang CS, Lu LS, Liu CA, Chan MT, Charng YY (2003) Over-expression of Arabidopsis thaliana heat shock factor gene (AtHsfA1b) enhances chilling tolerance in transgenic tomato. Bot Bull Acad Sinica 44:129–140

Limm EB, Simonin KA, Bothman AG, Dawson TE (2009) Foliar water uptake: a common water acquisition strategy for plants of the redwood forest. Oecologia 161:449–459

Liu MS, Li HC, Chang YM, Wu MT, Chen LF (2011) Proteomic analysis of stress-related proteins in transgenic broccoli harboring a gene for cytokinin production during postharvest senescence. Plant Sci 181:288–299

Liu MS, Li HC, Lai YM, Lo HF, Chen LF (2013) Proteomics and transcriptomics of broccoli subjected to exogenously supplied and transgenic senescence-induced cytokinin for amelioration of postharvest yellowing. J proteomics 20:133–144

Markus V, Lurie S, Bravdo B, Stevens MA, Rudich J (1981) High temperature effects on RuBP earboxylase and carbonic anhydrase activity in two tomato cultivars. Physiol Plant 53:407–412

Miller SR, McGuirl MA, Carvey D (2013) The evolution of RuBisCO stability at the thermal limit of photoautotrophy. Mol Biol Evol 30:752–760

Mittler R (2006) Abiotic stress, the field environment and stress combination. Trends Plant Sci 11:15–19

Mittler R, Vanderauwera S, Gollery M, Van Breusegem F (2004) Reactive oxygen gene network of plants. Trends Plant Sci 9:490–498

Mittler R, Vanderauwera S, Suzuki N, Miller G, Tognetti VB, Vandepoele K, Gollery M, Shulaev V, Van Breusegem F (2011) ROS signaling: the new wave? Trends Plant Sci 16:300–309

Moller IM, Kristensen BK (2004) Protein oxidation in plant mitochondria as a stress indicator. Photochemical and photobiological sciences. Photoch Photobio Sci 3:730–735

Ohara N, Naito M, Miyazaki C, Matsumoto S, Tabira Y, Yamada T (1997) HrpA, a new ribosome-associated protein which appears in heat-stressed Mycobacterium bovis bacillus Calmette-Guerin. J Bacteriol 179:6495–6498

Pucciariello C, Banti V, Perata P (2012) ROS signaling as common element in low oxygen and heat stresses. Plant Physiol Biochem 59:3–10

Rizhsky L, Liang H, Shuman J, Shulaev V, Davletova S, Mittler R (2004) When defense pathways collide. The response of Arabidopsis to a combination of drought and heat stress. Plant Physiol 134:1683–1696

Salvucci ME (2008) Association of Rubisco activase with chaperonin-60b: a possible mechanism for protecting photosynthesis during heat stress. J Exp Bot 59:1923–1933

Seki M, Narusaka M, Ishida J, Nanjo T, Fujita M, Oono Y, Kamiya A, Nakajima M, Enju A, Sakurai T, Satou M, Akiyama K, Taji T, Yamaguchi-Shinozaki K, Carninci P, Kawai J, Hayashizaki Y, Shinozaki K (2002) Monitoring the expression profiles of 7000 Arabidopsis genes under drought, cold and high-salinity stresses using a full-length cDNA microarray. Plant J 31:279–292

Simova-Stoilova L, Demirevska K, Kingston-Smith A, Feller U (2012) Involvement of the leaf antioxidant system in the response to soil flooding in two Trifolium genotypes differing in their tolerance to waterlogging. Plant Sci 183:43–49

Tabira Y, Ohara N, Yamada T (2000) Identification and characterization of the ribosome-associated protein, HrpA, of Bacillus Calmette-Guerin. Microb Pathogen 29:213–222

Uchida A, Jagendorf AT, Hibino T, Takabe T, Takabe T (2002) Effects of hydrogen peroxide and nitric oxide on both salt and heat stress tolerance in rice. Plant Sci 163:515–523

Wang G, Kong F, Zhang S, Meng X, Wang Y, Meng Q (2015) A tomato chloroplast-targeted DnaJ protein protects Rubisco activity under heat stress. J Exp Bot. doi:10.1093/jxb/erv102

Wei MY, Chao YY, Kao C (2012) NaCl-induced heme oxygenase in roots of rice seedlings is mediated through hydrogen peroxide. Plant Growth Regul 69:1–6

Xu Q, Xu X, Shi Y, Xu J, Huang B (2014) Transgenic tobacco plants overexpressing a grass *ppexp1* gene exhibit enhanced tolerance to heat stress. PLoS ONE 9:e100792

6

Effects of hot-water extracts from *Ganoderma lucidum* residues and solid-state fermentation residues on prebiotic and immune-stimulatory activities in vitro and the powdered residues used as broiler feed additives in vivo

Yuh-Hwa Liu[1,2†], Yin-Shiou Lin[3†], Kuan-Ling Lin[3], Yeh-Lin Lu[4], Chao-Hsiang Chen[3,5], Mei-Yin Chien[5], Huey-Fang Shang[6], Shyr-Yi Lin[2*] and Wen-Chi Hou[3*]

Abstract

Background: Large amounts of *Ganoderma lucidum* (GL) commercial products are provided in the worldwide market such as powders, tea bags, or capsules as dietary supplements which contained triterpenoids and/or polysaccharides. Therefore, it was estimated that several thousand tons of GL residues (GLR) are produced and discarded. For recycling uses, the aim of this study was to evaluate the benefits of two hot-water extracts from GLR (HWP_GLR) and solid-state fermentation GLR inoculated with GL mycelia (HWP_GLRF) on the growths of *Lactobacillus rhamnosus* and *Bifidobacterium longum*. The RAW264.7 cells were used to investigate the effects of HWP_GLR and HWP_GLRF on nitric oxide productions, phagocytic activities against FITC-labeled *E. coli*, and to lower lipopolysaccharide (LPS)-binding capacities. The powders of GLR and GLRF were used as additives in the commercial feeds for feeding broiler chicks in vivo to evaluate the immune-stimulatory and prebiotic activities.

Results: HWP_GLR and HWP_GLRF with molecular size 5 to 8 kDa were showed to stimulate growths of *L. rhamnosus* and *B. longum*. It was found that in the presence of polymyxin B HWP_GLR and HWP_GLRF could stimulate nitric oxide productions, elevate phagocytic activities against FITC-labeled *E. coli*, and to lower lipopolysaccharide-binding capacities in RAW264.7 cells. The broiler chicks were selected for feedings in vivo. The 1-day-old chicks were fed commercial feeds for 1 week, and then were fed without or with 4 or 8 % of GLR and GLRF additives for 3 weeks. There was no significant weight difference among feeding groups. However, the phagocytosis and natural killer cytotoxicity in the peripheral bloods, and prebiotic activities of bifidobacteria in feces of GLR and/or GLRF groups were significantly different compared to the control ($P < 0.05$).

Conclusions: The GLR, GLRF, and their hot-water extracts with beneficial activities could be processed as feed additives which could increase the waste-recycling.

Keywords: Arbor Acres broiler chicks; *Ganoderma lucidum* residues (GLR); Phagocytic; Natural killer cytotoxicity; Prebiotic activity; Probiotics

* Correspondence: sylin@tmu.edu.tw; wchou@tmu.edu.tw
†Equal contributors
[2]Department of General Medicine, Taipei Medical University, Taipei, Taiwan
[3]Graduate Institute of Pharmacognosy, Taipei Medical University, Taipei, Taiwan
Full list of author information is available at the end of the article

Background

Ganoderma lucidum (GL), known as "Ling-Zhi", is one of valuable Traditional Chinese Medicines, which has long been used as a tonic and to prevent and treat various diseases in China and other Asian countries for over 2000 years (Lee et al. 2012a). GL possessed a broader range of biological activities, such as antitumor, hypoglycemic, immunomodulatory, antioxidant, anti-diabetic, anti-atherosclerotic, anti-inflammatory, and anti-HIV activities (Wachtel-Galor et al. 2004; Boh et al. 2007; Mahajna et al. 2008; Sanodiya et al. 2009). Further, the clinical investigations also revealed that GL polysaccharide extracts could significantly enhance immune responses in patients with advanced-stage cancer (Gao et al. 2003). Moreover, the bioactive components of GL have been identified as polysaccharides, proteins, alkaloids, steroids, nucleosides, and triterpenoids etc. (Kino et al. 1989; Shiao 1992; Lee et al. 2012a). In additions, several literatures have shown that the GL polysaccharide-rich fractions exhibited abilities to stimulate the productions of IL-1, IL-6, IL-12, IFN-γ, tumor necrosis factor-α, IL-6, colony stimulating factor, and nitric oxide, which resulted in the stimulation of splenocyte proliferation and macrophage phagocytosis (Ohno et al. 1998; Mao et al. 1999; Chen et al. 2004).

Recently, mushroom with abundances of chitins, hemicelluloses, and glucans (Manning and Gibson 2004) has attracted much attentions as potential sources of prebiotics (Aida et al. 2009). The prebiotics are termed as "nondigestible food ingredient that beneficially affects the host by selectively stimulating the growth and/or activity of one or a limited number of bacteria in the colon, and thus improves host health", and intakes of prebiotics can significantly modulate the colonic microbiota by increasing the number of specific bacteria and thus changing the composition of the microbiota (Gibson and Roberfroid 1995). It is reported that lactobacilli, such as *Lactobacillus acidophilus*, *L. casei*, *L. delbruekii*, and bifidobacteria, such as *Bifidobacterium longum*, are commonly recognized as health-promoting bacteria in guts and can be stimulated growth by prebiotics (Gibson and Roberfroid 1995; Roberfroid et al. 2010). These health-promoting effects may include the modulation of the intestinal microflora, prevention of rotavirus-induced or antibiotic-associated diarrhea, beneficial effects on gastrointestinal tract inflammation, modulate the mucosal immune responses (Roberfroid et al. 2010; Gourbeyre et al. 2011). The water-soluble and alkali-soluble extracts from cultivated oyster mushroom which contained branched 1,3-β-D-glucans, 1,6-β-D-glucans, and linear 1,3-α-D-glucans showed the growth stimulation toward probiotics (Synytsya et al. 2009). It was reported that a large amount of GL commercial products are provided in the worldwide market such as powders, tea bags, or capsules as dietary supplements which contained triterpenoids and/or polysaccharides. Therefore, it was estimated that several thousand tons of GL residues (GLR) are produced and discarded. For recycling uses, the aim of this study was to evaluate the benefits of two hot-water extracts from GLR (HWP_GLR) and solid-state fermentation GLR inoculated with GL mycelia (HWP_GLRF) on the growths of *Lactobacillus rhamnosus* and *Bifidobacterium longum*. The RAW264.7 cells were used to investigate the effects of HWP_GLR and HWP_GLRF on nitric oxide productions, phagocytic activities against *E. coli*, and to lower lipopolysaccharide (LPS)-binding capacities. The GLR and GLRF were used as additives in the commercial feeds for feeding broiler chicks in vivo to evaluate the immune-stimulatory and prebiotic activities.

Methods

Materials

The *G. lucidum* BCRC 36041 and RAW 264.7 were obtained from the Bioresource Collection and Research Center (BCRC), Food Industry Research and Development Institute (Hsinchu, Taiwan). LSCC-RP9 cells (avian leukosis virus-transformed B-lymphoblastoid cell line from chickens) were kindly provided by Avian Disease and Oncology Laboratory, Agricultural Research Service, US Department of Agriculture (MI, USA). LPS, sulfanilamide, sodium nitrite, lipopolysaccharide-fluorescein isocyanate (FITC), polymyxin B-agarose resin, 3-(4, 5-Dimethylthiazol-2-yl)-2,5-diphenyltetrazolium bromide (MTT), inulin, trypan blue, and propidium iodide (PI) were all purchased from Sigma Chemical Co. (St. Louis, MO, USA). Difco™ Lactobacilli MRS broth, Endo broth, and reinforced clostridial broth were from Becton Dickinson (USA). Polymyxin B, kanamycin, iodoacetic acid, nalidixic acid, and 2,3,5-triphenyltetrazolium chloride, and vancomycin were from Abbott Lab. (IL, USA). Agar Bacteriological was from Oxoid Ltd. (Basingstoke, England). The GLR and GLRF were provided by Ko Da Pharmaceutical Co. Ltd (Tao-Yuan, Taiwan). The GL raw materials were extracted by 12-fold boiling water (*w/v*) to produce commercial products, and the residues were then dried as GLR. The *G. lucidum* BCRC 36041 was maintained on potato dextrose agar slants, and the activated mycelia of *G. lucidum* BCRC 36041 were obtained in liquid medium following the previous method (Hsieh and Yang 2004). For GLRF production, the activated mycelia of *G. lucidum* BCRC 36041 in the potato dextrose broth were homogenized and inoculated onto the autoclaved GLR in the flask [about 10 % inoculum for GLR (*v/w*) in each container] and placed at 25 to 30 °C for solid-state fermentations.

The HWP_GLR and HWP_GLRF preparations and β-glucan content determinations

Each GLR and GLRF (about 60 g) was weighted and placed in the bottle with screw cap and 600 ml distilled water was added, the mixture was autoclaved at 121 °C for 30 min. After filtration, each filtrate was lyophilized as HWP. The proximate compositions of GLR and GLRF were assayed by TÜV Rheinland AIMEX Ltd. (Pintung, Taiwan). The β-glucan contents of HWP_GLR and HWP_GLRF were determined according to the manufacturer's instructions (K-BGLU 07/11, Megazyme International, Ireland) which were hydrolyzed by lichenase (a specific endo-(1–3)(1–4)-β-D-glucan 4-glucanohydrolase) and β-glucosidase and the released glucose was then determined by glucose oxidase/peroxidase reagents.

HWP_GLR and HWP_GLRF molecular weight determinations

The molecular weights of HWP_GLR and HWP_GLRF were determined by using a Breeze HPLC system (Waters Co., Milford, MA) equipped with a 10 μl loop, a G4000PW$_{XL}$ size exclusion column (7.8 × 300 mm, TSK-GEL, Tosoh Corp., Tokyo, Japan) at 55 °C, and a RI detector (oven, 50 °C). An aliquot of 10-μl (10 μg/μl) sample solution was injected and eluted with 0.1 M NaN_3 aqueous solution (Karnjanapratum et al. 2012) at a flow rate of 0.5 ml/min. The calibration curve was prepared from dextran standards (1400, 670, 270, 50, 12 and 5 kDa).

Effects of HWP_GLR and HWP_GLRF on the cell viability, nitric oxide production, phagocytosis, and LPS binding inhibitory activity in RAW264.7 cells

The RAW 264.7 macrophages were cultured in Dulbecco's modified Eagle medium (DMEM), 10 % fetal bovine serum (FBS), and penicillin/streptomycin solution (10,000 units/ml of penicillin, 10 mg/ml of streptomycin). The cells were incubated at 37 °C in a humidified atmosphere with 5 % CO_2. For cell viability assay, 4×10^5/ml RAW 264.7 cells were seeded onto a 96-well microtiter plate, and various concentrations of HWP_GLR and HWP_GLRF (0.25, 0.5, 0.75 and 1 mg/ml) were added in the present of polymyxin B (50 μg/ml). Then, the cells were incubated at 37 °C in a humidified atmosphere with 5 % CO_2 for 24 h. After the end of incubation, RAW 264.7 cells were incubated with 250 μg/ml of MTT for cell viability assay (Liu et al. 2007). For nitric oxide production assay, RAW 264.7 cells were seeded onto a 96-well microtiter plate and various concentrations of HWP_GLR and HWP_GLRF (0.25, 0.5, 0.75 and 1 mg/ml) in the presence of polymyxin B or LPS (600 ng/ml) were added with or without polymyxin B additions. The nitric oxide production was assayed by Griess reagents and the sodium nitrite (0–100 μM) was used to plot the calibration curve (Liu et al. 2008). For stimulated phagocytic activity of RAW264.7 cells toward FITC-labeled *E. coli*, the different concentration of HWP_GLR and HWP_GLRF (0.25, 0.5, 0.75 and 1 mg/ml) or LPS (600 ng/ml) were added to the cultured RAW264.7 cells in the presence or absence of polymyxin B following the previous method (Liu et al. 2007). For LPS binding inhibitory activity in RAW264.7 cells, each reagent used was passed through polymyxin B-agarose column and HWP_GLR and HWP_GLRF solutions were pre-mixed with polymyxin B-agarose resins in advance to remove LPS contaminations, then, HWP_GLR and HWP_GLRF (0.1 and 1 mg/ml) were added into RAW 264.7 cells for 15 min and then FITC-labeled LPS (200 ng/ml) were added for 1 h at 37 °C and analyzed by a BD FACSCanto II flow cytometry (BD Bioscience, USA) which the median fluorescence intensity was measured, and FITC-LPS binding to the RAW264.7 cells was recognized as 100 % (An et al. 2002).

Bacterial enumeration after HWP_GLR and HWP_GLRF treatments in vitro

The 100 μl of each bacterial culture (grown in each selective broth) was pre-mixed with different concentrations of HWP_GLR and HWP_GLRF (1, 5, 10 mg/ml), diluted serially and then spread onto selective agar plate, and then incubated in an anaerobic chamber (Becton Dickinson, USA) at 37 °C for 48 h (*L. rhamnosus* GG and *B. longum*) or 24 h (*E. coli*). The selective growth medium for *L. rhamnosus* GG was MRS medium (Difco™ Lactobacilli MRS broth plus Agar Bacteriological. After being autoclaved, the filtered vancomycin was added and poured onto petri dishes for further uses); the selective growth medium for *B. longum* was BIM-25 medium (reinforced clostridial broth plus Agar Bacteriological. After being autoclaved, the filtered polymyxin B, kanamycin, iodoacetic acid, nalidixic acid and 2,3,5-triphenyltetrazolium chloride each was serially added and poured onto petri dishes for further uses); the selective growth medium for *E. coli* was Endo medium (Endo broth and Agar Bacteriological. After being autoclaved, the medium was poured onto petri dishes for further uses). The inulin (40 mg/ml) was used as a positive control. The bacterial colonies were counted and calculated, and expressed as growth stimulation index as following: (colony numbers of inulin or HWP_GLR or HWP_GLRF)/(colony numbers of the blank).

Broiler chick feeding experiments in vivo

The feeding experiment was performed at Ko Da Pharmaceutical Co. Ltd (Tao-Yuan, Taiwan). All animal experimental procedures followed the published guidelines of National Science Council, Taipei, Taiwan (1994). The newborn Arbor Acres broiler chicks ($N = 30$, 1-day-old) were purchased from Arbor Acres Taiwan Co., Ltd. (Taipei, Taiwan) and had free access to newborn broiler chick feeds (Fwusow Industry Co. Ltd, Shalu, Taichung) and water for 1 week. These broiler chicks were randomly divided into five groups ($N = 6$, two chicks in one cage), the control group fed with medium-sized broiler chick feeds (Fwusow Industry Co. Ltd, Shalu, Taichung) and four groups fed with functional feeds, 4 or 8 % (*w/w*) of GLR and GLRF powders added in the medium-sized feeds, for 3 weeks. The broiler chicks were weighed every week. The feces of each group were collected at day 28 (the end of the feeding experiments). For lactobacilli and bifidobacteria enumeration determinations, the MRS and BIM-25 selective agar plates were used, respectively. The feces (about 0.05 g) were suspended in the autoclaved broth and diluted serially and spread onto selective agar plate, and then incubated in an anaerobic chamber at above-mentioned conditions. The bacterial colonies were counted and calculated, and expressed as log colony forming unit (CFU)/g feces. To evaluate the effects of feed additives on phagocytic activities of polymorphonuclear (PMN) cells and the natural killer (NK) activities of peripheral blood mononuclear cells (PBMC) in the peripheral blood, the blood samples were drawn from the chicken wings at the last week of the feeding experiments.

Phagocytosis of polymorphonuclear cells in the peripheral blood

The FITC-labeled *E. coli* (Molecular Probes, USA) powder (5 mg) was suspended in 0.5 ml of Hank's balanced salt solution (HBSS) and used for phagocytic analysis by flow cytometry (Liu et al. 2007; Shang et al. 2007). Eighty microliters of blood from the chicken wing were mixed with 20 μl of FITC-labeled *E. coli* solution at 37 °C for 10 min. The Falcon tube was immersed in an ice bath to stop the phagocytosis. Eighty microliters of trypan blue (1.25 mg/ml) were added to quench the residual FITC-labeled *E. coli*. After lysis and washing, 5 μl PI (2 mg/ml) were added for 10 min, and the phagocytosis of PMN cells was determined by flow cytometry (Becton Dickinson FACS Calibur™, CA).

NK cytotoxicity of peripheral blood mononuclear cells in the peripheral blood

The LSCC-RP9 cells (avian leukosis virus-transformed B-lymphoblastoid cell line from chickens) were used as target cells of PBMC in the peripheral blood for NK cytotoxic activity assay by a flow cytometry using the DiOC/18 membrane dye (Molecular Probes, Eugene, OR) to stain alive LSCC-RP9 cells and PI nuclear dye to stain the dead cells (Kushima et al. 2003). The chicken PBMC were obtained by centrifuging the whole blood and a PBS mixture on Ficoll–Hypaque (2.4:1, V/V) by the density-gradient centrifugation. The cultural medium (2 l) for LSCC-RP9 cells included 11.9 g McCoy's 5A medium modified (Sigma Co., USA), 13.8 g Leibowitz L-15 medium (Sigma Co., USA), 20 % chicken serum, 10 % fetal bovine serum, and 5 % tryptose phosphate broth. For NK cytotoxic activity assay, the LSCC-RP9 cells were adjusted to 2×10^6 cell/ml by the cultural medium. Ten microliter of 3 mM $DiOC_{18}$ were added into 1000 μl of target cells at 37 °C for 20 min and then were suspended in cultural medium for further uses. The PBMC (effector cells) were mixed with target cells at ratios of 20:1 and then were co-cultured in 5 % CO_2 humidified incubator at 37 °C for 2 h. The supernatants were removed and the same volume of PI solution (0.2 mg/ml) was added. The NK cytotoxicity was determined by flow cytometry (Becton Dickinson FACS Calibur™, CA).

Statistical analyses

All data were calculated as means ± SD and multiple group comparisons were performed using one-way ANOVA, followed by the post hoc Tukey's test for comparisons. Values not sharing the same alphabetic letter were significantly different ($P < 0.05$). The statistical analysis was performed using the GraphPad Prism Software 5.0 (San Diego, CA, USA).

Results

The general properties of GLR, GLRF, and the hot-water extracts

The proximate compositions of GLR and GLRF were showed in Table 1. The GLR showed high amounts of crude fiber contents (41 %) and the nitrogen free extracts (NFE, 31.9 %). The crude fiber contents in GLRF were 22.4 %, and the NFE were calculated to be 46.1 %. The GLR and GLRF were suspended in distilled water and autoclaved at 121 °C for 30 min, filtered and lyophilized to get hot water extracts, HWP_GLR and HWP_GLRF, which the recovery was 4/100 g and 6.6/100 g, respectively. The Fig. 1 showed the size exclusion HPLC profiles of (a) HWP_GLR and (b) HWP_GLRF, and the percentage distributions of the molecular size in each HWP were showed in the Table 2. There were three peaks in HWP_GLR, 8144, 5688 and 5044 Da which accounted for 82.2, 9.6 and 8.3 % of total area, respectively; four peaks in HWP_GLRF, 8225, 6461, 5733 and 5136 Da which accounted for 58.9, 24.7, 6.4 and 10 % of total area. It seemed that parts of higher molecular size in HWP_GLR (such as 8.144 kDa) was shifted to medium size one (6461 Da) in HWP_GLRF after solid-state fermentations. Using enzymatic hydrolysis to determine the β-glucan contents in HWP, it was found that the HWP_GLRF was about 6-folds higher that of HWP_GLR.

Table 1 The proximate compositions of GLR and GLRF

Sample	Crude protein (g/100 g)	Crude fat (g/100 g)	Crude fiber (g/100 g)	Moisture (g/100 g)	Ash (g/100 g)	NFE[a] (g/100 g)
GLR	10.9	0.6	41.0	13.0	2.6	31.9
GLRF	13.2	0.6	22.4	7.6	10.1	46.1

[a]*NFE* (nitrogen free extracts) is calculated by 100 % – (crude protein + crude fat + crude fiber + ash + moisture)%

Effects of HWP_GLR and HWP_GLRF on the cell viability, nitric oxide production, phagocytosis, and LPS binding inhibitory activity of RAW264.7 cells

The HWP_GLR and HWP_GLRF showed no significant cytotoxicity among 0.25 to 1.0 mg/ml (Fig. 2a). The LPS showed significantly to stimulate the nitric oxide productions ($P < 0.05$) compared to the medium only (as the control), and this effect was abolished in the presence of polymyxin B (Fig. 2b) and showed no significant difference ($P > 0.05$) compared to the medium only (the control). Under non-toxic concentrations, HWP_GLRF at 0.75 and 1.0 mg/ml and HWP_GLR at 1 mg/ml in the presence of polymyxin B were showed to stimulate nitric oxide productions and significantly different to the LPS in the presence of polymyxin B ($P < 0.05$) (Fig. 2b). The HWP_GLR and HWP_GLRF (0.5, 0.75 and 1.0 mg/ml) exhibited dose-dependently to elevate phagocytic activities of the RAW264.7 cells against *E. coli* and showed significant differences compared to the control (medium only) ($P < 0.05$) (Fig. 2c). Using the FITC-labeled LPS as the positive controls, the pretreatments of HWP_GLR and HWP_GLRF at concentrations of 0.1 and 1.0 mg/ml could reduce LPS binding to RAW264.7cells and showed significant difference ($P < 0.05$) to the control (Fig. 2d).

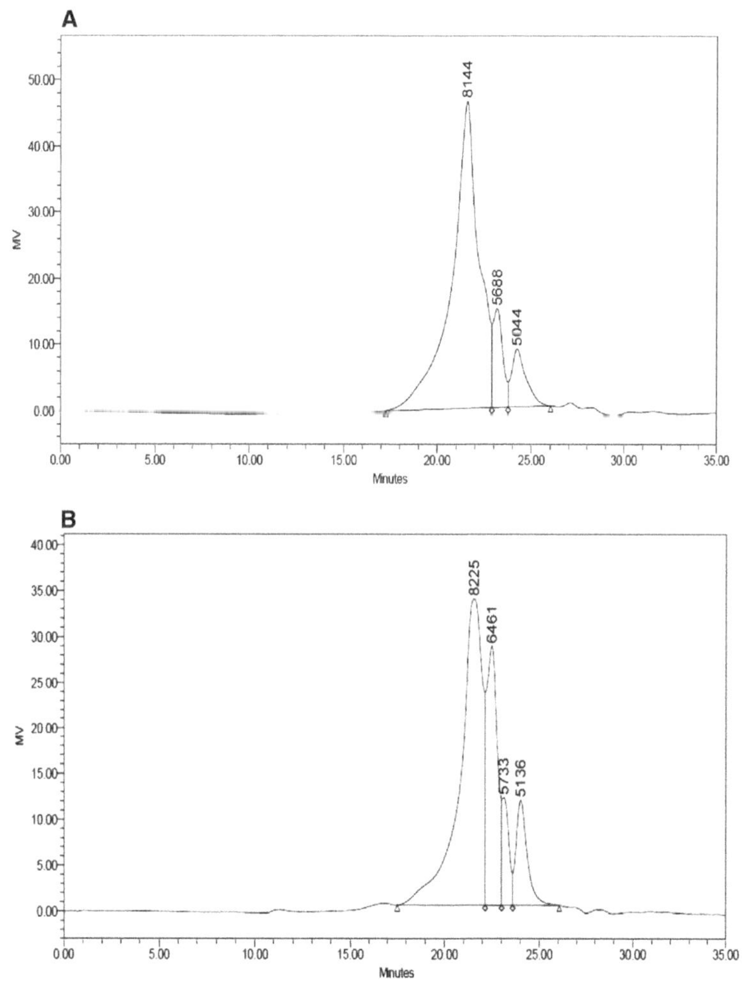

Fig. 1 The size exclusion HPLC profiles of (**a**) HWP_GLR and (**b**) HWP_GLRF. The molecular sizes were determined by using a Breeze HPLC system (Waters Co., Milford, MA) equipped with a 10 μl loop, a G4000PW$_{XL}$ size exclusion column (7.8 × 300 mm, TSK-GEL, Tosoh Corp., Tokyo, Japan) at 55 °C, and a RI detector (oven, 50 °C). An aliquot of 10-μl (10 μg/μl) sample solution was injected and eluted with 0.1 M NaN_3 aqueous solution at a flow rate of 0.5 ml/min

Table 2 The β-glucan contents and molecular weight distributions of HWP_GLR and HWP_GLRF

Parameter	HWP_GLR	HWP_GLRF
β-glucan contents (mg/g)	2.10	12.84
Molecular weight (distributions, area %)[a]	8144 Da (82.2 %)[a]	8225 Da (58.9 %)
	5688 Da (9.57 %)	6461 Da (24.7 %)
	5044 Da (8.27 %)	5733 Da (6.38 %)
		5136 Da (10.0 %)

[a]Calculated from Fig. 1

Prebiotic activity of HWP_GLR and HWP_GLRF in vitro

The effects of HWP_GLR and HWP_GLRF (5 and 10 mg/ml) on growths of *L. rhamnosus* GG, *B. longum*, and *E. coli* were showed in Table 3 and expressed as growth stimulation index (the blank was recognized as 1). Compared to the positive control of inulin (40 mg/ml), the HWP_GLR and HWP_GLRF showed better prebiotic activities and significant difference ($P < 0.05$) than those of inulin in the stimulatory growths of probiotics, *L. rhamnosus* GG and *B. longum*, but not the pathogenic *E. coli*, especial at the concentration of 10 mg/ml. Though HWP_GLRF showed higher β-glucan contents than HWP_GLR did (Table 2), the latter seemed to have higher capacity in stimulatory growths of *L. rhamnosus* GG and *B. longum* (Table 3).

Broiler chick feeding experiments in vivo

Therefore, the 4 or 8 % of powdered GLR and GLRF added in the commercial medium-sized broiler chick feeds were used for 3-week feeding experiments. During feeding periods, the weights of Arbor Acres broiler chickens among groups were not shown significantly different (Fig. 3), however, the phagocytic activity of PMN in GLR or GLRF groups (4 or 8 % additives) was elevated and showed significantly different compared to the control ($P < 0.05$) (Fig. 4a). For the 4 % additives, the

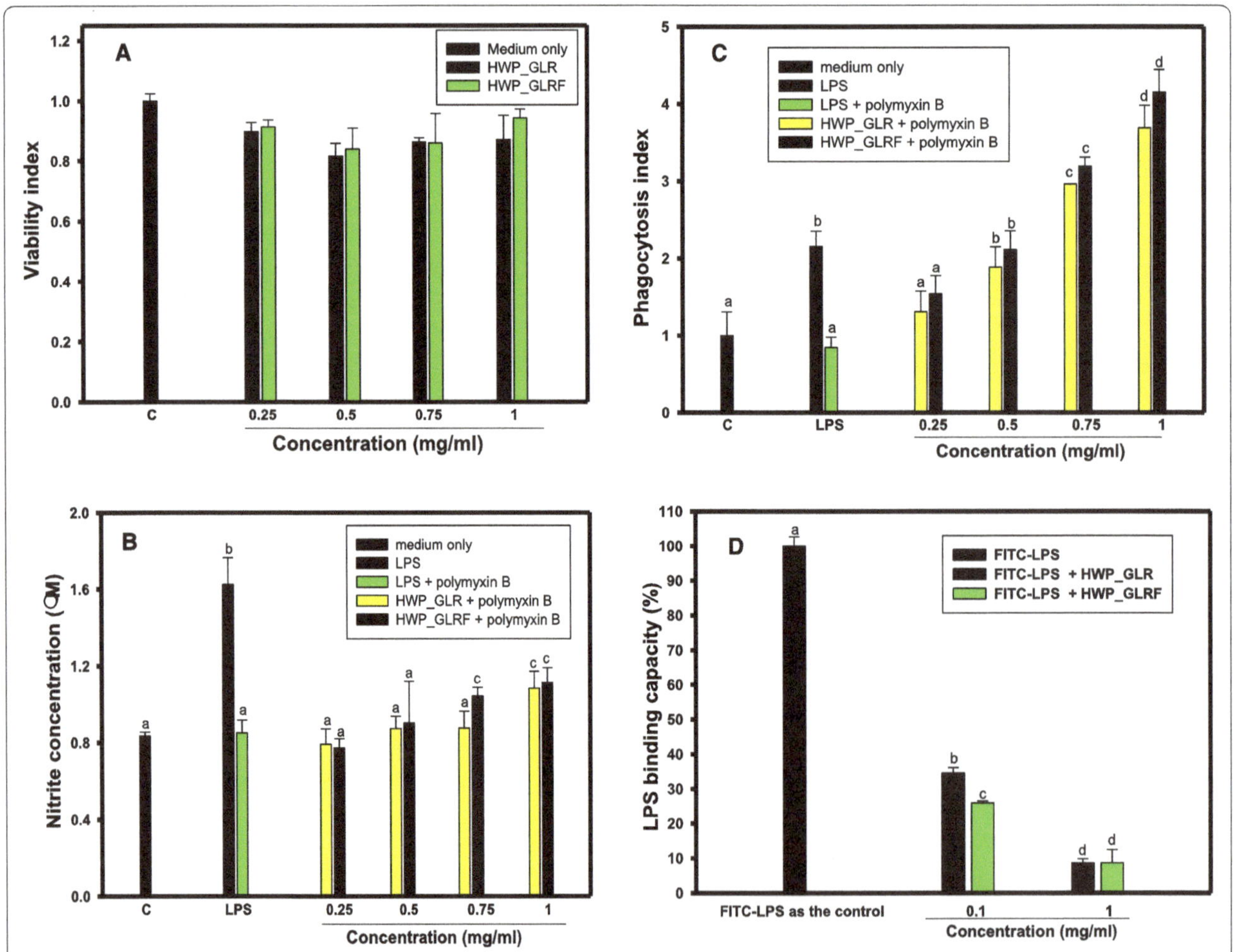

Fig. 2 Effects of HWP_GLR and HWP_GLRF (0.25, 0.5, 0.75 and 1 mg/ml) on (**a**) the viability; (**b**) nitric oxide productions; (**c**) the phagocytosis; and (**d**) FITC-labeled LPS binding capacities of RAW 264.7 cells in the presence of polymyxin B (50 μg/ml). LPS with or without polymyxin B additions were used as the controls. Data were means ± SD and analyzed using one-way ANOVA, followed by the post hoc Tukey's test for comparisons. Values not sharing the same *alphabetic letter* were significantly different ($P < 0.05$)

Table 3 Effects of HWP_GLR, HWP_GLRF, and inulin on bacterial growths

Treatment	Growth stimulation index		
	Lactobacillus rhamnosus GG	*Bifidobacterium longum*	*Escherichia coli*
Blank	1.0 ± 0.4 a[#]	1 ± 0.07 a	1.0 ± 1.3 a
Inulin (40 mg/ml)	3.0 ± 0.2 b	3.3 ± 0.2 b	N.D.
HWP_GLR			
5 mg/ml	11.1 ± 4.5 c	0.8 ± 0.7 a	1.1 ± 1.1 a
10 mg/ml	83.2 ± 9.3 f	100.3 ± 22.0 d	3.4 ± 1.1 a
HWP_GLRF			
5 mg/ml	38.9 ± 7.0 d	5.1 ± 1.6 b	0.6 ± 0.6 a
10 mg/ml	67.2 ± 1.0 e	63.7 ± 9.9 c	2.2 ± 1.0 a

N.D. not detected
[#]All data were calculated as means ± SD and multiple group comparisons were performed using one-way ANOVA, followed by the post hoc Tukey's test for comparisons. Values not sharing the same alphabetic letter were significantly different ($P < 0.05$)

PMN in GLR group showed higher phagocytic activity than that of GLRF group ($P < 0.05$). For the 8 % additives, the PMN in both groups showed similar phagocytic activities. It was found that only 8 % GLRF additives could elevate NK activities of PBMC and showed significantly different compared to the control ($P < 0.05$) (Fig. 4b). The present results showed that GLR or GLRF additives in the feeds could stimulate the innate immunity of Arbor Acres broiler chickens against microbial invasions.

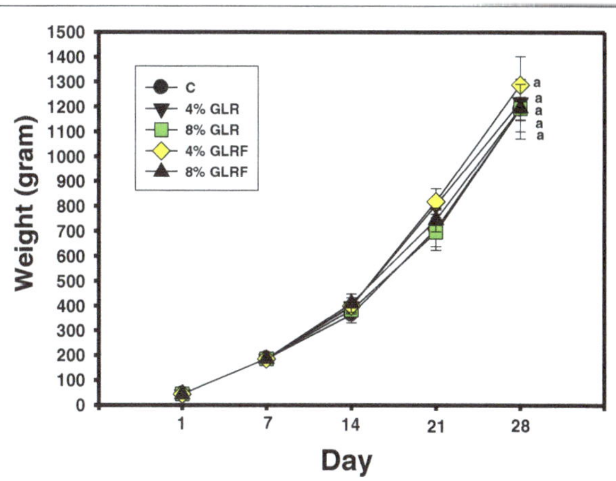

Fig. 3 Effects of GLR or GLRF additives (4 or 8 %) in the commercial feeds on weights of Arbor Acres broiler chickens. The newborn Arbor Acres broiler chicks ($N = 30$, 1-day-old) were purchased from Arbor Acres Taiwan Co., Ltd. and had free access to newborn broiler chick feeds and water for 1 week. These broiler chickens were randomly divided into five groups, the control group fed with medium-sized broiler chick feeds and four groups fed with functional feeds, 4 or 8 % (*w/w*) of GLR and GLRF powders added in the medium-sized feeds, for 3 weeks

The effects of GLR or GLRF additives on the growths of probiotics were showed in the Fig. 5. The MRS selective agar plate was used to count the lactobacilli (Fig. 5a) and the BIM-25 selective agar plate was used to count the bifidobacteria (Fig. 5b). It was found that the bifidobacteria, but not the lactobacilli, showed higher counts and significant differences compared to the control ($P < 0.05$) of the 4 % GLRF or 8 % GLR additives in the feeds.

Discussion

The large amounts of triterpenoid-contained or polysaccharide-contained or enriched GL commercial products with an annual global value over $1.5 billion were available in the markets (Liu et al. 2010). It was estimated that several thousand tons of GLR were discarded. It was previously reported that the processed product (sacchachitin) from residues of *G. tsugae*, a copolymer of 60 % β-1,3-glucan and 40 % poly(*N*-acetylglucosamine), was applied as skin substitutes (Su et al. 1997). In the present study, the wastes of GLR or the solid-state fermented GLR (GLRF) as broiler chick feed additives (such as 4 % additives) exhibited prebiotic (Fig. 4) and innate immune-stimulatory activities (Fig. 5) in vivo which might increase the recycling uses in the future. Chang and Lu (2004) reported that HWP from GL fruiting bodies (designed as GLG, GLPS-2, and GLPS-3 in eluted orders with 0.3 M NaN_3 by size exclusion HPLC equipped with the RI detector) contained mainly glucose and galactose and less amounts of mannose constituents. The GLG contained the aniline blue-positive β-glucans with average molecular mass of 2080 kDa; the GLPS-2 and GLPS-3 contained less amount of aniline blue-positive β-glucans with average molecular mass of 230 kDa and 120 kDa, respectively. These three fractions (GLG, GLPS-2, and GLPS-3) showed immune-stimulatory activities toward TNF-α secretions in human blood mononuclear cells in orders of GLG > > GLPS-2 > > GLPS-3. Liu et al. (2010) reported two low-molecular-weight glucans (GLP_L1 and GLP_L2) isolated from HWP of GL fruiting bodies by ion exchange chromatography and gel filtration with average molecular weight of 5.2 and 15.4 kDa, respectively. The GLP_L1 contained glucose with four major and one minor linkages in the percentage ratio of 21.9:20.3:23.7:24.0:3.7 of (a) → 3)-Glcp-(1→, (b) → 4)-Glcp-(1→, (c) → 6)-Glcp-(1→, (d) → 3,6)-Glcp-(1→, and (e) → 4,6)-Glcp-(1→. In the present results, using the size exclusion HPLC equipped with the RI detector and the same NaN_3 eluent, a smaller molecular size of 5 to 8 kDa HWP were found in GLR and GLRF (Fig. 1) which were apparently different from that of Chang and Lu (2004) reported, but closed to GLP_L1 of Liu et al. (2010) reported from fruiting bodies of *G. lucidum*. The inulin (β-2,1-fructan) was recognized as prebiotic substances (Aida et al. 2009) and used in the present

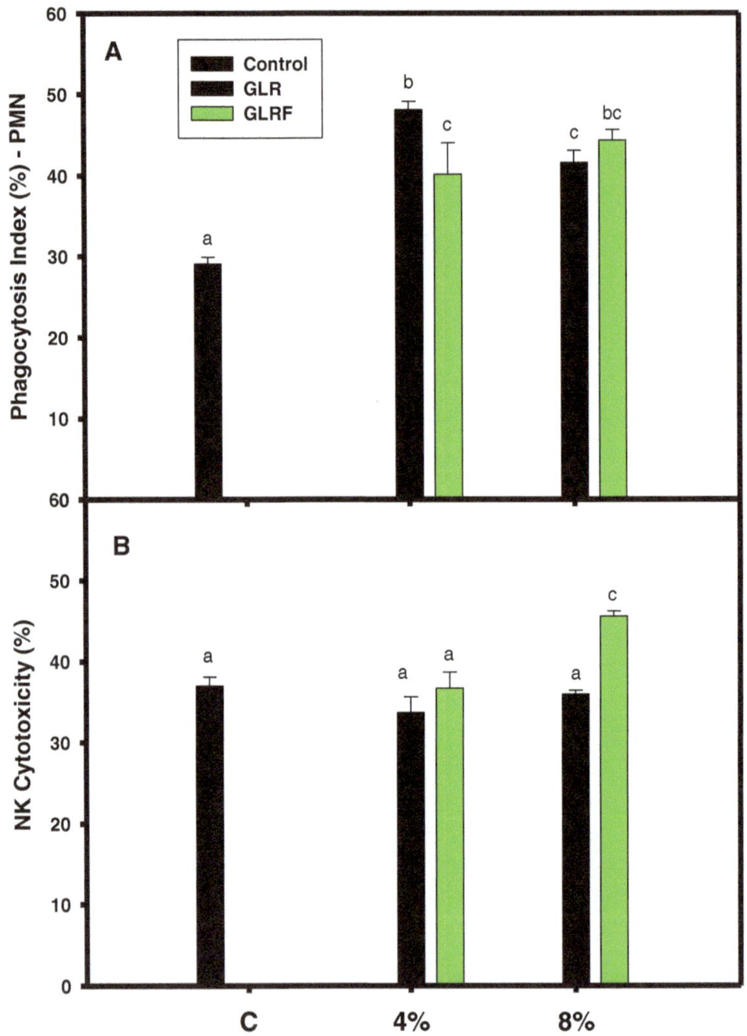

Fig. 4 Effects of powders of GLR or GLRF additives (4 or 8 %) on the (**a**) phagocytic activities of polymorphonuclear (PMN) cells and (**b**) natural killer (NK) activities of peripheral blood mononuclear cells in the peripheral blood. The blood samples were drawn from the Arbor Acres broiler chicken wings at the last week of the feeding experiments. Data were means ± SD and analyzed using one-way ANOVA, followed by the post hoc Tukey's test for comparisons. Values not sharing the same *alphabetic letter* were significantly different ($P < 0.05$)

research as the positive control (Table 3). The HWP_GLR and HWP_GLRF (10 mg/ml) showed much better than inulin (40 mg/ml) in stimulatory growths on *L. rhamnosus* GG and *B. longum* (Table 3). The β-glucan content in the present assay was based on lichenase and β-glucosidase hydrolysis for mixed-linkage of [(1–3)(1–4)]-β-D-glucan determinations. The amounts of detected β-glucan in HWP_GLRF were higher than HWP_GLR (Table 2), however, both of the stimulatory growths toward *L. rhamnosus* GG and *B. longum* (Table 3), and nitric oxide productions (Fig. 2b), phagocytic activities (Fig. 2c), and reduced LPS binding capacities (Fig. 2d) in RAW264.7 cells in the presence of polymyxin B in vitro were similar. It meant that the HWP_GLRF showed to increase the hot-water soluble β-glucan contents (Table 2), however, these increased glucans might be biological inactive in RAW264.7 cells and needed further investigations. Macrophages released several mediators, including inflammatory cytokines, IL-1, IL-6 and TNF-α, and nitric oxide which could induce the activation and differentiation of lymphocytes and the proliferation of granulocytes (Nathan 1987). In mammals, phagocytosis was a very important defense against pathogen invasions and apoptotic cell scavenging which were performed by phagocytes like macrophages, dendritic cells, and granulocytes (Stuart and Ezekowitz 2005). The LPS might bind to the Toll-like receptor 4 in the macrophages to activate downstream signaling pathway, such as NF-κB-mediated cytokine expression including IL-1, IL-6, TNF-α, and IFN-γ inflammatory cytokines (Akira 2003), and the polymyxin B could interact with LPS to retard binding to Toll-like receptor 4 and further inflammation. Therefore, the reduced LPS-binding capacities (Fig. 2d),

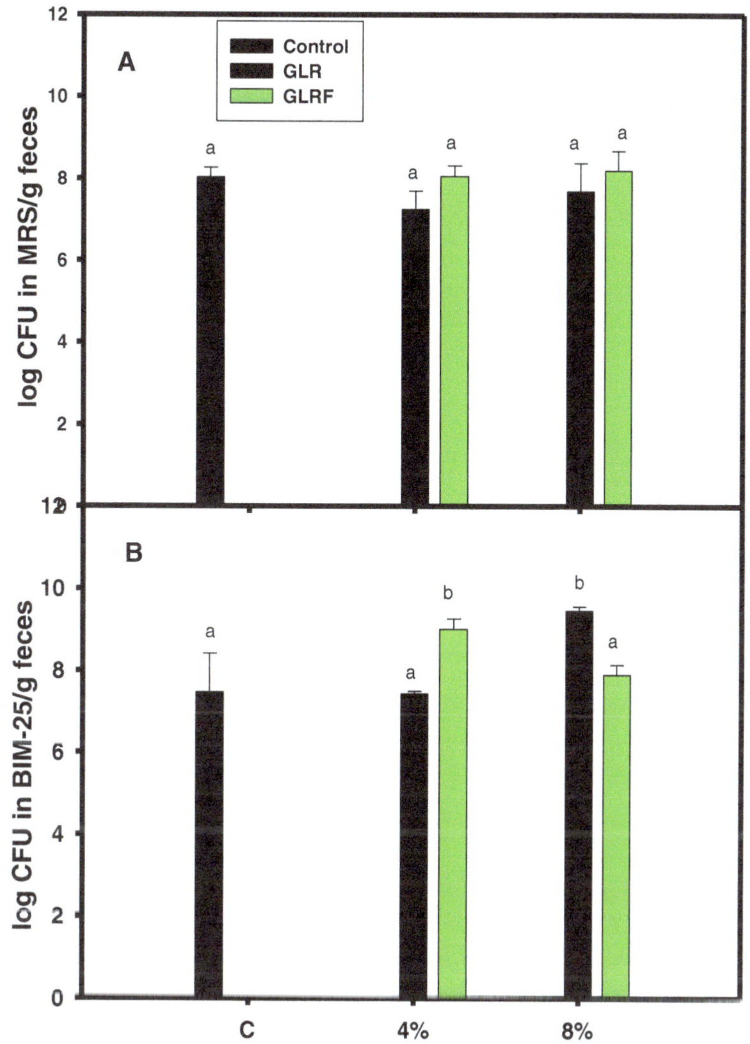

Fig. 5 Effects of powders of GLR or GLRF additives (4 or 8 %) on the growths of probiotics in vivo. (**a**) Lactobacilli counts by uses of MRS selective agar plates and (**b**) bifidobacteria counts by uses of BIM-25 selective agar plates. The feces (about 0.05 g) were suspended in the autoclaved broth and diluted serially and spread onto selective agar plate, and then incubated in an anaerobic chamber. The bacterial colonies were counted and calculated, and expressed as log colony forming unit (CFU)/g feces. Data were means ± SD and analyzed using one-way ANOVA, followed by the post hoc Tukey's test for comparisons. Values not sharing the same *alphabetic letter* were significantly different ($P < 0.05$)

elevated nitric oxide productions (Fig. 2b) and phagocytic activities (Fig. 2c) of HWP_GLR and HWP_GLRF were beneficial to innate immune responses.

Some reports were investigated on the improvement of feed properties in order to elevate the growth and/or meat qualities of chicken, such as the mushroom stalk residue additives in feeds could improve the oxidative status and meat quality of broiler chickens (Lee et al. 2012b), and the *Echinacea purpurea* L. additives in feeds could improve the oxidative status and meat quality in Arbor Acres broilers (Lee et al. 2013). The weights of broiler chicken among feeding groups did not show significant difference (Fig. 3), however, the phagocytosis and NK cytotoxicity (Fig. 4) of the innate immune in the peripheral blood and the counted probiotics of bifidobacteria fed with GLR and/or GLRF additives were apparently elevated compared to the commercial feeds (Fig. 5). It meant that the 4 or 8 % GLR and/or GLRF additives in the commercial feeds did not apparently influence the outcomes of chicken weights during developments, but might enhance some physiological properties, such as prebiotic activities and immune responses.

Conclusion

In conclusion, from the present in vitro and in vivo experiments, the GLR, GLRF, and their hot-water extracts with prebiotic and immune-stimulatory activities could be processed as feed additives which could increase the waste-recycling. It might be possible to add anti-bacterial properties of herbal medicines with GLR or GLRF to develop the new functional feeds with prebiotic, immune-stimulatory and anti-bacterial activities to reduce the uses of antibiotics and needed further investigations.

Competing interests
The authors declare that they have no competing interests.

Authors' contributions
YHL, SYL, and WCH participated the discussion and concepts of experimental designs, MS writing and revision; YSL, KLL, YLL, and HFS performed the analytical experiments; CHC and MYC designed the feeding experiments; KLL performed the feeding experiments. All authors read and approved the final manuscript.

Acknowledgments
The authors want to thank the Council of Agriculture, Republic of China and Ko Da Pharmaceutical Co. Ltd (97AS-1.1.3-AD-U4), and Shin Kong Wu Ho-Su Memorial Hospital (SKH-TMU-103-06) for financial supports. Authors want to thank the Avian Disease and Oncology Laboratory, Agricultural Research Service, US Department of Agriculture for providing LSCC-RP9 cells.

Author details
[1]Division of Gastroenterology, Shin Kong Wu Ho-Su Memorial Hospital, Taipei, Taiwan. [2]Department of General Medicine, Taipei Medical University, Taipei, Taiwan. [3]Graduate Institute of Pharmacognosy, Taipei Medical University, Taipei, Taiwan. [4]Department of Pharmacy, Taipei Medical University, Taipei, Taiwan. [5]Ko Da Pharmaceutical Co. Ltd, Tao-Yuan, Taiwan. [6]Department of Microbiology and Immunology, Taipei Medical University, Taipei, Taiwan.

References
Aida FMNA, Shuhaimi M, Yazid M, Maaruf AG (2009) Mushroom as a potential source of prebiotics: a review. Trends Food Sci Technol 20:567–575

Akira S (2003) Toll-like receptor signaling. J Biol Chem 278:38105–38108

An SJ, Pae HO, Oh GS, Choi BM, Jeong S, Jang SI, Oh H, Kwon TO, Song CE, Chung HT (2002) Inhibition of TNF-α, IL-1β, and IL-6 productions and NF-κB activation in lipopolysaccharide-activated RAW264.7 macrophages by catalposide, an iridoid glycoside isolated from *Catalpa ovata* G. Don (Bignoniaceae). Int Immunopharmacol 2:1173–1181

Boh B, Berovic M, Zhang J, Lin ZB (2007) *Ganoderma lucidum* and its pharmaceutically active compounds. Biotechnol Ann Rev 13:265–301

Chang YW, Lu TJ (2004) Molecular characterization of polysaccharides in hot-water extracts of *Ganoderma lucidum* fruiting bodies. J Food Drug Anal 12:59–67

Chen HS, Tsai YF, Lin S, Lin CC, Khoo KH, Lin CH, Wong CH (2004) Studies on the immuno-modulating and anti-tumor activities of *Ganoderma lucidum* (Reishi) polysaccharides. Bioorg Med Chem 12:5595–5601

Gao Y, Zhou S, Jiang W, Huang M, Dai X (2003) Effects of ganopoly (a *Ganoderma lucidum* polysaccharide extract) on the immune functions in advanced-stage cancer patients. Immunol Invest 32:201–215

Gibson GR, Roberfroid MB (1995) Dietary modulation of the colonic microbiota:introducing the concept of prebiotics. J Nutr 125:1401–1412

Gourbeyre P, Denery S, Bodinier M (2011) Probiotics, prebiotics and synbiotics: impact on the gut immune system and allergic reactions. J Leukoc Biol 89:685–695

Hsieh C, Yang FC (2004) Reusing soy residue for the solid-state fermentation of *Ganoderma lucidum*. Biores Technol 91:105–109

Karnjanapratum S, Tabarsa M, Cho M, You S (2012) Characterization and immunomodulatory activities of sulfated polysaccharides from *Capsosiphon fulvescens*. Int J Biol Macromol 51:720–729

Kino K, Yamashita A, Yamaoka K, Watanabe J, Tanaka S, Ko K, Shimizu K, Tsunoo H (1989) Isolation and characterization of a new immunomodulatory protein, Ling Zhi-8 (LZ-8), from *Ganoderma lucidium*. J Biol Chem 264:472–478

Kushima K, Fujita M, Shigeta A, Horiuchi H, Matsuda H, Furusawa S (2003) Flow cytometoric analysis of chicken NK activity and its use on the effect of restraint stress. J Vet Med Sci 65:995–1000

Lee KH, Morris-Natschke SL, Yang X, Huang R, Zhou T, Wu SF, Shi Q, Itokawa H (2012a) Recent progress of research on medicinal mushrooms, foods, and other herbal products used in traditional Chinese medicine. J Tradit Complement Med 2:84–95

Lee TT, Ciou JY, Chiang CJ, Chao YP, Yu B (2012b) Effect of *Pleurotus eryngii* stalk residue on the oxidative status and meat quality of broiler chickens. J Agric Food Chem 60:11157–11163

Lee TT, Ciou JY, Chen CL, Yu B (2013) Effect of *Echinacea purpurea* L. on oxidative status and meat quality in Arbor Acres broilers. J Sci Food Agric 93:166–172

Liu YW, Shang HF, Wang CK, Hsu FL, Hou WC (2007) Immunomodulatory activity of dioscorin, the storage protein of yam (*Dioscorea alata* cv. Tainong No.1) tuber. Food Chem Toxicol 45:2312–2318

Liu YH, Lin SY, Lee CC, Hou WC (2008) Antioxidant and nitric oxide production inhibitory activities of galacturonyl hydroxamic acid. Food Chem 109:159–166

Liu W, Wang H, Pang X, Yao W, Gao X (2010) Characterization and antioxidant activity of two low-molecular-weight polysaccharides purified from fruiting bodies of *Ganoderma lucidum*. Int J Biol Macromol 46:451–457

Mahajna J, Dotan N, Zaidman BZ, Petrova RD, Wasser SP (2008) Pharmacological values of medicinal mushrooms for prostate cancer therapy: the case of *Ganoderma lucidum*. Nutr Cancer 61:16–26

Manning TS, Gibson GR (2004) Prebiotics. Best Pract Res Clin Gastroenterol 18:287–298

Mao T, van de Water J, Keen C, Stern J, Hackman R, Gershwin M (1999) Two mushrooms, *Grifola frondosa* and *Ganoderma lucidum*, can stimulate cytokine gene expression and proliferation in human T lymphocytes. Int J Immunother 15:13–22

Nathan CF (1987) Secretary products of macrophages. J Clin Invest 79:319–326

Ohno N, Miura N, Sugawara N, Tokunaka K, Kirigaya N, Yadomae T (1998) Immunomodulation by hot water and ethanol extracts of *Ganoderma lucidum*. Pharm Pharmacol Lett 8:174–177

Roberfroid M, Gibson GR, Hoyles L, McCartney AL, Rastall R, Rowland I, Wolvers D, Watzl B, Szajewska H, Stahl B, Guarner F, Respondek F, Whelan K, Coxam V, Davicco MJ, Léotoing L, Wittrant Y, Delzenne NM, Cani PD, Neyrinck AM, Meheust A (2010) Prebiotic effects: metabolic and health benefits. Br J Nutr 104:S1–S63

Sanodiya BS, Thakur GS, Baghel RK, Prasad G, Bisen P (2009) *Ganoderma lucidum*: a potent pharmacological macrofungus. Curr Pharm Biotechnol 10:717–742

Shang HF, Cheng HC, Liang HJ, Liu HY, Liu SY, Hou WC (2007) Immunostimulatory activities of yam tuber mucilages. Bot Stud 48:63–70

Shiao MS (1992) Triterpenoid natural products in the fungus *Ganoderma lucidum*. J Chin Chem Soc 39:669–674

Stuart LM, Ezekowitz RAB (2005) Phagocytosis: elegant complexity. Immunity 22:539–550

Su CH, Sun CS, Juan SW, Hu CH, Ke WT, Sheu MT (1997) Fungal mycelia as the source of chitin and polysaccharides and their application as skin substitutes. Biomaterials 18:1169–1174

Synytsya A, Míčková K, Jablonský I, Spěváček J, Erban V, Kovářiková E, Čopíková J (2009) Glucans from fruit bodies of cultivated mushrooms *Pleurotus ostreatus* and *Pleurotus eryngii*: structure and potential prebiotic activity. Carbohydr Polym 76:548–556

Wachtel-Galor S, Szeto YT, Tomlinson B, Benzie IFF (2004) *Ganoderma lucidum* ('Lingzhi'); acute and short-term biomarker response to supplementation. Int J Food Sci Nutr 55:75–83

7

Antimicrobial activity and mechanisms of *Salvia sclarea* essential oil

Haiying Cui, Xuejing Zhang, Hui Zhou, Chengting Zhao and Lin Lin*

Abstract

Background: Nowadays, essential oils are recognized as safe substances and can be used as antibacterial additives. *Salvia sclarea* is one of the most important aromatic plants cultivated world-wide as a source of essential oils. In addition to being flavoring foods, *Salvia sclarea* essential oil can also act as antimicrobials and preservatives against food spoilage. Understanding more about the antibacterial performance and possible mechanism of *Salvia sclarea* essential oil will be helpful for its application in the future. But so far few related researches have been reported.

Results: In our study, *Salvia sclarea* oil showed obvious antibacterial activity against all tested bacterial strains. Minimum inhibitory concentration (MIC) and minimum bactericide concentration (MBC) of seven pathogens were 0.05 and 0.1 % respectively. In addition, *Salvia sclarea* oil also exhibited a significant inhibitory effect on the growth of *Escherichia coli* (*E. coli*) in phosphate buffer saline (PBS) and meats. After treated with *Salvia sclarea* oil, Scanning Electron Microscope (SEM) images can clearly see the damage of cell membrane; the intracellular ATP concentrations of *E. coli* and *S. aureus* reduced 98.27 and 69.61 % respectively, compared to the control groups; the nuclear DNA content of *E. coli* and *S. aureus* was significantly reduced to 48.32 and 50.77 % respectively. In addition, there was massive leakage of cellular material when *E. coli* and S. aureus were exposed to *Salvia sclarea* oil.

Conclusions: *Salvia sclarea* essential oil damaged the cell membrane and changed the cell membrane permeability, leading to the release of some cytoplasm such as macromolecular substances, ATP and DNA. In general, the antimicrobial action of *Salvia sclarea* essential oil is not only attributable to a unique pathway, but also involves a series of events both on the cell surface and within the cytoplasm. Therefore, more experiments need to be done to fully understand the antimicrobial mechanism of *Salvia sclarea* essential oil.

Keywords: *Salvia sclarea*, Essential oil, Antibacterial activity; Antibacterial mechanisms

Background

The genus *Salvia* is one of the largest and the most important aromatic and medicinal genera of the *Lamiaceae* family and comprises about 900 species, widespread throughout the world (Russo et al. 2013). Since ancient times, *Salvia sclarea* is one of the most appreciated medicinal herbs native to Mediterranean countries, and widely used in medicine and cooking (Durling et al. 2007), as well as in cosmetics, perfumery and the pharmaceutical industry (Kong et al. 2010). Plants from this genus are renowned for their biological activities such as antibacterial, antioxidant, antitumor, antidiabetic, antimicrobial, anxiolytic, sedative and anti-inflammatory activities (Rajabia et al. 2014).

In the world, the serious consequences caused by food-borne disease every year make food safety issues become the focus of attention by people and the priority issues of global public health (Cao et al. 2013). As a kind of the necessary food in our life, meat and its products are rich in protein, lipids and have suitable moisture content, which makes them to be the 'natural media' of microorganisms (Álvarez-Fernández et al. 2013), and they can be easily contaminated by *E. coli* during animal evisceration after slaughter, through contact with tainted water or during meat handling (Newell et al. 2010). So in spite of modern improvements in slaughter hygiene and food production techniques, the food safety of meat products is still an increasingly important public health issue (Burt 2004). To minimize meats degradation by

* Correspondence: linl@ujs.edu.cn
School of Food & Biological Engineering, Jiangsu University, Zhenjiang 212013, P. R. China

microorganisms and extend the shelf life of meats, legislation allows the application of antibacterial additives (Guillard et al. 2009). However, excessive use of antibacterial additives encourages the emergence of resistant bacteria. Bacteria contaminated nearly half of U.S. meat samples in a recent analysis, with 96 % of bacteria showing resistance to at least one type of drug (Fadli et al. 2012). So, due to the potential risk with synthetic antibacterial agents, there is a growing interest in the use of natural antimicrobials, particularly those that are derived from plant sources, for example, essential oils.

Essential oils (EOs) are aromatic oily liquids obtained from various plants (Marianne 2011). They are a group of secondary metabolites, complex mixtures of terpenoids, with characteristic flavor and fragrance properties (Stefanakis et al. 2013). So far, EOs play an important role in the protection of food safety as antibacterials, antivirals, antifungals, and antioxidants (Pandey et al. 2014). Nowadays essential oils are recognized as safe substances by the Food and Drug Administration and some contain compounds which can be used as antibacterial additives (Tian et al. 2014). A large number of reports concerning the antioxidant and the antimicrobial ability of essential oils have already been published. EOs become increasingly popular as natural antimicrobial and antioxidant agents that may be used in food preservation (Hossain et al. 2014; Jallali et al. 2014). Among many EOs, *Salvia sclarea* is an aromatic herb and thus was considered mainly for its essential oil content (Taarit et al. 2009; Alves-Silva et al. 2013). The essential oil is the main bioactive ingredient of *Salvia sclarea*. In addition to being flavouring foods, *Salvia sclarea* essential oil can also act as antimicrobials and preservatives against food spoilage (Kozics et al. 2013).

Understanding more about the antibacterial performance and possible mechanism of *Salvia sclarea* essential oil in meats will be helpful for its application in the future. But so far few related researches have been reported. Therefore, the aims of this study were a) to evaluate the antibacterial activity of the *Salvia sclarea* essential oil, and b) investigate the antibacterial mechanisms of *Salvia sclarea* essential oil.

Methods

Bacterial strains

The following bacterial strains were selected: *Escherichia coli* ATCC 25922, *Staphyloccocus aureus* ATCC 25923, *Bacillus pumilus* ATCC 27142, *Klebsiella pneumonia* ATCC 13883, *Bacillus subtilis* IFO 3457, *Salmonella typhi*murium B11, *Pseudomonas aeruginosa* ATCC 27853. The strains were cultured in Nutrient Broth (NB) at 35 °C for 24 or 48 h, and stored at −80 °C in NB.

Salvia sclarea essential oil

The *Salvia sclarea* essential oil tested in this study was supplied by J.E International (French). linalyl acetate (74.562 %), linalol (12.326 %) and germacrene D (1.931 %) are the main constituents.

Evaluation of antimicrobial activity

The antimicrobial activity of Salvia sclarea oil

Salvia sclarea essential oil was added into tubes containing NB to obtain concentrations of 0.1, 0.05, 0.025, 0.0125 and 0.00625 % respectively. Subsequently, the tubes were inoculated with the freshly prepared bacterial suspension in order to maintain initial bacterial concentration 10^3–10^4 CFU/mL, and then incubated in a rotary shaker at 150 rpm and 37 °C. The final treatment time of *Escherichia coli, Bacillus subtilis, Pseudomonas Aeruginosa, Bacillus pumilus* and *Salmonella typhimurium* is 24 h; *Staphylococcus aureus* and *Klebsiella pneumonia* is 48 h. The lowest concentration of essential oil showing growth inhibition (as seen visually) was considered as the minimum inhibitory concentration. The MBC was recorded as the lowest concentration of essential oil that showed no growth on Nutrient Agar (NA) plates after spot inoculation and incubation at 35 °C. The incubation time of *Escherichia coli, Bacillus subtilis, Pseudomonas Aeruginosa, Bacillus pumilus* and *Salmonella typhimurium* is 24 h; *Staphylococcus aureus* and *Klebsiella pneumonia* is 48 h (Bolou et al. 2011; Ruparelia et al. 2008). Each assay in this experiment was repeated three times.

Kill time analysis of *Salvia sclarea* oil

The plate colony-counting method was used to analyze the kill time of *Salvia sclarea* oil. *Salvia sclarea* oil was diluted into tubes with PBS to obtain concentration of 0.1 %. Subsequently, the tubes were inoculated with the freshly prepared bacterial suspension in order to maintain initial bacterial concentration 10^5 CFU/mL and cultured at 150 rpm and 37 °C. Numbers of viable bacteria were enumerated at 0, 0.5, 1, 2, 4 and 8 h by counting the number of bacterial colonies grown on the plate. As a control, the bacterial suspension in sterile PBS without *Salvia sclarea* oil was also tested (Petersen et al. 2006; Nakayama et al. 2012).

Antimicrobial activity testing in meats

Chicken, pork and beef were bought from the local supermarket. The aseptic sample with an amount of 10 and 90 g water was blended into slurry. Subsequently, mixture was inoculated with 1 % (v/w) of *E. coli* suspension (1×10^4 CFU/mL) and put into a sterilized 100 ml glass bottle. The samples with (0.1 v/v) or without *Salvia sclarea* oil were incubated at 30 °C for 24 and 48 h. Finally, the plate count method was carried out to determine the number of viable cells in the agar plate (Liu and Yang 2012).

The bactericidal mechanism of *Salvia sclarea* oil

The integrity of the cell membrane

The size and morphology of the bacteria, *E. coli* and *S. aureus*, were examined by Scanning electron microscopy (SEM, JSM-7001F, JEOL, Japan). Firstly, *Salvia sclarea* oil was diluted into tubes with PBS to obtain concentration of 0.05 %, and then added to two test tubes that containing *E. coli* and *S. aureus*. The tubes were cultured at 150 rpm and 37 °C for 24 h. Then the bacteria cultures were fixed with 2.5 % glutaraldehyde for at least 2 h. The samples washed twice with PBS each for 20 min. The washed samples were put into 1 % osmium tetroxide for 4 h, followed by washing twice with PBS each 20 min. The samples were then dehydrated using sequential exposure per ethanol concentrations ranging from 50 to 100 % and washed twice with PBS each 20 min. The washed sample was then put onto a stub for air drying, coating with gold followed by microscopic examinations (Zhang et al. 2007).

Loss of 260 nm-absorbing material

The release of cytosolic material absorbing at 260 nm from E. coli and S .aureus treated with *Salvia sclarea* oil at different concentrations, 0.25 MIC, 0.5 MIC, 1 MIC and 2 MIC, were performed on the bacteria suspension (10^8 CFU/mL in PBS). After treatment at 37 °C for 20 h, cells were centrifuged at 4000 rpm for 15 min at 4 °C and the supernatant were filtered by microporous membrane filter. The absorbency of the filtrate at 260 nm was detected by ultraviolet spectrophotometer (UV-1801, Beijing, China). As a control group, a bacterial suspension in sterile PBS without *Salvia sclarea* oil was tested as well (Chami et al. 2005; Sharma et al. 2013).

Measurement of cellular ATP concentrations

The broth of *E. coli* and *S. aureus* (10^8 CFU/mL) were centrifuged for 10 min at 8000 rpm. The cell pellets were washed three times and resuspended in the buffer to make up 10^8 CFU/mL suspensions of bacteria. After 0.05 % *Salvia sclarea* oil was added, the samples were cultured at 150 rpm and 37 °C for 30 min. Then, the samples were centrifuged for 10 min at 8000 rpm, and the cell pellets were collected. Finally, the cellular ATP concentrations of samples were determined by the Clean Sense TM Surface Hygiene Test Kit (LEYU Biotechnology, Shanghai, China). As a control group, the samples without *Salvia sclarea* oil were tested (Finger et al. 2012).

Observed nucleic acids with fluorescent staining method

S. aureus was cultured at 37 °C with gentle agitation for 48 h, and *E. coli* for 24 h. Subsequently, the cells were collected and resuspended in PBS, then added in 0.05 % (v/v) *Salvia sclarea* oil and cultured at 37 °C for 24 h. The nucleic acids were observed using the diluted 4′6-diamidino-2- phenylindole (DAPI) staining method. Ten microliter of the sample solution treated with *Salvia sclarea* oil was dropped on a glass slide. Three times volume of diluted DAPI (10 μg/mL, Roche Diagnostics GmbH, Germany) was added to the glass slide and kept in the dark for 10 min. The nucleic acids were observed with fluorescent microscope (Leica DMI4000B). As a control group, the bacteria without *Salvia sclarea* oil were also detected (Wang et al. 2010).

Measurement of the quantification of DNA

Centrifuge the suspensions at 4000 rpm for 15 min at 4 °C and the pellets were harvested. The pellets were washed three times with PBS and resuspended in the buffer containing 0.05 % (v/v) *Salvia sclarea* oil to make up bacterial suspensions of 10^8 CFU/mL. Eight-hundred micro liters were removed from the suspensions at 0 h hand 24 h and immediately added to three times the volume of diluted DAPI and kept in the dark for 10 min. The fluorescence intensity of DNA was estimated using fluorescence spectrophotometer (Cary Eclpise, VARIAN, America) with the excitation wavelength of 364 nm. As control, the bacterial suspension in sterile PBS without *Salvia sclarea* oil was tested (Wang et al. 2010).

Results and discussion

Evaluation of antimicrobial activity

Determination of Minimum Inhibitory Concentration (MIC) and Minimum Bactericidal Concentration (MBC)

Obtained MIC and MBC values of the investigated isolates were summarized in Table 1. As can be seen, *Salvia sclarea* oil showed the strongest antibacterial activity against the investigated bacteria. Meanwhile *Salvia sclarea* oil showed equal bacteriostatic and bactericidal activity against gram-positive bacteria (*Staphylococcus aureus* and *Bacillus subtilis*) and gram-negative bacteria (*Escherichia coli, Salmonella typhimurium, Klebsiella pneumonia, Pseudomonas Aeruginosa, Bacillus pumilus*).

Table 1 MICs and MBCs of *Salvia sclarea* oil against 7 pathogens

Test strain	*Salvia sclarea* oil	
	MIC (%)	MBC (%)
Escherichia coli	0.05	0.1
Staphylococcus aureus	0.05	0.1
Bacillus subtilis	0.05	0.1
Salmonella typhimurium	0.05	0.1
Klebsiella pneumonia	0.05	0.1
Pseudomonas Aeruginosa	0.05	0.1
Bacillus pumilus	0.05	0.1

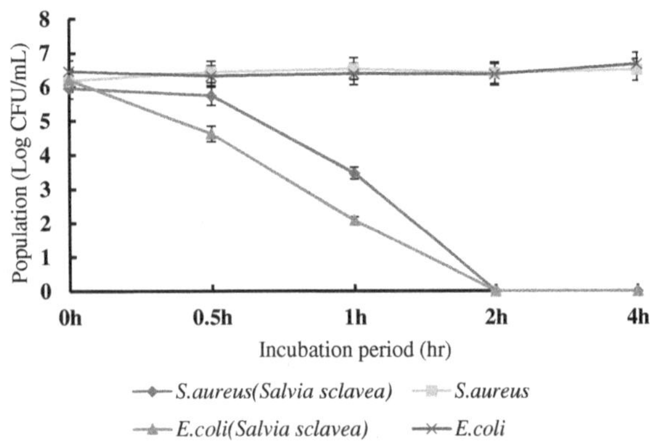

Fig. 1 The antimicrobial activity of *Salvia sclarea* oil of *E. coli* and *S. aureus*

The MICs of *Salvia sclarea* oil were both 0.05 %, and the MBCs were both 0.1 %. Those all indicated that *Salvia sclarea* oil was an effective bacterial inhibitor and bactericide with a broad antibacterial spectrum. Sepahvand et al. (2014) has demonstrated that *S. sclarea* essential oil did have different anti-bacterial effects with the magnitude *S. aureus* = *K. pneumonia* > *P. aeruginosa*. There are several reasons for the discrepancies of my observation and the previous report: (1) the difference among different species was obvious; (2) the antibacterial activities of *S. sclarea* essential oil were variable with different material and extraction methods; (3) there are differences between strains of the same origin. The difference between the strains of the same origin is resulted from long adaption to ecological environment, artificial selection and cross breeding. So the drug resistances of different strains can't be exactly the same; (4) different experiment methods may lead to different results.

The antimicrobial activity of *Salvia sclarea* oil testing in agar media

Due to equal bacteriostatic and bactericidal activity of *Salvia sclarea* oil against gram-negative bacteria and gram-positive bacteria, as the most representative gram-negative and gram-positive bacterium, *E. coli* and *S. aureus* were selected to further study to discuss the antibacterial activity and bactericidal mechanism of *Salvia sclarea* oil. From the Fig. 1, *Salvia sclarea* oil showed a good antimicrobial effect against both *E. coli* and *S. aureus*. About 99.99 and 99.9 % reduction in population were observed in *E. coli* and *S. aureus*

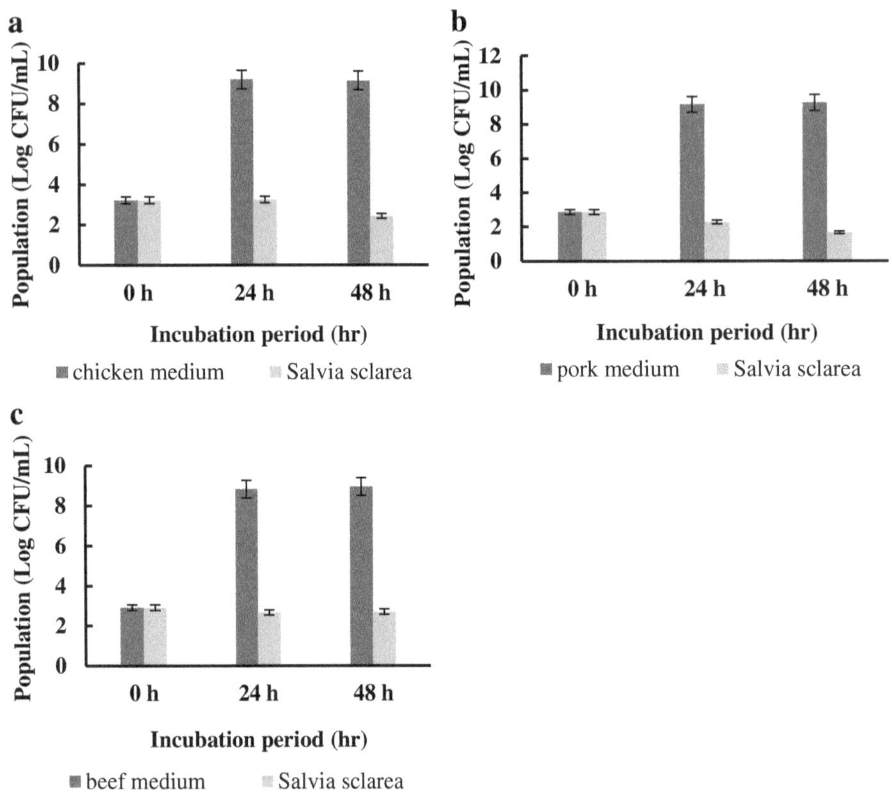

Fig. 2 a The antibacterial activity of *Salvia sclarea* oil of *E. coli* on chicken medium (**b**) The antibacterial activity of *Salvia sclarea* oil of *E. coli* on pork medium (**c**) The antibacterial activity of *Salvia sclarea* oil of *E. coli* on beef medium

Fig. 3 SEM images of *E. coli* and *S. aureus* before (**a**, **b**) and after (**c**, **d**) *Salvia sclarea* oil treatment. *Scale bar*: *E. coli* (1 μm); *S. aureus* (2 μm)

respectively after 1 h of *Salvia sclarea* treatment. And after 2 h treatment, almost 99.9999 % reductions in population were achieved both in *E. coli* and *S. aureus*. In summary, *Salvia sclarea* oil could act as promising antimicrobials because of its short antimicrobial time.

The antimicrobial activity of *Salvia sclarea* oil testing in meats

Escherichia coli (*E. coli*) are common in the normal microflora of the intestinal tract of humans and animals and are often used as an indicator of faecal contamination. Hence, the antibacterial activity of *Salvia sclarea* oil against *E. coli* on meat was investigated. As shown in Fig. 2, *Salvia sclarea* oil exhibited great antimicrobial activity on meat. After 48 h, almost 99.99999, 99.99999 and 99.9999 % reduction in population was observed on chicken medium, pork medium and beef medium respectively after treated by *Salvia sclarea* oil. The results indicated that *Salvia sclarea* oil had a marked effect on the inhibition of *E. coli* on meat. Therefore, *Salvia sclarea* oil can be regarded as a natural and efficient antiseptic of the pathogens which used on meat.

The bactericidal mechanism of *Salvia sclarea* oil

Scanning Electron Microscope (SEM)

The surface morphology of *E. coli* and *S. aureus* cells were evaluated by SEM analysis. The electron micrographs of both control and *Salvia sclarea* oil treated microbial cells were presented in Fig. 4. In control groups, the untreated cells had the typical structure, showing a striated wall for *E. coli* (Fig. 3a) and a smooth wall for *S. aureus* (Fig. 3b). In contrast, obvious harmful effects on the morphology of cell membranes were presented when strains were treated with *Salvia sclarea* oil. Deformed, incomplete and pitted shapes were observed in treated *E. coli* and *S. aureus* (Fig. 3c and d)

Loss of 260 nm-absorbing material

Changes in membrane integrity by *Salvia sclarea* oil cause the release of intracellular components. Small ions such as potassium and phosphate tend to flow out first, followed by macromolecular substances such as DNA and others (Lee et al. 2014). As we can see

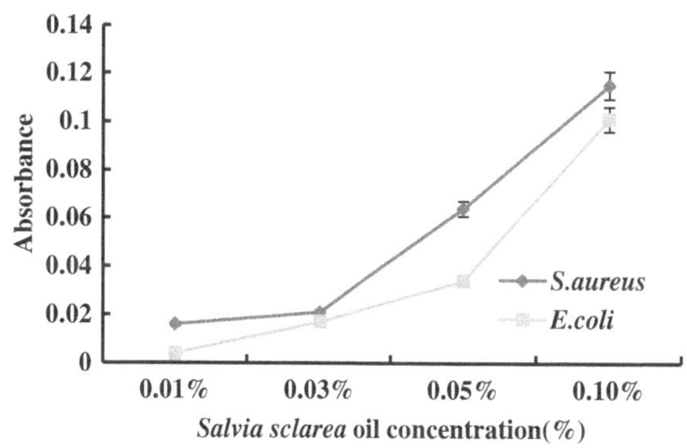

Fig. 4 Changes in optical density at 260 nm of supernatant from *E. coli* and *S. aureus* cell suspension treated with *Salvia sclarea* oil

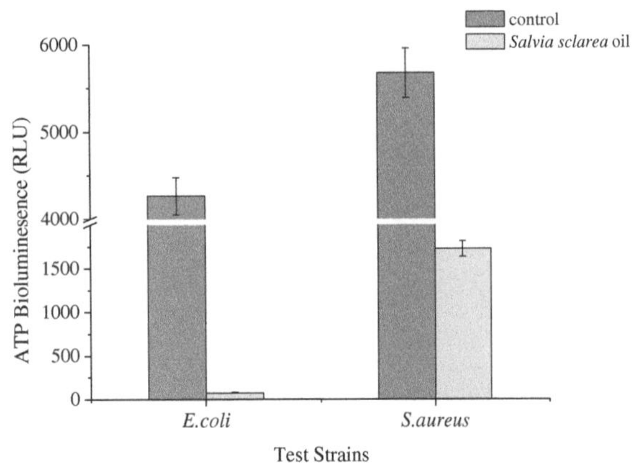

Fig. 5 The cellular ATP concentrations of *E. coli* and *S. aureus* before and after *Salvia sclarea* oil treatment

from Fig. 4, there was massive leakage of cellular material when *E. coli* and *S.* aureus were exposed to *Salvia sclarea* oil. Higher concentration of *Salvia sclarea* oil resulted in higher cell leakage, associated with loss of macromolecular substances. The level of cellular material reached a maximum after treatment with 0.1 % *Salvia sclarea* oil. And the results indicated that at higher concentration, the antibacterial activity of *Salvia sclarea* oil might be due to bactericidal damage to the membrane.

Measurement of cellular ATP concentrations

To observe the level of *E. coli* and *S. aureus* membrane damage caused by *Salvia sclarea* oil, the amount of cellular ATP was measured by ATP bioluminescence assay (Fig. 5). As is shown above, cellular ATP concentrations of *E. coli* and *S. aureus* reduced after treated with *Salvia sclarea* oil. Results of ATP bioluminescence assay indicated the intracellular ATP concentrations of *E. coli* and *S. aureus* reduced 98.27 and 69.61 % respectively, compared to the control groups. The results suggest that *Salvia sclarea* oil affects membrane permeability and results in the release of ATP. Moreover, it is possible that *Salvia sclarea* oil acts against different bacterial enzymes, such as ATPase (Nazzaro et al. 2013). And the results also showed that the gram negative strains of bacteria were more sensitive than gram positive strains to *Salvia sclarea* oil.

Nucleic acid staining

DAPI is a fluorescent dye that could penetrate into the bacteria cells and integrate with DNA. The Fig. 6 showed that the fluorescence intensities of *E. coli* and *S. aureus* treated with *Salvia sclarea* oil (Fig. 6b and d) were obviously lower than control group. Similarly, the fluorescence spectrophotometer measurements indicated that the DNA content in *S. aureus* was significantly reduced to 48.32 % compared to the control group (Fig. 7a), and

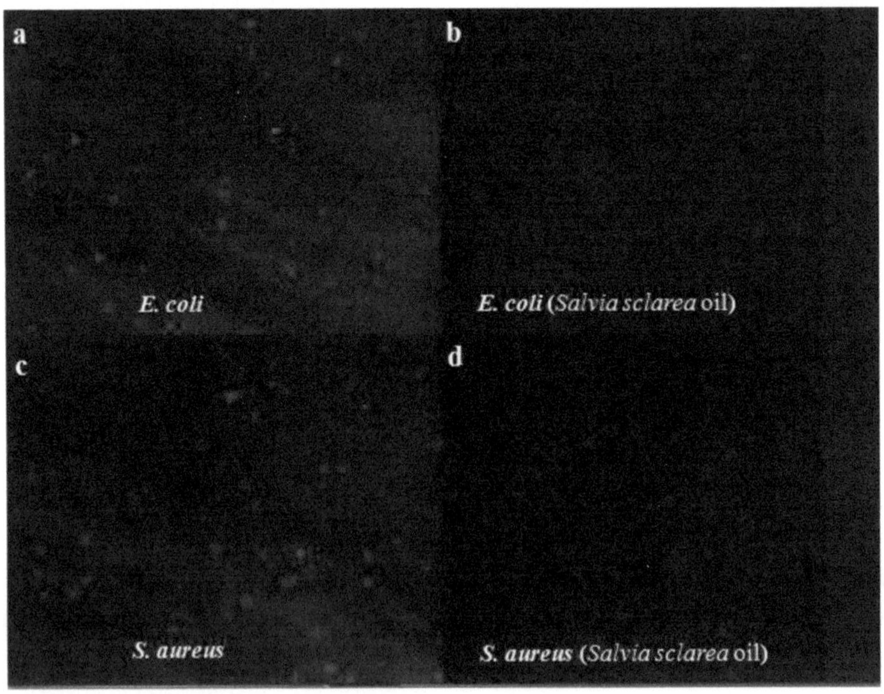

Fig. 6 DNA fluorescence images of *E. coli* and *S. aureus* cells before (**a, c**) and after (**b, d**) *Salvia sclarea* oil treatment

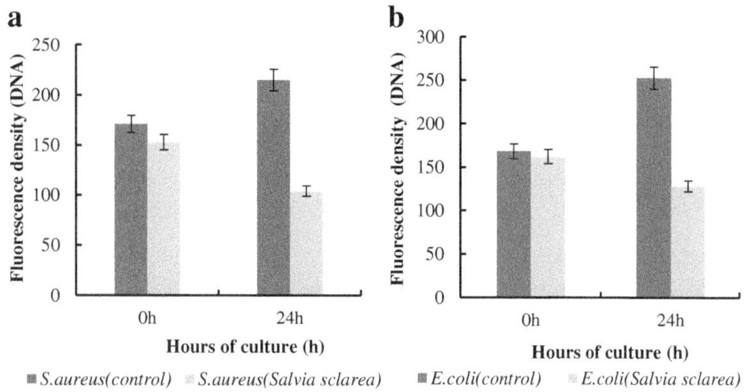

Fig. 7 **a** The fluorescence density of DNA of *S. aureus*. **b** The fluorescence density of DNA of *E. coli*

the DNA content in *E. coli* was reduced to 50.77 % (Fig. 7b). The results are the same as the fluorescent microscopic figures of *E. coli* and *S. aureus* cells treated with *Salvia sclarea* oil in Fig. 6. This implied that *Salvia sclarea* oil led to the decrease of DNA. Based on the results of the SEM images, this may be due to the interaction between *Salvia sclarea* oil and the bacterial cell membrane.

Conclusion

The results of the study revealed that *Salvia sclarea* essential oil exhibited significant antibacterial activity against all seven tested bacterial strains in vitro. Moreover, the similar antibacterial performances were observed in different meats. It indicated that *Salvia sclarea* essential oil has strong broad-spectrum antimicrobial activity. *Salvia sclarea* essential oil damaged the cell membrane and changed the cell membrane permeability, leading to the release of the material inside the cell such as macromolecular substances, ATP and DNA. In general, the antimicrobial action of *Salvia sclarea* essential oil is not only attributable to a unique pathway, but also involves a series of events both on the cell surface and within the cytoplasm. Therefore, more experiments need to be done to fully understand the antimicrobial mechanism of *Salvia sclarea* essential oil. In addition, these studies provided an experimental basis of practical application of *Salvia sclarea* essential oil as a natural antibacterial agent.

Abbreviations

MIC: Minimum inhibitory concentration; MBC: Minimum bactericide concentration; PBS: Phosphate buffer saline; NA: Nutrient Agar; SEM: Scanning electron microscope; DAPI: Diluted 4'6-diamidino-2-phenylindole.

Competing interests

The authors declare that they have no competing interests.

Authors' contributions

CHY and ZXJ designed the antibacterial mechanism measurement protocol, conducted all sampling and statistical analyses. ZH and ZCT designed the antibacterial experiments. LL served as the boss for all tests and established the permanent plots. All authors wrote and revised the manuscript and have read and approved the final manuscript.

Acknowledgments

The authors acknowledge the financial support from National Natural Science Foundation of China (grant no. 31301573), National Science Foundation for Post-doctoral Scientists of China (grant no.12M511223), Natural Science Foundation of Jiangsu Province (grant no. BK20130493), Jiangsu University research foundation (grant no. 11JDG050) and Project Funded by the Priority Academic Program Development of Jiangsu Higher Education Institutions.

References

Álvarez-Fernández E, Cancelo A, Díaz-Vega C, Capita R, Alonso-Calleja C (2013) Antimicrobial resistance in *E. coli* isolates from conventionally and organically reared poultry: a comparison of agar disc diffusion and Sensi Test Gram-negative methods. Food Control 30:227–234

Alves-Silva JM, Dias SM, Pintado ME (2013) Chemical composition and in vitro antimicrobial, antifungal and antioxidant properties of essential oils obtained from some herbs widely used in Portugal. Food Control 32:371–378

Bolou GEK, Bagré I, Ouattara K, Djaman AJ (2011) Evaluation of the antibacterial activity of 14 menicinal plants in Côte d'Ivoire. Trop J Pharm Res 10:335–340

Burt S (2004) Essential oils: their antibacterial properties and potential applications in foods—a review. Int J Food Microbiol 94:223–253

Cao YM, Gu WG, Zhang JJ, Chu Y, Ye XQ, Hu YQ, Chen JC (2013) Effects of chitosan, aqueous extract of ginger, onion and garlic on quality and shelf life of stewed-pork during refrigerated storage. Food Chem 141:1655–1660

Chami F, Chami N, Bennis S, Bouchikhi T, Remmal A (2005) Oregano and clove essential oils induce surface Alteration of *Saccharomyces Cerevisiae*. Phytother Res 19:405–408

Durling NE, Catchpole OJ, Grey JB, Webby RF (2007) Extraction of phenolics and essential oil from dried sage (*Salvia officinalis*) using ethanol-water mixtures. Food Chem 101:1417–1424

Fadli M, Saad A, Sayadi S, Chevalier J, Mezrioui N (2012) Antibacterial activity of *Thymus maroccanus* and *Thymus broussonetii* essential oils against nosocomial infection–bacteria and their synergistic potential with antibiotics. Phytomedicine 9:464–471

Finger S, Wiegand C, Buschmann HJ, Hipler UC (2012) Antimicrobial properties of cyclodextrin–antiseptics-complexes determined by microplate laser nephelometry and ATP bioluminescence assay. Int J Pharmaceut 436:851–856

Guillard V, Issoupov V, Redl A, Gontard N (2009) Food preservative contentreduction by controlling sorbic acid release from a superficial coating. Innov Food Sci Emerg 10:108–115

Hossain MA, Harbi SR, Weli AM (2014) Comparison of chemical constituents and antimicrobial activities of three essential oils from three different brands' clove samples collected from Gulf region. Asian Pac J Trop Dis 4:262–268

Jallali I, Zaouali Y, Missaoui I, Smeoui A (2014) Variability of antioxidant and antibacterial effects of essential oils and acetonic extracts of two edible halophytes: *Crithmum maritimum* L. and *Inula crithmoïdes* L. Food Chem 145:1031–1038

Kong BH, Zhang HY, Xiong YL (2010) Antioxidant activity of spice extracts in a liposome system and in cooked pork patties and the possible mode of action. Meat Sci 85:772–778

Kozics K, Klusova V, Srancikova A (2013) Effects of Salvia officinalis and Thymus vulgaris on oxidant-induced DNA damage and antioxidant status in HepG2 cells. Food Chem 141:2198–2206

Lee SY, Kim KB, Lim SI, Ahn DH (2014) Antibacterial mechanism of *Myagropsis myagroides* extract on *Listeria monocytogenes*. Food Control 42:23–28

Liu TT, Yang TS (2012) Antimicrobial impact of the components of essential oil of *Litsea cubeba* from Taiwan and antimicrobial activity of the oil in food systems. Int J Food Microbiol 156:68–75

Marianne (2011) U.S. meats tainted with bacteria. http://news.discovery.com/human/drug-resistant-bacteria-common-in-us-meats-110415.htm. Accessed 15 April 2011

Nakayama M, Shigemune N, Tsugukuni T, Jun H, Matsushita T, Mekada Y, Kurahachi M, Miyamoto T (2012) Mechanism of the combined anti-bacterial effect of green tea extract and NaCl against *Staphylococcus aureus* and *Escherichia coli* O157:H7. Food Control 25:225–232

Nazzaro F, Fratianni F, Martino LD, Coppola R, Feo VD (2013) Effect of essential oils on pathogenic bacteria. Pharmaceuticals 6:1451–1474

Newell DG, Koopmans M, Verhoef L, Duizer E, Aidara-Kane A, Sprong H, Opsteegh M, Langelaar M, Threfall J, Scheutz F, Giessen J, Kruse H (2010) Food-borne diseases—the challenges of 20 years ago still persist while new ones continue to emerge. Int J Food Microbio 139:S3–S15

Pandey AK, Mohan M, Singh P, Palni UT, Tripathi NN (2014) Chemical composition, antibacterial and antioxidant activity of essential oil of *Eupatorium adenophorum Spreng*. from Eastern Uttar Pradesh, India. Food Biosci 7:80–87

Petersen PJ, Labthavikul PC, Jones H, Bradford PA (2006) In vitro antibacterial activities of tigecycline in combination with other antimicrobial agents determined by chequerboard and time-kill kinetic analysis. J Antimicrob Chemoth 57:573–576

Rajabia Z, Ebrahimia M, Farajpoura M, Mirzac M, Ramshini H (2014) Compositions and yield variation of essential oils among and within nine Salvia species from various areas of Iran. Ind Crop Prod 61:233–239

Ruparelia JP, Chatterjee AK, Duttagupta SP, Mukherji S (2008) Strain specificity in antimicrobial activity of silver and copper nanoparticles. Acta Biomater 4:707–716

Russo A, Formisano C, Rigano D (2013) Chemical composition and anticancer activity of essential oils of Mediterranean sage (*Salvia officinalis L.*) grown in different environmental conditions. Food Chem Toxicol 55:42–47

Sepahvand R, Delfan B, Ghanbarzadeh S, Rashidipour M, Gholam GH, Yadegari JG (2014) Chemical composition, antioxidant activity and antibacterial effect of essential oil of the aerial parts of Salvia sclareoides. Asian Pac J Trop Med 7:S491–S496

Sharma A, Bajpai VK, Baek KH (2013) Determination of antibacterial mode of action of allium sativum essential oil against foodborne pathogens using membrane permeability and surface characteristic parameters. J Food Saf 33:197–208

Stefanakis MK, Touloupakis E, Anastasopoulos E, Ghanotakis D, Katerinopoulos HE, Makridis P (2013) Antibacterial activity of essential oils from plants of the genus Origanum. Food Control 34:539–546

Taarit MB, Msaadaa K, Hosnia K (2009) Plant growth, essential oil yield and composition of sage (*Salvia officinalis* L.) fruits cultivated under salt stress conditions. Ind Crop Prod 30:333–337

Tian J, Zeng XB, Zhang S (2014) Regional variation in components and antioxidant and antifungal activities of *Perilla frutescens* essential oils in China. Ind Crop Prod 59:69–79

Wang Q, Wang H, Xie MJ (2010) Antibacterial mechanism of soybean isoflavone on *Staphylococcus aureus*. Arch Microbiol 192:893–898

Zhang LL, Jiang YH, Ding YL, Povey A, York D (2007) Investigation into the antibacterial behaviour of suspensions of ZnO nanoparticles (ZnO nanofluids). J Nanopart Res 9:479–489

8

Scale-dependent effects of habitat fragmentation on reproduction in the annual *Circaeaster agristis*, a narrow endemic and threatened species

Jie-Cai Zhao[1], Jun Luo[1], Chun-Ping Yang[1] and Guo-Xing Cao[1,2*]

Abstract

Background: Habitat fragmentation and the resulting decline in population size and density commonly reduce the reproduction of rare and threatened species. We investigated the impacts of population size and density on reproduction in more than 30 populations of *Circaeaster agristis*, a narrow endemic and threatened species, in 2010 and 2011. We also examined the effects of NND (nearest neighbor distance) and LNS (local neighbor size), within radii of 0.1 m, 0.2 m and 0.3 m, on reproduction in two of the populations in 2011.

Results: Population size did not affect fruit (seed) number and fruit set in either year studied. Population density had an indirect negative effect on fruit number and fruit set as a consequence of a negative effect on plant size in 2010, but had no effect on fruit number and fruit set in 2011. Within populations, individual fruit number did not change, and individual fruit set increased independent of plant size, in response to increasing NND. Both individual fruit number and individual fruit set increased, independent of plant size, with increases in LNS within a 0.1 m radius, but did not change with increases in LNS within radii of between 0.1 m and 0.2 m radii or between 0.2 m and 0.3 m.

Conclusions: The effect of habitat fragmentation on reproduction of *C. agristis* is scale-dependent. In contrast to the generally accepted idea that fragmentation reduces plant reproduction, reproductive success may increase in sparse populations or increase in response to decreases in LNS in *C. agristis*.

Keywords: *Circaeaster agristis*; Habitat fragmentation; Population size; Population density; Reproduction; Scale-dependency

Background

Habitat fragmentation is a worldwide phenomenon, and is considered to be one of the major threats to the persistence and viability of plant populations (Eriksson and Ehrlén 2001; Oostermeijer, 2003). Habitat fragmentation often causes a decline in the size and density of populations, thereby altering abiotic and biotic environmental conditions (Wilcove et al., 1986; Saunders et al., 1991), all of which can affect the reproductive success of plants.

Experimental studies and field observations have shown that plants in small and sparse populations, because of their smaller display size and lower supply of rewards, often receive lower visitation rates (Ågren, 1996; Kunin, 1997; Weber and Kolb 2013), and experience smaller pollen loads (Bosch and Waser, 1999; Waites and Ågren, 2004; Jakobsson et al., 2009), resulting in a limitation of pollen quantity and reduced production of fruits and seeds. Furthermore, plants in small and sparse populations may witness higher within-plant movement of pollinators (Antonovics and Levin, 1980; Klinkhamer and de Jong, 1990; Hermansen et al., 2014), or may be more likely to cross with related individuals (Barrett and Kohn, 1991; Glémin et al. 2008; Suarez-Gonzalez and Good, 2014). In self-incompatible plants, within-plant pollinator movements or crossing with related individuals decrease compatible pollen receipt,

* Correspondence: cgxing@sicau.edu.cn
[1]Department of Forestry, Sichuan Agricultural University, Yaan 625014, China
[2]Xishuangbanna Tropical Botanical Garden, Chinese Academy of Sciences, Mengla 666303, China

which can reduce reproductive success (Waites and Ågren, 2004). In self-compatible plants, within-plant pollinator movements or crossing with related individuals lead to greater self-fertilization rates (van Treuren et al., 1993; Karron, 1995) or biparental inbreeding (Jones and Comita, 2008), potentially decreasing offspring quantity or quality. Finally, if the population size and density are positively correlated with habitat quality, plant size and reproductive success may increase in response to increasing population size and density (Leimu et al., 2006). Literature reviews on this topic have indicated that a positive correlation between reproductive success and population size or density is a common phenomenon in rare and endangered species (Ghazoul, 2005; Leimu et al., 2006).

Notwithstanding the above points, pollination and reproductive success may not always be curtailed in small and sparse populations. For example, plants in such populations may experience lower competition for pollinator visits than those in large and dense populations (Steven et al., 2003). If fruit or seed set is pollinator-limited and pollinators visit a smaller proportion of flowers per plant in large and dense rather than small and sparse populations, then reproductive success may decline in large and dense populations, particularly in self-incompatible species (Johnson et al., 2012; Stein et al., 2013; Ward et al., 2013). Plants facing chronic pollinator scarcity and pollen limitation may evolve self-compatibility for reproductive assurance (Baker, 1967), and are expected to be less subject to reduced pollinator visitation and the effects of fragmentation (Aguilar et al., 2006). Furthermore, reproductive success in plants in sparse populations may benefit from lower competition for resources, such as light, soil nutrients and water. As a result, plant size and reproductive success may decrease with plant density (Harper, 1977; Antonovics and Levin, 1980; Weiner, 1982; Mustajärvi et al., 2001).

When assessing the effects of habitat fragmentation on reproductive success, some researchers have focused on the population level: population size (number of individuals) and/or mean population density (Ågren, 1996; Morgan, 1999; Hermansen et al., 2014). On the other hand, others have focused on the individual level: nearest neighbor distance (NND) (Allison, 1990; Metcalfe and Kunin, 2006; Caraballo-Ortiz et al., 2011; Lawes et al., 2013) or local neighbor size (LNS) (Roll et al., 1997; Mitchell and Ankeny, 2001; Jakobsson et al., 2009; Weber and Kolb 2013). Although the perspective from either level can provide valuable insight into the effects of habitat fragmentation, few studies have investigated the effects of habitat fragmentation on reproductive success at both levels within a single species (Wagenius, 2006; Gunton and Kunin, 2007; Spigler and Chang, 2008).

In this study, we explored the effects of habitat fragmentation on reproductive success in the endangered annual *Circaeaster agristis*, at both population (population size and density) and individual (NND and LNS) levels, in 2010 and 2011. The data obtained here may assist in understanding the effects of fragmentation on plant reproductive success, and provide useful guidelines for the management and conservation of *C. agristis*.

Methods

Study species

Circaeaster agristis Maxim. is the only species of the genus *Circaeaster*, which is a member of Circaeasteraceae together with another monotypic genus, *Kingdonia* (Tian et al., 2007). *Circaeaster agristis* is an annual alpine herb, narrowly distributed in southwestern and northwestern China. This species primarily grows in humus-rich forest soils. Its population is declining and its distribution range is shrinking because of deforestation and habitat fragmentation, and has consequently been listed as a rare and endangered species in the Chinese red list (Fu and Jin, 1992). It reproduces exclusively by seeds. Individual plants of *C. agristis* produce a single stem which is, on average, 7.2 ± 2.2 cm (mean ± SD, range = 4.3-12.9 cm, $n = 60$, unpublished data) tall. The flowers are green, hermaphroditic and about 1 mm long, have two or three tepals, one or two stamens, and one carpel (Hu et al., 1990). Flowering occurs from late May to early June, and fruits mature in August. The fruit is an elliptic achene, and is 6–12 mm long when mature. Each achene contains one seed. The upper surface of the fruit exhibits hooked trichomes.

Study site

Our experiment was conducted in the Wanglang Nature Reserve (32°49′-33°02′ N, 103°55′-104°10′ E, 2300–4980 m above sea level), southwestern China. The vertical distribution of vegetation types in this region includes mixed forests of conifers and broadleaf trees, and broadleaf deciduous forest (2300–2600 m), fir forest (dominated by *Abies faxoniana*) and spruce-cypress forest (dominated by *Picea purpurea* and *Sabina saltuaria*, 2600–3500 m), subalpine shrubs and meadow (3500–4400 m), and sparse alpine vegetation (4400–4900 m).

Population size and density

In 2010, we identified 34 *C. agristis* populations (Table 1). Three populations disappeared in 2011, but another four populations were identified in this year (Table 1). The populations selected for analysis were present at between 2503 and 2645 m altitude. All selected populations were present in a forest and were separated from one another by at least 30 m. Natural *C. agristis* populations are patchy and discrete, and are easy to delimit. In early to mid-August of each year, we measured the perimeter of each population, calculating its area, and then counted the number of individual plants (population size). We then estimated the mean density of each population by dividing its population size by its area. The log-transformed mean

Table 1 Summary data for study populations including location, population size and population density

Population	Location		Population size		Population density (number of plants/m^2)	
	Latitude	longitude	2010	2011	2010	2011
1	32°58′32"	104°04′33"	312	93	118	38
2	32°58′35"	104°04′35"	15434	32647	230	211
3	32°58′05"	104°04′38"	25	7	71	28
4	32°58′44"	104°04′43"	18433	29277	146	122
5	32°58′42"	104°04′50"	8062	13961	129	122
6	32°58′42"	104°04′55"	715	253	63	21
7	32°58′42"	104°04′59"	138	—	139	—
8	32°58′39"	104°05′13"	1138	7761	27	117
9	32°58′14"	104°05′18"	257	31	612	20
10	32°58′34"	104°05′19"	5166	1905	118	43
11	32°58′28"	104°05′21"	2137	557	161	40
12	32°58′25"	104°05′24"	43495	22049	185	87
13	32°58′23"	104°05′26"	10224	7522	238	139
14	32°58′24"	104°05′28"	231	975	27	103
15	32°58′26"	104°05′30"	7231	1879	272	77
16	32°58′11"	104°05′37"	137	302	117	122
17	32°58′14"	104°05′43"	156	134	158	61
18	32°58′14"	104°05′48"	450	379	58	175
19	32°58′15"	104°05′54"	17	—	57	—
20	32°58′16"	104°05′59"	1315	7942	33	83
21	32°58′40"	104°05′08"	53	428	7	229
22	32°58′16"	104°06′09"	3860	247	283	11
23	32°58′16"	104°06′12"	37519	96071	309	469
24	32°58′09"	104°06′13"	3774	3340	165	178
25	32°58′12"	104°06′16"	8138	8961	134	36
26	32°58′07"	104°06′17"	301	—	1158	—
27	32°58′14"	104°06′17"	4912	4417	120	115
28	32°58′10"	104°06′20"	85897	91925	134	142
29	32°58′09"	104°06′24"	1537	147	422	21
30	32°58′09"	104°06′29"	13657	19962	184	89
31	32°58′02"	104°06′46"	84500	40357	285	133
32	32°57′59"	104°06′56"	18084	46203	239	275
33	32°58′00"	104°07′02"	247	554	30	71
34	32°57′58"	104°07′24"	426	1069	12	34
35	32°58′11"	104°03′18"	—	41389	—	88
36	32°58′46"	104°03′44"	—	91669	—	109
37	32°58′45"	104°03′57"	—	17157	—	70
38	32°58′29"	104°04′20"	—	70989	—	64

— the population disappeared in 2011 or not found in 2010

population density was significantly correlated to the log-transformed population size in 2010 ($r_p = 0.371$, $P = 0.031$, $n = 34$) and 2011 ($r_p = 0.614$, $P < 0.001$, $n = 35$).

Sampling on a population basis

From early to mid-August 2010 and 2011 in each population, 30 individuals were randomly marked, except in populations with less than 30 individuals, in which case all of them were marked. The number of leaves, flowers (flower scars) and mature fruits were counted on each of the marked individuals.

Sampling on an individual basis

In August 2011, we randomly marked 35 focal individuals in each of populations 2 (intermediate population density) and 37 (low population density), with the focal plants separated from each other by more than 1 m. For each focal plant, the number of leaves, flowers (flower scars) and mature fruits were counted. The effect of NND was quantified by measuring the distance from each focal plant to the nearest conspecifics. The effect of LNS was quantified by counting the number of conspecifics within radii of 0.1 m, 0.2 m, and 0.3 m around each focal plant. We included areas with multiple radii because the effects of LNS can be scale-dependent (Roll et al., 1997).

Statistical analyses

Effects of population size and density on reproduction

To understand the effects of population size and mean population density, we first calculated the population means of leaf number per plant (plant size), flower number per plant, fruit number per plant, and fruit set (fruits/flowers) per plant, in both 2010 and 2011. We first used linear regression analysis to evaluate the effects of population size and mean population density on mean population fruit number or fruit set for each year separately. Because the population means of plant size were highly correlated with mean population flower number (2010: $r_p = 0.865$, $P < 0.001$, $n = 34$; 2011: $r_p = 0.861$, $P < 0.001$, $n = 35$), fruit number (2010: $r_p = 0.782$, $P < 0.001$, $n = 34$; 2011: $r_p = 0.709$, $P < 0.001$, $n = 35$) and fruit set in 2010 ($r_p = 0.583$, $P < 0.001$, $n = 34$), we used a multiple regression analysis to analyze whether population size and mean population density had an influence on mean population fruit number in 2010 and 2011 or on fruit set in 2010, when accounting for mean population plant size. All variables except mean population fruit set were natural-log-transformed to stabilize variances. For all analyses, we examined the residuals for departures from normality (Shapiro-Wilk test).

Effects of NND and LNS on reproduction

We first used analyses of covariance to analyze fruit number and fruit set of selected individuals as a function of population (random factor) and NND (covariate). Because individual leaf number (plant size) was highly correlated with individual flower number (population 2: $r_p = 0.750$, $P < 0.001$, $n = 35$; population 37: $r_p = 0.829$, $P < 0.001$, $n = 35$), individual fruit number (population 2: $r_p = 0.818$, $P < 0.001$, $n = 35$; population 37: $r_p = 0.795$, $P < 0.001$, $n = 35$) and individual fruit set (population 2: $r_p = 0.531$, $P = 0.001$, $n = 35$; population 37: $r_p = 0.570$, $P < 0.001$, $n = 35$), we then included individual plant size as a covariate in each of the analyses to evaluate whether NND had effects on fruit number or fruit set when differences in individual plant size were accounted for. All variables except fruit set were natural-log-transformed to stabilize variances. For each analysis, initial models included all possible two-way or three-way interactions involving covariates. Non-significant interactions were dropped from the model using backwards elimination. For all analyses, we examined residuals for departures from normality.

Similarly, we used analyses of covariance to analyze fruit number and fruit set as a function of population (random factor) and 0.1 m LNS (covariate), 0.2 m LNS (covariate) and 0.3 m LNS (covariate). We considered 0.2 m LNS and 0.3 m LNS as the number of additional individuals beyond 0.1 m LNS within 0.2 m, and the number of additional individuals beyond 0.2 m LNS within 0.3-m, respectively. Multicollinearity was assessed by inspection of variance inflation factors, which was always ≤ 2.1, indicating that the level of collinearity was not problematic (Quinn and Keough, 2002). We then included individual plant size as a covariate in each of the analyses to assess whether LNS had effects on fruit number or fruit set when differences in individual plant size were accounted for. Subsequent manipulations of the statistical model were conducted as described previously for NND. All analyses were conducted using SPSS 17.0.

Results

Effects of population size and density on reproduction

Population means for leaf number per plant varied from 6.5 to 13.0 in 2010 and 7.1 to 11.7 in 2011; population means for flower number per plant varied from 6.5 to 22.2 in 2010 and 9.4 to 32.8 in 2011; population means for fruit number per plant varied from 1.7 to 13.1 in 2010 and 3.0 to 11.9 in 2011. Finally, population means for fruit set per plant varied from 0.14 to 0.59 in 2010 and 0.19 to 0.57 in 2011. All measured traits differed significantly among populations within the same year ($P < 0.001$; all variables except fruit set were log-transformed). All correlations were conducted using these log-transformed variables, where applicable.

Mean population fruit number was not correlated to population size in either 2010 ($F_{1,32} = 0.036$, $P = 0.850$) or 2011 ($F_{1,33} = 1.697$, $P = 0.202$) (Fig. 1a). When mean population plant size was included in the regression analysis of each year, mean population fruit number was not related to population size in either 2010 ($t_{31} = -0.626$, $P = 0.536$) or 2011 ($t_{32} = 0.066$, $P = 0.948$). Meanwhile, mean population fruit number was negatively correlated to mean population density in 2010 ($b \pm s_b = -0.192 \pm 0.075$, $R^2 = 0.169$, $F_{1,32} = 6.500$, $P = 0.016$), and was not

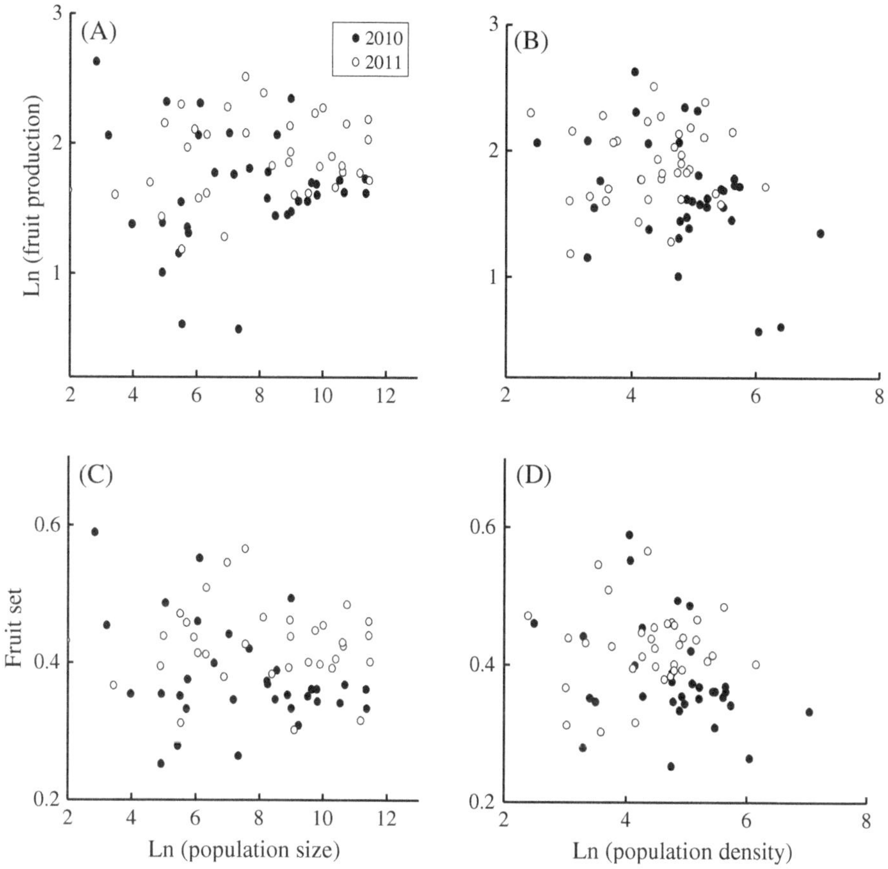

Fig. 1 Effects of population size and density on reproduction in *C. agristis*. Relationships between population size and (**a**) mean population fruit production, (**c**) mean population fruit set, and between mean population density and (**b**) mean population fruit production and (**d**) mean population fruit set. Effects of plant size were not accounted for

significantly correlated to mean population density in 2011 ($F_{1,33} = 0.076$, $P = 0.785$) (Fig. 1b). When mean population plant size was included in the regression analysis for each year, mean population fruit production was not correlated to mean population density in 2010 ($t_{31} = -0.785$, $P = 0.439$) or 2011 ($t_{32} = 0.321$, $P = 0.750$). Furthermore, mean population plant size was negatively correlated to mean population density) in 2010 ($b \pm s_b = -0.064 \pm 0.024$, $R^2 = 0.180$, $F_{1,32} = 7.037$, $P = 0.012$).

Mean population fruit set was not related to population size in either 2010 ($F_{1,32} = 1.483$, $P = 0.232$) or 2011 ($F_{1,33} = 0.355$, $P = 0.555$) (Fig. 1c). When mean population plant size was included in the regression analysis for 2010, mean population fruit set was not correlated to population size ($t_{31} = -1.700$, $P = 0.099$). Mean population fruit set was negatively correlated to mean population density in 2010 ($b \pm s_b = -0.038 \pm 0.014$, $R^2 = 0.186$, $F_{1,32} = 7.305$, $P = 0.011$), and not correlated to mean population density in 2011 ($F_{1,33} = 0.701$, $P = 0.408$) (Fig. 1d). When mean population plant size was included in the regression analysis for 2010, the correlation between mean population fruit set and mean population density became non-significant ($t_{31} = -1.436$, $P = 0.161$).

Effects of NND

Individual plants of populations 2 and 37 respectively produced an average of 10.3 (SD = 2.6, range = 7 – 18) and 10.6 (SD = 2.5, range = 7 – 17) leaves, and 10.5 (SD = 5.6, range = 2 – 29) and 11.8 (SD = 7.9, range = 3 – 29) fruits. Mean fruit set of individual plants was 0.45 (SD = 0.13, range = 0.25 – 0.70) and 0.47 (SD = 0.12, range = 0.23 – 0.71) in populations 2 and 37 respectively. All of these measured traits did not differ significantly between populations ($P > 0.5$ for all comparisons). Mean NND was 0.020 m (SD = 0.019, range = 0.001 – 0.081 m) and 0.044 m (SD = 0.056, range = 0.001 – 0.230 m) in populations 2 and 37 respectively, and differed significantly between populations ($F_{1,68} = 4.001$, $P = 0.049$).

Individual fruit number did not vary with NND irrespective of whether or not individual plant size was included in the analysis of covariance (Table 2A-B, Fig. 2a). Individual fruit set was positively correlated with NND ($b \pm s_b = 0.029$

Table 2 Individual fruit production and fruit set as a function of population, NND and plant size

	Effect	Fruit production	Fruit set
A	Population	$F_{1,67} = 0.006$	$F_{1,67} = 0.020$
	NND	$F_{1,67} = 0.512$	$F_{1,67} = 4.807^{*}$
	Model R^2	n.s.	0.069
B	Population	$F_{1,66} = 0.500$	$F_{1,66} = 0.063$
	NND	$F_{1,66} = 3.015$	$F_{1,66} = 9.664^{**}$
	Plant size	$F_{1,66} = 125.026^{***}$	$F_{1,66} = 33.551^{***}$
	Model R^2	0.657	0.383

n.s. = not significant, *$P < 0.01$, **$P < 0.01$, ***$P < 0.001$

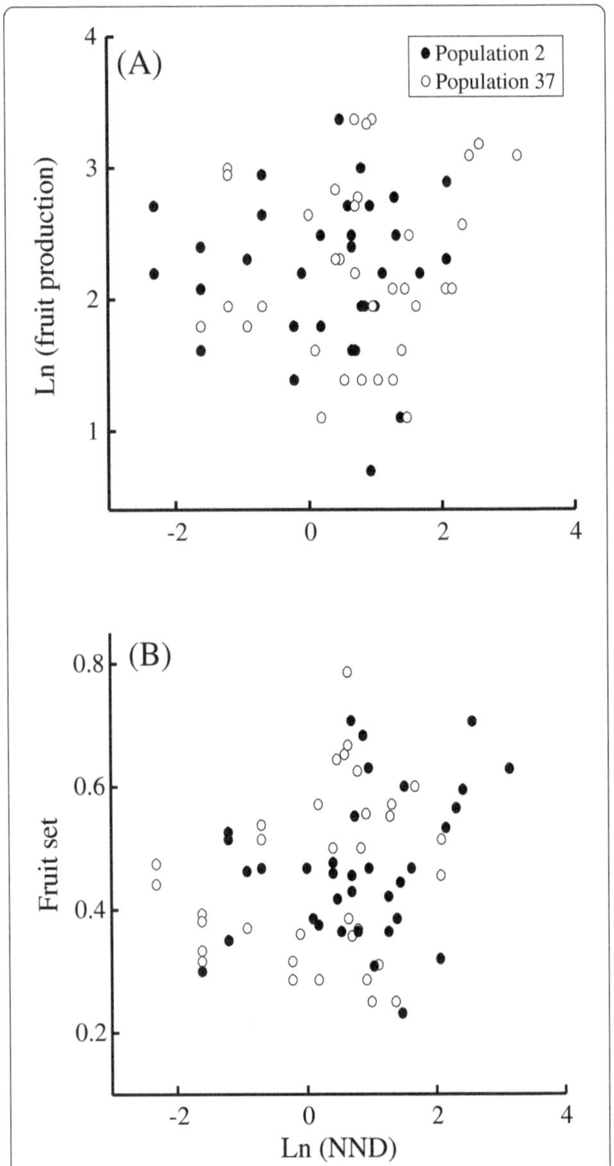

Fig. 2 Effects of NND on individual reproduction within populations of *C. agristis*. Relationships between NND and (**a**) individual fruit production and (**b**) individual fruit set. Effects of plant size were not accounted for

± 0.013, Table 2A, Fig. 2b). When individual plant size was included in the analysis of covariance, a positive effect of NND on individual fruit set remained ($b \pm s_b = 0.032 \pm 0.011$, Table 2B). As expected, individual plant size affected individual fruit number and individual fruit set (Table 2B). Furthermore, individual plant size was not significantly correlated to NND ($F_{1,67} = 0.150$, $P = 0.699$).

Effects of LNS

In populations 2 and 37, the mean 0.1 m LNS was 15.8 (SD = 10.6, range = 2–40) and 9.5 (SD = 10.0, range = 1–56), respectively. Similarly, the mean 0.2 m LNS was 24.3 (SD = 18.8, range = 3–72) and 12.3 (SD = 11.3, range = 1–45), respectively, while the mean 0.3 m LNS was 28.3 (SD = 28.0, range = 4–137) and 18.6 (SD = 24.5, range = 1–131), respectively. All pairwise contrasts between populations were significant ($P < 0.05$).

Individual fruit production was negatively correlated to 0.1 m LNS ($b \pm s_b = -0.269 \pm 0.075$, Table 3A, Fig. 3a), and was not correlated to 0.2 m LNS or 0.3 m LNS (Table 3A). When individual plant size was included in the analysis of covariance, individual fruit production was again negatively correlated to 0.1 m LNS ($b \pm s_b = -0.127 \pm 0.056$, Table 3B), and was not correlated to 0.2 m LNS or 0.3 m LNS (Table 3B). Furthermore, individual plant size was negatively correlated to 0.1 m LNS ($b \pm s_b = -0.096 \pm 0.032$, $F_{1,65} = 9.133$, $P = 0.004$), and was not correlated to 0.2 m LNS ($F_{1,65} = 3.775$, $P = 0.056$) or 0.3-m LNS ($F_{1,65} = 0.033$, $P = 0.856$).

Individual fruit set was negatively correlated to 0.1 m LNS ($b \pm s_b = -0.068 \pm 0.015$, Table 3A, Fig. 3b), and was not correlated to 0.2 m LNS or 0.3 m LNS (Table 3A). When individual plant size was included in the analysis of covariance, individual fruit set was again negatively

Table 3 Individual fruit production and fruit set as a function of population, LNS and plant size

	Effect	Fruit production	Fruit set
A	Population	$F_{1,65} = 0.555$	$F_{1,65} = 1.391$
	0.1 m LNS	$F_{1,65} = 15.213^{***}$	$F_{1,65} = 16.637^{***}$
	0.2 m LNS	$F_{1,65} = 2.807$	$F_{1,65} = 0.162$
	0.3 m LNS	$F_{1,65} = 1.019$	$F_{1,65} = 0.691$
	Model R^2	0.196	0.244
B	Population	$F_{1,64} = 2.227$	$F_{1,64} = 2.289$
	0.1 m LNS	$F_{1,64} = 5.252^{*}$	$F_{1,64} = 7.445^{**}$
	0.2 m LNS	$F_{1,64} = 0.055$	$F_{1,64} = 0.468$
	0.3 m LNS	$F_{1,64} = 1.924$	$F_{1,64} = 0.731$
	Plant size	$F_{1,64} = 102.676^{***}$	$F_{1,64} = 23.627^{***}$
	Model R^2	0.691	0.448

*$P < 0.05$, **$P < 0.01$, ***$P < 0.001$

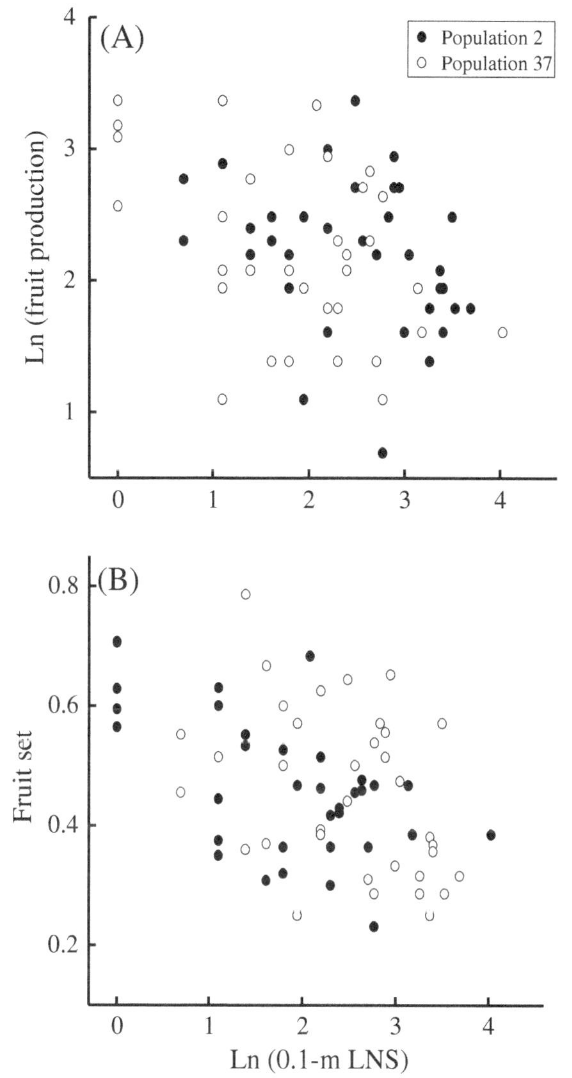

Fig. 3 Effects of 0–0.1 m LNS on individual reproduction within populations of *C. agristis*. Relationships between 0.1-m LNS and (**a**) individual fruit production and (**b**) individual fruit set. Effects of plant size were not accounted for

correlated to 0.1 m LNS ($b \pm s_b = -0.042 \pm 0.015$, Table 3B), and was not correlated to 0.2 m LNS or 0.3 m LNS (Table 3B).

Discussion

Effects of habitat fragmentation on reproduction at the population level

The effects of population size on reproduction have been widely studied (Ghazoul, 2005; Aguilar et al., 2006; Leimu et al., 2006). There is a large body of evidence indicating that plants of small populations have lower fruit or seed production than those of large populations because of insufficient pollen quantity and poor pollen quality (e.g. Ågren 1996; Tomimatsu and Ohara, 2002; Brys et al., 2004; Waites and Ågren, 2004; Hermansen et al., 2014). However, we did not detect such a pattern in *C. agristis* in either of the years studied, possibly because pollen quantity and quality are not related to its population size. No relationships between fruit or seed number per plant and population size have been reported in some studies (e.g. Alexandersson and Ågren, 1996; Molano-Flores et al. 1999; Spigler and Chang, 2008), suggesting that positive relationships between population size and either pollination or reproduction maybe not the rule. Alternatively, a lack of population size effects might arise if studies examining the effects of population size require the sampling of a wide range of population sizes. This explanation cannot be applied to our results, given that our study included, in both years of study, both populations with only a few individuals and populations with about 100, 000 individuals (Table 1).

In contrast to the vast majority of studies reporting a positive correlation between fruit or seed number per plant and population density (Ghazoul, 2005; Wagenius, 2006; Brys et al., 2008; Feldman, 2008), the effects of population density on mean population fruit number and fruit set in *C. agristis* varied between years, from negative in 2010 to neutral in 2011. As indicated in Table 1, Fig. 1b and Fig. 1d, some populations (9, 26 and 29) surveyed in 2010 were very dense, and it is possible that a negative correlation between population density and mean population fruit number or fruit set for *C. agristis* may be detectable only when dense populations are surveyed. Our findings support the idea that a 1-year study may not provide a realistic view of the effects of fragmentation on plant reproduction (Hobbs and Yates, 2003). However, when population plant size was accounted for, no relationship between population density and mean population fruit production or fruit set was evident in either year. These results suggest that the effects of population density on mean population fruit production and fruit set in 2010 were exerted indirectly via a negative effect on population plant size, and that pollination was independent of population density. A negative relationship between population density and mean population plant size might arise if competition for resources increases with density, reducing resource availability for plants in populations of high density (Harper, 1977).

Effects of habitat fragmentation on reproduction at the individual level within populations

Many studies have shown that the reproductive success of individual plants may decrease in response to increasing NND or decreasing LNS due to pollen limitation (Allison, 1990; Roll et al. 1997; Jakobsson et al., 2009; Caraballo-Ortiz et al., 2011). Contrary to these studies, individual fruit set of *C. agristis* in our study increased in response to increasing NND and decreased with increasing LNS at the 0.1 m spatial scale, and these relationships remained even when individual plant size was accounted for. These results are highly suggestive of intra-specific competition for resources: competition for resources decreased with increasing NND and increased with increasing LNS at the 0.1 m spatial scale during fruit maturation. Individual fruit production in *C. agristis* decreased with increasing LNS at the 0.1 m spatial scale but did not change in response to NND when individual plant size was accounted for; this indicates that competition for resources between the focal individual and the nearest conspecific individual may have been too weak to affect fruit production in the focal plants. Similarly, individual plant size of *C. agristis* decreased with increasing LNS at the 0.1 m spatial scale but did not change in response to NND, suggesting that intra-specific competition for resources between neighboring plants during growth was not strong enough to be detected when only the nearest conspecific was considered. These results also suggest that plants of *C. agristis* with fewer neighbors within a 0.1 m radius would benefit from not only a direct negative effect of LNS on reproductive success, but also from an indirect negative effect of LNS on reproductive success, namely via its effect on plant size. This is because large individuals typically have a higher fruit set than small individuals. However, these effects were not evident at the 0.2 m and 0.3 m spatial scales, suggesting that competition for resources in *C. agristis* occurs on very local scales.

Conclusions

The results of our study indicate that the reproductive success of *C. agristis* does not respond negatively to habitat fragmentation. Although previous studies on both agricultural and experimental populations have shown that high plant density may reduce fruit or seed production through competition and/or plant size (Harper, 1977; Antonovics and Levin, 1980; Weiner, 1982; Mustajärvi et al., 2001), our study is an instance of negative density-dependent reproduction occurring in a rare and endangered species in the wild. Furthermore, our findings highlight the view that any studies of the effects of fragmentation on reproduction should consider multiple spatial scales (Wagenius, 2006; Gunton and Kunin, 2007; Spigler and Chang, 2008).

Our study has implications for the conservation and management of *C. agristis*. First, our finding that *C. agristis* populations are distributed patchily on forest floors after fragmentation suggests that the maintenance of *C. agristis* habitat should be a priority. Second, small and sparse populations are as important as large and dense populations from a conservation perspective, because the former will be at least as successful, if not more, at reproduction, compared to large and dense populations. Third, a program aimed at re-establishing populations of this species should consider increasing the spacing between plants (compared to populations in the wild), which may reduce competition for resources and thereby increase fruit production.

Competing interests

The authors declare that they have no competing interests.

Authors' contributions

CGX conceived and designed the experiments. ZJC, LJ, YCP performed the experiments. CGX analyzed the data. CGX, ZJC wrote the manuscript. All authors read and approved the final manuscript.

Acknowledgements

This research was supported by the National Natural Science Foundation of China (Project no. 30870388, 31370436).

References

Ågren J (1996) Population size, pollinator limitation, and seed set in the self-incompatible herb Lytrhum salicaria. Ecology 77:1779–1790
Aguilar R, Ashworth L, Galetto L, Aizen MA (2006) Plant reproductive susceptibility to habitat fragmentation: review and synthesis through a meta-analysis. Ecol Lett 9:968–980
Alexandersson R, Ågren J (1996) Population size, pollinator visitation and fruit production in the deceptive orchid Calypso bulbosa. Oecologia 107:533–540
Allison TD (1990) Pollen production and plant density affect pollination and seed production in Taxus canadensis. Ecology 71:516–522
Antonovics J, Levin DA (1980) The ecological and genetic consequences of density-dependent regulation in plants. Annu Rev Ecol Syst 11:411–452
Baker HG (1967) Support for Baker's law-as a rule. Evolution 21:853–856
Barrett SCH, Kohn JR (1991) Genetic and evolutionary consequences of small population size in plants: implication for conservation. In: Falk DA, Holsinger KE (eds) Genetics and Conservation of Rare Plants. Oxford University Press, New York
Bosch M, Waser NM (1999) Effects of local density on pollination and reproduction in Delphinium nuttallianum and Aconitum columbianum (Ranuculaceae). Amer J Bot 86:871–879
Brys R, Jacquemyn H, Endels P, Van Rossum F, Hermy M, Triest L, de Bruyn L, Blust GDE (2004) Reduced reproductive success in small populations of the self-incompatible Primula vulgaris. J Ecol 92:5–14
Brys R, Jacquemyn H, Hermy M (2008) Pollination efficiency and reproductive patterns in relation to local plant density, population size, and floral display in the rewarding Listera ovata (Orchidaceae). Bot J Linn Soc 157:713–721
Caraballo-Ortiz MA, Santiago-Valentín E, Carlo TA (2011) Flower number and distance to neighbours affect the fecundity of Goetzea elegans (Solanaceae). J Trop Ecol 27:521–528
Eriksson O, Ehrlén J (2001) Landscape fragmentation and the viability of plant populations. In: Silvertown J, Antonovics J (eds) Integrating Ecology and Evolution in a Spatial Context. Blackwell Science, Oxford
Feldman TS (2008) The plot thickens: does low density affect visitation and reproductive success in a perennial herb, and are these effects altered in the presence of a co-flowering species? Oecologia 156:807–807
Fu LK, Jin JM (1992) China Plant Red Data Book: Rare and Endangered Plants. Science Press, Beijing
Ghazoul J (2005) Pollen and seed dispersal among dispersed plants. Biol Rev 80:413–443
Glémin S, Petit C, Maurice S, Mignot A (2008) Consequences of low mate availability in the rare self-incompatible species Brassica insularis. Conserv Biol 22:216–221
Gunton RM, Kunin WE (2007) Density effects at multiple scales in an experimental plant population. J Ecol 95:435–445
Harper JL (1977) Population Biology of Plants. Academic, New York
Hermansen TD, Ayre DJ, Minchinton TE (2014) Effects of stand size on pollination in temperate populations of the mangrove Avicennia marina. Plant Ecol 215:1153–1162

Hobbs RJ, Yates CJ (2003) Impacts of ecosystem fragmentation on plant populations: generalizing the idiosyncratic. Aust J Bot 51:471–488
Hu ZH, Yang J, Jing RQ, Dong ZM (1990) Morphology studies on Circaeaster agrestis II. Morphology and anatomy of flower, fruit and seed. Cathaya 2:77–88
Jakobsson A, Lázaro A, Totland Ø (2009) Relationships between the floral neighborhood and individual pollen limitation in two self-incompatible herbs. Oecologia 160:707–719
Johnson SD, Hollens H, Kulhmann M (2012) Competition versus facilitation: conspecific effects on pollinator visitation and seed set in the iris Lapeirousia oreogena. Oikos 121:545–550
Jones FA, Comita LS (2008) Neighbourhood density and genetic relatedness interact to determine fruit set and abortion rates in a continuous tropical tree population. Proc R Soc Lond B 275:2759–2767
Karron JD (1995) The influence of population density on outcrossing rates in Mimulus ringens. Heredity 75:175–180
Klinkhamer PGL, de Jong TJ (1990) Effects of plant size, plant density and sex differential nectar reward on pollinator visitation in the protandrous Echium vulgare (Boraginaceae). Oikos 57:399–405
Kunin WE (1997) Population size and density effects in pollination: pollinator foraging and plant reproductive success in experimental arrays of Brassica kaber. J Ecol 85:225–234
Lawes MJ, Taplin P, Bellairs SM, Franklin DC (2013) A trade-off in stand size effects in the reproductive biology of a declining tropical conifer Callitris intratropica. Plant Ecol 214:169–174
Leimu R, Mutikainen P, Koricheva J, Fischer M (2006) How general are positive relationships between plant population size, fitness and genetic variation? J Ecol 94:942–952
Metcalfe DB, Kunin WE (2006) The effects of plant density upon pollination success, reproductive effort and fruit parasitism in Cistus ladanifer L. (Cistaceae). Plant Ecol 185:41–47
Mitchell RJ, Ankeny DP (2001) Effects of local conspecific density on reproductive success in Penstemon digitalis and Hesperis matronalis. Ohio J Sci 101:22–27
Molano-Flores B, Hendrix SD, Heard SB (1999) The effect of population size on stigma pollen load, fruit set, and seed set in Allium stellatum ker. (Liliaceae). Int J Plant Sci 160:753–757
Morgan JW (1999) Effects of population size on seed production and germinability in an endangered, fragmented grassland plant. Conserv Biol 13:266–273
Mustajärvi K, Siikamäki P, Rytkönen S, Lammi A (2001) Consequences of plant population size and density for plant–pollinator interactions and plant performance. J Ecol 89:80–87
Oostermeijer JGB (2003) Threats to rare plant persistence. In: Brigham CA, Schwartz MW (eds) Population Viability in Plants: Conservation, Management, and Modeling of Rare Plants. Springer, Berlin
Quinn GP, Keough MJ (2002) Experimental design and data analysis for biologists. Cambridge University Press, Cambridge
Roll J, Mitchell RJ, Cabin RJ, Marshall DL (1997) Reproductive success increases with local density of conspecifics in a desert mustard (Lesquerella fenleri). Conserv Biol 11:738–746
Saunders DA, Hobbs RJ, Margules CR (1991) Biological consequences of ecosystem fragmentation: a review. Conserv Biol 5:18–32
Spigler RB, Chang SM (2008) Effects of plant abundance on reproductive success in the biennial Sabatia angularis (Gentianaceae): spatial scale matters. J Ecol 96:323–333
Stein K, Templin M, Hensen I, Fischer M, Matthies D, Schleuning M (2013) Negative effects of conspecific floral density on fruit set of two neotropical understory plants. Biotropica 45:325–332
Steven JC, Rooney TP, Boyle OD, Waller DM (2003) Density dependent pollinator visitation and self-incompatibility in upper Great Lakes populations of Trillium grandiflorum. J Torrey Bot Soc 130:23–29
Suarez-Gonzalez A, Good SV (2014) Pollen limitation and reduced reproductive success are associated with local genetic effects in Prunus virginiana, a widely distributed self-incompatible shrub. Ann Bot 113:595–605
Tian XH, Zhao L, Ren Y, Zhang XH (2007) Number of floral organs in Circaeaster agrestis (Circaeasteraceae) and possible homeosis among floral organs. Plant Syst Evol 265:259–265
Tomimatsu H, Ohara M (2002) Effects of forest fragmentation on seed production of the understory herb trillium camschatcense. Conserv Biol 16:1277–1285
Van Treuren R, Bulsma R, Ouborg NJ, Van Delden W (1993) The effects of population size and plant density on outcrossing rates in locally endangered Salvia pratensis. Evolution 47:1094–1104
Wagenius S (2006) Scale dependence of reproductive failure in fragmented Echinacea populations. Ecology 87:931–941
Waites AR, Ågren J (2004) Pollinator visitation, stigmatic pollen loads, and among-population variation in seed set in Lythrum salicaria. J Ecol 92:512–526
Ward M, Johnson SD, Zalucki MP (2013) When bigger is not better: intraspecific competition for pollination increases with population size in invasive milkweeds. Oecologia 171:883–891
Weber A, Kolb A (2013) Local plant density, pollination and trait–fitness relationships in a perennial herb. Plant Biol 15:335–343
Weiner J (1982) A neighborhood model of annual-plant interference. Ecology 63:1237–1241
Wilcove DS, McLellan CH, Dobson AP (1986) Habitat fragmentation in the temperate zone. Conserv Biol 6:237–256

Investigation of metabolites accumulation in medical plant *Gentiana rigescens* during different growing stage using LC-MS/MS and FT-IR

Yu Pan[1,2], Ji Zhang[2], Yan-Li Zhao[2], Yuan-Zhong Wang[2*] and Heng-Yu Huang[1*]

Abstract

Background: *Gentiana rigescens*, an important medicinal plant in China, has been widely cultivated in Yunnan province, China. Previous studies were focused on analysis and determination of the metabolites isolated from this species, the accumulation of these metabolites during growth period are not yet clear. In this study, samples for the experiments were obtained by tissue culture. FT-IR and LC-MS/MS method were performed to distinguish the variation on the major metabolites in *G. rigescens* during growing stage when combined with chemometrics.

Results: Methodology validations were all within the required limits. The metabolites were visually different in tissue culture samples and mature plants. The diversity of metabolites increased proportionally with plant growth. The quantitative analysis showed the content of gentiopicroside was significantly vary during different growing stage. The highest content of gentiopicroside (122.93 ± 7.01 mg/g) was detected in leaf of regenerated plantlet, whereas its content in root significantly increased along with underground parts growth. Moreover, flavonoids mainly distributed in aerial parts showed potential competitive relationship during plant growth.

Conclusion: The distribution and accumulation of metabolites are associated with different parts and plant growth, which provide potential evidences for the rational application and exploitation of *G. rigescens*.

Keywords: *Gentiana rigescens*; Tissue culture; Metabolites variation; LC-MS/MS; FT-IR

Background

Gentiana rigescens Franch belonging to Gentianaceae is an important medicinal plant in China for treatment of hepatitis, jaundice and dysentery. The root and rhizome of this plant has been officially documented in Chinese Pharmacopoeia as one of raw materials of Gentianae Radix et Rhizoma (Longdan in Chinese), a hepatoprotective agent (State Pharmacopoeia Commission 2010). According to ethnobotanical information, its aerial parts are also used as folk medicine for treatment of fever and rheumatic arthritis or as antiophidica, when prepared with vegetable oil. The phytochemistry and bioactivities of *G. rigescens* have been intensively studied. More than 100 secondary metabolites with different activities like hepatoprotective, anti-inflammatory, antioxidant and neuritogenic growth have been isolated from this plant (Gao et al. 2010a, 2010b; Xu et al. 2006, 2007, 2009a; Wang et al. 2012). Among them, iridoid glycosides is the most abundant components especially gentiopicroside which content is more than 4.5 % and serve as major active ingredient and standard for quality control (Jiang et al. 2005; Pan et al. 2015a; Wang et al. 2012).

As one of well-known traditional Chinese medicine with remarkable medicinal functions, the wild resources have been under heavy threat owing to human activities and environmental pollution. Although *G. rigescens* have been extensively planted in Yunnan, some disadvantages on cultivation such as continuous cropping obstacle, time-consuming and laborintensive, etc. result in decline of production and quality. Fortunately, plant tissue culture is conducive to the accumulation of biomass and metabolites, in particular with individual metabolites, which amount is multifold higher than control group when treated with appropriate elicitors (Chuang et al. 2014; Huang et al. 2014; Kuzovkina et al. 2014; Kumari

* Correspondence: boletus@126.com; hhyhhy96@163.com
[2]Institute of Medicinal Plants, Yunnan Academy of Agricultural Sciences, Kunming, China
[1]College of Traditional Chinese Medicine, Yunnan University of Traditional Chinese Medicine, Kunming, China

et al. 2015; Marsh et al. 2014; Su et al. 2014). However, the efficacy of medical plant, to a large extend, are attributed to synergistic effect of a number of metabolites. The amount of individual metabolites is significantly increased, which might lead to different pharmacological activities and therapeutic effects.

On the other hand, previous studies (Jiang et al. 2005; Pan et al. 2014, 2015a; Wang et al. 2012) on metabolites of *G. rigescens* were most focused on quality assessment, phytochemistry, pharmacology, etc. However, metabolites are not only the efficacious properties for maintaining human health, but also play an important role for resistance to abiotic and biotic threats during plant growth (Hall et al. 2008; Wink 2003). To our best knowledge, the accumulation and variation of metabolites in *G. rigescens* have not yet clear.

Currently, analysis of metabolites based on separation technologies such as gas chromatography–mass spectrometry (Hu et al. 2014), liquid chromatography coupled with photodiode array detector (Yu et al. 2014), mass spectrometry detector (Won et al. 2014) and nuclear magnetic resonance spectroscopy (Hilbert et al. 2015) can rapidly provide complex chemical information and clarify the similarities and differences of bio-samples when combined with chemometrics. However, comprehensive chemical information on metabolites cannot be analyzed in a single chromatogram. Fourier transform infrared spectroscopy (FT-IR) enables to rapid reflect holistic molecular structure-analyte relationships, which is considered as a well-established and non-destructive method for analysis of bio-sample, whereas it fails to recognise the variation of specific compound in sample due to the limited specificity and sensitivity (Karoui et al. 2010; Lohumi et al. 2014; Zhao et al. 2014).

In this study, the variation on distribution and accumulation of metabolites in *G. rigescens* are investigated based on plant tissue culture. Individual parts of sample during different growing stage are subjected to targeted and non-targeted analysis using FT-IR and liquid chromatography tandem mass spectrometry (LC-MS/MS). Moreover, the biosynthetic pathway of iridoid glycosides is also discussed. The combinative comparison approach can reflect the overall chemical difference during different growing stage, which may provide the useful information for reasonable utilization of resources.

Methods

Materials and chemicals

Tissue culture materials provide by Dr. Heng-Yu Huang (College of Traditional Chinese Medicine, Yunnan University of Traditional Chinese Medicine) were established using leaves of *G. rigescens.* The operations of tissue culture were completed on a super-clean bench. The plantlets were grown on Murashige-Skoog (MS) medium supplemented different concentration of indoleacetic acid (IAA), zeatin(ZT), dichlorphenoxyacetic acid (2,4-D), kinetin (KT) and 6-benzyladenine (BA) during different stage. The processed of tissue culture is present in Fig. 1. The culturing conditions were incubated at 23 ± 2 °C under cool white fluorescent light at 1500–2000 lx under 10 h per day.

KBr (specpure) was purchased from Tianjin FengChuan Fine Chemical Research Institute (Tianjin, China). Methanol and formic acid (assigned purity > 98 %) were LC grade and purchased from Thermo Fisher Scientific (USA) and Dikmapure (USA), respectively. Water was purified to 18.25 MΩ using Milli-Q system from Millipore (USA). All other chemicals for extraction were analytical grade. The standard compounds (**1.** loganic acid, **2.** swertiamarin, **3.**

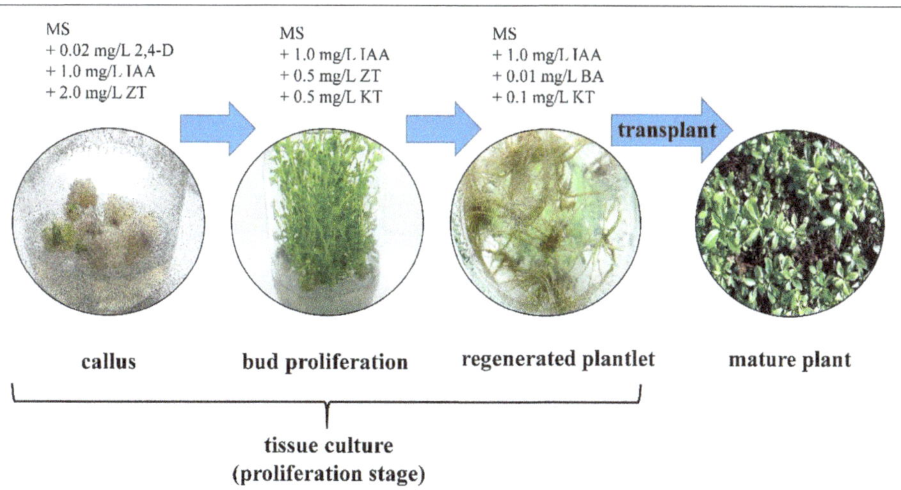

Fig. 1 Tissue culture flowchart of *G. rigescens*

gentiopicroside, **4**. sweroside, **5**. isoorientin and **6**. isovitexin) (Fig. 2) were provided by Chinese National Institute for the Control of Pharmaceutical and Biological Products (Beijing, China). Their assigned purity were all >98 %.

Apparatus

Tablet press (YP-2, Shanghai Shanyue instrument Inc) was used to press powder into thin samples. The FT-IR spectrometer (PerkinElmer, USA) equipped with a DTGS detector. IR spectra were recorded from the accumulation of 16 scans in 4000–400 cm^{-1} range with a resolution of 4 cm^{-1}.

Separation, quantitation and quantification of metabolites were performed on a Shimadzu Nexera UHPLC tandem mass spectrometry (LCMS-8030, Shimadzu, Japan) equipped with a Shim-pack XR-ODS III (75 × 2.0 mm, 1.6 μm) column, UV detection and triple quadrupole mass spectrometer via an electrospray ionization (ESI) interface. The mobile phase consisted of 0.1 % formic acid in water (A) and methanol (B) was applied at a flow rate of 0.35 mL/min with gradient as follow: initial, 13 % B; 0.31-7.00 min 20 % B linear, 7.01-13.00 min 46 % B linear, 13.01-16.50 min 83 % B linear; followed by a final increase to 90 % in 1 min. After a 3 min wash, the column was reconditioned at 13 % B for 3 min to prepare for the next injection. Injection volume and column temperature were 1 μL and 40 °C, respectively. The detection wavelength was set at 242 nm, where all the standards and UPLC profiling showed a satisfactory performance.

High-resolution electrospray ionization mass spectrometry was performed using a Agilent G6550 QTOF (Agilent technologies Santa Clara, CA, USA) equipped with an ESI inter-face. Mass spectra were acquired in both positive and negative modes over the range *m/z* 100–1000. The capillary voltages were set at 3000 V (positive mode) and 2700 V (negative mode), respectively, and nozzle voltage was 300 V. Sheath gas and drying gas were nitrogen at a flow rate of 3.0 and 14.0 L/min, respectively, nebulizer pressure was 20 psi. The precise molecular mass was determined by the accurate-mass data of the TOF analyzer within a reasonable degree of measurement error, normally with mass errors below 5 ppm in routine analysis, which was sufficient to verify the elemental compositions of the known constituents in *G. rigescens*.

The quantification of the targeted compounds with low concentration was carried out on multiple reaction monitoring (MRM). The settings of MRM were auto-optimized by Labsolutions software (Shimadzu, Japan). The triple quadrupole mass spectrometer parameters were set as follows: nebulizing gas and drying gas were nitrogen at a flow rate of 3.0 and 15.0 L/min, respectively; the interface voltage was set to 4.5 kV; desolvation line (DL) temperature was 250 °C and the heat block temperature was 400 °C. Reference solutions containing ions of *m/z* 503.15 and 1004.60 were continuously introduced into the MS system during the analysis procedure to ensure the accuracy of the measured mass.

Sample preparation

Fresh samples were dried at 60 °C and ground into fine powder. Then, 2 mg sample was blended with 200 mg KBr powder, ground again and pressed into a tablet. The FT-IR spectra of all samples were collected three times after by subtraction of KBr pellet background.

Sample preparation was based on our previous work (Pan et al. 2015b). An accurately weighed sample powder

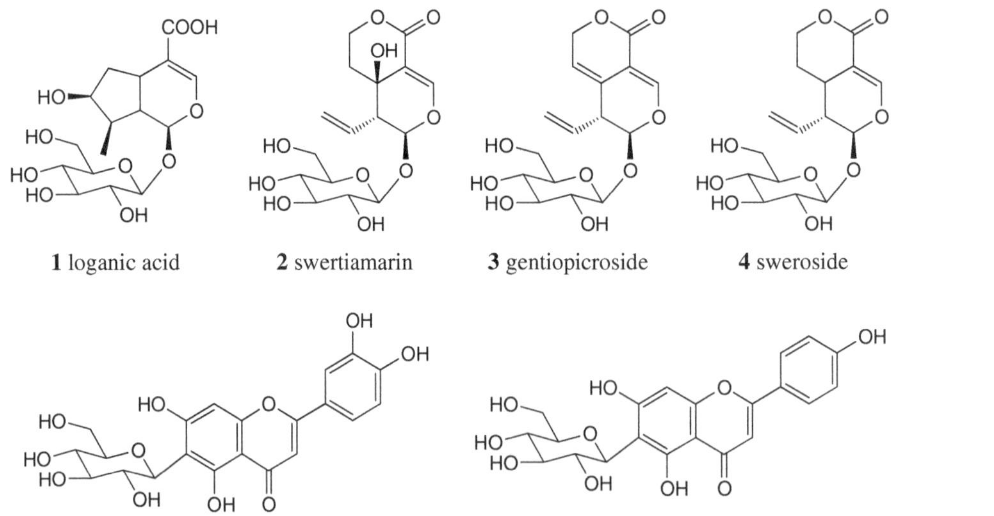

1 loganic acid **2** swertiamarin **3** gentiopicroside **4** sweroside **5** isoorientin **6** isovitexin

Fig. 2 Index compounds of *G. rigescens*. (1. loganic acid; 2. swertiamarin; 3. gentiopicroside; 4. sweroside; 5. isoorientin; 6. isovitexin)

(0.1 g) was extracted by ultrasonication with 7 mL 80 % methanol for 35 min. The extract solution were filtered through a paper filter. Then, the filtrate were stored at 4 °C and filtered through a 0.22 μm membrane filter before injection into the LC system for analysis. Injection volume was 1 μL.

Data analysis

The raw FT-IR spectroscopy were processed by Omnic 8.0 (Thermo Fisher Scientific, USA). The peaks of UV and mass data were picked and filtered by Labsolutions software (Shimadzu, Japan). Principal component analysis (PCA), an unsupervised chemometric approach for classification, was exploited to optimize the complex data set for reflecting relationships among different samples. The PCA was performed by software SIMCA-P^{+}10.0 (Umetrics AB, Sweden).

Results

Comparative analysis of *G. rigescens* in different growing stage

FT-IR and LC-MS/MS were used to investigate the metabolites variation in both integrity and detail on chemical information in different plant part and growing stage. There are few visual differences in averaged FT-IR spectra of different sample. In order to explore the relationships between metabolites and plant growth, the 1800–600 cm^{-1} region of the FT-IR spectral data without the interferences of CO_2 and H_2O were subjected to PCA. A two-dimension (2D) scores plot (PC1 × PC2) was constructed from a data matrix (623 × 54) by PCA which could visually reflect the similarity between IR spectra and samples in this plot where the closer the points, the more similar the spectral data. In PCA, the first and second principal components cumulatively accounted for 92.8 % of the total variance, which suggested that the former two principal components could explain the proportion of the experimental data. In Fig. 3, sample of mature plant and samples during proliferation stage were explicitly separated into two groups. However, samples in the corresponding groups were crossed and could not be classified according to their plant parts or growing stage. These results implied that the whole metabolome based on FT-IR spectra were significant different, whereas these the detailed variation failed to be monitored especially for content of individual metabolites.

Moreover, a targeted method based on LC-MS/MS was designed for monitoring variations of iridoid glycosides and flavonoids. The 80 % MeOH extracts of *G. rigescens* were injected into LC-MS/MS system for analysis. As shown in Fig. 4a-i, peak numbers and their peak areas were visually different in samples, especially for peaks a, 3, 5, 6, b and c, which indicated the accumulation of metabolites varied with plant growth. For example, peak b and c were only found in leaf of mature plant. Furthermore, peak a is one of characteristic marker in chromatogram of callus, whereas it could not detected in chromatogram of mature plant.

Except for peak 1–6 which can be unambiguously identified by standard compounds, other peaks in chromatogram are still unknown. In order to further clarify metabolites variations during growing stage, peak a-g (Additional file 1: Table S1), characteristic markers in chromatogram of samples, were tentatively identified via matching the mass data (high revolution data and MS/MS spectra) with published works on chemical compounds isolated from *G. rigescens*. According to comparison of mass data of standard compounds, peak

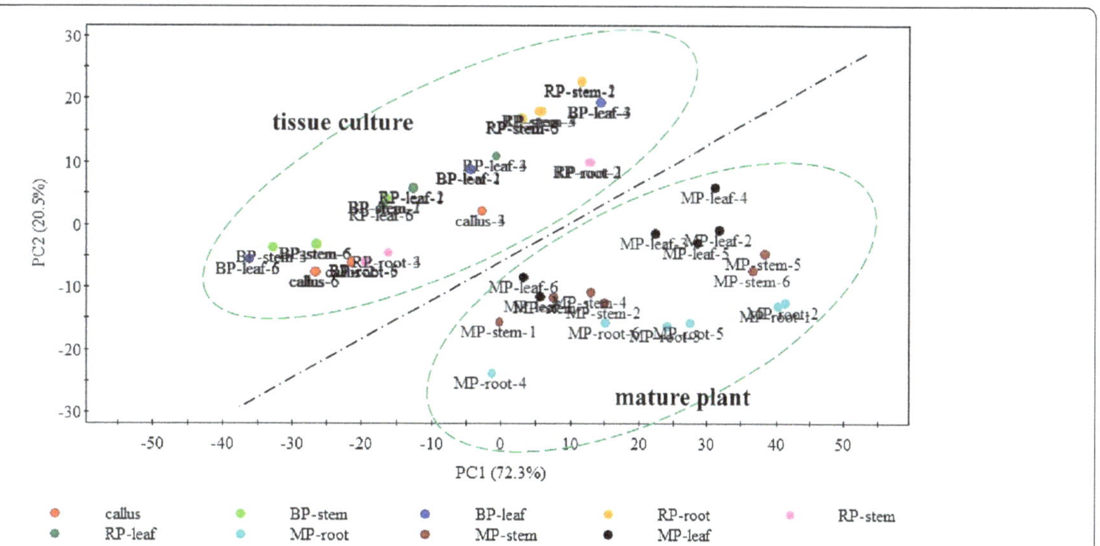

Fig. 3 The PCA scores plot of samples during different stage based on FT-IR

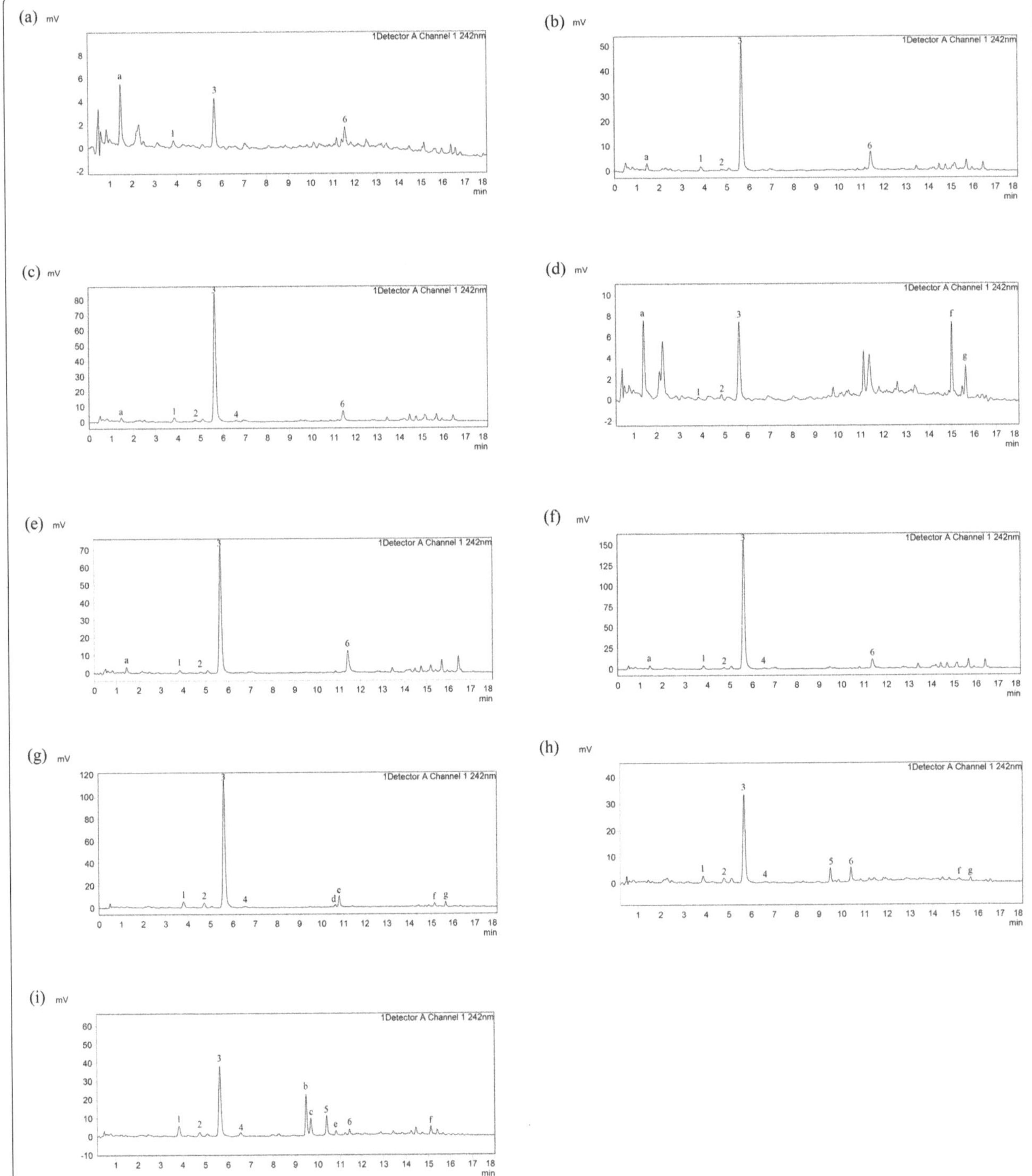

Fig. 4 UPLC profiling of *G. rigescens.* (1. loganic acid; 2. swertiamarin; 3. gentiopicroside; 4. sweroside; 5. isoorientin; 6. isovitexin; **a**: callus; **b**: BP-stem; **c**: BP-leaf; **d**: RP-root; **e**: RP-stem; **f**: RP-leaf; **g**: MP-root **h**: MP-stem; **i**: MP-leaf)

b and c could be tentatively assigned as flavonoids while peak d-g were iridoid glycosides. In mass spectra of isoorientin and isovitexin (Additional file 1: Figure S1, see supplement data), the characteristic neutral loss of 90 and 120 Da are correspond to *C*-glycosidic structure, which could be considered as diagnosed markers. Interestingly, the mass data of peak b and c (Additional file 1: Figure S1) were highly similar with isoorientin and isovitexin, respectively. Besides, the neutral loss of 162 Da, the feature of glucosyl group, were detected in both peak b and c. Therefore, peak b (m/z 593.1513 $[M\text{-}H]^-$, $C_{27}H_{30}O_{15}$) and c (m/z 611.1503 $[M\text{-}H]^-$ $C_{27}H_{30}O_{16}$) were the *O*-glycosidic derivatives of isoorientin and isovitexin, respectively.

Furthermore, a neutral loss of 136 Da which may correspond to the loss of a dihydroxy benzoyl group via a classical McLafferty-type rearrangement (Tan et al. 1996; Xu et al. 2009b) was observed in mass spectrum of peak d-g. By further analyzing the mass data, peak d-g were dihydroxy benzoyl iridoid glycosides. Their mass spectrum and fragmentation pattern are shown in Additional file 1: Figure S2. In mass spectrum of peak a, the product ions at m/z 153 and 109 were correspond to losing a glucose (162 Da) and carboxyl group (44 Da), indicating peak a (m/z 315.0832 $[M\text{-}H]^-$ $C_{13}H_{16}O_9$) were tentatively assigned as *O*-glucosyl-dihydroxy benzoyl acid (Additional file 1: Figure S3) (Xu et al. 2009a). All compounds but peak 1–6 were identified by matching corresponding mass data in previous studies.

Method validation and quantification

Standard solutions of each compounds with seven different concentrations were prepared individually in methanol and were injected into LC-MS/MS system for generation of external standard calibration curves. Calibration curves for each compound were performed by plotting the peak area (y) against the concentrations (x, μg/ml). The correlation coefficients (R^2) of each calibration curve were more than 0.9992. The limits of detection (LOD) and quantification (LOQ), S/N (signal-to-noise ratio) of 3 and 10, were determined by serial dilution of each standard solution using the described conditions. The LOD and LOQ for UV detector and mass analyzer, together with standard calibration curves of each standards, were listed in Table 1.

Precision was evaluated by intra- and inter-day variation which determined by analyzing mixed standard solutions with known concentration six times within a day and on three consecutive days in triplicate. The intra- and inter-day precision of peak area (expressed in terms of %RSD) were in the range of 0.76–2.27 %. Accuracy was validated by recovery test performed by accurately adding three different amounts (low, medium and high spike) of each standard to the crude sample. The recovery rates of six standards were ranged from 97.3-102.6 % and their RSD values were less than 3 %. Then, six independently samples analysed by repeating the described produce of sample preparation under this chromatographic condition were used to investigate the repeatability. These results are displayed in Table 2.

In this study, the identification of standards in the chromatogram was confirmed by retention times and ion pairs determined by MRM. Quantification depended on the external standard method. The contents of the six standards in *G. rigescens* during different rowing stage are listed in Table 3. The results showed that the content of the six standards differed greatly in different stage. The lowest contents of all standards were found in callus. Gentiopicroside was the highest yield compound in the whole growth stage. The highest gentiopicroside yield was found in leaf of regenerated plantlet (122.93 ± 7.01 mg/g), followed by root of mature plant (96.78 ± 8.54 mg/g). Interestingly, gentiopicroside yield was reduced and tended to accumulate in root after regenerated plantlet. Furthermore, root did not contain the two flavonoids. The highest content of the two flavonoids were observed in leaf. For isoorientin, the highest content was found in mature plants. On the contrary, the highest isovitexin yield was detected in proliferation stage.

Table 1 Linear regression and MRM parameters of standards

Analytes	Regression equation	Linearity range (μg/mL)	R^2	LOD (μg/mL)	LOQ (μg/mL)	Ions pairs	CE (eV)
Loganic acid	y = 2099.0x + 19518.7	50-500	0.9996	0.34	1.01	421 > 375	27
Loganic acid[a]	y = 200150.0x + 2801.4	0.05-5	0.9997	0.004	0.022	421 > 213	22
Swertiamarin	y = 2151.2x + 11429.0	10-150	0.9998	0.27	0.92	419 > 179	18
Swertiamarin[a]	y = 169057.0x + 3753.2	0.1-5	0.9995	0.005	0.027	419 > 141	23
Gentiopicroside	y = 1494.4x + 21722.0	5-1500	0.9995	0.23	1.03	401 > 197	14
-	-	-	-	-	-	401 > 89	21
Sweroside[a]	y = 292679.0x + 5034.6	0.5-10	0.9992	0.008	0.037	359 > 197	−31
-	-	-	-	-	-	359 > 127	−10
Isoorientin	y = 2945.4x-6172.4	5-250	0.9993	0.57	2.33	447 > 327	27
Isoorientin[a]	y = 278957.0x + 4532.8	0.05-5	0.9994	0.007	0.041	447 > 357	23
Isovitexin	y = 1979.7x + 19518.7	5-250	0.9998	0.63	1.97	431 > 311	24
Isovitexin[a]	y = 409813.0x + 1523.7	0.05-5	0.9996	0.007	0.037	431 > 331	22

-:not mentioned
[a]:determination by MRM

Table 2 Precision accuracy and repeatability (%RSD) of this method

Analytes	Intra-day RSD%			Inter-day RSD%	Accuracy ($n = 9$)		Repeatability ($n = 6$)
	Day 1	Day 2	Day 3				
	P_a	P_a	P_a	P_a	Mean recovery	RSD%	RSD%
UV detector							
Loganic acid	1.13	2.09	2.36	1.58	102.6 %	1.93	1.79
Swertiamarin	1.53	1.72	1.97	1.46	101.8 %	2.62	1.88
Gentiopicroside	1.74	1.85	2.27	1.38	98.7 %	1.77	1.69
Isoorientin	1.22	2.21	2.33	2.25	102.2 %	2.56	1.71
Isovitexin	1.27	1.85	1.76	1.64	97.3 %	2.13	2.07
MRM							
Loganic acid	1.26	1.58	1.71	1.57	99.3 %	1.99	1.86
Swertiamarin	1.45	1.31	1.67	1.78	100.9 %	1.86	1.22
Sweroside	1.48	1.37	1.94	0.76	101.4	1.78	1.85
Isoorientin	0.95	1.08	1.88	1.21	98.7 %	2.14	1.29
Isovitexin	1.66	1.33	1.92	1.59	98.9 %	1.98	2.24

P_a: peak areas

Discussions

According to the results of comparative analysis, the distribution and accumulation of metabolites are associated with plant growth. The whole metabolome in different stage are significantly different. The reason might be attributed to (1) elicitors in medium and (2) environment conditions such as soil, sun exposure time and rainfall, which could also result in fluctuation on the distribution and accumulation of metabolites after transplant (Manukyan 2011; Marsh et al. 2014; Xie et al. 2011). Additionally, the proportion among metabolites vary with plant growth. The proportion of peak a (*O*-glucosyl-dihydroxy benzoyl acid) in chromatogram gradually reduce with the increase of other metabolites. Based on the identification of peak d-g, *O*-glucosyl-dihydroxy benzoyl acid may transformed into dihydroxy benzoyl iridoid glycosides via esterification of iridoid glycosides. The obvious differences are observed in chromatogram of leaf during different stage especially for peak b and c (*O*-glycosidic derivatives of isoorientin and isovitexin) which only is detected in mature plant. These result implied that the synthesis and transform of metabolites with complex molecular structure tend to be in mature plant with more abundant substance when compared with plant during in proliferation stage.

Gentiopicroside is the characteristic compound with the highest amount in *G. rigescens*, which serve as standard for quality control of *G. rigescens* (State Pharmacopoeia Commission 2010; Wang et al. 2012). Combined with previous study on *Gentiana scabra* (Huang et al. 2014), another raw materials of Gentianae Radix et Rhizoma, 1.8 higher-fold gentiopicroside content was observed in sample of hair root culture than plants grown in greenhouse. Although leaf in regenerated plantlet contains the highest gentiopicroside yield in this study, it can only be used for industrial extraction of gentiopicroside rather than medical application because the efficacy of herb medicine, to large extend, is derived from synergistic effect of metabolites.

Table 3 Mean Contents (mg/g) of six standards ($n = 6$)

Analytes	Loganic acid	Swertiamarin	Gentiopicroside	Sweroside	Isoorientin	Isovitexin
Callus	0.053 ± 0.002^a	0.017 ± 0.001^a	0.35 ± 0.25	0.05 ± 0.001^a	0.013 ± 0.002^a	0.075 ± 0.01^a
BP-stem	3.12 ± 0.17	1.66 ± 0.22	45.11 ± 2.21	0.06 ± 0.001^a	0.016 ± 0.002^a	2.25 ± 0.62
BP-leaf	3.73 ± 0.28	2.78 ± 0.44	85.13 ± 3.93	0.81 ± 0.04^a	0.021 ± 0.002^a	2.17 ± 0.58
RP-root	0.034 ± 0.005^a	0.038 ± 0.005^a	0.55 ± 0.03	0.03 ± 0.007^a	-	-
RP-stem	3.66 ± 0.02	2.65 ± 0.02	62.13 ± 5.66	0.36 ± 0.03^a	0.036 ± 0.001^a	2.87 ± 0.72
RP-leaf	4.55 ± 0.72	2.83 ± 0.14	122.93 ± 7.01	0.41 ± 0.04^a	0.047 ± 0.002^a	3.07 ± 0.33
MP-root	8.52 ± 0.59	6.16 ± 0.72	96.78 ± 8.54	0.27 ± 0.07^a	-	-
MP-stem	4.90 ± 0.11	3.06 ± 0.25	17.58 ± 3.21	0.46 ± 0.02^a	$1.58 \pm 0.32.$	0.83 ± 0.02
MP-leaf	5.51 ± 0.21	4.11 ± 0.21	26.05 ± 4.33	1.16 ± 0.13^a	2.84 ± 0.52	0.85 ± 0.22

[a]:Determination by MRM

The variation on the accumulation of gentiopicroside in different stage is present in Fig. 5a. An interesting phenomenon that gentiopicroside content in root is significant raised with evidently decrease in leaf and stem, when the root start to regenerate. During this stage, root growth rate is far higher than in stem and leaf with the occurrence of gentiopicroside growth, which could be consistent with the results of specific-tissue analysis where the distribution of secondary metabolites vary in different tissue. Moreover, gentiopicroside in aerial parts may translocate into root or transform into other metabolites when the root start to regenerate.

In leaf, a significant negative correlation is found between isoorientin and isovitexin (Fig. 5b). From the biosynthetic pathway of view, it can be explain that these two flavonoids are competitive relationship in a common biosynthetic pathway. Additionally, isovitexin can also be considered as a precursor of isoorientin and *O*-glycosidic isovitexin in this pathway. This may explain why the isovitexin first increase during proliferation stage, and then decrease.

Conclusions

In the present study, the combination use of FT-IR, LC-UV-MS/MS and chemometrics was designed for investigation of the variation on metabolites during different growing stage of *G. rigescens*. For whole metabolome, the molecular structure-analyte relationships are significantly different between plants during proliferation stage and mature plants according to FT-IR analysis. Combined with LC-UV-MS/MS, mature plants contains more abundant secondary metabolites than plants during proliferation stage, whereas the higher content of some characteristic metabolites like gentiopicroside and peak a are observed in plants during proliferation. Moreover, the distribution and accumulation of metabolites, together with biosynthetic pathway, are associated with plant growth and significantly vary during different growing stage. In practical application, the root in mature plants with rich chemical components could be of better quality for medicinal application, whereas leaf in regenerated plantlet would be used for industrial extraction

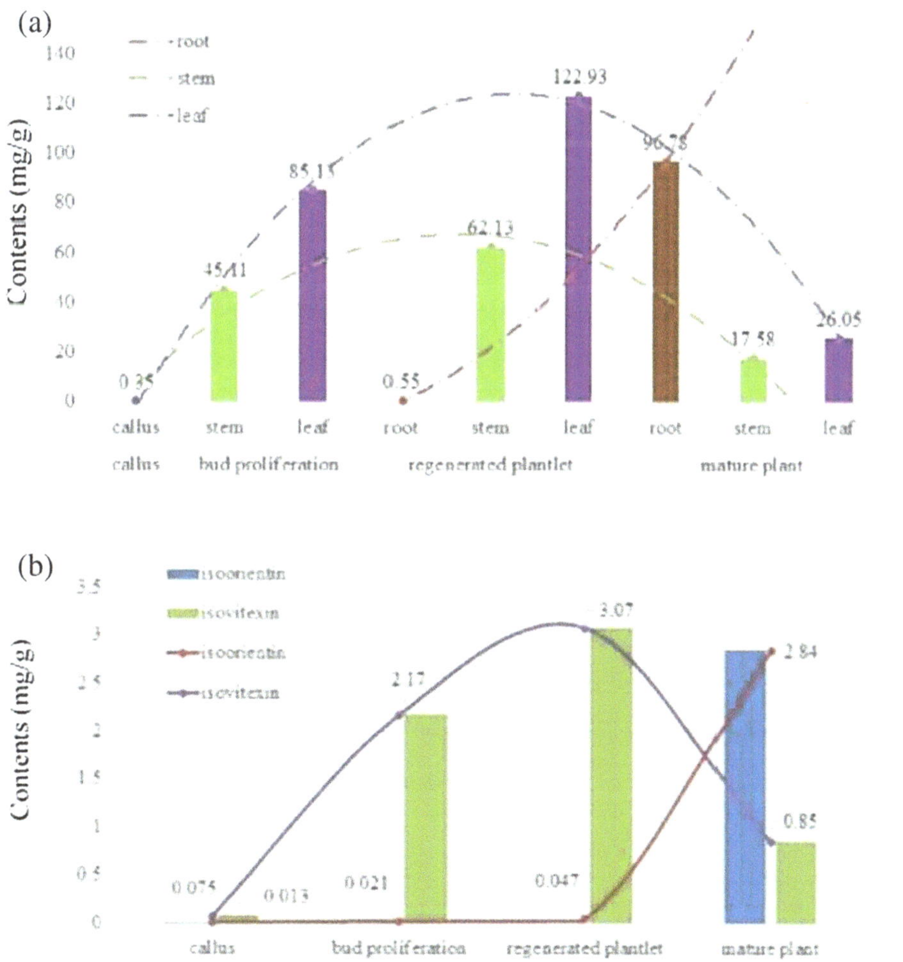

Fig. 5 a: Determination of gentiopicroside in different sample. **b**: Determination of isoorientin and isovitexin in leaf during different stage

of gentiopicroside. These results provide evidence for reasonable exploitation and distinct usage of *G. rigescens* during different growing stage.

Abbreviations

BP: Bud proliferation; FT-IR: Fourier transform infrared spectroscopy; LOD: The limits of detection; LOQ: The limits of quantification; LC-MS/MS: Liquid chromatography tandem mass spectrometry; MP: Mature plant; MRM: Multiple reaction monitoring; RP: Regenerated plantlet.

Competing interests

The authors declare that they have no competing interests.

Authors' contributions

PY conducted all laboratory experiments, data analysis and partial interpretation of results under the guidance of HH-Y and WY-Z, who designed the study. ZJ and Z Y-L are responsible for the grammar checking of the paper and finalized the manuscript. All authors have read and approved the final manuscript.

Acknowledgments

This work was supported by the National Natural Science Foundation of China (81260608 and 31260077), Key Project of Yunnan Provincial Natural Science Foundation (2013FC006), and the Yunnan Provincial Natural Science Foundation (2013FZ150 and 2013FD066).

References

Chuang YK, Yang IC, Lo YM, Tsai CY, Chen S (2014) Integration of independent component analysis with near-infrared spectroscopy for analysis of bioactive components in the medicinal plant *Gentiana scabra* Bunge. J Food Drug Anal 22:336–344

Gao L, Li J, Qi J (2010a) Gentisides A and B, two new neuritogenic compounds from the traditional Chinese medicine *Gentiana rigescens* Franch. Bioorgan Med Chem 18:2131–2134

Gao L, Xiang L, Luo Y, Wang G, Li J, Qi J (2010b) Gentisides C–K: Nine new neuritogenic compounds from the traditional Chinese medicine *Gentiana rigescens* Franch. Bioorgan Med Chem 18:6995–7000

Hall RD, Brouwer ID, Fitzgerald MA (2008) Plant metabolomics and its potential application for human nutrition. Physiol Plantarum 132:162–175

Hilbert G, Temsamani H, Bordenave L, Pedrot E, Chaher N, Cluzet S, Delaunay JC, Ollat N, Delrot S, Mérillon JM, Gomès E, Richard T (2015) Flavonol profiles in berries of wild Vitis accessions using liquid chromatography coupled to mass spectrometry and nuclear magnetic resonance spectrometry. Food Chem 169:49–58

Hu Y, Kong W, Yang X, Xie L, Wen J, Yang M (2014) GC–MS combined with chemometric techniques for the quality control and original discrimination of Curcumae longae rhizome: Analysis of essential oils. J Sep Sci 37:404–411

Huang SH, Vishwakarma RK, Lee TT, Chan HS, Tsay HS (2014) Establishment of hairy root lines and analysis of iridoids and secoiridoids in the medicinal plant *Gentiana scabra*. Bot Stud 55:17–25

Jiang RW, Wong KL, Chan YM, Xu HX, But PPH, Shaw PC (2005) Isolation of iridoid and secoiridoid glycosides and comparative study on Radix Gentianae and related adulterants by HPLC analysis. Phytochemistry 66:2674–2680

Karoui R, Downey G, Blecker C (2010) Mid-infrared spectroscopy coupled with chemometrics: A tool for the analysis of intact food systems and the exploration of their molecular structure-quality relationships-a review. Chem Rev 110:6144–6168

Kumari T, Sharma C, Bajpai V, Kumar B, Srivastava M, Arya KR (2015) Qualitative determination of bioactive metabolites through Q-TOF LC/MS in different parts and undifferentiated cultures of *Ulmus wallichiana* Planchon. Plant Growth Regul 75:331–340

Kuzovkina IN, Prokof'eva MY, Umralina AR, Chernysheva TP (2014) Morphological and biochemical characteristics of genetically transformed roots of *Scutellaria andrachnoides*. Russ J Plant Physiol 61:697–706

Lohumi S, Lee S, Lee WH, Kim MS, Mo C, Bae H, Cho BK (2014) Detection of Starch Adulteration in Onion Powder by FT-NIR and FT-IR Spectroscopy. J Agri Food Chem 62:9246–9251

Manukyan A (2011) Effect of growing factors on productivity and quality of lemon catmint, lemon balm and sage under soilless greenhouse production: I. drought stress. Med Aromat Plant Sci Biotechnol 5:119–125

Marsh Z, Yang T, Nopo-Olazabal L, Wu S, Ingle T, Joshee N, Medina-Bolivar F (2014) Effect of light, methyl jasmonate and cyclodextrin on production of phenolic compounds in hairy root cultures of *Scutellaria lateriflora*. Phytochemistry 107:50–60

Pan Y, Shen T, Pan J, Xiao D, Li Z, Li W, Wang YZ (2014) Development and validation of a UPLC-MS/MS method for the simultaneous determination and detection of four neuritogenic compounds in different parts of *Gentiana rigescens* Franch using multiple reaction monitoring and precursor ion scanning. Anal Methods 6:1782–1787

Pan Y, Shen T, Zhang J, Zhao YL, Wang YZ, Li WY (2015a) Simultaneous determination of six index constituents and comparative analysis of four ethnomedicines from genus *Gentiana* using a UPLC-UV-MS method. Biomed Chromatogr 29:89–96

Pan Y, Zhang J, Shen T, Zuo ZT, Jin H, Wang YZ, Li WY (2015b) Optimization of ultrasonic extraction by response surface methodology combined with UFLC-UV method for determination of four iridoids in *Gentiana rigescens*. J Food Drug Anal doi: 10.1016/j.jfda.2014.11.002

State Pharmacopoeia Commission (2010) Chinese Pharmacopoeia, 2010th edn. Chemistry and Industry Press, Beijing, pp 89–90

Su P, Cheng Q, Wang X, Cheng X, Zhang M, Tong Y, Li F, Gao W, Huang LQ (2014) Characterization of eight terpenoids from tissue cultures of the Chinese herbal plant, *Tripterygium wilfordii*, by high-performance liquid chromatography coupled with electrospray ionization tandem mass spectrometry. Biomed Chromatogr 28:1183–1192

Tan RX, Wolfender JL, Zhang LX, Ma WG, Fuzzati N, Marston A, Hoste'itman K (1996) Acyl secoiridoids and antifungal constituents from *Gentiana macrophylla*. Phytochemistry 42:1305–1313

Wang YM, Xu M, Wang D, Zhu HT, Zhang YJ, Yang CR (2012) Review on "Long-Dan", one of the traditional Chinese medicinal herbs recorded in Chinese Pharmacopoeia. Nat Prod Bioprospect 2:1–10

Wink M (2003) Evolution of secondary metabolites from an ecological and molecular phylogenetic perspective. Phytochemistry 64:3–19

Won TH, Liao L, Kang SS, Shin J (2014) Simultaneous analysis of furfural metabolites from Rehmanniae radix preparata by HPLC-DAD-ESI-MS. Food Chem 142:107–113

Xie CX, Suo FM, Jia GL, Song JY, Huang LF, Chen S (2011) Correlation between ecological factors and ginsenosides. Acta Ecol Sin 31:7551–7563

Xu M, Wang D, Zhang YJ, Yang CR (2006) A new secoiridoidal glucoside from *Gentiana rigescens* (Gentianaceae). Acta Botanica Yunnanica 28:669–672

Xu M, Wang D, Zhang YJ, Yang CR (2007) Dammarane triterpenoids from the roots of *Gentiana rigescens*. J Nat Prod 70:880–883

Xu M, Yang CR, Zhang YJ (2009a) Minor antifungal aromatic glycosides from the roots of *Gentiana rigescens* (Gentianaceae). Chinese Chem Lett 20:1215–1217

Xu M, Zhang M, Zhang YJ, Yang CR (2009b) New acylated secoiridoid glucosides from *Gentiana straminea* (Gentianaceae). Helv Chim Acta 92:321–327

Yu C, Wang CZ, Zhou CJ, Wang B, Han L, Zhang CF, Wu XH, Yuan CS (2014) Adulteration and cultivation region identification of American ginseng using HPLC coupled with multivariate analysis. J Pharmaceut Biomed Anal 99:8–15

Zhao Y, Zhang J, Yuan T, Shen T, Li W, Yang S, Hou Y, Wang YZ, Jin H (2014) Discrimination of wild *Paris* based on near infrared spectroscopy and high performance liquid chromatography combined with multivariate analysis. PloS One 9, e89100

Antibacterial mechanism of the action of *Enteromorpha linza* L. essential oil against *Escherichia coli* and *Salmonella Typhimurium*

Jayanta Kumar Patra[†], Gitishree Das[†] and Kwang-Hyun Baek[*]

Abstract

Background: Identification of natural antibacterial agents from various sources that can act effectively against disease causing foodborne bacteria is one of the major concerns throughout the world. However, the natural antibacterial agents identified to date are primarily effective against Gram positive bacteria, but less effective against Gram negative bacteria. In the present study, *Enteromorpha linza* L. essential oil (EEO) was evaluated for antibacterial activity against *Escherichia coli* and *Salmonella Typhimurium* along with the mode of their antibacterial action.

Results: The chemical composition of EEO revealed high amounts of acids (54.6 %) and alkenes (21.1 %). EEO was effective against both *E. coli* and *S. Typhimurium*. The minimum inhibitory concentration (MIC) and minimum bactericidal concentration (MBC) values of EEO for both pathogens were 12.5 mg/ml and 25.0 mg/mL, respectively. EEO at the MIC acted on the loss in viability of *E. coli* ATCC 43890, which was used as the model system for evaluation of the antibacterial mode of action of EEO against Gram negative bacteria. Significant increase in relative electrical conductivity and K^+ concentration were recorded with respect to time, indicating the disruption of tested *E. coli* cells owing to the controlling effect of EEO. Alternation of the morphology of the cell surface, increase in the release of 260 nm absorbing materials and loss of high salt tolerance were observed.

Conclusions: The results suggest that EEO induced a bactericidal effect via structural membrane damage caused by deposition of EEO in the cytosol or through enzymatic degradation of bacterial intracellular enzymes that resulted in cellular lysis. Accordingly, EEO can be used as a strong natural antibacterial agent against Gram negative foodborne pathogens such as *E. coli* and *S. Typhimurium*.

Keywords: Cell membrane; *Enteromorpha linza*; *Escherichia coli*; Essential oil; *Salmonella Typhimurium*; Seaweed

Background

Foodstuffs of animal and plant origin along with sea products are always subjected to spoilage and decaying phenomena that occur due to various microorganisms, mainly bacteria (Kim et al. 1995; Faid 2011). These phenomena often lead to heavy loss of food products and raw materials, ultimately resulting in health problems caused by the occurrence of food poisoning and infections (Kim et al. 1995). Contamination of food by different types of microorganisms is a matter of great concern for this world with increasing human population day by day (Acharya et al. 2011). Various foodborne microorganisms negatively affect food via two possible pathway, direct growth in food products and the release of microbial toxins while handling the food before processing. However, it is always difficult to find solutions to control food contamination.

The use of chemicals to protect meat, fruits, vegetables and drinks can be harmful to human health and the environment (Acharya et al. 2011; Khwaldia 2011). However, the application of chemical antimicrobials has been approved by different medical authorities worldwide to maintain food security. Owing to increased awareness, the use of synthetic chemicals has been rejected by consumers because of their adverse side effects that in many cases led to deadly diseases (Goni et al. 2009). As a result, the demand for natural preservatives in food has increased, leading many food industries to search for

* Correspondence: khbaek@ynu.ac.kr
[†]Equal contributors
School of Biotechnology, Yeungnam University, Gyeongsan, Gyeongbuk 712-749, Republic of Korea

natural antimicrobial agents for use in packaging and preservation. Indeed, the demand for natural antimicrobial agents is high as they are more effective and generally recognized as safe (GRAS) (Acharya et al. 2011).

More than 150,000 algae and seaweed species exist in the ocean, many of which possess components with antibacterial, antifungal and antiviral properties (Salvador et al. 2007; Souhaili et al. 2008; Lategan et al. 2009; Plaza et al. 2010; Faid 2011). Several chemical compounds from marine seaweeds such as agar, carrageenan, alginates and pigments like beta carotene have been used widely in food industries and pharmacology. Marine algae/seaweeds have become a part of the human diet in many countries, and many ingredients from these seaweeds have also being used in drinks, snacks, soups and beverages (Faid 2011). Various natural antimicrobial agents that have the potential for food processing and preservation are effective against Gram positive foodborne bacteria, but they are very less effective and sometimes not effective against the Gram negative foodborne bacteria (Trombetta et al. 2005; Bassole and Juliani 2012; Sfeir et al. 2013; Nazzaro et al. 2013).

Enteromorpha linza (L.) is a common edible seaweed that is abundant in coastal areas of Asian and European countries (Say et al. 1986) and consumed as food by people worldwide (Demirel et al. 2011). Therefore, in this study, we evaluated the antibacterial potential of essential oil extracted from *E. linza* against two dominant Gram negative foodborne pathogenic bacteria, *Escherichia coli* and *Salmonella Typhimurium*, as well as its mode of antibacterial action.

Methods

Extraction of essential oil from *E. linza*

Fresh seaweed (*E. linza* L.) was purchased from a local market at Gyeongsan, Republic of Korea, washed thoroughly and put into a specially designed glass container. Five liters of water were then added to a glass container containing 500 g seaweed and subjected to hydrodistillation to extract essential oils using a micro-wave assisted extraction machine manufactured by KMD Engineering (Paju, Republic of Korea) for 4 h. The temperature of the chamber was maintained by a thermo controller (oven power capacity of 40 W and frequency of 15 gkH). Approximately 500 mL of the distillate was then collected from the collecting nozzle of the apparatus with a conical flask, mixed with an equal volume of dichloromethane in a separating funnel, vigorously shaken, and allowed to stand until the two layers separated. Next, the lower layer was taken and concentrated using a rotary evaporator (N-1110, Eyela, Tokyo Rikakikai Co., Ltd., Japan) at 40 °C. The extracted *E. linza* L. essential oil (EEO) was then kept in a tightly closed vial at 4 °C until further use.

Chemical analysis of EEO

A detailed analysis of the chemical composition of EEO was conducted using a gas chromatography–mass spectrometry system (JMS 700 Mstation, Jeol Ltd., Japan) equipped with an Agilent 6890 N GC DB-5 MS fused silica capillary column (Agilent Technologies, Santa Clara, USA) according to the method described by Patra et al. (2015).

Determination of antibacterial activity of EEO

The antibacterial potential of EEO was evaluated against four foodborne bacteria belonging to two species, *S. Typhimurium* ATCC 19586 and ATCC 43174 and *E. coli* ATCC 43889 and ATCC 43890. The pathogens were obtained from the American Type Culture Collection (ATCC, Manassas, VA, USA). Prior to use, EEO was diluted two times in 5 % dimethylsulphoxide (DMSO), then filter-sterilized using a 0.22 μm micro syringe filter (Chemco Scientific, Chungbuk, Republic of Korea). The standard disc diffusion method described by Diao et al. (2013) was employed to evaluate the antibacterial activity of EEO. EEO and rifampicin (Duchefa Biochemie, Haarlem, The Netherlands) at 25 mg/disc and at 20 μg/disc, respectively, were used as the test sample and the positive control, while 5 % DMSO was used as the negative control. The inhibition zone was measured in mm for each pathogen.

The minimum inhibitory concentration (MIC) and minimum bactericidal concentration (MBC) of EEO against the four pathogens was determined by the two-fold dilution method described by Kubo et al. (2004), with slight modification. The lowest concentration of EEO without any visible growth of test organism was taken as the MIC value, and the concentration not showing any growth of bacterial colony on nutrient agar (NA) was selected as the MBC. Both the MIC and MBC were expressed in mg/mL. All experiments were repeated three times.

Effect of EEO on the viability of *E. coli*

The effects of EEO on the viability of Gram negative foodborne bacteria were investigated according to the procedure described by Joray et al. (2011). Only one foodborne bacterium, *E. coli* ATCC 43890, was selected as the model organism for all the subsequent assays. Overnight grown bacterial culture treated with EEO at the MIC served as the treatment, while those treated with 5 % DMSO served as the control. Both treatment and control samples were incubated at 37 °C for 8 h, during which time samples were collected every 2 h, appropriately diluted in phosphate buffer saline, spread on the surface of NA plates and incubated further at 37 °C for 24 h. After incubation, the number of bacterial colonies in terms of colony forming unit (cfu) was determined by multiplying it with the appropriate dilution factor. The results were expressed in terms of log_{10} (cfu/mL).

Effect of EEO on surface morphology of *E. coli*

The effects EEO treatment on the morphology of *E. coli* ATCC 43890 were determined by scanning electron microscopy (SEM). Both a control treated with 5 % DMSO and sample treated with EEO were prepared according to the procedure described by Bajpai et al. (2009). The specimens were then sputter-coated with platinum in an ion coater for 120 s and observed under SEM (S-4100, Hitachi, Japan).

Effect of EEO on cell membrane permeability of *E. coli*

The effect of EEO on the cell membrane permeability of *E. coli* ATCC 43890 were evaluated by the standard procedure (Kong et al. (2008). The loss in membrane permeability of the bacterial cell was measured using a conductivity meter (Con 6, LaMotte, Maryland, USA) and calculated in terms of its relative electrical conductivity according to the following formula:

$$\text{Relative conductivity } (\%) = [(L2 - L1)/L0] \text{ X } 100$$

where, L_0 was the electrical conductivity of the bacteria in 5 % glucose killed by being treated in boiling water for 5 min, L_1 was the electrical conductivity of the bacteria in 5 % glucose with the EEO at the MIC and L_2 was the electrical conductivity of the EEO-bacteria mixtures in 5 % glucose collected at 2 h intervals during 8 h of incubation.

Effect of EEO on *E. coli* cell membrane integrity

The effect of EEO on the cell membrane integrity of *E. coli* ATCC 43890 were measured by the standard procedure described by Carson et al. (2002). Both the control treated with DMSO and the sample treated with EEO at the MIC were incubated at 37 °C. After 60 and 120 min, samples were centrifuged at 3500 rpm for 10 min and the absorbance of the supernatant at 260 nm was measured using a spectrophotometer (ASP 3700, ACTGene Inc., NJ, USA).

Effect of EEO on the leakage of potassium ion from *E. coli*

The leakage of potassium ion (K^+) in the suspension culture of *E. coli* ATCC 43890 treated with EEO at the MIC was measured according to the method described by Lee et al. (2002) using a Kalium/Potassium kit (Quantofix, Macherey-Nagel GmbH & Co., Germany). All measurements were carried at 2 h time intervals during 8 h of incubation and expressed as the amount of free K^+ ion (mg/L) in the bacterial suspension at each time interval. Bacterial culture treated with DMSO was used as a control.

Effect of EEO on salt tolerance capacity of *E. coli*

The effect of EEO on the salt tolerance capacity of *E. coli* ATCC 43890 were determined by growing the bacterial cultures on NA plates supplemented with different concentrations of NaCl (0, 2.5, 5.0 and 10.0 %) (Miksusanti et al. 2008). Bacterial cultures amended with 5 % DMSO and EEO at the MIC were plated on NA plates and incubated at 37 °C for 24 h. After determining the CFU, the results were expressed in terms of Log_{10} (cfu/mL).

Statistical analysis

The results of all experiments were expressed as the mean ± standard deviation (SD). Statistical analyses consisted of one way ANOVA and Duncan's multiple range tests, with $P < 0.05$ considered to indicate significance. Statistical Analysis Software (SAS) version 9.4 (SAS Inc., Cary, USA) was used for all analyses.

Results and discussion

For years, pathogenic bacteria have been known to be the primary cause of foodborne illness and food poisoning throughout the world (Kim et al. 1995). The use of natural antibacterial agents in food processing and food preservation has played a vital role in controlling the adverse effect of bacteria in food. However, the natural antibacterial agents currently used as preservatives are primarily effective against Gram positive foodborne pathogens, whereas they are not much effective against the Gram negative foodborne pathogens (Trombetta et al. 2005; Bassole and Juliani 2012; Sfeir et al. 2013; Nazzaro et al. 2013). The reason for the failure of most natural antibacterial agents against Gram negative pathogens is that they are unable to penetrate the strong cell wall of a Gram negative bacterium and thus do not affect it (Trombetta et al. 2005).

There have been many reports of the bioactive potential of marine algae and seaweeds, and many novel compounds have been isolated from them (Kamat et al. 1992; Crasta et al. 1997; Bhosale et al. 2002; Salvador et al. 2007; Lategan et al. 2009; Patra et al. 2009; Plaza et al. 2010). However, there have been limited reports on the extraction and evaluation of bioactive potential of essential oils from marine algae and seaweeds on antibacterial capacity. In this study, we evaluated the marine resource *E. linze* as a source of essential oil for use as a strong antibacterial agent that can efficiently control against the Gram negative bacteria *E. coli* and *S. Typhimurium*.

The composition of essential oil obtained from hydrodistillation of *E. linze* was analyzed by GC-MS analysis and individual compounds present in the EEO were described in detail in our earlier publication (Patra et al. 2015). All of the identified compounds representing 94 % of the oil were categorized into six different groups (Table 1), 30 % of which had antimicrobial potential. Acids (54.6 %) and alkenes (21.1 %) were found in higher quantities, followed by alcohols (4.5 %), aldehydes (3.7 %) and ketones (2.8 %) (Table 1). The compounds with antimicrobial potential as previously reported include n-hexanal, 2-chloro-5-methyl-4-(2-thienyl) pyrimidine, tetradecanoic acid, pentadecanoic acid, (z,z)-6,9-cis-3,4-epoxy-nonadecadiene, 13-octadecenal and azetidine (Gutierrez et al. 2002; Agoramoorthy et al. 2007; Deep et al. 2012; Jemaa 2014; Sharma et al. 2014).

The antibacterial potential of EEO against four foodborne bacterial pathogens is presented in Table 2. The result showed that EEO was most effective against *E. coli* ATCC 43889, with a 13.33 mm inhibition zone, as well as against all tested *S. Typhimurium* strains, with a 10.66 mm inhibition zone. Rifampicin, a standard antibiotic showed significantly high controlling activity against all four strains of the tested bacteria, whereas DMSO, a negative control) exerted no inhibitory activity.

Table 1 Percentage composition of functional groups present in *Enteromorpha linza* essential oil

Functional group	Active compounds	Percentage composition
Acid	Cyclopropanecarboxylic acid	54.6
	Tetradecanoic acid	
	Pentadecanoic acid	
	Pentadecanoic acid	
	Hexadecanoic acid	
Alkene	1-heptadecene	21.1
	(z,z)-6,9-cis-3,4-epoxy-nonadecadiene	
Alcohol	2-Hexanol	4.5
	Tridecanol	
Aldehyde	n-Hexanal	3.7
	2,4 heptadienal	
	2 Octenal	
	n-nonanal	
	13-Octadecenal	
Ketone	2-heptanone	2.8
	Alpha-Ionone	
	Trans-beta-Ionone	
	2(4H)-Benzofuranone	
Others	Dimethyl sulfone	6.9
	2-chloro-5-methyl-4-(2-thienyl) pyrimidine	
	2-pentadecen-4-yne	
	3-heptadecen-5-yne	
	Azetidine	

Table 2 Efficacy of *Enteromorpha linza* essential oil (EEO) against four strains of the two Gram negative foodborne pathogens, *E. coli* and *S. Typhimurium*

Foodborne bacterial pathogen	Zone of inhibition (mm)	
	EEO[#]	Rifampicin[##]
E. coli ATCC 43889	13.33 ± 0.58[a]	32.33 ± 1.52[a]
E. coli ATCC 43890	10.00 ± 0.00[b]	32.66 ± 0.57[a]
S. Typhimurium ATCC 19586	10.66 ± 1.15[ab]	21.50 ± 4.94[b]
S. Typhimurium ATCC 43174	10.66 ± 0.57[ab]	31.66 ± 2.08[a]

Values are expressed as the mean ± standard deviation. Values with different superscript letters in the same column are significantly different at $P < 0.05$
[#]EEO at 25 mg/disc; [##]Rifampicin as a standard antibiotic at 20 µg/disc

The MIC values of the EEO ranged from 12.5 to 25 mg/mL, whereas it had an MBC of 25 mg/mL against all four test pathogens (Table 3).

Only a few studies have investigated the antibacterial effects of algal essential oil on pathogenic bacteria (Koz et al. 2009; Demirel et al. 2011; Gressler et al. 2011). However, the results of the present study clearly demonstrated that EEO exerted antibacterial effects on Gram negative bacteria. The outer membrane of Gram negative bacteria is always impermeable to various antibacterial agents owing to their strong lipolysaccharide molecules (Bezic et al. 2003; Sharma et al. 2013). In this study, however, the positive results of antibacterial activity of EEO against the two Gram negative bacteria suggested an hypothesis that, although essential oil is hydrophobic in nature, it might have slowly entered into the bacterial cell through the general bacterial porins owing to its small size, after which it might have affected the bacterial enzymes, causing cellular lysis and death (Nikaido 1996; Kotzekidou et al. 2008). The antibacterial activity of the EEO against the tested pathogenic bacteria might be due to the presence of various active compounds such as n-hexanal, 2-chloro-5-methyl-4-(2-thienyl) pyrimidine, tetradecanoic acid, pentadecanoic acid, (z,z)-6,9-cis-3,4-epoxy-nonadecadiene, 13-octadecenal and azetidine in the EEO that might have individually of synergistically altogether inhibited the growth of the pathogenic bacteria.

Among the two investigated pathogens, EEO was more effective against *E. coli* (Tables 2 and 3); hence, *E. coli* ATCC 43890 was selected for further investigation of its antibacterial mode of action. The antibacterial mode of action of EEO against *E. coli* ATCC 43890 was evaluated by

Table 3 MIC and MBC values of *Enteromorpha linza* essential oil against four strains of two Gram negative foodborne pathogens

Foodborne bacterial pathogen	MIC (mg/mL)	MBC (mg/mL)
E. coli ATCC 43889	12.5	25.0
E. coli ATCC 43890	12.5	25.0
S. Typhimurium ATCC 19586	25.0	25.0
S. Typhimurium ATCC 43174	12.5	25.0

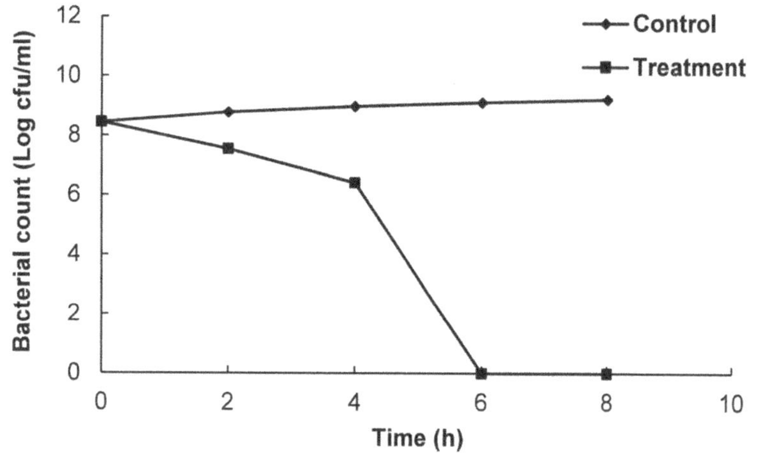

Fig. 1 Effect of *Enteromorpha linza* essential oil on the cell viability of *E. coli* ATCC 43890. The control and treatment were *E. coli* ATCC 43890 treated with 5 % DMSO and *Enteromorpha linza* essential oil at the MIC, respectively. Values are expressed as the mean ± SD

several assays, including cell viability, observation by SEM, effect on cell membrane leakage and loss of salt tolerance. The effects of EEO on the viability of *E. coli* ATCC 43890 are shown in Fig. 1. EEO at the MIC gradually decreased the viability of the bacterial cells by 4 h of incubation; however, it drastically decreased the number of CFU between 4 and 6 h, and completely inhibited cell growth after 6 h of incubation. In contrast, there was a continuous increase in the number of CFUs on the control plates.

Disruption of the morphology of *E. coli* ATCC 43890 by EEO was also observed by SEM analysis (Fig. 2). When compared with the uniform and smooth surface found in the control *E. coli* treated with 5 % DMSO, a ruptured and rough surface with an elongated morphology was observed in bacterial cells treated with EEO at the MIC. As shown in the mode of action for other essential oils, EEO might have induced a bactericidal effect through membrane damage and enzymatic degradation that resulted in cellular lysis (Ghannoum 1988; Bajpai et al. 2009).

The effect of EEO and DMSO on the membrane permeability of *E. coli* ATCC 43890 are presented based on the relative conductivity, optical density and free K^+ contents (Figs. 3, 4 and 5). The control sample treated with DMSO showed no increase in the relative electrical conductivity with respect to time, whereas *E. coli* ATCC 43890 treated with EEO showed increased relative electrical conductivity with respect to time (Fig. 3). Especially, after 6 h of incubation, there was a sharp increase in the relative electrical conductivity (Fig. 3). The effect of EEO at the MIC on the release of 260 nm absorbing materials from *E. coli* ATCC 43890 are presented in Fig. 4. *E. coli* ATCC 43890 treated with EEO at the MIC displayed a continuous increase in the optical density; however, the control treated with 5 % DMSO showed no remarkable change in optical density (Fig. 4). *E. coli* ATCC 43890 treated with EEO at the MIC displayed a continuous increase in the leakage of K^+ ions from bacteria with respect to incubation time. An especially sharp

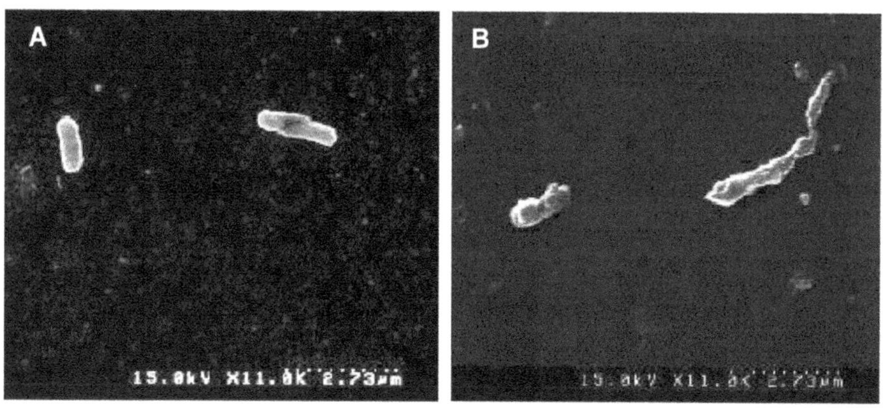

Fig. 2 Photograph of the scanning electron microscopy image of *E. coli* ATCC 43890. **a**: *E. coli* treated with 5 % DMSO; **b** *E. coli* treated with *Enteromorpha linza* essential oil at the MIC

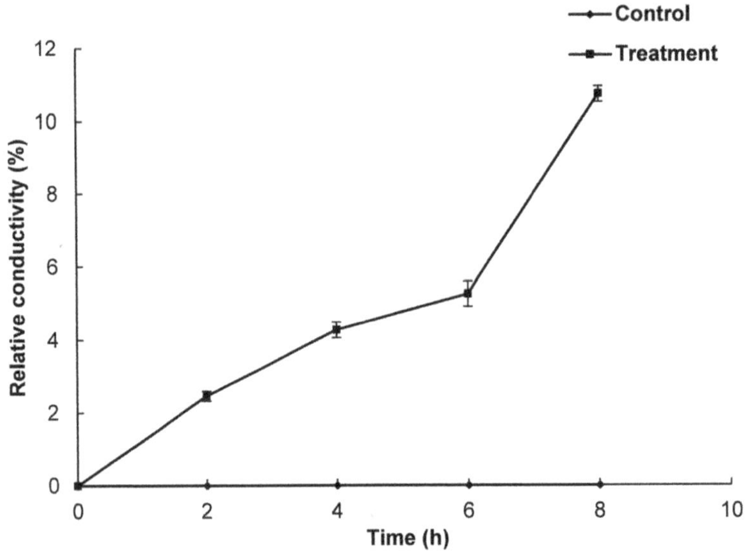

Fig. 3 Effect of *Enteromorpha linza* essential oil at the MIC on the membrane permeability of *E. coli* ATCC 43890. The control and treatment were *E. coli* ATCC 43890 treated with 5 % DMSO and *Enteromorpha linza* essential oil at the MIC, respectively. Values are expressed as the mean ± SD

increase in the leakage of K^+ was observed after 6 h of incubation (Fig. 5), corresponding to the complete loss of cell viability and a sharp increase in relative electrical conductivity in *E. coli* ATCC 43890 treated with EEO at the MIC after 6 h of incubation (Figs. 1 and 3).

The increase in the relative electrical conductivity, leakage of 260 nm absorbing materials and leakage of K^+ ions to the incubating culture indicated that EEO after being escaped into the bacterial cells through the general porins might have been accumulated in the cytosol, resulting in loss of plasma membrane integrity. This caused the cells to become permeable to protons and ions from outside, which eventually caused cellular lysis and cell death (Bajpai et al. 2013). The loss of 260 nm absorbing materials along with the leakage of K^+ ions to the outside of the bacteria might have been due to the structural damage caused by the action of EEO (Cox et al. 2001). In addition, the loss of membrane permeability, release of 260 nm absorbing material, and leakage of K^+ ions to the outside can be sensitive indicators of

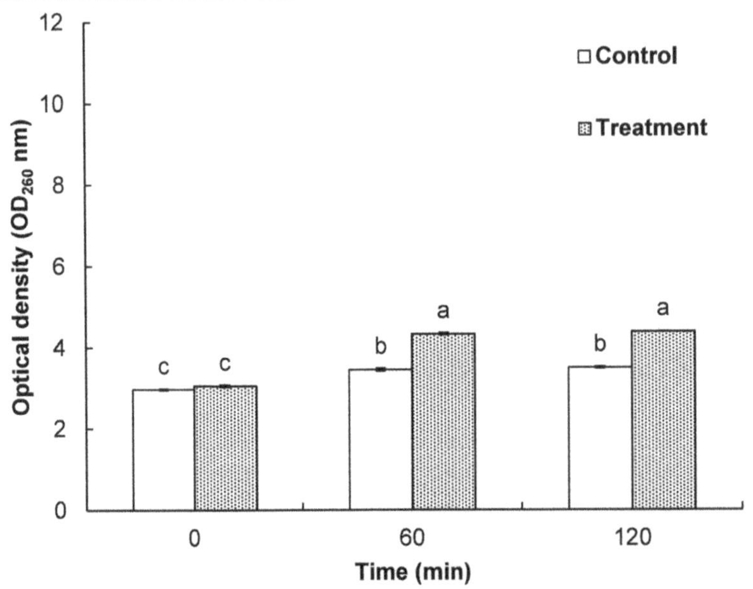

Fig. 4 Effect of *Enteromorpha linza* essential oil at the MIC on the release rate of 260 nm absorbing material from *E. coli* ATCC 43890. The control and treatment were *E. coli* ATCC 43890 treated with 5 % DMSO and *Enteromorpha linza* essential oil at the MIC, respectively. Data are expressed as the mean ± SD. Values with different superscript letters are significantly different at $P < 0.05$

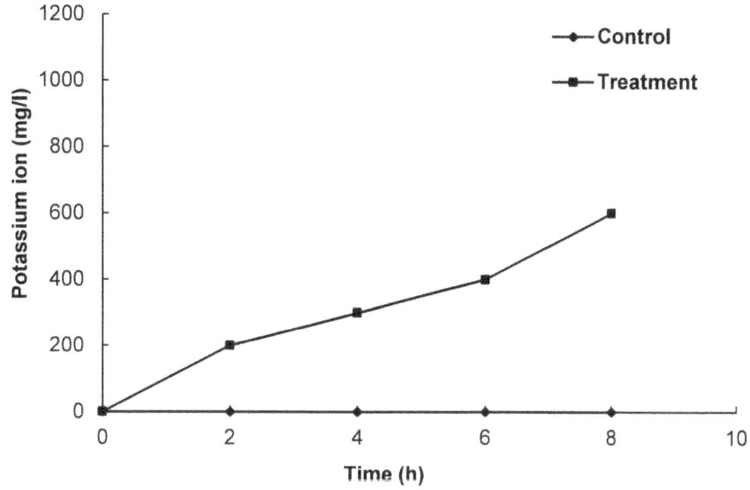

Fig. 5 Effect of *Enteromorpha linza* essential oil at the MIC on the leakage of potassium ions from *E. coli* ATCC 43890. The control and treatment were *E. coli* ATCC 43890 treated with 5 % DMSO and *Enteromorpha linza* essential oil at the MIC, respectively

the disruption of membrane integrity (Cox et al. 2001). Any type of minor changes to the structural integrity of the cell membrane could adversely affect the normal metabolic functioning of the bacterial cell and result in complete lysis (Hugo and Snow 1981; Davidson and Branen 2005).

Loss of the ability to tolerate high salt concentrations was observed in *E. coli* ATCC 43890 treated with EEO at the MIC (Fig. 6). When the *E. coli* strain pretreated with EEO at the MIC was incubated on NA supplemented with different concentrations of NaCl (0, 2.5 and 5 %), its concentration decreased significantly relative to bacterium pretreated with 5 % DMSO (Fig. 6). The underlying mechanism for the loss of salt-tolerance capacity might have been related to the loss of membrane permeability and degradation of important enzymes inside the bacterial cells. Additionally, EEO might have caused reduced osmoregulatory ability in high salt concentrations, resulting in their loss of salt tolerance (Miksusanti et al. 2008).

Conclusions

EEO from the edible marine seaweed *E. linza* was found to have a potential inhibitory effect on two Gram negative foodborne bacteria. This effect of EEO was associated with its ability to disrupt the bacterial membrane and probably the bacterial enzymes by accumulation inside the cytosol, causing loss of membrane integrity and leakage of 260 nm absorbing material along with K^+ ions. Since EEO is extracted from the edible seaweed

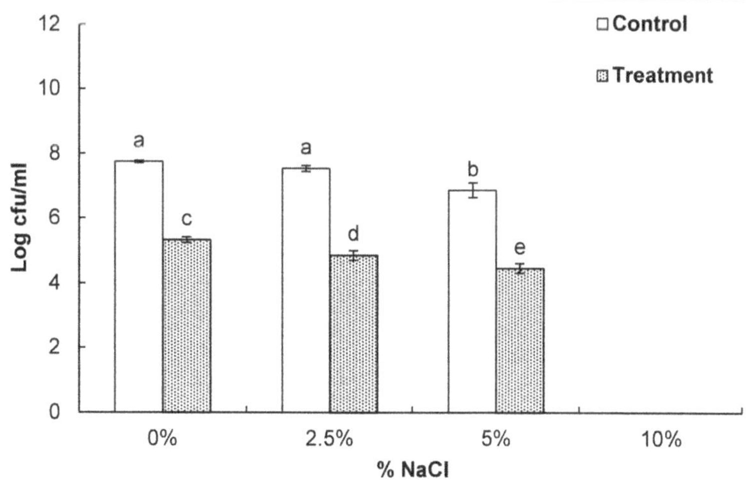

Fig. 6 Effect of *Enteromorpha linza* essential oil at the MIC on the loss of salt tolerance capacity of *E. coli* ATCC 43890. The control and treatment were *E. coli* ATCC 43890 treated with 5 % DMSO and with *Enteromorpha linza* essential oil at the MIC, respectively. Data are expressed as the mean ± SD. Values with different superscript letters are significantly different at $P < 0.05$

widely available around the coastal belt of many countries worldwide, it can be an inexpensive but valuable source of natural mixtures of antibacterial compounds that exhibit the potential for use in food systems. EEO could prevent the growth of foodborne bacteria, particularly Gram negative bacteria, and extend the shelf life of the processed food.

Competing interests
The authors declare that they have no competing interests.

Authors' contributions
JKP and GD conducted the experiment and wrote the manuscript. JKP and KB planned and designed the study and analyzed data. All authors participated in drafting the manuscript and have read and approved the final manuscript.

Authors' information
Jayanta Kumar Patra and Gitishree Das were the combined first authors to the manuscript.

Acknowledgements
This work was supported by a grant from the Systems and Synthetic Agro-biotech Center through the Next-Generation BioGreen 21 Program (PJ011117012015), Rural Development Administration, Korea.

References
Acharya D, Rios JL, Rai M (2011) Naturally occurring biocides in the food industry. In: Rai M, Chikindas M (eds) Natural antimicrobials in food safety and quality. CAB International, United Kingdom

Agoramoorthy G, Chandrasekaran M, Venkatesalu V, Hsu MJ (2007) Antibacterial and antifungal activities of fatty acid methyl esters of the blind-your-eye mangrove from India. Braz J Microbiol 38:739–42

Bajpai VK, Al-reza SM, Choi UK, Lee JH, Kang SC (2009) Chemical composition, antibacterial and antioxidant activities of leaf essential oil and extracts of *Metaseqoia glyptostroboides* Miki ex Hu. Food Chem Toxicol 47:1876–83

Bajpai VK, Sharma A, Baek KH (2013) Antibacterial mechanism of action of *Texus cuspidate* stem essential oil against selected foodborne pathogens. J Food Safety 33:348–59

Bassole IHN, Juliani HR (2012) Essential oils in combination and their antimicrobial properties. Molecules 17:3989–4006

Bezic N, Skocibusic M, Dinkic V, Radonic A (2003) Composition and antimicrobial activity of *Achillea clavennae* L. essential oil. Phytother Res 17:1037–40

Bhosale SH, Nagle VI, Jagtap TG (2002) Antifouling potential of some marine organisms from Indian species of *Bacillus* and *Pseudomonas*. Marine Biotechnol 4:111–8

Carson CF, Mee BJ, Riley TV (2002) Mechanism of action of *Melaleuca alternifolia* (tea tree) oil on *Staphylococcus aureus* determined by time-kill, lysis, leakage, and salt tolerance assays and electron microscopy. Antimicrob Agent Chemother 46:1914–20

Cox SD, Mann CM, Markhan JL, Gustafson JE, Warmington JR, Wyllie SG (2001) Determining the antimicrobial action of tea tree oil. Molecules 6:87–91

Crasta PJ, Raviraja NS, Sridhar KR (1997) Antimicrobial activity of some marine algae of southwest coast of India. Indian J Marine Sci 26:201–5

Davidson PM, Branen AL (2005) Food antimicrobials - an introduction. In: Davidson PM, Sofos JN, Branen AL (eds) Antimicrobial in food. CRC Press, Boca Raton, FL

Deep A, Phogat P, Kumar M, Kakkar S, Mittal SK, Malhotra M (2012) New tetradecanoic acid hydrazones in the search for antifungal agents: synthesis and *in vitro* evaluations. Acta Poloniae Pharmaceut Drug Res 69:129–33

Demirel Z, Yilmaz-Koz FF, Karabay-Yavasoglu NU, Ozdemir G, Sukatar A (2011) Antimicrobial and antioxidant activities of solvent extracts and the essential oil composition of *Laurencia obtusa* and *Laurencia obtusa* var. *pyramidata*. Romanian Biotechnol Lett 16:5927–36

Diao WR, Hu QP, Feng SS, Li WQ, Xu JG (2013) Chemical composition and antibacterial activity of the essential oil from green huajiao (*Zanthoxylum schinifolium*) against selected foodborne pathogens. J Agric Food Chem 61:6044–9

Faid M (2011) Antimicrobials from marine algae. In: Rai M, Chikindas M (eds) Natural antimicrobials in food safety and quality. CAB international, United Kingdom

Ghannoum MA (1988) Studies on the anticandidal mode of action of *Allium sativum* (garlic). J Gen Microbiol 134:2917–24

Goni P, Lopez P, Sanchez C, Gomez-Lus R, Becerril R, Nerin C (2009) Antimicrobial activity in the vapour phase of a combination of cinnamon and clove essential oils. Food Chem 116:982–9

Gressler V, Stein EM, Dorr F, Fujii MT, Colepicolo P, Pinto E (2011) Sesquiterpenes from the essential oil of *Laurencia dendroidea* (Ceramiales, Rhodophyta): isolation, biological activities and distribution among seaweeds. Brazilian J Pharmacog 21:248–54

Gutierrez A, Rio JCD, Martinez-Inigo MJ, Martinez MJ, Martinez AT (2002) Production of new unsaturated lipids during wood decay by *Ligninolytic basidiomycetes*. Appl Environ Microbiol 68:1344–50

Hugo W, Snow G (1981) Biochemistry of antibacterial action. Chapman & Hall, London

Jemaa JMB (2014) Essential oil as a source of bioactive constituents for the control of insect pests of economic importance in Tunisia. Med Aromatic Plants 3:1–7

Joray MB, del-Rollan MR, Ruiz GM, Palacios SM, Carpinella MC (2011) Antibacterial activity of extracts from plants of central Argentina- isolation of an active principle from *Achyrocline satureioides*. Plant Med 77:95–100

Kamat SY, Wahidulla S, D'Souza L, Naik CG, Ambiye V, Bhakuni DS et al (1992) Bioactivity of marine organisms-VI. Antiviral evaluation of marine algal extracts from the Indian coast. Bot Mar 35:161–4

Khwaldia K (2011) Antimicrobial films and coatings from milk proteins. In: Rai M, Chikindas M (eds) Natural antimicrobials in food safety and quality. CAB International, United Kingdom

Kim J, Marshall MR, Wei C (1995) Antibacterial activity of some essential oil components against five foodborne pathogens. J Agric Food Chem 43:2839–45

Kong M, Chen XG, Liu CS, Meng XH, Yu LJ (2008) Antibacterial mechanism of chitosan microspheres in a solid dispersing system against *E. coli*. Colloid Surface B 65:197–202

Kotzekidou P, Giannakidis P, Boulamatsis A (2008) Antimicrobial activity of some plant extracts and essential oils against foodborne pathogens in vitro and on the fate of inoculated pathogens in chocolate. LWT- Food Sci Technol 41:119–27

Koz FFY, Yavasoglu NUK, Demirel Z, Sukatar A, Ozdemir G (2009) Antioxidant and antimicrobial activities of *Codium fragile* (Suringar) Hariot (Chlorophyta) essential oil and extracts. Asian J Chem 21:1197–209

Kubo I, Fujita K, Kubo A, Nihei K, Ogura T (2004) Antibacterial activity of coriander volatile compounds against *Salmonella choleraesuis*. J Agric Food Chem 52:3329–32

Lategan C, Kellerman T, Afolayan AF, Mann MG, Antunes EM, Smith PJ et al (2009) Antiplasmodial and antimicrobial activities of South African marine algal extracts. Pharm Biol 47:408–13

Lee JY, Kim YS, Shin DH (2002) Antimicrobial synergistic effect of linolenic acid and monoglyceride against *Bacillus cereus* and *Staphylococcus aureus*. J Agric Food Chem 50:2193–9

Miksusanti JB, Priosoeryanto B, Syarief R, Rekso G (2008) Mode of action Temu kunci (*Kaempferia pandurata*) essential oil on *E. coli* K1.1 cell determined by leakage of material cell and salt tolerance assays. Hayati J Biosci 15:56–60

Nazzaro F, Fratianni F, Martina LD, Coppola R, De-Feo V (2013) Effect of essential oils on pathogenic bacteria. Pharmaceuticals 6:1451–74

Nikaido H (1996) Outer membrane. In: Neidhardt FC (ed) *Escherichia coli* and *Salmonella*: cellular and molecular biology. ASM Press, Washington, DC

Patra JK, Kim SH, Baek KH (2015) Antioxidant and free radical scavenging potential of essential oil from *Enteromorpha linza* L. prepared by microwave assisted hydrodistillation. J Food Biochem 39:80–90

Patra JK, Patra AP, Mohapatra NK, Das S, Thatoi H, Sahu RK et al (2009) Biomolecular constituents and antibacterial activity of some marine algae from Chilika Lake (Orissa, India). Int J Algae 11:222–35

Plaza M, Santoyo S, Jaime L, Garcia-Blairsy Reina G, Herrero M, Senorans MJ et al (2010) Screening for bioactive compounds from algae. J Pharma Biomed Anal 51:450–5

Salvador N, Garreta AG, Lavelli L, Ribera MA (2007) Antimicrobial activity of Iberian macroalgae. Scientia Marina 71:101–13

Say PJ, Burrows IG, Whitton BA (1986) Enteromorpha as a monitor of heavy metals in estuarine and coastal intertidal waters: A method for the sampling, treatment and analysis of the seaweed Enteromorpha to monitor heavy metals in estuaries and coastal waters. Northern Environmental Consultants Ltd, Durham DH8 6SS, UK, pp 8–9

Sfeir J, Lefrancois C, Baudoux D, Derbre S, Licznar P (2013) *In vitro* antibacterial activity of essential oils against *Streptococcus pyogenes*. Evid based Complement Altern Med 2013:1–9

Sharma A, Bajpai VK, Baek KH (2013) Determination of antibacterial mode of action of *Allium sativum* essential oil against foodborne Pathogens using membrane permeability and Surface characteristic parameters. J Food Safety 33:197–208

Sharma V, Chitranshi N, Agarwal AK (2014) Significance and biological importance of pyrimidine in the microbial world. Int J Med Chem 2014:1–31

Souhaili Z, Mohammadi H, Habti N, Faid M (2008) Lethal effect of the aqueous extract of the brown marine algae (*Cystoseira tamariscifolia*) on the mouse and the myelom cells. Afrique Sci 4:580–90

Trombetta D, Castelli F, Sarpietro MG, Venuti V, Cristani M, Daniele C et al (2005) Mechanisms of antibacterial action of three monoterpenes. Antimicrob Agents Chemother 49:2474–8

11

Effects of light-emitting diode supplementary lighting on the winter growth of greenhouse plants in the Yangtze River Delta of China

Xue Li[1], Wei Lu[1], Guyue Hu[1], Xiao Chan Wang[1,2*], Yu Zhang[1,2], Guo Xiang Sun[1,2] and Zhichao Fang[1]

Abstract

Background: The winter in the Yangtze River Delta area of China involves more than 1 month of continuous low temperature and poor light (CLTL) weather conditions, which impacts horticultural production in an unheated greenhouse; however, few greenhouses in this area are currently equipped with a heating device. The low-cost and long-living light-emitting diode (LED) was used as an artificial light source to explore the effects of supplementary lighting during the dark period in CLTL winter on the vegetative characteristics, early yield, and physiology of flowering for pepper plants grown in a greenhouse without heating. Two LED lighting sets were employed with different light source to provide 65 μmol m^{-2} s^{-1} at night: (1) LED-A: red LEDs (R, peak wavelength 660 nm) and blue LEDs (B, peak wavelength 460 nm) with an R:B ratio of 6:3; and (2) LED-B: R and B LEDs at an R:B ratio of 8:1. Plants growth parameters and chlorophyll fluorescence characteristics were compared between lighting treatments and the control group.

Results: Plants' yield and photosynthesis ability were improved by LED-A. Pepper grown under the LED-A1 strategy showed a 303.3 % greater fresh weight of fruits and a 501.3 % greater dry mass compared with the control group. Plant leaves under LED-A1 showed maximum efficiency of the light quantum yield of PSII, electron transfer rate, and the proportion of the open fraction of PSII centers, with values 113.70, 114.34, and 211.65 % higher than those of the control group, respectively, and showed the lowest rate constant of thermal energy dissipation of all groups. LED-B was beneficial to the plant height and stems diameter of the pepper plants more than LED-A.

Conclusions: These results can serve as a guide for environment control and for realizing low energy consumption for products grown in a greenhouse in the winter in Southern China.

Keywords: LED, Supplementary lighting, Greenhouse, Winter, Pepper

Background

Light and temperature are the most important environmental factors affecting plant survival, dry mass distribution, and crop yield (Janda et al. 2014; Zoratti et al. 2014). Light is the direct energy source for plant photosynthesis to determine the distribution of organic materials, the morphogenesis of plant and subsequent yield (Blom and Ingratta 1984; Fan et al. 2008). The temperature state can affect the activity of enzymes in plant cells to exert physiological changes (Chen and Tang 2013). Because of this close relationship between temperature and physiological reactions, temperature and lighting cues provide vital information for plants to assure optimal development (Franklin et al. 2014). Under the condition of light with a low red/far-red (R/FR) ratio and a comfortable temperature, the plant leaf area and dry mass are reduced (Patel et al. 2013); however, at lower temperatures, lighting with a low R/FR ratio can significantly promote

*Correspondence: wangxiaochan@njau.edu.cn
[1] College of Engineering, Nanjing Agricultural University, Nanjing, Jiangsu, People's Republic of China
Full list of author information is available at the end of the article

the plant-soluble sugar content and the expression of C-repeat-binding factor, a regulator of cold-acclimation genes (Franklin and Whitelam 2007). Furthermore, Marcelis (1993a, b) found that plant fruits dry mass were improved with increasing irradiance and decreasing temperature, demonstrating that when plants are in a relatively low temperature environment (or suboptimal temperature), adjusting the lighting condition exerts positive effects on plant production.

A light-emitting diode (LED) can emit the monochromatic light required for plant growth. The combination of red LEDs with blue LEDs can form a spectral absorption peak suitable for plant photosynthesis and morphogenesis with an 80–90 % light energy utilization rate and a remarkable energy-saving effect (Yang. 2008). Some reports have indicated that the use of a single red or blue LED light source or their combination could improve the efficiency of photosynthesis to promote plant production and regulate morphogenesis (Barta et al. 1992; Fang and Jao 1996; Nhut et al. 2003; Ding et al. 2005; Xiao et al. 2013; Shen et al. 2014).

In the Yangtze River Delta region of southern China, the low temperature and poor light (CLTL) environment always lasts for more than 1 month in winter (Wu 2011). Moreover, greenhouses in this area are scarcely heated owing to the high energy consumption and costs. Consequently, this suboptimal weather in southern China is quite detrimental to plant growth and production. Most of the research conducted in this area has focused on seedling growth in response to CLTL conditions or has examined the influence of only a single factor on plant growth in order to select plants that are resistant to the harsh temperature conditions (Wang et al. 2001; Ren et al. 2002; Hu and Yu 2003). Little studies have been conducted to study how to improve the production of flowering plants grown in greenhouse under the CLTL conditions. Thus, the aim of this study was to evaluate the effects of LED supplementation on flowering plants grown in a greenhouse without heating equipment during the dark period in winter. These results should serve as a guide for environmental control and for achieving low energy consumption with high productivity in plants grown in greenhouses in the winter in southern China.

Methods

Plant material and experimental design

Experiments were conducted from December 2014 to January 2015 in a greenhouse located in the College of Engineering in Nanjing Agricultural University (Nanjing, Jiangsu Province, China, 118°46′ N, 32°03′ E), which is characterized by a sub-tropical monsoon climate. The greenhouse is equipped with internal and external shading, skylights, a drip irrigation system, and a spray system, which are all computer-controlled. The Eco-Watch ecological environment monitoring system (Dynamax, USA) was used to monitor the micro-environmental parameters inside the greenhouse. Pepper (*Capsicum frutescens* L., Sujiao No. 5) was chosen as the experimental material. The seed of pepper were sown in nutrient soil (Galaku Pty Ltd., Australia) on September 10, 2014 and transplanted into plots after the fourth true leaf was fully mature. The medium was a coconut shell (Galaku Pty, Ltd.). The Yamazaki nutrient (Table 1) solution was used to supply nutrients for pepper growth. Chlorophyll fluorescence parameters were measured with the Mini-Pam II system (Walz, Germany), quantum photon density and light homogeneity were tested by a PAR sensor equipped by Mini-Pam II.

The LED lighting sets were established for each plant individually. Each light set consisted of 144 red (R, peak wavelength 660 nm) or blue (B, peak wavelength 460 nm) LEDs with ultra-high brightness at an R:B ratio of 6:3 (LED-A group) or 8:1 (LED-B group) (Hogewoning et al. 2010). The lighting area was 555 × 22.8 mm. The plant photosynthetic quantum flux density, measured 15 cm above the plants, was 65 μmol m^{-2} s^{-1}.

Plants in anthesis with similar growth trends were chosen for the experimental treatment. Supplementary lighting began on December 1, 2014 and finished on January 15, 2015. The humidity arranged from 60 to 70 %. Temperature was the environment temperature as well as the light. The experiment included seven treatments: six experimental groups and the control group, with six plants in per treatment. The control group was plants grown without light supplementation at night. The supplementary lighting duration and light source for each treatment are given in Table 2.

Parameter measurements

The greenhouse environmental parameters were collected by the Eco-Watch system at 15-min intervals. All

Table 1 Elements composition of Yamazaki nutrient in 1 L deionized water

Elements composition	Dosage (mg L^{-1})
$Ca(NO_3)_2·4H_2O$	354
KNO_3	607
$NH_4H_2PO_4$	96
$MgSO_4·7H_2O$	185
Na_2Fe-EDTA	25
H_3BO_3	2.13
$MnSO_4·4H_2O$	2.86
$ZnSO_4·7H_2O$	0.22
$CuSO_4·5H_2O$	0.08
$(NH_4)_6MO_7O_2·4H_2O$	0.02

Table 2 The light quality and light supplementation arrangement

Light source	R/B ratio	Lighting duration	Serial number
LED-A	6:3	18:00–00:00 (6 h)	LED-A1
	6:3	18:00–22:00 (4 h)	LED-A2
	6:3	18:00–20:00 (2 h)	LED-A3
LED-B	8:1	18:00–00:00 (6 h)	LED-B1
	8:1	18:00–22:00 (4 h)	LED-B2
	8:1	18:00–20:00 (2 h)	LED-B3
–	–	–	Control group

pepper parameters were measured after 2 weeks of light supplementation at night period. The chlorophyll fluorescence parameters of the pepper plants were tested every 7 days, with healthy canopy leaves chosen for measurement, ensuring that each measure was taken at the same location as much as possible. The chlorophyll fluorescence parameters would be measured for 5 weeks in all. The actinic light intensity was four and the SAT-Plus intensity was 10, the duration of induction curve was 20 s with the length of 12. The measuring time was from 17:00–21:00 after 25 min of dark adaptation for the leaves, and three leaves were tested of each plant with the averages used for statistical analysis. Three plants were randomly tested in each treatment. At the end of the experiment, the following vegetative characteristics of the plants were measured: plant height, stem diameter, and plant canopy width. The dry mass of the roots, stems, leaves, and fruits were also recorded.

Statistical analysis

Excel 2010 (Chinese version) and Origin 7.5 (Chinese version) were used for data processing and graph generation. Analysis of variance (ANOVA) was used to evaluate statistical significance among groups, and the Duncan method was used for multiple comparisons evaluated at $p < 0.05$ or $p < 0.01$ with SPSS 17.0 (Chinese version).

Results

Environmental parameters in the greenhouse during winter

Figure 1 showed the temperature and solar radiation fluctuations during daytime hours in the greenhouse. The daily minimum temperature was 7.67 °C, the average temperature was 15.8 °C, and the highest temperature was 22.18 °C. With respect to solar radiation, the minimum value was 5.35 W m^{-2}, the average value was 101.5 W m^{-2}, and the maximum value was 178.86 W m^{-2}, indicating that the peppers grown in the greenhouse were in a CLTL environment.

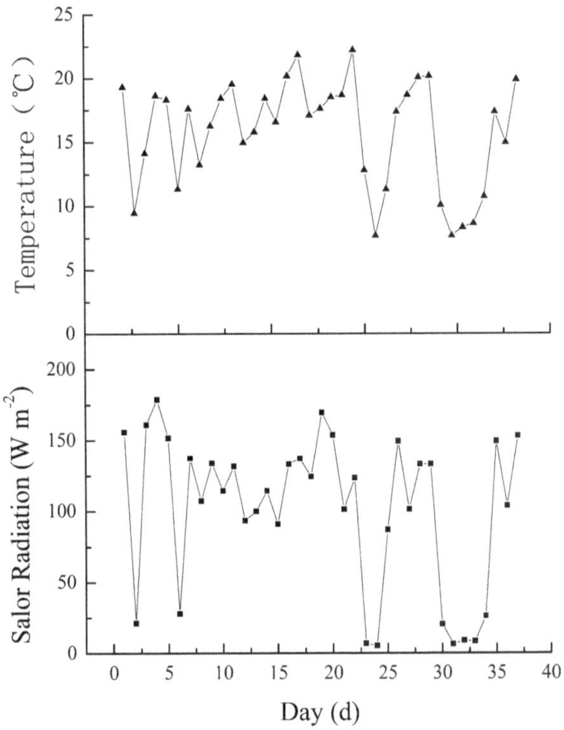

Fig. 1 Daily microclimate conditions (air temperature and solar radiation) inside the greenhouse. The data were the average of temperature and solar radiation from 6:00 to 17:00

Plant growth and development in cltl with supplementary led lighting during dark periods in winter

Impacts of LED supplementary lighting on plant vegetative characteristics during the dark period

When the lighting was applied for 2 h during the dark period, plants grown under the LED-A treatment showed the highest plant height, which was 13.94 % higher than that of the control group, representing a significant difference ($p < 0.05$; Fig. 2a). The stem diameter under the LED-B treatment was significantly higher than that of the control with an increase of 5.56 % ($p < 0.05$; Fig. 2b). No significant difference in plant canopy width or plant dry mass was observed among the groups.

When the LED lighting was applied for 4 h, the plants grown under the LED-B treatment showed the highest plant height and stem diameter, which were 20.44 and 20.83 %, respectively, greater than those of the control group (Fig. 2a, b). Plants grown under LED-A showed significantly increased plant dry mass of 81.86 % compared to the control group ($p < 0.05$; Fig. 2d). There was no significant difference among the groups on plant canopy width.

When the lighting was applied for 6 h in CLTL winter, the plants height under the LED-B treatment was significantly higher than that of the LED-A and control groups

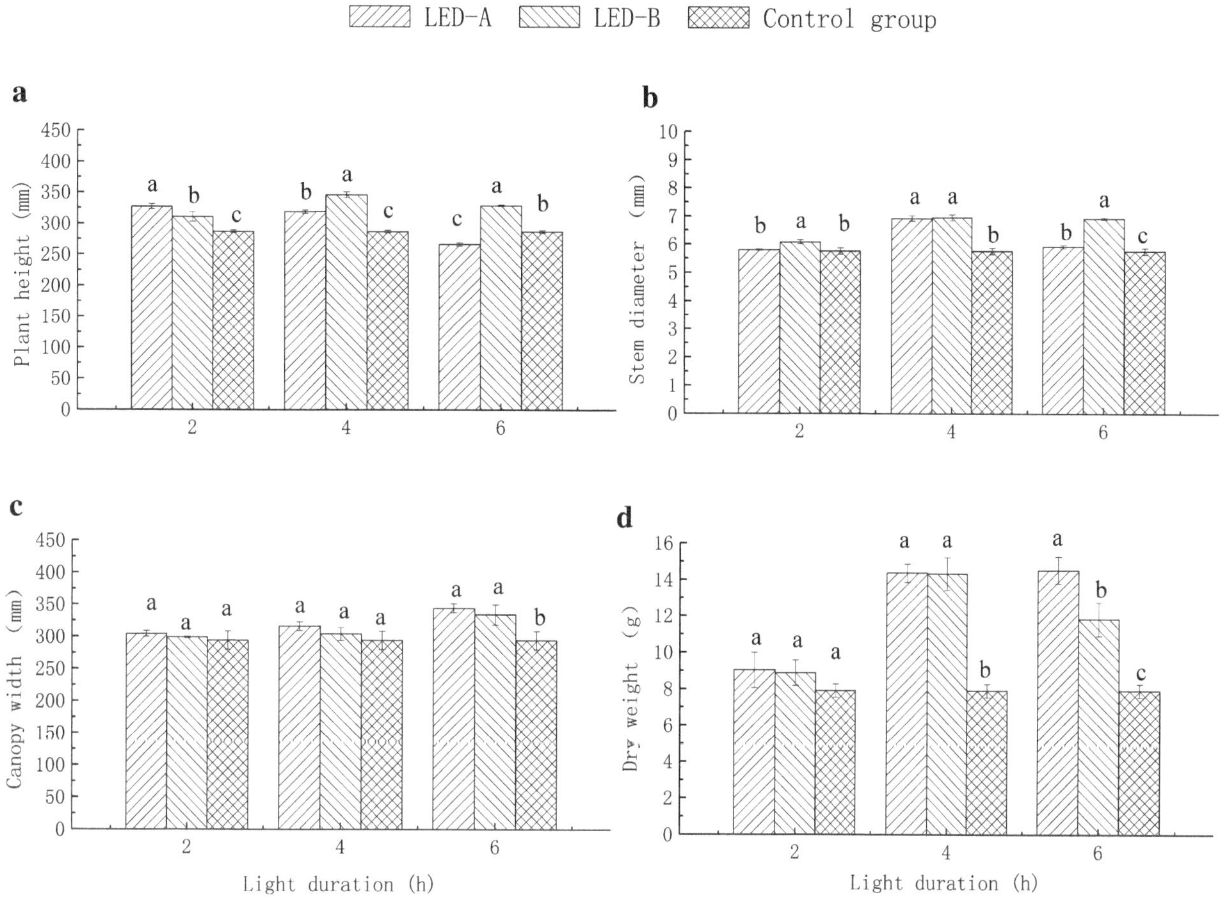

Fig. 2 Vegetative characteristics and dry mass of peppers grown under LED light supplementation during dark period. *Different letters* indicate a significant difference at $p < 0.05$. The R:B ratios of LED-A and LED-B were 6:3 and 8:1, respectively, and the control group received no light. **a** illustrates the plant height under different lighting treatments; **b** illustrates the stem diameter of plants under different lighting treatments; **c** explains the canopy width of plants under different lighting treatments; **d** explains the whole dry weight of plants under different lighting treatments

($p < 0.05$), with a 14.63 % increase compared to the control (Fig. 2a). The stem diameter of plants grown under the LED-B treatment was significantly greater than that of plants under the LED-A treatment, and was increased by 20.31 % compared to the control (Fig. 2b). The plant canopy width under LED-A was 17.01 % greater than that of the control (Fig. 2c; $p < 0.05$). There was no significant difference in plant canopy width between the LED-A and LED-B groups. Plants grown under the LED-A treatment showed the largest plant dry mass, which was 83.96 % greater than that of the control, which showed the lowest level of all groups; the difference among groups was statistically significant ($p < 0.05$, Fig. 2d).

Effects of different lighting strategies on early fruit yield and plant dry mass distribution

The dry mass distribution ratio can reflect the effects of lighting on the dry mass distribution to all of the pepper plant organs, whereas the shoot-to-root ratio reflects the effects on the dry mass distribution of the underground parts relative to the over-ground parts. As shown in Table 3, the root growth of plants under LED-B was higher than the LED-A and the control, while the stem growth of plants grown under LED-B was significant greater than the control as well as the LED-A ($p < 0.05$). Plants grown under LED-A1 showed the lowest leaf dry mass ratio, which was 24.32 % lower than that of the control. Compared with the control, plants grown under the LED-B treatment showed no significant difference in the leaf dry mass ratio and fruit dry mass ratio, whereas lighting strategy LED-A1 appeared to be the most conducive for dry weight accumulation of fruits, with a 200.00 % increase compared to that in the control groups. There was no significant difference in the shoot-to-root ratio among treatments excluded the LED-B2.

As shown in Table 4, LED-A promoted the greatest yield of plants in CLTL conditions of winter. In addition, the fruit fresh and dry weights per plant grown under LED-A were

Table 3 Dry mass distribution ratio of pepper under different lighting strategies

Lighting strategy	Dry mass distribution ratio of pepper				
	Root	Stem	Leaf	Fruit	Shoot/Root
LED-A1	0.14 ± 0.01^{ab}	0.25 ± 0.01^{e}	0.28 ± 0.02^{d}	0.33 ± 0.02^{a}	6.42 ± 1.07^{ab}
LED-B1	0.16 ± 0.01^{a}	0.35 ± 0.02^{bc}	0.35 ± 0.01^{b}	0.14 ± 0.02^{c}	5.27 ± 0.40^{b}
LED-A2	0.14 ± 0.02^{ab}	0.30 ± 0.01^{d}	0.30 ± 0.01^{c}	0.26 ± 0.03^{b}	6.38 ± 0.86^{ab}
LED-B2	0.11 ± 0.02^{b}	0.39 ± 0.01^{a}	0.38 ± 0.01^{a}	0.12 ± 0.01^{cd}	7.67 ± 1.19^{a}
LED-A3	0.16 ± 0.02^{a}	0.35 ± 0.01^{bc}	0.37 ± 0.01^{a}	0.12 ± 0.01^{cd}	5.27 ± 0.21^{b}
LED-B3	0.15 ± 0.02^{a}	0.37 ± 0.01^{a}	0.38 ± 0.01^{a}	0.10 ± 0.00^{d}	5.73 ± 0.83^{b}
Control group	0.16 ± 0.01^{a}	0.35 ± 0.01^{bc}	0.37 ± 0.01^{a}	0.11 ± 0.01^{cd}	5.13 ± 0.21^{b}

Data are the mean ± SD of each group. Different letters in same column indicate a significant difference at $p < 0.05$

Table 4 Early fruits yield of pepper under different LED lighting strategies

Lighting strategy		Average fresh weight (g $plant^{-1}$)	Average dry weight (g $plant^{-1}$)	Average fruit number
LED-A	LED-A1	32.67 ± 2.00^{a}	4.75 ± 0.38^{a}	9.3 ± 0.7^{a}
	LED-A2	26.50 ± 1.68^{ab}	3.79 ± 0.52^{b}	6.0 ± 0.7^{b}
	LED-A3	22.98 ± 4.44^{bc}	1.05 ± 0.15^{d}	6.5 ± 0.7^{b}
LED-B	LED-B1	16.15 ± 5.85^{cd}	1.68 ± 0.50^{c}	6.0 ± 0.7^{b}
	LED-B2	14.4 ± 7.10^{de}	1.72 ± 0.20^{c}	5.6 ± 0.7^{b}
	LED-B3	8.3 ± 0.4^{e}	0.92 ± 0.08^{d}	6.3 ± 0.7^{b}
Control group		8.1 ± 1.40^{e}	0.79 ± 0.04^{d}	5.7 ± 0.7^{b}

Data are the mean ± SD of each group. Different letters in same column indicate a significant difference at $p < 0.05$

significantly higher than those of the LED-B and control groups, with increases of 238.02 and 305.06 % compared to the control, respectively. Together, these results demonstrate that supplementary lighting augmented plant dry mass production (Fig. 2d), and promoted partitioning of the dry mass to the fruit (Tables 3 and 4). Moreover, LED-A supplementation could reduce the distribution of the dry mass of plants to the stems and leaves, resulting in an effective increase in fruit dry mass accumulation to ultimately increase plant yield. Specifically, plants grown under the LED-A1 strategy showed the highest fruit yield per plant, which was significantly higher than that of control group with a 303.33 and 501.27 % increase in fresh weight and dry mass of fruits, respectively.

Chlorophyll fluorescence characteristics under different lighting strategies during the dark period

The Fv/Fm ratio is the potential maximum photosynthetic ability, which is steadily maintained at 0.8–0.85 in healthy plants, and decreases in stress conditions; thus, Fv/Fm is an importance index for studying the impact of various environmental stresses on plant photosynthesis (Björkman and Demmig 1987; Pathre and Shirke 2003). Table 5 shows that the CLTL environment in winter did not destroy the photosynthetic function of peppers plants, given that the Fv/Fm ratio remained basically stable at 0.8–0.85. Furthermore, the trends in Fv/Fm dynamics were similar among the LED-A, LED-B, and control group (Fig. 3), showing an increase trend, indicating that all of the pepper plants became increasingly stronger and healthier and that the initial light energy conversion efficiency was maintained. Peppers had the highest Fv/Fm under the LED-A1 condition (Table 5).

The parameter Y(II) is used to directly estimate the efficiency of the actual captured energy of the PSII photosynthetic reaction center (Bilger and Björkman 1990), reflecting the amount of excitation energy reaching the PSII reaction center (Schreiber et al. 1995). The electron transfer rate (ETR) is used to reflect the photosynthesis rate under the actual light intensity (Kramer et al. 2004). As shown in Table 5, the Y(II) and ETR values in the experimental groups were higher than those in the control group. The Y(II) and ETR values of plants under LED-A1 strategy were significantly higher than those in the control group and other treatment groups ($p < 0.05$); the maximum values observed in the LED-A1 group were 113.70 and 114.34 % higher than the control group, respectively. This indicated that the LED-A1 strategy is the most helpful for improving the photoreaction rate of the PSII reaction center, which in turn promotes the electron transfer ability of the PSII to ultimately increase the assimilatory power (ATP and NADPH) in PSI and improves the dry mass accumulation of peppers.

Table 5 Effects of LED supplementation on chlorophyll fluorescence parameters of pepper

Lighting strategy	Y(II)	ETR	qL	NPQ	Fv/Fm
LED-A1	0.312 ± 0.090[a]	11.81 ± 3.51[a]	0.321 ± 0.130[a]	1.419 ± 0.256[d]	0.831 ± 0.014[a]
LED-B1	0.259 ± 0.065[b]	9.89 ± 2.59[b]	0.216 ± 0.085[b]	1.605 ± 0.210[ab]	0.820 ± 0.013[cd]
LED-A2	0.201 ± 0.066[c]	7.02 ± 2.96[c]	0.131 ± 0.072[c]	1.570 ± 0.249[bc]	0.829 ± 0.015[a]
LED-B2	0.176 ± 0.049[c]	7.40 ± 1.50[c]	0.135 ± 0.039[c]	1.700 ± 0.187[a]	0.827 ± 0.011[ab]
LED-A3	0.178 ± 0.073[c]	6.76 ± 2.81[c]	0.126 ± 0.073[c]	1.622 ± 0.183[ab]	0.821 ± 0.019[bc]
LED-B3	0.146 ± 0.053[d]	5.65 ± 1.93[d]	0.104 ± 0.047[c]	1.439 ± 0.270[d]	0.818 ± 0.012[cd]
Control group	0.146 ± 0.059[d]	5.51 ± 2.24[d]	0.103 ± 0.063[c]	1.479 ± 0.286[cd]	0.815 ± 0.012[d]

Data are the mean ± SD of each group; Different letters in same column indicate a significant difference at $p < 0.05$

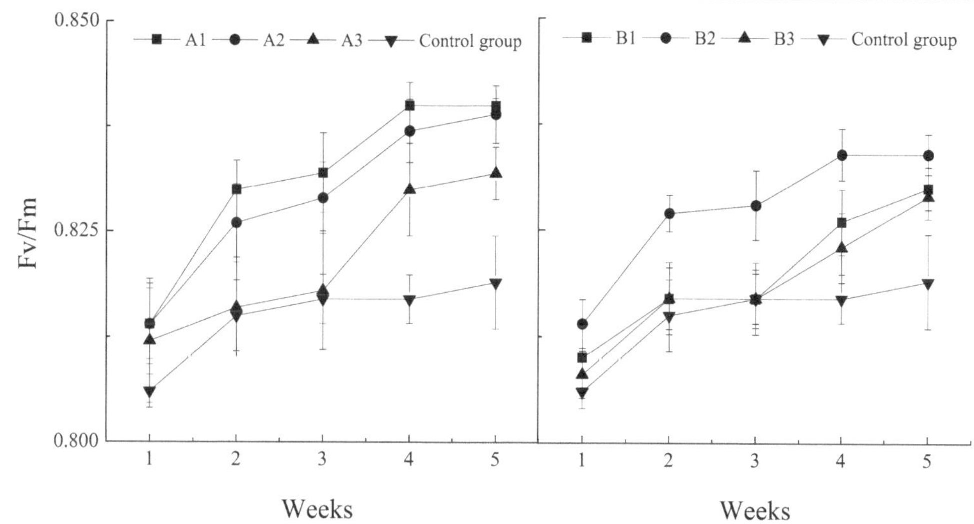

Fig. 3 Fv/Fm curve of pepper leaves for different weeks under different light supplementation strategies at night. Data are the mean ± SD of each group

The parameter qL reflects the proportion of light absorbed by the antenna pigment in PSII for the photochemical reaction (Kramer et al. 2004), and NPQ is proportional to the rate constant of thermal energy dissipation, which is related to the plant self-protection mechanism of the photosynthetic apparatus in a stressful environment (Zribi et al. 2009). Plants grown under the LED-A1 strategy had the highest qL, which was significantly higher ($p < 0.05$) than that of other treatments with a 211.65 % increase compared to the control. In addition, there was a significantly correlated relation among qL, Y(II) and ETR ($p < 0.01$), indicating that these parameters were to reflect the actual redox degree of photoelectrons in PSII. The NPQ of plants under the LED-B treatment was significantly higher than that of other groups; especially LED-B2 was the highest one with a 14.94 % increase compared to the control, indicating that the light absorbed by the plant's antenna pigments was dissipated as heat to a greater extent in the LED-B light quality, and that the plants' self-protection mechanisms of the photosynthetic apparatus in PSII was strengthened. The NPQ of plants under the LED-A1 strategy was the lowest, significant lower than the groups and with a 4.06 % increase compared to the control, indicating that more light could be used for the photochemical reaction under this condition (Table 5).

Effects of light quality and lighting duration on the chlorophyll fluorescence parameters in winter

As shown in Table 6, the light source of LEDs had a significant effect on Y(II), qL, Fv/Fm, and ETR values ($p < 0.01$). The lighting duration had extreme significant influence on all chlorophyll fluorescence parameters ($p < 0.01$). Light source and lighting duration had a significant impact on Fv/Fm and ETR ($p < 0.05$) and a highly significant effect on qL and NPQ ($p < 0.01$).

Discussion

China's Yangtze River Delta region is the most concentrated area of horticulture cultivation in the country, and its economic benefits directly reflect the productivity

Table 6 Two-way ANOVA results (F-values) for effects of supplementary light source and duration on the chlorophyll fluorescence parameters of pepper plants

Source	Y(II)	QL	NPQ	Fv/Fm	ETR
Light source	20.128**	18.393**	2.274	10.089**	7.848**
Light duration	85.719**	104.381**	7.046**	7.986**	80.325**
Light source × Light duration	1.012	12.104**	15.821**	3.138*	4.604*

* and ** indicate significant difference at $p < 0.05$ and $p < 0.01$, respectively

level of horticulture cultivation of China. However, due to its geographical location and meteorological conditions, the area is subject to more than 1 month of a harsh CLTL environment in winter, with severe consequences for production in plants grown in a greenhouse without a heater. Therefore, studies focused on improving greenhouse production in winter in this region should be a priority on the basis of plant physiology combined with agricultural engineering technology theory. In this study, LED artificial supplementary lighting technology was exposed to pepper plants grown in a greenhouse under CLTL winter conditions during dark periods to detect the influences of different light supplementation strategies on the growth and development of pepper plants.

Hao and Papadopoulos (1999) found that plant dry mass production could be increased through lighting supplementation, and dry mass allocation was increased to the fruit and decreased to the stem. In this study, the total and fruits dry mass of plants with light supplementation was significantly higher than the control, however, though the stem dry mass ratio under LED-B was significant higher than the control but it was decreased under LED-A, plant height and stem diameter were higher under LED-B than under LED-A and control lighting conditions, the stem and leaf dry mass ratio of plants under the LED-B strategy were higher than those under LED-A and control conditions, indicating that red light plays an important role in promoting elongation of the plant stem and dry matter accumulation, in order to guide assimilation of the mass to the stem and leaf (Moe et al. 2002). Blue light facilitates leaf expansion and helps to assimilate products to the fruit (Chu et al. 1999; Ni et al. 2009). Thus, the stem dry mass ratio of plants under LED-A, in this study, was lower than the control. While, the pepper leaf width was slightly higher under LED-A than that under LED-B and the control, and the fruit dry mass ratio and yield were significantly higher under LED-A than those of LED-B plants.

The duration of supplementary lighting directly affects the photoperiod of the plant, which determines the duration of plant photosynthesis, thereby affecting carbohydrate metabolism and the absorption and transformation of nutrients (Jin et al.2014). In this study, plants under an LED-A1 and LED-B1 strategies with an 18-h photoperiod showed the largest fruit dry mass ratio as well as fruit fresh weight and dry mass per plant, followed by the LED-A2 and LED-B2 strategies with a 14-h photoperiod; the LED-A3 and LED-B3 treatments showed the lowest values with respect to productivity.

The photosynthesis ability of plant leaves is directly proportional to the R/B ratio of LEDs (Hogewoning et al. 2010). While the results of this study showed that both light source and light duration had a significant impact on ETR (can be used to evaluate the photosynthetic rate of plant leaves), qL (the open ratio of photochemical reaction in PSII), Fv/Fm and NPQ. Therefore, in this study the R/B ratio of the LEDs is only one aspect that contributes to photosynthetic ability. In addition, the Fv/Fm of plants under LED-A was higher than that under LED-B when the lighting duration was the same, and was significantly increased compared to that of the control group. These results suggest that when using supplementary lighting sources with different R/B ratios, an LED with a higher amount of blue would increase the Fv/Fm, which could improve the original capturing ability of light and enhance the health of peppers. However, the optimal R/B ratio requires further study as well as the duration of supplementary lighting.

Conclusions

The results of the present study showed several benefits of LED supplementary lighting during the dark period for pepper plants grown in greenhouse in a CLTL environment. First, LED-A is particularly beneficial for promoting the production of peppers grown in the winter in a greenhouse without a heater, whereas LED-B is helpful to increase the height and stem of plants. Second, the photosynthesis ability of plants was influenced by both light source and the lighting duration of the supplementary lighting setting. In addition, a lighting setting with a high blue-to-red light ratio is more favorable for enhancing the pepper plant's natural ability to capture light energy and can improve the photosynthetic ability of the plants as well.

Abbreviations

CLTL: continuous low temperature and poor light; h: hour; PSII: photosystem II; Y(II): the light quantum yield of PSII; ETR: electron transfer rate; qL: fraction of PSII centers that are open; NPQ: rate constant of non-photochemical quenching; Fv/Fm: maximum photochemical efficiency of PSII.

Authors' contributions

XL conceived the experiments, analyzed the data and drafted the manuscript. XL, WL, GYH and ZCF performed the experiment. XCW and YZ approved the manuscript. We also thank GXS for his input during the experiment. All authors read and approved the final manuscript.

Author details
[1] College of Engineering, Nanjing Agricultural University, Nanjing, Jiangsu, People's Republic of China. [2] Jiangsu Province Engineering Lab for Modern Intelligent Facilities of Agriculture Technology & Equipment, Nanjing, Jiangsu, People's Republic of China.

Acknowledgements
This study was supported by the Natural Science Foundation of China (Project No. 61273227) and the Natural Science Foundation of Jiangsu Province (Project No. BK20150686).

Competing interests
The authors declare that they have no competing interests.

References

Barta DJ, Tibbitts TW, Bula RJ, Morrow RC (1992) Evaluation of lighting-emitting diodes characteristics for a space-based plant irradiation source. Adv Space Res 12:141–149
Bilger W, Björkman O (1990) Role of the xanthophylls cycle in photoprotection elucidated by measurements of light induced absorbance changes, fluorescence and photosynthesis in leaves of Hedera canariensis. Photosynthesis Res 25:173–185
Björkman O, Demmig B (1987) Photon yield of O2 evolution and chlorophyll fluorescence characteristics at 77 K among vascular plants of diverse origins. Planta 170:489–504
Blom TJ, Ingratta FJ (1984) The effect of high pressure sodium lighting on the production of tomatoes, cucumbers and roses. Acta Horticulturae 148:905–914
Chen M, Tang YL (2013) Chlorophyll fluorescence characteristics of Amaranthus tricolor L. under high temperature stress. Chin J Ecol 7:1813–1818
Chu ZX, Tong Z, Feng LJ, Zhang Q, Wen XG, Song ST, Zhu XF (1999) Effect of different light quality on photosynthetic characteristics of cucumber leaves. Acta Botanica Sinica 8:867–870
Ding JJ, Yang ZC, Wang PB, Geng FZ, Gao B (2005) Influence of LED light intensity on growth and photosynthetic characteristics of non-heading Chinese cabbage. J Northwest A and F Univ (Nat.) 3:113–118
Fan ZQ, Yu X, Wang J (2008) Effect of light intensity on leave physiological activity of *Brassica campestris* L. Anhui Agric Sci Bull 20(26–27):39
Fang W, Jao RC (1996) Simulation of light environment with fluorescent lamps and design of a movable light mounting fixture in a growing room. Acta Horticulturae 440:181–186
Franklin KA, Whitelam GC (2007) Light-quality regulation of freezing tolerance in *Arabidopsis thaliana*. Nat Genet 11:1410–1413
Franklin KA, Ortiz GT, Pyott DE, Halliday KJ (2014) Interaction of light and temperature signaling. J Exp Bot 11:2859–2871
Hao XM, Papadopoulos AP (1999) Effects of supplemental lighting and cover materials on growth, photosynthesis, biomass partitioning, early yield and quality of greenhouse cucumber. Sci Horticulturae 80:1–18
Hogewoning SW, Trouwborst G, Maljaars H, Poorter H, van Leperen W, Harbinson J (2010) Blue light dose-responses of leaf photosynthesis, morphology, and chemical composition of *Cucumis sativus* grown under different combinations of red and blue light. J Exp Bot 11:3107–3117
Hu HW, Yu JQ (2003) Effect of low temperature and low light on the growth, development and physiological functions of tomato plants. Chin J Ecoagric 3:55–57
Janda T, Majláth I, Szalai G (2014) Interaction of temperature and light in the development of freezing tolerance in plants. J Plant Growth Regul 30:460–469
Jin ZY, WANG SS, Liu N, Gong B, Liu SQ, Shi QH (2014) Effect of different light supplement time and method on bulbing enlargement and nutritional quality of greenhouse-grown garlic. Acta Horticulturae Sinica 12:2411–2418
Kramer DM, Johnson G, Kiirats O, Edwards GE (2004) New fluorescence parameters for the determination of QA redox state and excitation energy fluxes. Photosynth Res 79:209–218
Marcelis LFM (1993a) Fruit growth and biomass allocation to the fruits in cucumber 1. Effect of fruit load and temperature. Sci Horticulturae 54:1078–1121
Marcelis LFM (1993b) Fruit growth and biomass allocation to the fruits in cucumber 2. Effect of irradiance. Sci Horticulturae 54:123–130
Moe R, Morgan L, Grindal G (2002) Growth and plant morphology of *Cucumis sativus* and fuchsia hybrids are influenced by light quality during the photoperiod and by temperature alternations. Acta Horticulturae 580:229–234
Nhut DT, Takamura T, Watanabe H, Tanaka M (2003) Efficiency of a novel culture system by using light-emitting diode (LED) on in vitro and subsequent growth of micro propagated banana plantlets. ISHS Acta Horticulturae 616:121–127
Ni JH, Chen XH, Chen CH, Xu Q (2009) Effects of supplemental different light qualities on growth, photosynthesis, biomass partition and early yield of greenhouse cucumber. Scientia Agricultura Sinica 7:2615–2623
Patel D, Basu M, Hayes S, Majláth I, Hetherington FM, Tschaplinski TJ, Franklin KA (2013) Temperature-dependent shade avoidance involves the receptor like kinase ERECTA. Plant J 73:980–992
Pathre UV, Shirke PA (2003) Diurnal and seasonal changes in photosynthesis and photosystem2 photochemical efficiency in *prosopis juliflora* leaves subjected to natural environmental stress. Photosynthetica 41:173–185
Ren ZH, Huang W, Zhang MF (2002) Effects of low temperature and poor light on some physiological parameters of tomato. J Chin Agric Univ 1:95–101
Schreiber U, Bilger W, Neubauer C (1995) Chlorophyll fluorescence as a non-invasive indicator for rapid assessment of in vivo photosynthesis. In: Schulze ED, Caldwell MM (eds) Ecophysiology of Photosynthesis. Springer, Berlin Heidelberg, pp 49–70
Shen BY, Li YN, Zhao SQ, Ding WM, Hui N, Li J (2014) Effect of dark period lighting regulation on cucumber seedling morphology and comprehensive evaluation analysis and comprehensive evaluation. Transactions of the Chinese Society of Agricultural Engineerin. 22:201–208
Wang YJ, Zhang HY, Zhang F, Xu Y, Cao WH, Kang GB (2001) Effects of low temperature and low light intensity stress on Photosynthesis in seedlings of different cucumber varieties. Acta Horticulturae Sinica 3:230–234
Wu CC (2011) The analysis and regionalization of climate for vegetable in protected cultivation in China. Chinese Master's Theses Full-text Database: S1, Agriculture Science and Technology. 2: 1–55
Xiao YX, Gao XJ, Zhou FM (2013) Effect of supplemental lighting on growth and development of Tobacco. Tianjin Agric Sci 11:85–87
Yang QC (2008) Application and prospect of Light Emitting Diode (LED) in agriculture and bio-industry. J Agric Sci Technol 6:42–47
Zoratti L, Karppinen K, Escobar AL, Häggman H, Jaakola L (2014) Light-controlled flavonoid biosynthesis in fruits. Front Plant Sci 5:1–16
Zribi L, Fatma G, Fatma R, Salwa R, Hassan N, Ne´jib RM (2009) Application of chlorophyll fluorescence for the diagnosis of salt stress in tomato "Solanumly copersicum (variety Rio Grande)". Sci Horticulturae 120:367–372

12

Effects of volatile organic compound ether on cell responses and gene expressions in *Arabidopsis*

Yun-Ting Tseng and Kuo-Chih Lin*

Abstract

Background: The volatile organic compound ether is widely used as an industrial solvent and easily released to the environment. Our previous research indicated that ether triggers reactive oxygen species (ROS) production and activates ethylene biosynthetic genes and defense gene expressions in tomato. In the present study, we investigated the effect of ether on cell responses and gene expressions in *Arabidopsis* and compared the ROS and phytohormones produced in *Arabidopsis* and tomato plants in response to different air pollutants (O_3 vs. ether).

Results: Ether induced the sequential production of superoxide anion and hydrogen peroxide in *Arabidopsis*. Ether also triggered expressions of ethylene, salicylic acid and jasmonic acid biosynthetic genes. The temporal expression patterns of MAP kinase and protein phosphatase genes are in good accordance with those of the ethylene and salicylic acid biosynthetic genes, suggesting that induction of these phytohormone biosynthesis were through signaling pathways including both phosphorylation and/or dephosphorylation. By contrast, expression pattern of protein phosphatase PP2A3&4 coincided well with the expression of jasmonic acid biosynthetic gene LOX4, suggesting that induction of jasmonic acid biosynthesis is through PP2A3&4. However, the production of ROS and temporal expression patterns of phytohormone biosynthetic genes in *Arabidopsis* in response to ether were different from those to O_3 and were different from those in tomato as well.

Conclusions: Different plants have different strategies to respond to the same abiotic stress, and each plant species possesses its own unique signaling pathways that regulate the responding process.

Keywords: Ether, Volatile organic compound, Ozone, Reactive oxygen species, Pollutant

Background

The deleterious effects of industrial air pollution on plants have been an issue due to the massive increase in fossil fuel combustion. Among the air pollutants, tropospheric ozone (O_3) is considered the most phytotoxic, and there are many literature reports document the adverse effects of tropospheric O_3 on crop yield, forest growth, and the metabolism and physiology of plants (Karnosky et al. 2007). In addition to ozone, the role of volatile organic compounds (VOC) as air pollutants has drawn much attention (Cape 2003). VOCs are carbon-based compounds that may contain oxygen, nitrogen, bromide and chloride atom and easily evaporate into the environment. The major sources of anthropogenic airborne VOCs are industrial processes, oil refining, fuel combustions, and vehicle exhaust emissions. Experimental studies have shown that some VOCs (e.g. acetate, formaldehyde, peroxyacetyl nitrate, and trichloroethylene.) are detrimental to vegetation, although most experiments investigated short-term exposures at high concentrations as opposed to ambient air conditions (Cape 2003). Cape (2003) prepared a review article, summarizing several studies on the effects of VOCs on vegetation at low concentrations, and long exposure times. VOCs may cause detrimental effects on leaves, flowering, seed production, protein content, plant metabolism, and total dry weight. Some countries publish emission inventories of

*Correspondence: kclin@mail.ndhu.edu.tw
Department of Life Science, National Dong Hwa University, Hualien County 974, Taiwan

VOCs, and take efforts to minimize VOC release into the environment. Despite a large body of research on the effects of O_3 on plants at molecular and cellular levels, few reports describe the effects of VOCs on plants at the molecular and cellular levels.

For a period of time, the plants in our laboratory were polluted by diethyl ether (ether) that came from nearby chemical laboratory. Ether, a VOC, is widely used as an industrial solvent for waxes, fats, oils, perfumes, alkaloids, gums, and nitrocellulose, and also a solvent for hormone extraction from plant and animal tissues (Howard 1993). During its use as an industrial solvent and extractant, ether would easily release into the environment and evaporate into the atmosphere. Once in the vapor phase, it has very low chemical reactivity. Not absorbing light with wavelength longer than 290 nm, ether is not decomposed through photolysis (Howard 1993). However, ether can be photooxidized via vapor-phase reaction with photochemically produced hydroxyl radicals at a half-life of 29 h. Ether is not hydrolyzed or biodegraded in environmental media, so is persistent if released to soil or water. The compound is detectable in drinking water at water treatment plants, in the tap water in three out of ten North American cities, in drinking-water groundwater wells located downstream of municipal and industrial solid-waste landfill, and in the effluent of sewage treatment plants and chemical industries (Howard 1993). The detected concentrations range from 2.5–1000 μg/L.

In our previous research, we discovered that ether triggers a sequential production of superoxide anion ($^{\cdot}O_2^-$) and hydrogen peroxide (H_2O_2), and specifically induces tomato 1-aminocyclopropane-1-carboxylic acid oxidase (LeACO1) and LeACO4 genes through distinct signaling pathways (Lin et al. 2011). While the LeACO1 gene was induced via protein phosphorylation (e.g. via MAP kinase), the LeACO4 was induced via protein dephosphorylation (e.g. via protein phosphatase). Furthermore, induction of ACO gene expression occurred simultaneously with ROS accumulation and coincided with the occurrence of cell death. Our study results suggest that the cellular responses of tomato plants to ether are different from the plant responses to O_3, and that tomato plants respond to different air pollutants through different perceptions and downstream signaling pathways (Lin et al. 2011). These findings unveiled the complex signaling networks of plant response to different air pollutants.

The purpose of the present study is to elucidate how the model plant *Arabidopsis* responds to ether in the context of ROS production and phytohormone biosynthetic gene expressions. The results will be helpful in understanding how differently plants (tomato vs. *Arabidopsis*) respond to the same type of molecule, ether, and how plants respond to different air pollutants (O_3 vs. ether).

Methods

Plant materials

Arabidopsis seeds (*A. thaliana* cv Columbia) were soaked in ddH_2O at 4 °C for 3 days and sown in wet soil. Seven days later, the seedlings were transferred to pots and grown in a temperature- and lighting-controlled growth chamber at 22 °C at 16/8 h light/dark cycles. Twenty-five- to thirty-day old plants were used for all the experiments.

Ether fumigation and histochemical staining of $^{\cdot}O_2^-$, H_2O_2, and dead cells

For ether fumigation, twenty-five- to thirty-day old plants were exposed to ether (Sigma, MO, USA) with 500 μL/L in a 4 L can for various time intervals. Histochemical staining of $^{\cdot}O_2^-$, H_2O_2, and dead cells were performed as described by Lin et al. (2011). Two *Arabidopsis* plants were exposed to 500 μL/L ether fumigation for each time point, and all the mature leaves were harvested for histochemical staining. The $^{\cdot}O_2^-$ production was detected by nitroblue tetrazolium (NBT) staining. The leaves were vacuum-infiltrated with 50 mL staining buffer [10 mM NaN_3, 10 mM potassium phosphate buffer, pH7.8, and 0.1 % NBT] (Sigma, MO, USA) for 1 min and with a total of four times. The leaves were then incubated in the staining buffer in dark for 30 min. After staining, the leaves were cleared in boiling 70 % ethanol for 15 min. Production of $^{\cdot}O_2^-$ was directly visualized by forming blue formazan precipitate. The H_2O_2 production was detected by the 3,3′-diaminobenzidine tetrahydrochloride (DAB) staining. The whole plants were removed from soil and soaked in 100 mL DAB staining buffer [10 mM 2-(*N*-morpholino)-ethane sulfonic acid (MES), pH6.5, and 0.1 % DAB] (Sigma, MO, USA) for 10 h in the dark. Following staining, the leaves were cleared in boiling 70 % ethanol for 15 min. Production of H_2O_2 was visualized as a reddish-brown coloration. Dead cells in the mesophyll tissues were assessed by the Evans blue staining. The leaves were vacuum-infiltrated with 50 mL of 1 % (w/v) Evans blue (Sigma, MO, USA) staining buffer for 1 min with a total of four times and then incubated in the staining buffer in dark for 30 min. After staining, the leaves were cleared in boiling 70 % ethanol for 5 min. Dead cells were unable to exclude Evans blue and were stained deep blue. The samples were mounted on glass slides with 30 % glycerol and examined with a fluorescence microscope (Eclipse E600, Nikon, Japan). The mechanically wounded plants that were punched with a brush leaving 0.5–1 mm holes spaced 2 mm apart were used a positive control. The density of the brush hole is approximately 25 holes/cm^2. The wounded plants were then incubated in growth chamber for 3 h before staining.

RNA extraction and real-time reverse transcription PCR (RT-PCR)

For RNA extraction, two plants were used for each time point and five to six mature leaves were collected. Total RNA was isolated from treated leaves using TRIzol™ reagent (Ambion, CA, USA) according to the manufacture's protocol. Briefly, approximately 100 mg of leaves were homogenized with 1 mL of TRIzol reagent with a pestle. Two hundred microliters of chloroform were added to separate the RNA from cell pellets. The RNA was precipitated with 0.5 mL isopropanol. The RNA precipitates were washed with 70 % ethanol, air-dried, and resuspended with 20 μL DEPC-treated ddH_2O. The concentration of RNA was determined by spectrophotometer. For cDNA synthesis, reverse transcription was performed in a 30 μL reaction volume containing 3 μg of the total RNA, 100 units of MMLV reverse transcriptase (Promega, WI, USA), 1X MMLV reaction buffer, 0.1 mM each of dNTPs (Thermo Scientific, MA, USA), and 1 μM oligo dT_{18} primer (Protech, Taipei, Taiwan). The reaction was carried out at 37 °C for 2 h. Real-time RT-PCR amplification of the cDNA clones were performed in a 10 μL reaction volume containing 2 μg of cDNA, 1X GoTaq™ qPCR Master Mix reagent (Promega, WI, USA), 1X CXR reference dye, and 0.2 μM of each primer (Additional file 1: Table S1). Thermocycling conditions were 95 °C for 2 min followed by 39 cycles of 95 °C for 15 s, 55 °C for 45 s, and 72 °C for 30 s, and a final 5 min at 72 °C on a CFX96™ real-time PCR detection system (Bio-Rad, USA). The amplified *Arabidopsis Actin2* (ACT2) was used as an internal control. Relative gene expression levels were calculated with the $2^{-\triangle\triangle Ct}$ method. Each value was the mean ± standard deviation of three independent experiments. The results were analyzed using Student's *t* test. Differences of relative fold increase between the 0 min the other time points were considered statistically significant if $P < 0.05$ (*).

Results

Detection of cell death and ROS generation in response to ether

In an attempt to understand *Arabidopsis* plant response to acute exposure to ether, twenty-five- to thirty-day old *Arabidopsis* plants were exposed to ether fumigation for 24 h. No obvious cell death occurred during the 24 h of ether exposure (Fig. 1). To elucidate whether ether could induce cellular ROS (e.g. $^{\cdot}O_2^-$ and H_2O_2) production, *Arabidopsis* plants were exposed to ether for various time intervals and histochemically stained with nitroblue tetrazolium (NBT) and 3,3′-diaminobenzidine tetrahydrochloride (DAB) to detect the accumulation of $^{\cdot}O_2^-$ and H_2O_2, respectively. We detected a burst of $^{\cdot}O_2^-$ production at 30 min after initiation of ether fumigation, and the amount decreased at 1 h and kept decreasing to a scarce level at 2 h (Fig. 2). No $^{\cdot}O_2^-$ production was detectable thereafter. We detected a trace amount of H_2O_2 at 30 min, and a burst of H_2O_2 production was observed during 1–2 h of exposure (Fig. 3). The accumulation of H_2O_2 declined at 4 h, and the H_2O_2 levels remained low till 12 h. However, a small rise of H_2O_2 production was detectable after 24 h exposure to ether.

Expressions of ROS producing and phytohormone biosynthetic genes in response to ether

To further investigate the oxidative burst reactions in *Arabidopsis* in response to ether, we studied the temporal expression patterns of genes involved in ROS production. The selected genes were the NADPH oxidase (RBOHD), a gene responsible for catalyzing O_2 to $^{\cdot}O_2^-$, copper/zinc superoxide dismutase 1 (CSD1), a gene responsible for reducing $^{\cdot}O_2^-$ to H_2O_2, and L-ascorbate peroxidase 1 (APX1) which involves in H_2O_2 reduction reaction (Overmyer et al. 2003). The results in Fig. 4 showed that the transcript levels of RBOHD increased at 15 min and then quickly decreased to the basal level at 30 min and thereafter. The transcript levels of CSD1 increased at

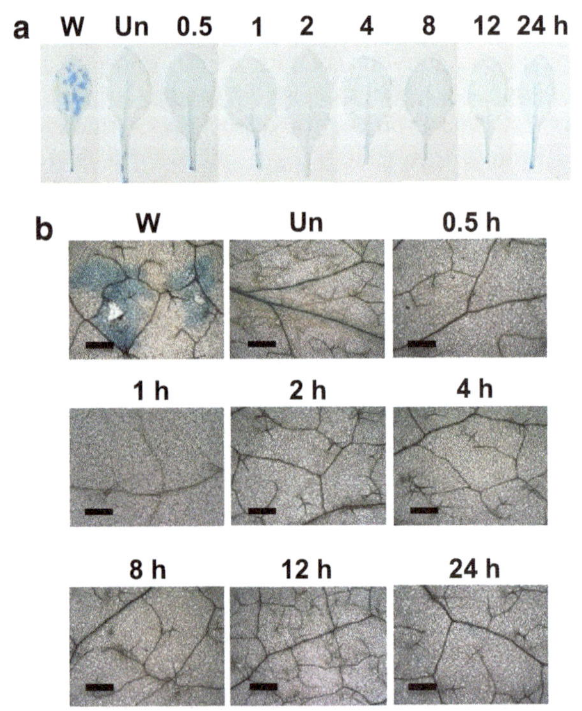

Fig. 1 *Evans blue* assay for cell death caused by ether on *Arabidopsis* leaves. *Arabidopsis* plants were fumigated with 500 μL/L of ether for various time intervals (**a**). Panel (**b**) is the close-up pictures of the *Evans blue* staining. Three replicates for each treatment were performed, and representative leaves were presented. *W* wounding; *Un* untreated plants. *Scale bars* = 300 μm

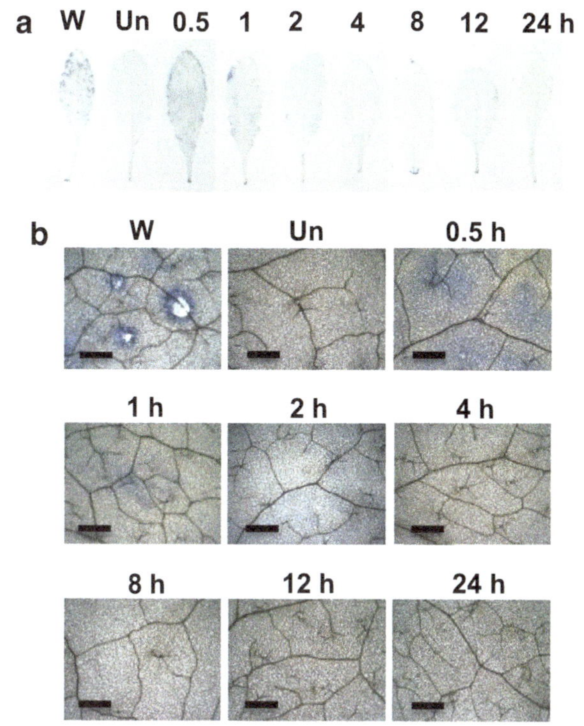

Fig. 2 NBT assay for O_2^- production induced by ether on *Arabidopsis* leaves. *Arabidopsis* plants were fumigated with 500 μL/L of ether for various time intervals (**a**). Panel (**b**) is the close-up pictures of the NBT stainings. Three replicates for each treatment were performed, and representative leaves were presented. *W* wounding; *Un* untreated plants. *Scale bars* = 300 μm

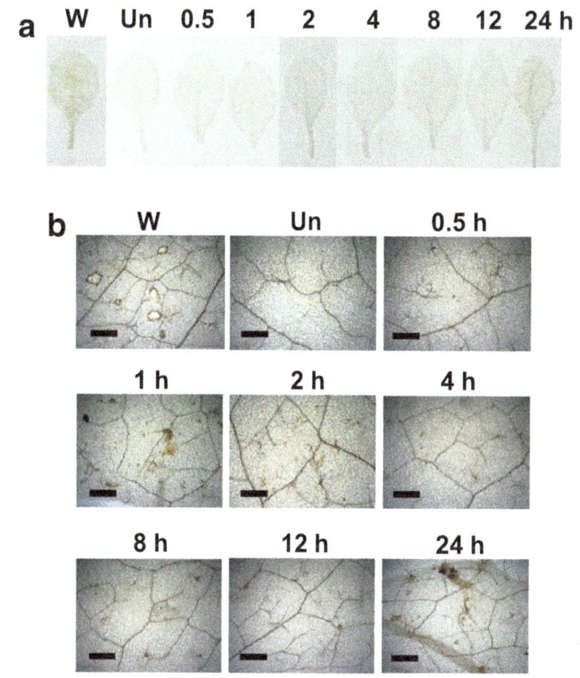

Fig. 3 DAB assay for H_2O_2 production induced by ether on *Arabidopsis* leaves. *Arabidopsis* plants were fumigated with 500 μL/L of ether for various time intervals (**a**). Panel (**b**) is the close-up pictures of the DAB staining. Three replicates for each treatment were performed, and representative leaves were presented. *W* wounding; *Un* untreated plants. *Scale bars* = 300 μm

15–60 min and then decreased to the basal level thereafter. The transcript levels of APX1 showed a decreasing pattern over time.

To investigate the phytohormone biosynthesis in *Arabidopsis* in response to ether fumigation, we studied the temporal expression patterns of genes involved in biosynthesis of ethylene, salicylic acid, and jasmonic acid (Fig. 5). The selected genes were 1-aminocyclopropane-1-carboxylate synthase (ACS), 1-aminoocyclopropane-1-carboxylate oxidase (ACO), isochorismate synthase (ICS), and lipoxygenase (LOX) (Tamaoki 2008). ACS and ACO are responsible for ethylene biosynthesis. ACS converts S-adenosyl-methionine to 1-aminocyclopropane-1-carboxlic acid, and ACO subsequently converts 1-aminocyclopropane-1-carboxlic acid to ethylene. ICS, a key enzyme in salicylic acid biosynthesis, converts chorismate to isochorismate. LOX is a gene responsible for converting linolenic acids into 13-hydroperoxy-octadecatrienoic acid in the jasmonic acid biosynthetic pathway. Two ACS gene members, ACS2 and ACS6, which are activated in *Arabidopsis* in response to O_3 were analyzed (Tosti et al. 2006). We detected no ACS2 transcript signal from ether-treated plants (data not shown). By contrast, the ACS6 transcripts greatly accumulated at 15 min, fell to the basal level at 60 min, and decreased thereafter. We designed gene-specific primer sets for analyzing the temporal expression patterns of ACO1–ACO5 genes in response to ether. Of these genes, we detected no signal for the ACO1 and ACO5 transcripts, and the transcript levels of ACO2 showed a decreasing pattern over time (data not shown). However, the ACO3 transcripts rapidly accumulated at 15 min and then quickly decreased to the basal level at 30 min. The ACO4 transcripts promptly accumulated at 15 min and then gradually decreased to basal level at 240 min. The transcript levels of ICS increased at 15 min and reached a maximum by 30 min. The transcript levels then started decreasing during the rest of exposure period. We chose LOX4 and LOX5, each belongs to the lipoxygenase gene family of distinct enzymatic activity groups, for analyzing their expression upon ether treatment (Bannenberg et al. 2008). The accumulation level of LOX4 transcripts exhibited a biphasic pattern. The transcript levels showed a quick rise at 15 min, reached a peak at 60 min, and declined to basal level by 120 min. The second peak was detected at 240 min post-exposure. We detected no signal for the LOX5 transcripts upon ether exposure (data not shown).

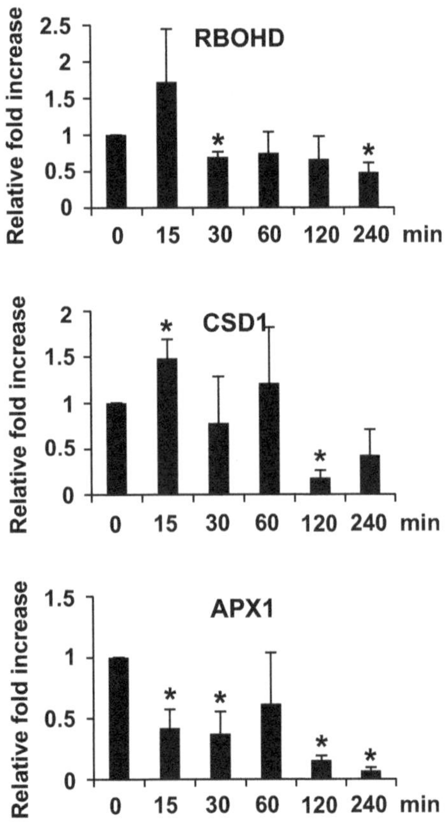

Fig. 4 Real-time RT-PCR assay of ROS metabolizing gene expressions in response to ether. The relative fold increase of each gene was the average of three independent repeats

Meanwhile, we determined the expression patterns of the pathogenesis-related protein (PR) 1, PR2, PR3, PR4, and PR5 in response to ether. We detected no PR1 and PR3 transcript signal from ether-treated plants (data not shown). The transcript levels of PR2 increased at 15 min, reached a maximal level by 30 min, and then decreased to basal level at 60 min (Fig. 6). The transcript levels of PR4 and PR5 increased at 15 min and then quickly decreased to basal level at 30 min after exposure.

Determination of signaling pathways in response to ether

To elucidate whether protein phosphorylation and/or dephosphorylation might be involved in ether-induced signaling pathway, we investigated the temporal expression patterns of protein kinase and protein phosphatase genes in response to ether fumigation by real time RT-PCR. The protein kinase genes include mitogen-activating protein kinase (MPK) 1, MPK3, MPK4, and MPK6 which are activated in response to O_3 and other abiotic stresses (Ahlfors et al. 2004; Mishra et al. 2006). The results in Fig. 7 revealed that MPK1 and MPK6 exhibited similar transcript accumulation patterns, which are quickly accumulated at 15 min and dropped to basal level at 30 min and thereafter. Increasing levels of MPK3 transcripts appeared at 15–60 min, and then dropped to basal level thereafter. The transcript levels of MPK4 increased

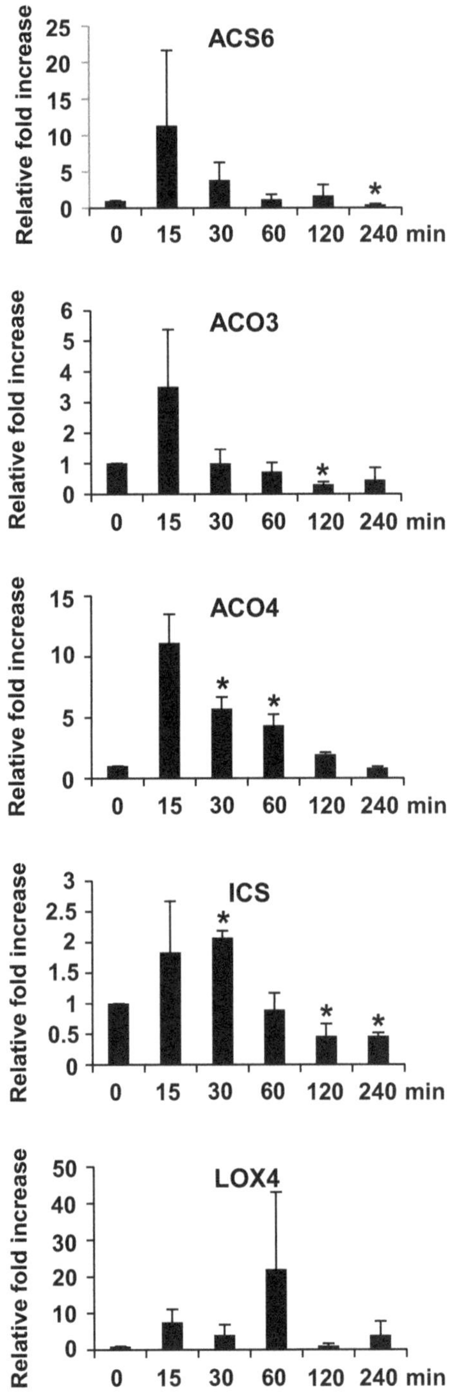

Fig. 5 Real-time RT-PCR assay of hormone biosynthetic gene expressions in response to ether. The relative fold increase of each gene was the average of three independent repeats

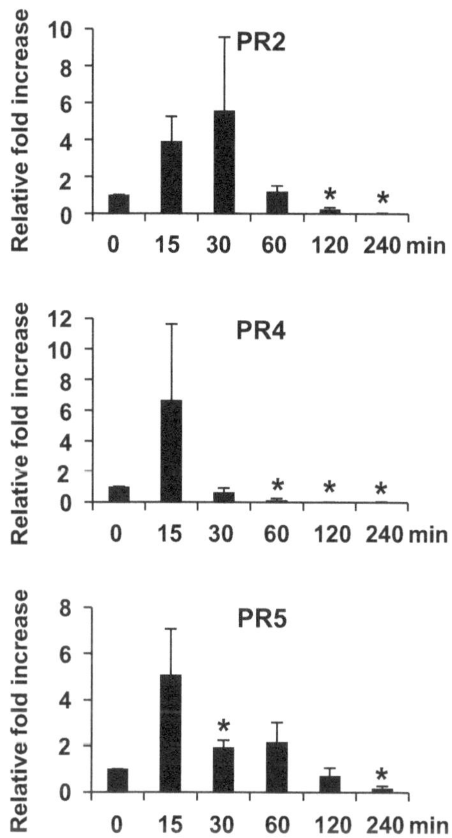

Fig. 6 Real-time RT-CR assay of pathogenesis-related protein gene expressions in response to ether. The relative fold increase of each gene was the average of three independent repeats

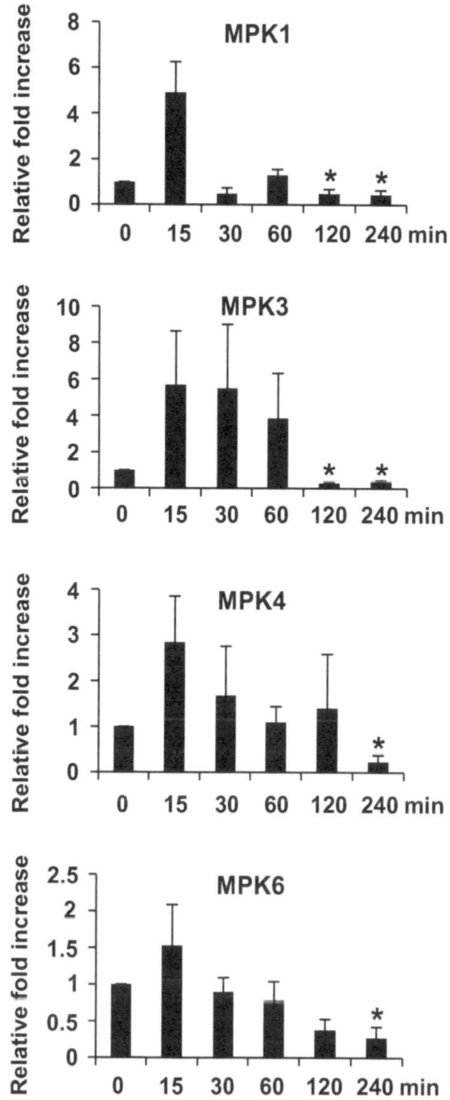

Fig. 7 Assay for ether responsive phosphorylation signaling pathways with real-time RT-PCR. The relative fold increase of each gene was the average of three independent repeats

at 15–30 min and dropped to basal level at 60 min and thereafter.

We studied the expression patterns of protein phosphatase genes in response to ether by real time RT-PCR. These genes include MAP kinase phosphatase 1 (MKP1), protein phosphatase (PP) 2A1 (PP2A1), PP2A2, PP2A3&4, and PP2C-type phosphatase AP2C1 (AP2C1). These phosphatases are involved in modulating phytohormone biosynthesis in response to ozone or developmental cues (Ludwików et al. 2014; Schweighofer et al. 2007; Skottke et al. 2011). The results in Fig. 8 revealed that the transcript accumulation patterns of MKP1, AP2C1, and PP2A1 were similar. Their transcript levels accumulated at 15 min and then gradually decreased to basal level by 30 or 120 min post-exposure. PP2A2 showed a straight decreasing pattern during the ether fumigation period. By contrast, the transcript accumulation of PP2A3&4 exhibited a biphasic pattern. The first peak was present at 15 min and declined to basal level by 60 min. The second peak was detectable at 120 min post-exposure and returned to basal level by 240 min.

Discussion

Base on the observations of ROS production, ROS and phytohormone biosynthetic gene expressions, PR gene expressions and signaling gene expressions, we proposed a model for depicting the responses of *Arabidopsis* to ether fumigation (Fig. 9).

Oxidative burst is a general response of plants to different air pollutants

Oxidative burst is a general response of sensitive plants to air pollutant O_3 (Wohlgemuth et al. 2002). In tomato and tobacco, H_2O_2 is the major ROS produced in response to O_3, and ethylene is involved in the amplification of

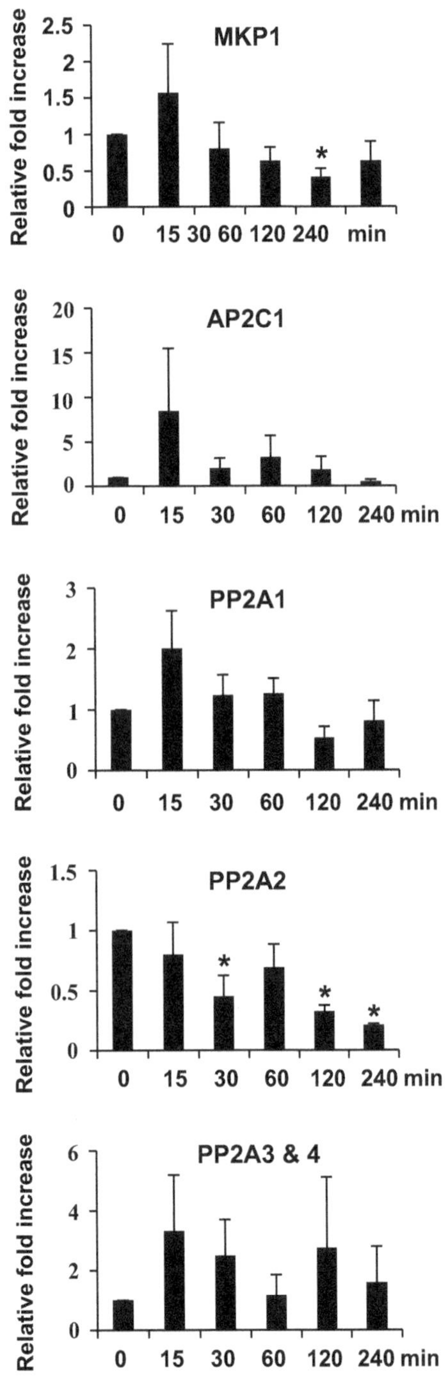

Fig. 8 Assay for ether responsive dephosphorylation signaling pathways with real-time RT-PCR. The relative fold increase of each gene was the average of three independent repeats

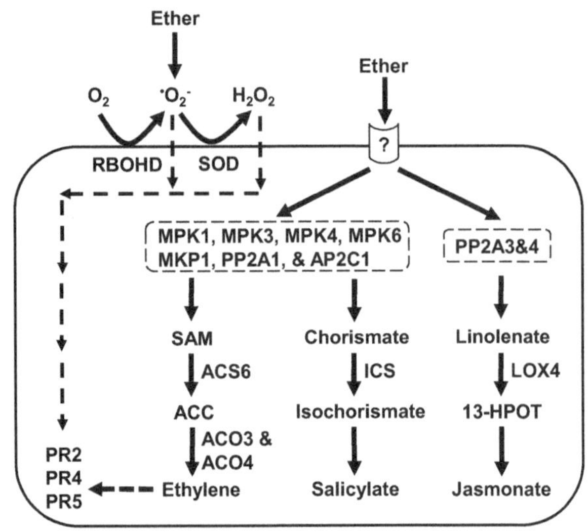

Fig. 9 A schematic model illustrating the responses and signaling pathways for activation of gene expressions in response to ether. The *dashed boxes* indicate possible kinase and phosphatase involved. The *dashed lines* indicate the possible involvement of ethylene and ROS on PR protein gene induction

ROS production and regulation of cell death (Moeder et al. 2002; Schraudner et al. 1998; Castagna et al. 2007). By contrast, $^{\cdot}O_2^-$ is the major ROS produced in *Arabidopsis* in response to O_3, and salicylic acid is responsible for cell death (Rao and Davis 1999). Our previous study showed that acute exposure of tomato plants to ether causes cell death and that $^{\cdot}O_2^-$ was the major ROS produced in response to ether (Lin et al. 2011). In the present study, we detected a transient, sequential production of $^{\cdot}O_2^-$ and H_2O_2 in *Arabidopsis* in response to ether, and no cell death observed, suggesting that *Arabidopsis* is not as susceptible to ether treatment as tomato is (Figs. 1, 2, 3). Expression patterns of ROS metabolizing genes such as RBOHD and CSD1 corresponded well with the histochemical observations of ROS accumulation. Production of RBOHD transcripts at 15 min after initiation of ether fumigation coincided well with the accumulation of $^{\cdot}O_2^-$ at 30 min post-exposure (Figs. 2, 4). Diminution of $^{\cdot}O_2^-$ at 30–60 min was accompanied by accumulation of H_2O_2 at 30–120 min (Figs. 2, 3). Furthermore, induction of CSD1 at 15–60 min was also in good accordance with the production of H_2O_2 (Figs. 3, 4). All these results suggested that $^{\cdot}O_2^-$ and H_2O_2 were sequentially produced by RBOHD and CSD, respectively, in *Arabidopsis* in response to ether, and oxidative burst is a general response of plants to different air pollutants (Fig. 9).

Ascorbate is believed to provide important protection in the apoplast against O_3-derived ROS (Castagna and Ranieri, 2009). In addition to direct reaction with ROS, ascorbate also acts as a reducing substrate for ascorbate peroxidase (APX), which scavenges H_2O_2 and thus affects the propagation of the O_3 stress. However, in this study no induction of APX1 gene expression in ether-treated *Arabidopsis* suggested that ether-derived ROS were more likely to be quickly quenched by direct reaction with ascorbate or other antioxidant systems rather than by APX1 (Fig. 4).

Induction of salicylate and ethylene accumulation by ether does not suppress jasmonate biosynthesis

In O_3-treated *Arabidopsis*, salicylic acid is the key signal in antioxidant response and in initiating O_3-induced cell death in *Arabidopsis* (Rao and Davis 1999; Tamaoki 2008). O_3 stimulates the production of ethylene via salicylic acid. Both salicylic acid and ethylene suppress jasmonic acid biosynthesis and act in concert to induce high levels of $^{\cdot}O_2^-$ or $^{\cdot}O_2^-$ together with H_2O_2, and the production of ROS finally leads to cell death (Rao and Davis 1999; Wohlgemuth et al. 2002). By contrast, in O_3-treated tomato ethylene is the key factor that triggers the progression of O_3-induced lesions (Castagna et al. 2007). Furthermore, O_3 induce ethylene and jasmonate biosynthesis, although with different timings and to different extents. In the present study, ether stimulated expressions of salicylic biosynthetic gene, ICS, transcripts from 15–30 min, and the amount returned to basal level by 60 min post-exposure (Fig. 5). The transcript levels of ethylene biosynthetic genes, ACS6, ACO3, and ACO4, were also promptly accumulated at 15 min and began to decrease at 30 min (Fig. 5). Occurrence of ROS production and ethylene and salicylic acid biosynthesis at the very early stage of ether treatment suggested that these three events simultaneously occurred in response to ether (Figs. 2, 3, 4, 5). Furthermore, no cell death occurred during this time period or thereafter, suggesting that production of ROS and ethylene and salicylic acid biosynthesis in response to ether did not promote cell death as they do in *Arabidopsis* in response to O_3.

We also detected an accumulation of jasmonic acid biosynthetic gene transcripts, LOX4, following exposure to ether for 15 min and maintaining at high level till 240 min, suggesting that ether also promptly induces jasmonic acid biosynthesis (Fig. 5). However, continuous induction of LOX4 gene at the time that ICS, ACS6, ACO3, and ACO4 gene expressions decreased in response to ether indicated that induction of salicylic acid and ethylene did not suppress the jasmonic acid biosynthesis as they do in response to O_3 (Rao and Davis 1999; Wohlgemuth et al. 2002). Biosynthesis of jasmonic acid may inhibit the spread of cell death (Tuominen et al. 2004). Therefore, it is possible that continuous production of jasmonic acid suppressed the occurrence and spread of cell death during ether exposure.

We assayed five PR genes and found that the expression patterns of PR2, PR4, and PR5 were similar to those of the ethylene and salicylic acid biosynthetic genes (Figs. 5, 6). Expressions of SA-dependent PR1 and ethylene-dependent PR4 genes in *Arabidopsis* have been reported to be induced under O_3 stress (Rao et al. 2002). The result of our previous study suggested that ethylene produced in tomato plants in response to ether activates expression of PR4 but not PR1 genes (Lin et al. 2011). Therefore, it is possible that expressions of these PR genes were induced by ethylene in *Arabidopsis* in response to ether (Fig. 9, dash lines). In addition, low dosage exposure to ROS may produce signaling effects on growth, development, stress acclimation, and pathogen responses (Foyer and Noctor 2005). Therefore, ether-generated ROS may play a role in signaling modulation and induce the expression of the PR genes (Fig. 9, dash lines). However, whether expressions of the PR genes were induced by ethylene or ROS signaling during ether fumigation remains unclear and needs further elucidation.

The ether responsive signaling pathways

The signal transduction pathways induced by O_3 exposure have been extensively elucidated in *Arabidopsis*, whereas rare reports, if any, are focused on tomato. MPK3 and MPK6 in *Arabidopsis* are activated within 30–60 min in response to O_3 (Ahlfors et al. 2004). Phosphorylation of ACS2 and ACS6 by MPK3 and MPK6 results in increased ethylene biosynthesis through protein stabilization (Liu and Zhang 2004; Han et al. 2010). Skottke et al. (2011) showed that PP2A mediates a finely tuned regulation of overall ethylene production by differentially affecting the stability of specific classes of ACS enzymes. Ludwików et al. (2014) showed that the *Arabidopsis* protein phosphatase type 2C is involved in the regulation of ethylene biosynthesis during ozone stress. In this study, the temporal expression patterns of MAP kinase and protein phosphatase genes were in good accordance with those of the hormone biosynthetic genes. The temporal expression patterns of ACS6, ACO3, ACO4 and ICS were in unanimity with the expression patterns of MAP kinase, MPK1, MPK3, MPK4, and MPK6, and protein phosphatase, MKP1, AP2C1, and PP2A1 (Figs. 5, 7, 8). They were all induced at 15–60 min, suggesting that MPK1, MPK3, MPK4, MPK6, MKP1, AP2C1, and PP2A1 were involved in protein phosphorylation/dephosphorylation leading to induction of ethylene and salicylic acid biosynthesis by ether (Fig. 9). Expression patterns of protein phosphatase PP2A3&4 coincided well with the expression of LOX4 (Figs. 5, 8). Both LOX4 and PP2A3&4 transcripts exhibited a biphasic accumulation pattern. Their transcript levels showed a quick rise at 15 min and reached a peak at 15–60 min, and the second peak was detected at 120–240 min post-exposure. The sequential expression of PP2A3&4 and LOX4 genes suggested that induction of jasmonic acid biosynthesis might be through phosphatase PP2A3&4 (Fig. 9). Although further experiments are required to confirm our model, ether induction of MAP kinase and protein phosphatase genes reinforces the idea that signaling events initiated by diverse biotic or abiotic stresses may converge at the point of MAP kinases and/or phosphatase (Farkas et al. 2007; Zhang et al. 2006).

Conclusion

In this study, $^{\cdot}O_2^{-}$ and H_2O_2 were sequentially produced in *Arabidopsis* in response to ether, indicating that oxidative burst is a general response of plants to different air pollutants. Induction of salicylic acid and ethylene did not suppress the jasmonic acid biosynthesis as they do in response to O_3 treatment, suggesting that plant responds differently to different air pollutants. The different ROS and hormones produced in *Arabidopsis* and tomato plants in response to different air pollutants (O3 vs. ether) suggested that different plants have different strategies to respond to the same abiotic stress, and each plant species possesses its own unique signaling pathways that regulate the responding process.

Authors' contributions

YTT designed and carried out the experiments, analyzed and interpreted data, and helped to draft the manuscript. KCL designed the experiments, analyzed and interpreted data, and drafted the manuscript. Both authors read and approved the final manuscript.

References

Ahlfors R, Macioszek V, Rudd J, Brosché M, Schlichting R, Scheel D, Kangasjärvi J (2004) Stress hormone-independent activation and nuclear translocation of mitogen-activated protein kinases in *Arabidopsis thaliana* during ozone exposure. Plant J 40:512–522

Bannenberg G, Martínez M, Hamberg M, Castresana C (2008) Diversity of the enzymatic activity in the lipoxygenase gene family of *Arabidopsis thaliana*. Lipids 44:85–95

Cape JN (2003) Effects of airborne volatile organic compounds on plants. Environ Pollut 122:145–157

Castagna A, Ranieri A (2009) Detoxification and repair process of ozone injury: from O_3 uptake to gene expression adjustment. Environ Pollut 157:1461–1469

Castagna A, Ederli L, Pasqualini S, Mensuali-Sodi A, Baldan B, Donnini S, Ranieri A (2007) The tomato ethylene receptor LE-ETR3 (NR) is not involved in mediating ozone sensitivity: causal relationships among ethylene emission, oxidative burst and tissue damage. New Phytol 174:342–356

Farkas I, Dombrádi V, Miskei M, Szabados L, Koncz C (2007) *Arabidopsis* PPP family of serine/threonine phosphatases. Trends Plant Sci 12:169–176

Foyer CH, Noctor G (2005) Oxidant and antioxidant signaling in plants: a re-evaluation of the concept of oxidative stress in a physiological context. Plant Cell Environ 28:1056–1071

Han L, Li GJ, Yang KY, Mao G, Wang R, Liu Y, Zhang S (2010) Mitogen-activated protein kinase 3 and 6 regulate *Botrytis cinerea*-induced ethylene production in *Arabidopsis*. Plant J 64:114–127

Howard PH (1993) Handbook of environmental fate and exposure data for organic chemicals. Diethyl ether, vol Vol. IV. Lewis Publishers, Florida, pp 221–229

Karnosky DF, Skelly JM, Percy KE, Chappelka AH (2007) Perspectives regarding 50 years of research on effects of tropospheric ozone air pollution on US forests. Environ Pollut 147:489–506

Lin KC, Sun PC, Lin PL (2011) Production of reactive oxygen species and induction of signaling pathways for the ACO gene expressions in tomato plants triggered by the volatile organic compound ether. Plant Cell Rep 30:599–611

Liu Y, Zhang S (2004) Phosphorylation of 1-aminocyclopropane-1-carboxylic acid synthase by MPK6, a stress-responsive mitogen-activated protein kinase, induces ethylene biosynthesis in *Arabidopsis*. Plant Cell 16:3386–3399

Ludwików A, Cieśla A, Kasprowicz-Maluśki A, Mituła F, Tajdel M, Gałgański L, Ziółkowski PA, Kubiak P, Małecka A, Piechalak A, Szabat M, Górska A, Dąbrowski M, Ibragimow I, Sadowski J (2014) *Arabidopsis* protein phosphatase 2C ABI1 interacts with type I ACC syntheses and is involved in the regulation of ozone-induced ethylene biosynthesis. Mol Plant 7:960–976

Mishra NS, Tuteja R, Tuteja N (2006) Signaling through MAP kinase networks in plants. Arch Biochem Biophys 452:55–68

Moeder W, Barry CS, Tauriainen AA, Betz C, Tuomainen J, Utriainen M, Grierson D, Sandermann H, Langebartels C, Kangasjärvi J (2002) Ethylene synthesis regulated by biphasic induction of 1-aminocyclopropane-1-carboxylic acid synthase and 1-aminocyclopropane-1-carboxylic acid oxidase genes is required for hydrogen peroxide accumulation and cell death in ozone-exposed tomato. Plant Physiol 130:1918–1926

Overmyer K, Brosché M, Kangasjärvi J (2003) Reactive oxygen species and hormonal control of cell death. Trends Plant Sci 8:335–342

Rao MV, Davis KR (1999) Ozone-induced cell death occurs via two distinct mechanisms in *Arabidopsis*: the role of salicylic acid. Plant J 17:603–614

Rao MV, Lee HI, Davis KR (2002) Ozone-induced ethylene production is dependent on salicylic acid, and both salicylic acid and ethylene act in concert to regulate ozone-induced cell death. Plant J 32:447–456

Schraudner M, Moeder W, Wiese C, Camp WV, Inzé D, Langebartels C, Sandermann H Jr (1998) Ozone-induced oxidative burst in the ozone biomonitor plant, tobacco Bel W3. Plant J 16:235–245

Schweighofer A, Kazanaviciute V, Scheikl E, Teige M, Doczi R, Hirt H, Schwanninger M, Kant M, Schuurink R, Mauch F, Buchala A, Cardinale F, Meskienea I (2007) The PP2C-type phosphatase AP2C1, which negatively regulates MPK4 and MPK6, modulates innate immunity, jasmonic acid, and ethylene levels in *Arabidopsis*. Plant Cell 19:2213–2224

Skottke KR, Yoon GM, Kieber JJ, DeLong A (2011) Protein phosphatase 2A controls ethylene biosynthesis by differentially regulating the turnover of ACC synthase isoforms. PLoS Genet 7:e1001370. doi:10.1371/journal.pgen.1001370

Tamaoki M (2008) The role of phytohormone signaling in ozone-induced cell death in plants. Plant Signal Behav 3:166–174

Tosti N, Pasqualini S, Borgogni A, Ederli L, Falistocco E, Crispi S, Paolocci F (2006) Gene expression profiles of O_3-treated *Arabidopsis* plants. Plant Cell Environ 29:1686–1702

Tuominen H, Overmyer K, Keinänen M, Kollist H, Kangasjärvi J (2004) Mutual antagonism of ethylene and jasmonic acid regulates ozone-induced spreading cell death in *Arabidopsis*. Plant J 39:59–69

Wohlgemuth H, Mittelstrass K, Kschieschan S, Bender J, Weigel HJ, Overmyer K, Kangasjärvi J, Sandermann H, Langebartels C (2002) Activation of an oxidative burst is a general feature of sensitive plants exposed to the air pollutant ozone. Plant Cell Environ 25:717–726

Zhang T, Liu Y, Yang T, Zhang L, Xu S, Xue L, An L (2006) Diverse signals converge at MAPK cascades in plant. Plant Physiol Biochem 44:274–283

13

Range extension of *Christisonia scortechinii* from mainland Southeast Asia into Borneo, and notes on the distinction between *Aeginetia* and *Christisonia* (Orobanchaceae)

Antony van der Ent[1,2] and K. M. Wong[3*]

Abstract

Background: *Christisonia* is a little-documented and poorly studied root-parasitic genus in the Orobanchaceae occurring in India, China, Indochina and part of the Malesian region. Recent collection of a *Christisonia* taxon in Kinabalu Park in Sabah, Borneo, taxonomically identical to earlier Sabah collections that have hitherto not been recorded in the literature, led to an assessment of the taxonomic identity of the species against *Christisonia scortechinii*, *C. siamensis*, *C. sinensis* and related species.

Results: Some taxa in China, Indochina, the Malay Peninsula, and the Philippines are morphologically identical to the Borneo taxon except in the number of calyx lobes, but differ by several distinctive characters from other well-distinguished species in the region. Studies of dried herbarium specimens, augmented by photographic images of different stages of fresh flowering material and a scrutiny of available descriptions confirmed that the calyx has two primary lobes in the bud that may separate into 3–5 portions, giving a variable number of apparent lobes in specimen material collected at different localities. This new scrutiny of the calyx also permitted an improved description of the calyx differences that separate *Christisonia* and the closely related *Aeginetia*, which have not been clearly elucidated in the past.

Conclusions: *Christisonia scortechinii* Prain (Orobanchaceae), the only species that was described as having an initially spathaceous calyx among species of this root-parasitic genus, is newly recorded for Borneo (including Kinabalu Park, where its presence has been overlooked). The range of the species in mainland Southeast Asia, previously extended from Peninsular Malaysia to Thailand and Vietnam, is here further extended to Laos and China. *Christisonia wightii* Elmer (relevant to the Philippines) and *C. sinensis* Beck (China) are reduced to synonymy.

Keywords: *Aeginetia*, Borneo, China, *Christisonia*, Kinabalu, Laos, New record, Orobanchaceae, Philippines, Ultramafics

Background

The root-parasitic genus *Christisonia* Gardn. has ca. 20 species distributed from India to China and the Malay Peninsula (Parnell 2001, 2008), Philippines, SW Celebes and Flores (Van Steenis 1967). *Christisonia* has hitherto not been documented for Borneo Island, in spite of pre-existing collections, although the closely related *Aeginetia* L. is recorded there (Beaman and Anderson 2004; under Scrophulariaceae). Both *Christisonia* (Quisumbing 1940) and *Aeginetia* (Kusano 1903, Lee and Goseco 1932) have been reported to parasitize the roots of sugarcane crops. In nature, *Christisonia* is also a root-parasite on other grasses, including bamboos (Barlow 1982; Parnell 2001), and other plant families, including Acanthaceae

*Correspondence: wkm2000@gmail.com
[3] Singapore Botanic Gardens, 1 Cluny Road, Singapore 259569, Singapore
Full list of author information is available at the end of the article

(Nandikar et al. 2013) and Vitaceae (Benniamin et al. 2012).

Christisonia and *Aeginetia* are resolved in the same clade, within the tribe Buchnereae, in the general molecular phylogeny of the Orobanchaceae (McNeal et al. 2013). However, a closer study of the phylogenetic relationship between these two genera with adequate taxon sampling is still elusive. *Christisonia* species have been distinguished by their distinctly lobed calyx from *Aeginetia*, also a root-parasitic genus, which has spatheate calyx (Parnell 2001). Prain (1904) had, however, also described the calyx of the Malayan *C. scortechinii* as "spathaceous rupturing into usually 2, occasionally 3–4 lobes in the fully opened flower", so causing uncertainty if the spatheate calyx of *Aeginetia* is a consistent character for distinguishing it from *Christisonia* (Parnell et al. 2014). Nevertheless, the presence of a covering of colourless slime over *Christisonia* flower buds (Fig. 1a) seems to provide another consistent distinction from *Aeginetia*, in which this slime is absent (Parnell et al. 2014). Also, most *Christisonia* spp. have largely white corollas, and most *Aeginetia* spp. lilac, purple, red or yellow corollas, although this is not mutually exclusive and a purplish corolla form has been reported for *C. scortechinii* (Yong 1989) (Fig. 1d) and species such as *A. indica* also have white corolla forms (Parnell 2012). Another general difference is that *Christisonia* flowers are typically on short pedicels borne on short or obscure stems that do not protrude much from the ground level, whereas *Aeginetia* species often have taller stems 10–40 cm long that carry the flowers clearly above ground level.

More recently, the distribution of several *Christisonia* species has become better documented. For example, *C. siamensis* Craib, once thought to be endemic to Thailand

Fig. 1 Flowering material of *Christisonia scortechinii* from Sabah (Kinabalu Park) (**a–c**) and a *purple*-tinged variant from Peninsular Malaysia (Sungei Wi, Cameron Highlands) (**d**). **a** Side view of flower showing the transparent slime covering. **b** Cluster of fresh flowers with short calyx lobes and a yellow band on the lip that continues down into the corolla tube. **c** Open flowers, with one showing the peltate discoid stigma in the throat. **d** *Christisonia* with unusual *purple*-tinged corolla. Photos: Sukaibin Sumail (**a**), A. van der Ent (**b, d**) and R. van Vugt (**c**)

(Parnell 2001) has been documented in Nagaland, NE India (Benniamin et al. 2012) and SW China (Tan 2013). Nevertheless, the identification of these root-parasites, the flowers of which are delicate and difficult to preserve (spirit collections are essential), remains challenging. Comparatively few collections of the various species and the poor state of preservation of key reference specimens of earlier documented species hamper comparison and reliable identification (Parnell 2001; Parnell et al. 2014). Only in a number of cases where photographic images of sufficient quantity are available can we elucidate the morphology and variation. This, fortunately, is becoming the case as amateur and professional botanists make their images more readily available on the World Wide Web.

This paper formally records *Christisonia* for Borneo through specimens collected recently from Kinabalu Park by Antony van der Ent (Fig. 1a–c), which began this investigation, as well as previous collections from there and elsewhere in Sabah.

Methods

Herbarium material of *Christisonia* for Borneo was studied at the Singapore Herbarium (SING) and the Sabah Parks Herbarium (SNP), and through images on JStor for the Global Plants Initiative, the latter including material from K and WU; herbarium acronyms used follow Thiers (2015). The SNP specimens collected by A. van der Ent in Kinabalu Park were preserved in FAA mixture. Photographic images of different stages of fresh flowering material taken by A. van der Ent and colleagues or which were available from various internet sources (as respectively acknowledged below) aided the interpretation of floral structures and development, especially that of the calyx.

Results and discussion

Christisonia newly recorded for Borneo

Cedric E. Carr, better known for his orchid collecting (Kiew 2003), was apparently the first person to collect a *Christisonia* from Mount Kinabalu and Borneo, although this was never recorded in any publication, including the recently completed *Plants of Mount Kinabalu* where Beaman and Anderson (2004) had missed Carr's number *SFN 26457* of this taxon in the Singapore Herbarium. Carr's locality was the Kinataki River (below the famous Marai Parai spur) at around 1400 m on the western flank of Mount Kinabalu.

Christisonia scortechinii (Fig. 1) was recorded by Antony van der Ent from two localities inside Kinabalu Park, the first at Marai Parai (near the Kinataki River where also Carr collected this species on Mount Kinabalu over 80 years earlier), and the second on Mount Tambuyukon in the north of Kinabalu Park. Both localities sit at 1500–1800 m asl in dense montane cloud forest on ultramafic soils and have remarkably similar vegetation in aspect and composition. Specimens were collected only from the Marai Parai site on Mount Kinabalu, while the Mount Tambuyukon occurrences were documented with photographs.

The *Christisonia* is root-parasitic on the bamboo *Racemobambus gibbsiae* (Poaceae) which is abundant at both sites. The parasitic plants occur as scattered typically "clumped" individuals in the permanently moist humus-rich soil. It appears to be flowering year-round as it has been recorded on a number of occasions during fieldwork. The Marai Parai site is a steep slope along the Kinataki River below the famed spur that is the type-locality for many of Kinabalu's enigmatic plant species, such as *Nepenthes rajah* and *N. edwardsiana* (Nepenthaceae). Mount Tambuyukon (2579 m asl) is the third highest mountain on the island of Borneo and the highest ultramafic mountain in the region. The vegetation on this mountain ranges from tall mixed dipterocarp forest on the foot slopes to stunted graminoid scrub near the summit. The *Christisonia* occurs between Musang Camp and General Camp on the east flank roughly 400 m below the summit.

Voucher material for *C. scortechinii* recorded on Mount Kinabalu include the following:

SABAH, Ranau, Mount Kinabalu, Kinataki River, ca. 4500 ft (1370 m), 7 Mar. 1933, *C.E. Carr SFN 26457* (SING), Marai Parai, near Kinataki River, 1630 m, 31 Mar. 2011, *A. van der Ent, S. Sumail & R.M. Karim SNP 23658* (SNP).

Christisonia scortechinii is also known from elsewhere in Sabah, also on ultramafic soils:

SABAH, Kinabatangan district. Telupid, Bukit Tawai, near summit, 1 Sep. 1999, *R. Kiew RK 4817* (SING), Sungai Ruku-Ruku Forest Reserve, 3 Sep. 1999, *Ali Ibrahim & R. Kiew AI 540* (SING).

The flora of Kinabalu Park is the richest globally in terms of the number of plant species per unit area with more than 5500 taxa recorded from an area roughly 1200 km^2 in extent (Beaman 2005, Van der Ent unpublished). Plants with alternative metabolism or nutrient-acquisition strategies in Kinabalu Park include insect and/or rodent scat trapping ('carnivorous') plants (Nepenthaceae, Droseraceae, Lentibulariaceae), myrmecophytes (Rubiaceae: *Hydnophytum*, *Myrmecodia*), myco-heterotrophic ('saprophytic') plants (mainly Burmanniaceae, Orchidaceae, Triuridaceae), and parasitic plants (principally Loranthaceae and Rafflesiaceae). Most of the hemi-parasitic plants (those that have an active photosynthesis system) in Kinabalu Park are in the order Santalales represented by approximately 60 species predominantly in the families Loranthaceae and Santalaceae (Beaman and Anderson 2004). Apart from the newly recorded *Christisonia*, holo-parasitic plants (which derive all carbohydrates exclusively from the host) include three species of *Balanophora* (*B. lowii, B. reflexa* and *B. papuana*) (Balanophoraceae), one species of *Rhizanthes* (*R. lowii*), one species of *Mitrastema* (*M. yamamotoi*) and two species of *Rafflesia* (*R. keithii* and *R. pricei*) (the latter three genera in Rafflesiaceae).

The local occurrence of ultramafic soils has been suggested as a driving force for speciation, in combination with recent uprising of the Mount Kinabalu Massif, isolation and other factors (Beaman and Beaman 1990). Soils developed from ultramafic (often called 'serpentine' by botanists and ecologists) bedrock induce an edaphic filter on the vegetation that includes nutrient deficiencies, major cation imbalances and metal toxicities (Proctor 2003). As a consequence, ultramafic vegetation in Sabah is often of lower stature, with distinct species composition compared to mixed dipterocarp forest on sedimentary soils. Ultramafic habitats in Borneo are known to harbour a multitude of rare and localised plant species (Wong 1998). Although ultramafic host a number of endemic taxonomic entities that could be evolutionarily derived from closely related taxa from adjacent environments, so-called "neo-endemics" (Van der Ent et al. 2014, 2015), we could not detect any consistent morphological differences between the *C. scortechinii* from the Sabah ultramafic sites and that from elsewhere.

Comparing Sabah *Christisonia* material with closely related taxa

At the outset we were able to determine that the Sabah taxon (Fig. 1a–c) is most related to *C. scortechinii* Prain, *C. siamensis* Craib and *C. sinensis* Beck, in that all these typically or frequently have glabrous white corollas with a single yellow patch that runs from the base of the lip to within the throat and peltate stigmas.

Christisonia scortechinii

An unnumbered collection of *C. scortechinii* at Kew (K000899781) represents potential type material against which to interpret the species in its original Malay Peninsula provenance, given the type sheet bearing Scortechini's number 2121 at Calcutta is in poor condition (Parnell et al. 2014). But even on the Kew sheet, it is difficult to make out intact calyx lobes and verify other characters because the dried material is in poor condition and mounted on a sheet. Thus the interpretation of *C. scortechinii* has depended in large part on direct visual assessment of fresh material from Peninsular Malaysia (Fig. 1d), other herbarium material available (see below), the original description by Prain (1904), as well as the sketch in Henderson (1954) and available photographic images, such as in http://www.parasiticplants.siu.edu/Orobanchaceae/images/Christisonia1.jpg and http://www.parasiticplants.siu.edu/Orobanchaceae/images/Christisonia3.jpg.

Christisonia siamensis

Christisonia siamensis is well distinguished from *C. scortechinii, C. sinensis* and the Sabah taxon in frequently having violet-purplish corolla coloration and a distinct hair tuft at the insertion of the anther that obstructs the corolla throat (Parnell et al. 2014), which is absent in the latter three taxa, which have quite glabrous filaments.

Christisonia sinensis

Christisonia sinensis, found in SW and SE China, was reduced to the synonymy of *C. hookeri* C.B.Clarke ex Hook. in the Flora of China (Zhang and Tzvelev 1998). However this may not be justified because, as pointed out by Parnell et al. (2014), the stigma was described as unequally bilobed in the type description of *C. hookeri* (Hooker 1885) whereas the Flora of China treatment describes a discoid stigma 4–6 mm in diameter, which fits the peltate stigma illustrated in photographic images from Chinese material, as shown in Fig. 2, for example.

Although Parnell et al. (2014) expressed doubt that Chinese material (with peltate stigmas) had been correctly identified as *C. hookeri* (described as having bilobed stigmas), they did not comment on the status of the name *C. sinensis.*

Identification of the Sabah *Christisonia*

Table 1 compares a number of characters among the Sabah taxon, *C. scortechinii, C. siamensis* and *C. sinensis*, as well as *C. calcarata* and *C. hookeri*, both of which are clearly differentiated from the others by their bilobed stigmas. Other very conspicuous and consistent characters also help differentiate consistently recognised species in Table 1, such as the slender pedicellate flowers of *C. calcarata*; a distinct hair tuft at the anther insertion in *C. siamensis*; and the densely pubescent corolla in *C. calcarata*. It would appear that while the predominantly white corolla is also important, occasional violet or purplish tinges do occur. The single band of yellow marking the lip and down into the corolla throat in *C. scortechinii* (Fig. 1), *C. siamensis* and *C. sinensis* (Fig. 2a) also distinguishes these species from *C. calcarata*, where the entire throat is yellow. These provide an idea of various specialised characters that clearly distinguish species.

It can be seen from Table 1 that, while *C. calcarata, C. hookeri* and *C. siamensis* are immediately distinguished from the rest as stated above, *C. scortechinii, C. sinensis* and the Sabah taxon remain rather similar, including in the variable number of calyx lobes, a character that had caused much hesitation in deciding whether or not *C.*

Fig. 2 *Christisonia sinensis* from Chinese material. **a** Flowers, with slime over young buds and (*center-right*) suturing of calyx lobes that incompletely separate. **b** Corolla cut open to show the peltate discoid stigma and two pairs of stamens; in the shorter stamens, the anther appendage has a short acute projection on the dorsal side. (Reproduced with the kind permission of Ming-I Weng, from: https://www.flickr.com/photos/mingiweng/4927639081/; http://www.flickriver.com/photos/mingiweng/4928219216/)

scortechinii from the type provenance (Peninsular Malaysia) and other provenances in Thailand and Vietnam should be conspecific.

In order to understand the nature of the calyx lobes better, we examined specimen material of *C. scortechinii* from Peninsular Malaysia (the type provenance) to see if they agreed with the original description of Prain (1904); and both dried preserved material (mostly expected to be in poor condition) and photographic images that could assist in interpreting this character for other similar taxa in China (as *C. sinensis* Beck or *C. hookeri* sensu Zhang & Tzvelev, non (C.B.Clarke) Hook.), Vietnam, and Thailand.

For material from the type provenance of *C. scortechinii*, the diagram given by Henderson (Henderson 1954: 335, Fig. 317) clearly shows calyces with 2 or 3 lobes. Earlier, Ridley (1923) had also described "Calyx spathaceous splitting into 2–4 lobes...", which is not different from what Prain (1904) stated. We also studied the following material:

PENINSULAR MALAYSIA, Perak, Bukit Kinta Forest Reserve, 28 Apr. 1987, *R. Kiew RK 2583* (KEP, SING); Perak, Ulu Batang Padang, no date, *Ridley 13693* (SING); Selangor, Ginting Simpah, top of pass, ca. 2000 ft (609 m), 20 Sep. 1947, *J. Reid s.n.* (SING).

In fact, Reid's specimen had helped the description in Henderson (1954), as evident from his letter to Henderson pasted on the specimen, which, among other things, gave the occurrence of the species as "sparsely distributed in small localised patches at moderate altitudes." All five calyces of young as well as open flowers on the Reid specimen sheet show 2 lobes; one young bud shows a single deep suturing across the top of the calyx (implying two lobes from a younger stage). The Ridley specimen has only one 'good' flower in the packet showing clearly two main lobes, each with a short median tear (which can be expected to give at least four lobes perceptible at a later stage). The Kiew specimen shows 2, 3, 4, or 5 calyx lobes, unequal in width as if from tears rather than from preformed equivalent calyx members.

Christisonia sinensis is very similar to *C. scortechinii*, including some Thai and Vietnamese (Central Mountains Region) material that was identified as the latter species (Parnell et al. 2014). The wider concept of *C. scortechinii* (Parnell et al. 2014) accepted that the calyx may be spatheate in the bud but becoming 2–4(–5)-lobed. Although herbarium material for the genus is admittedly poor (Parnell et al. 2014), the type of *C. sinensis*, *Handel-Mazzetti 9595*, Yunnan, 2850 m, 31 July 1916 (WU), clearly shows the short thick pedicels, short-lobed calyces and peltate stigmas that accord with the images of material from China as well as the Thai and Vietnam material mentioned. Moreover, the type also includes flowers where the calyx is variably 2–3-lobed or even 5-lobed, indicating that the calyx in *C. sinensis* does not always part into 5 lobes.

Table 1 The Sabah *Christisonia* compared with five other taxa

	Sabah *Christisonia*	*C. scortechinii*	*C. siamensis*	*C. sinensis*	*C. calcarata*	*C. hookeri*
Pedicel	Short-thick, ca. 1 cm long	Short-thick, ca. 1 cm long	Short-thick, ca. 1 cm long	Short-thick, ca. 1 cm long	*Slender, 2–3.5 cm long*	Short-thick, ca. 1 cm long
Calyx colour	White	White to purplish	White, yellowish to dark purple	White	Pale yellow	Violet
Calyx surface	Glabrous	Glabrous	Glabrous	Glabrous	Glabrous	Glabrous[a]
Calyx lobes/teeth	2–4 (–5)	2–4 (–5)	(2–) 3 (–4)	(2–) 3–4 (–5)	5	(4–) 5
Enclosure of corolla tube by calyx (herbarium material)	About half	Less than half	Less than half	About half	Less than half, to half	About half
Enclosure of corolla tube by calyx (fresh material)	About a quarter	A quarter to a third	A third to half	A quarter to a third	A quarter to a third	?
Corolla colour	White	White to tinged with violet	White tinged with violet	White	White	White with violet seams
Corolla pubescence	Glabrous	Glabrous	Glabrous	Glabrous	*Tube outer surface and both sides of lobes white-pubescent*	Glabrous
Corolla lobe & throat colour	White, with yellow-orange patch from base of lip down to inside the tube	White, with dark yellow patch from base of lip down to inside the tube	Bluish to violet, with yellow patch from base of lip down to inside the tube	White to slightly lilac, with yellow-orange patch from base of lip down to inside the tube	White to greyish or bluish, *bright yellow patch on all lobes to down inside the tube*	?
Hairs at point of attachment of anthers	Glabrous	Glabrous	*Distinct hair tuft* obstructing corolla throat	Glabrous	*Sparse scattered short hairs on filaments*	Glabrous
Anther appendage, dorsal side	With short acute projection	With "projecting acute posterior process" (Prain 1904)	With short acute projection	With short acute projection	*With a slender spur-like projection*	*Rounded, without acute projection*
Stigma	Peltate	Peltate	Peltate	Peltate	*Bilobed*	*Bilobed*
Known distribution	N Borneo (Sabah)	Peninsular Malaysia, Thailand, Vietnam[b]	Thailand (N NE, SW), India (Nagaland)	SW and SE China	India, Pakistan	India
Habitat	Montane sites over 2000 m	Montane sites ca. 650–1700 m	Mixed deciduous forest at 300–800 m	Montane forest, 1500–2000 m	Montane sites	Montane sites

[a] Not pubescent as stated in Benniamin et al. (2012)

[b] Material from Thailand (N, NE, SE, Peninsular, from ca. 1200–1655 m) and a more northerly latitude in C Vietnam (642 m) placed in *C. scortechinii* by Parnell et al. (2014)

Prain (1904) described that the lower pair of stamens of *C. scortechinii* had "a projecting acute posterior process", and *C. sinensis* material from China, as well as Thai and Vietnam material ascribed to *C. scortechinii* in Parnell et al. (2014), also develop a small acute or conical projection on the anther appendage. This can be seen in Chinese material (Fig. 2b) as well as the Sabah material (Fig. 3b). These vary slightly in prominence but are instantly distinguished from the slender elongate, spur-like projection from the dorsal side of the anther appendage developing in another species, *C. calcarata*.

Fresh Chinese material from Taiwan referable to *C. sinensis* beginning with two calyx lobes that sustain visible apical tears can be seen in images from the website: https://www.flickr.com/photos/mingiweng/4928220164/ and examples of calyx tearing into 3–5 lobes can be seen in: http://www.plant.csdb.cn/details?guid=photo:cfh@5c7f74b1-31df-4c72-a872-a2d446a8fa22.

Incompletely separating calyx lobes are clearly seen in Fig. 2a.

Even more instructive, Chinese (Taiwan) material showing a bud calyx with only a small apical opening, and other calyces tearing irregularly into several lobes are obvious in: http://blog.xuite.net/bibibear599/twblog/137870776.

Likewise, the variable number of calyx lobes that can be seen in web-images of fresh material from mainland China, Vietnam and Thailand has been discussed in detail by Parnell et al. (2014), including their Figs. 1c and 2a. Vietnam material with calyx tearing into 5 or more lobes is also available in: https://www.flickr.com/photos/biodivn/14351717859/.

To these we wish only to add a new record of the same taxon for Laos:

LAOS, Houa Phan Province, Vieng Thong District, Nam Et-Phou Louey Natural Protected Area, Nam Et side, near Ban Phon Song, 20°20′46″N, 103°22′04″E, 953 m, 8 Jun. 2013, *J. Leong-Skornickova, M. Rodda, Aung Thame, U. Souvannakhoummane, K. Phoutthavong, M. Norsaengsri & S. Sangvirotjanapat JLS 2407* (E, P, QBG, SING).

Most calyces from this specimen are torn, but one clearly shows 2 lobes, another in bud has the calyx showing only a single small median tear (or slit).

These observations imply that there is no consistent difference in the number of calyx lobes between *C. scortechinii* and *C. sinensis*, supporting a key premise on which Parnell et al. (2014) based their extension of the range of *C. scortechinii* to include sites in Thailand and Vietnam. Although for some other species, e.g., *C. calcarata*, the number of calyx lobes seems evenly formed and are consistently 5 in number, for other species such as *C. scortechinii* and *C. sinensis*, there is generally a variable number of calyx lobes. It would seem that the basic condition representative of the latter two taxa is a basically 2-lobed calyx, where the lobes are fused to only slightly separated in the bud stage, later separating into two clear

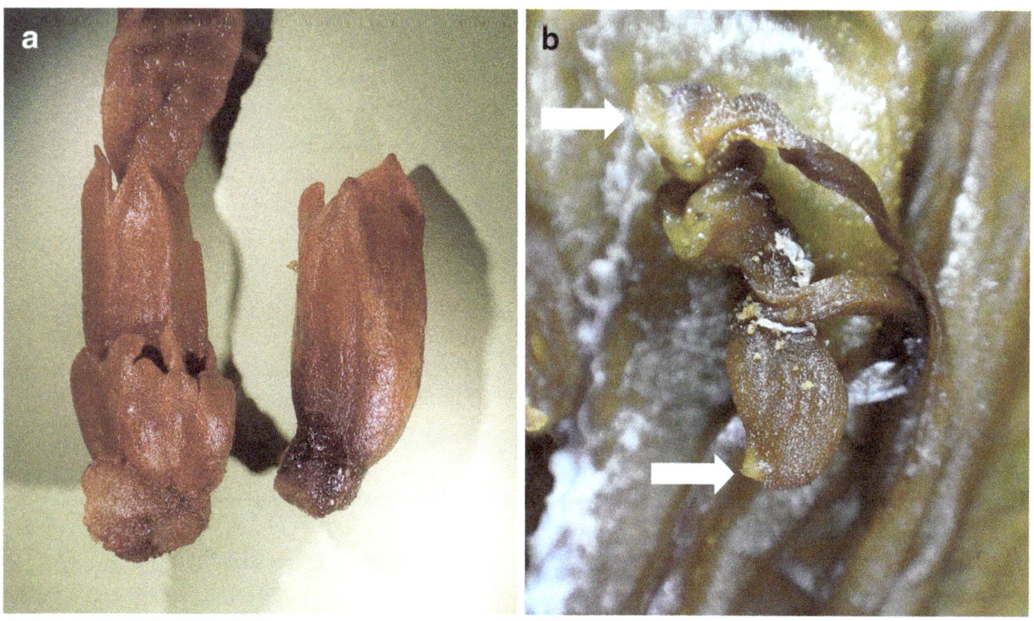

Fig. 3 *Christisonia scortechinii* from Kinabalu Park, showing calyces separating into 3 or more lobes (**a**) and detail of one set of stamens (**b**) showing anther of longer stamen (*upper arrow*) close to the anther of shorter stamen, which has a dorsal lobe with an acute process at its tip (*lower arrow*). From spirit-preserved material of *Van der Ent* et al. *SNP 23658* (SNP)

lobes, each of which can sustain a secondary median tear, in all giving 2, 3, or 4 lobes (as was recorded by Prain for typical *C. scortechinii* and not authoritatively recorded for *C. sinensis*), or even 5 lobes (as common in and recorded for *C. sinensis* and some material from Vietnam and Thailand). Other distinct species may also show a similar calyx character, e.g., the painting of *Christisonia thwaitesii* Trimen by H. DeAlwis in Trimen (1893–1898), t. 69, depicts calyces with 2 lobes, 3 lobes and irregularly tearing lobes: http://plantillustrations.org/illustration.php?id_illustration=157328&mobile=0.

In fact, the description of the calyx by Trimen (1895: 263) of *C. thwaitesii* is not only accurate but also describes the condition in *C. scortechinii*: "...quite closed in the bud (which is beaked), deeply 2-lipped, the lips irregularly or faintly 2- and 3-toothed...".

The material from Sabah also amply displays the same condition of basically two calyx lobes that may further divide (tear) to give 2, 3, or 4, even 5 lobes. Carr's material is rather well preserved, with one calyx clearly showing division into two main lobes, and another having two main lobes with apical tears. The recent material collected by van der Ent and colleagues also shows 3–4 calyx lobes (Fig. 3a).

A wider geographical range for *Christisonia scortechinii*

The impression from descriptions, dried herbarium specimens and photographic images of different stages of fresh flowering material in *C. scortechinii*, *C. sinensis* and the Sabah taxon is that the calyx is basically composed of two lobes that slightly to mostly cohere in the young flower bud, with an apical slit along which the main calyx lobes part and through which the corolla protrudes. In summary, there seem to be no significant differences between *C. scortechinii* and *C. sinensis* and our Sabah taxon. The Malayan name would take precedence.

Christisonia scortechinii is probably also distributed in the Philippines. Images by Leonardo L. Co identified as *C. wightii* Elmer from the Sierra Madre Mountain Range, Cagayan Province, Luzon Island, and the (ultramafic) Mount Victoria massif on Palawan Island (http://www.philippineplants.org/Families/Orobanchaceae.html), both show *Christisonia* flowers with a peltate discoid stigma, and appear to be the same as *C. scortechinii*. *Christisonia wightii* Elmer, described based on material from Negros, was the only species described for the Philippines, and of which the type is not found and probably not extant (likely at PNH and destroyed during WWII). Of the type specimen *Elmer 9510*, Elmer (1915) says "A specimen was however sent...to *Dr. Ridley* [italics his] for determination and who pronounced it a species of *Christisonia*." This collection is not at SING. We have examined the following material, however most calyces are fragmented:

PHILIPPINES, Mindanao, Camp Keelthler, Lake Loamao, Apr. 1907, *M.S. Clemens, s.n.* (SING).

Elmer (1915) described the stigma as "composed of 2 broad and fleshy lobes", which differs from the clearly peltate discoid stigma in *C. scortechinii*, otherwise the details he gives do not vary specially from the characters of *C. scortechinii* or *C. sinensis*. We regard that Elmer's description of the stigma may have been inaccurate and consider that his species is likely the same as *C. scortechinii*.

The material from China, Vietnam and Thailand (as detailed by Parnell et al. 2014), Laos, Peninsular Malaysia (Prain 1904; Ridley 1923) and Borneo (Sabah), probably also that from the Philippines which we identify as *C. scortechinii* all share the following characters: short thick flower pedicels barely 1 cm long; a calyx with 2 primary lobes in the bud that may separate into 3–5 lobes; a typically white, glabrous corolla with a distinctive yellow–orange patch on the lip that continues down into the corolla tube; also additionally, an anther appendage with a short acute-conical projection on the dorsal side (Figs. 2b, 3b). Furthermore, as with the Peninsular Malaysia, Sabah and Philippine records, the ecological distribution of the taxa here included into *C. scortechinii* is clearly montane: the Thai records were from 1200 to 1655 m, the Laos record was from 953 m, the Vietnam record from 642 m at Bi Doup in the Central Mountains Region (Parnell et al. 2014), and the Chinese account generally recorded 1500–2000 m for the species (as '*C. hookeri*') (Zhang and Tzvelev 1998).

A corollary by way of conclusion: *Christisonia* vs *Aeginetia*

Our current understanding of how the calyx structure varies within *C. scortechinii* allows us to recognise it as a species with the main part of its range from the Malay Peninsula to parts of Thailand, Vietnam, Laos and south China, and which is now recorded for Borneo (Sabah) and probably found also in the Philippines. We thus interpret the species to be vicariant across the South China Sea, between mainland Southeast Asia and the Philippine islands and Borneo. Correspondingly, *C. sinensis* and *C. wightii* should be regarded as synonyms of *S. scortechinii*:

Christisonia scortechinii Prain (1904: 205); Ridley (1923: 489); Henderson (1954: 335); Parnell et al. (2014: 16–23). Type: MALAY PENINSULA, Perak, no date, *Scortechini 2121* (holotype CAL, barcode 0000025048; possible isotype K, barcode K000899781).

Synonyms:

Christisonia wightii Elmer (1915: 2793). Type: PHILIPPINES, Negros, Oriental, Dumaguete (Cuernos Mountains), Mar. 1908, *Elmer 9510* (holotype PNH, not found, probably destroyed).

Christisonia sinensis Beck (1930: 314). Type: CHINA, Yunnan, 2850 m, 31 Jul. 1916, *Handel-Mazzetti 9595* (holotype WU, barcode WU0060576).

Christisonia hookeri sensu Zhang and Tzvelev (1998: 242), non Hooker (1885).

It was perhaps unfortunate that Prain (1904) had chosen the term "spathaceous" to describe the calyx of *C. scortechinii*. Use of this term has also led to some concern that an initially "spathaceous" calyx in *C. scortechinii*, the sole example of this in a genus where generally the species were described as having several (up to 5) discrete calyx lobes, would weaken the distinction between *Christisonia* and *Aeginetia* by their otherwise different calyx forms.

In fact, in *C. scortechinii* the two primary calyx lobes (which may remain intact or further tear to give more lobes) are subequal and so ensheath the emerging corolla tube at the beginning; this character is therefore rather different from the conspicuously spathaceous calyx in *Aeginetia*. In *Aeginetia*, there is only one (incomplete) split developing along the ventral median in the conspicuously curved calyx, through which the corolla emerges. Thus, the genera *Aeginetia* and *Christisonia* are still readily distinguished from their calyx form and the way it ruptures during floral development. We venture a distinction between these two otherwise very closely related genera as follows:

(1a) Flower buds covered in a translucent jelly-like slime; flowers typically on short pedicels borne on short or obscure stems that do not protrude much above ground; calyx not or only slightly curved, with 2-several distinct primary lobes, if only 2 then the lobes subequal but possibly dividing further through apical tears; corollas largely white outside, sometimes infused with pink or blue-violet to various degrees ….. *Christisonia*

(1b) Flower buds without a slimy covering; flowers carried on taller stems typically 10–40 cm above ground level; calyx markedly curved or unequally distended, developing only a single split along the ventral median and therefore effectively without distinct lobes; corollas lilac, purple, pink, red or yellow, less frequently white ….. *Aeginetia*

Authors' contributions

AVDE collected and photographed specimen material from Kinabalu Park and Peninsular Malaysia, and compiled the information on vegetation, habitat and host plants. KMW conducted the morphological studies and taxonomic comparisons, and initiated the manuscript draft. Both authors read and approved the final manuscript.

Author details

[1] Centre for Mined Land Rehabilitation, Sustainable Minerals Institute, The University of Queensland, St. Lucia, QLD 4072, Australia. [2] Laboratoire Sols et Environnement, Université de Lorraine-INRA, UMR 1120, 54000 Nancy, France. [3] Singapore Botanic Gardens, 1 Cluny Road, Singapore 259569, Singapore.

Acknowledgements

We thank Sabah Parks, especially Dr Jamili Nais and Rimi Repin, for research support and for the loan of the herbarium material. AVDE is grateful to Rositti Karim and Sukaibin Sumail for help during the fieldwork in Kinabalu Park, and Anthea Phillipps for an earlier identification of *Christisonia*. KMW acknowledges the support of the National Parks Board, Singapore and the Singapore Botanic Gardens.

Compliance with ethical guidelines

Competing interests

The authors declare that they have no competing interests.

References

Barlow HS (1982) A note on *Christisonia scortechinii*, a parasitic flowering plant of the Broom-Rape family (Orobanchaceae). Malay Nat J 36:140

Beaman JH (2005) Mount Kinabalu: Hotspot of plant diversity in Borneo. Biol Skr 55:103–127

Beaman JH, Anderson C (2004) The Plants of Mount Kinabalu. 5. Dicotyledon Families Magnoliaceae to Winteraceae. Natural History Publications (Borneo), Kota Kinabalu

Beaman JH, Beaman RS (1990) Diversity and distribution patterns in the flora of Mount Kinabalu. In: Baas P, Kalkman K, Geesink R (eds) The Plant Diversity of Malesia. luwer Academic Publishers: Dordrecht, The Netherlands

Beck von Mannagetta G (1930) Orobanchaceae. In: Engler A (ed) Das Pflanzenreich. regni vegetabilis conspectus, vol. 261, Part 96. W Engelmann, Leipzig

Benniamin A, Chaturvedi SK, Dey S, Moaakum (2012) Supplements to the root parasitic plant in India. A new recorded species *Christisonia siamensis* Craib (Orobanchaceae). Taiwania 57: 217–221

Elmer ADE (1915) Two Hundred Twenty Six New Species—II. Leafl Philipp Bot 8:2719–2884

Henderson MR (1954) Malayan wild flowers. Malay Nat Soc, Kuala Lumpur

Hooker JD (1885) Orobanchaceae. Flora of British India, vol 4. L. Reeve & Company, London, pp 319–328

Kiew R (2003) Kinabalu diary and orchid determinations. C.E. Carr's Kinabalu field diary. Gardenwise 20:8–12

Kusano S (1903) Notes on *Aeginetia indica*. Linn Bot Mag 17:81–84

Lee A, Goseco F (1932) Studies of the sugar-cane root parasite *Aeginetia indica*. Int Soc Sugar Cane Tech Congr San Juan Bull 101:1–12

McNeal JR, Bennett JR, Wolfe AD, Mathews S (2013) Phylogeny and origins of holoparasitism in Orobanchaceae. Am J Bot 100:971–983

Nandikar M, Singh R, Gurav R, Shinde R (2013) Lectotypification and notes on the identity of *Christisonia calcarata* (Orobanchaceae). Rheedea 23:26–29

Parnell J (2001) A revision of Orobanchaceae in Thailand. Thai For Bull (Bot) 29:72–80

Parnell J (2008) Orobanchaceae. In: Santisuk T, Larsen K (eds) Flora of Thailand, vol 9(2). The Forest Herbarium, National Park, Wildlife and Conservation Department, Bangkok, pp 142–147

Parnell J (2012) *Aeginetia flava*: a new and remarkable species of *Aeginetia*: Orobanchaceae from South-eastern Thailand. Kew Bull 67:81–84

Parnell J, Strijk JS, Maknoi C, Luu HT, Pooma R, Sirimongkol S, Rueangruea S, Tichai W, Ritphet N (2014) New records expand the ranges of *Christisonia siamensis* and *Christisonia scortechinii*, the latter species being new to Thailand. Thai For Bull (Bot) 42:16–23

Prain D (1904) Noviciae Indicae XXIV. Some new Indian plants. J As Soc Bengal 73(II, Natural History): 192–206

Proctor J (2003) Vegetation and soil and plant chemistry on ultramafic rocks in the tropical Far East. Perspect Plant Ecol Evol Syst 6:105–124

Quisumbing E (1940) On *Christisonia wightii*, Elmer, a parasite of sugarcane. Philipp J Plant Ind 11:397–400

Ridley HN (1923) The Flora of the Malay Peninsula: Gamopetalae, vol II. L. Reeve & Co., London

Tan Y-H (2013) *Christisonia siamensis* Craib, a newly recorded species of *Christisonia* (Orobanchaceae) from China. Guihaia 33:521–522

Thiers B (2015). Index Herbariorum: a global directory of public herbaria and associated staff. New York Botanical Garden's Virtual Herbarium. Available from: http://sweetgum.nybg.org/ih/. Accessed 1 July 2015 (continuously updated)

Trimen H (1893–1898). Plates in Illustration of A Hand-Book to the Flora of Ceylon. Dulau & Co., London

Trimen H (1895) A Hand-Book to the Flora of Ceylon, vol 3. Dulau & Co., London

Van der Ent A, Repin R, Sugau J, Wong KM (2014) The Ultramafic Flora of Sabah. An introduction to the plant diversity on ultramafic soils. Natural History Publications (Borneo). Kota Kinabalu

Van der Ent A, Repin R, Sugau J, Wong KM (2015) Plant diversity and ecology of ultramafic outcrops in Sabah (Malaysia). Aust J Bot 63:204–215

Van Steenis CGGJ (1967) Miscellaneous botanical notes XVIII. Blumea 15:145–155

Wong KM (1998) Patterns of plant endemism and rarity in Borneo and the Malay Peninsula. In: Peng C-I, Lowry PP II (eds) Rare, Threatened and Endangered Floras of Asia and the Pacific Rim. Institute of Botany, Taiwan

Yong HS (1989) Colour polymorphism in the flowers of the parasitic plant Malayan Bamboorape (*Christisonia scortechinii*). Nat Malays 14:100–103

Zhang Z-Y, Tzvelev NN (1998) Orobanchaceae. Flora of China vol 18. Science Press, Beijing & Missouri Botanical Garden, St. Louis, pp 229–243

14

Impatiens pandurata (Balsaminaceae), a new species from Yunnan, China

Yun-Hong Tan[1], Yan-Nan Liu[2], Hong Jiang[3], Xin-Xin Zhu[4], Wei Zhang[5] and Sheng-Xiang Yu[2*]

Abstract

Background: The species-rich genus *Impatiens* is mainly distributed throughout much of tropical Africa, India, southwest Asia, southern China and Japan. There are more than 270 species recorded in China, most of which are restricted to the southwest. An unknown species of *Impatiens* was collected from Yunnan, southwest China.

Results: *Impatiens pandurata* Y. H. Tan & S. X. Yu, a new species of Balsaminaceae from Jinping County and Malipo County, Yunnan, China is similar to *I. apalophylla* and *I. clavigera* in having racemose inflorescences, 4 lateral sepals, hammer-shaped capsules and ellipsoid seeds, but differs in having leaves with oblanceolate blades aggregated at the top of the stem, 3–5-flowered racemes, a yellow lower sepal without reddish patches, yellowish flowers, and a dorsal petal with stalks at the base. Molecular phylogenetic analyses of sequences from both nuclear ribosomal and plastid genes confirm that this new species is distinct from morphologically similar species previously recorded.

Conclusion: With the support of careful morphological studies and phylogenetic analysis, *I. pandurata* is a species new to science.

Keywords: Balsaminaceae, *Impatiens*, Morphology, New species, Phylogeny

Background

The genus *Impatiens* L. (Balsaminaceae), containing over 1000 species (Grey-Wilson 1980; Fischer 2004; Yu et al. 2015), is mainly distributed throughout much of tropical Africa, India, South-west Asia, southern China and Japan, with only a few species spreading into the north temperate zone of Europe, Russia and China as well as North America (Grey-Wilson 1980). *Impatiens* species occur in diverse habitats, from sea level to 4000 m in elevation, in forest understories, roadside ditches, valleys, abandoned fields, along streams and in seepage, usually in mesic or wet conditions, although some species can tolerate drier habitats (Yu et al. 2015). Because of its species diversity, the genus has been regarded as 'the dicot counterpart of the orchid' (Yuan et al. 2004).

Five diversity hotspots for *Impatiens* have been recognized, i.e. tropical Africa, Madagascar, southern India and Sri Lanka, the eastern Himalayas, and southeast Asia (Song et al. 2003; Yuan et al. 2004). *Impatiens* is notoriously difficult to classify morphologically (Hooker 1908; Grey-Wilson 1980) and the semi-succulent stems, fleshy leaves, and extremely fragile flowers make it challenging to prepare well-dried herbarium specimens. The publication of new species each year shows that the genus has been under-collected and under-studied (e.g. Narayanan et al. 2013; Utami 2013; Kuang et al. 2014; Luo et al. 2014).

There are more than 270 species of *Impatiens* recorded in China (Yu 2012; Chen 2001; Chen et al. 2007), most of them restricted to the southwest. During recent field expeditions in Yunnan, the authors collected several specimens with distinctive morphological characteristics. After careful consultation of the literature and specimens, we concluded that these specimens are morphologically distinct from any described species. After additional molecular phylogenetic analysis, we are confident that this species is new to science.

*Correspondence: yushengxiang@ibcas.ac.cn
[2] State Key Laboratory of Systematic and Evolutionary Botany, Institute of Botany, Chinese Academy of Sciences, Beijing 100093, People's Republic of China
Full list of author information is available at the end of the article

Methods

Morphology

Characteristics of the leaves, inflorescence and flowers were described and measured on both dried herbarium specimens (from HITBC and PE) and fresh specimens in the field.

Molecular methods

DNA sequences of 151 species of *Impatiens* were used and three species, *Hydrocera triflora* (L.) Wight & Arn. (Balsaminaceae), and *Marcgravia umbellata* L. and *Norantea guianensis* Aubl. (Marcgraviaceae), were included as outgroups, based on the results of Yuan et al. (2004), Janssens et al. (2006) and Yu et al. (2015). All sequences were downloaded from GenBank, except those of the new species, *I. pandurata*, which were newly generated for this study (Genbank accession numbers XXXX, XXXX for ITS, XXXX, XXXX for *atpB-rbcL* and XXXX, XXXX for *trnL-F*). Vouchers and GenBank accession numbers are listed in Additional file 1: Table S1.

Three molecular markers were used: ITS, *atpB-rbcL* and *trnL-F*. Total genomic DNA was extracted from silica gel-dried leaves using a modified CTAB protocol from Doyle and Doyle (1987). Primers and PCR protocols for ITS, *atpB-rbcL* and *trnL-F* are derived from White et al. (1990), Janssens et al. (2006) and Taberlet et al. (1991), respectively. PCR products were purified using a GFX™PCR DNA and Gel Band Purification Kit (Amersham Pharmacia Biotech, Piscataway, NJ, USA). Sequencing reactions were carried out using an ABI Prism Bigdye Terminator Cycle Sequencing Kit (Applied Biosystems, Foster City, CA, USA). Products were analyzed on an ABI3730xl automated DNA sequencer. Sequences were aligned using the default parameters in Clustal X v.1.83 (Thompson et al. 1997) and further adjusted manually in BioEdit v.7.0 (Hall 1999). Four difficult-to-align regions in *trnL-F* (encompassing 73 sites) and one difficult-to-align region in *atpB-rbcL* (encompassing 42 sites) were excluded from the analyses.

Maximum parsimony (MP) and Bayesian inference (BI) were used to analyze the ITS and plastid data sets. The MP analyses were carried out in PAUP* v.4.0b10 (Swofford 2003). Heuristic searches were conducted with 1000 replicates of random addition, one tree held at each step during stepwise addition, tree-bisection-reconnection (TBR) branch swapping, MulTrees in effect, and steepest descent off. Bootstrapping was conducted with 1000 replicates with 10 random taxon additions and heuristic research options. The BI analyses were carried out in MrBayes v.3.0b4 (Ronquist and Huelsenbeck 2003). Each of the three regions (ITS, *atpB-rbcL*, and *trnL-F*) was assigned its own model of nucleotide substitution, as determined by the Akaike information criterion (AIC) in Modeltest v.3.06 (Posada and Crandall 1998).

Results and discussion

Impatiens pandurata Y. H. Tan & S. X. Yu, sp. nov.

This species is similar to *I. apalophylla* and *I. clavigera* in having racemose inflorescences, 4 lateral sepals, hammer-shaped capsules and ellipsoid seeds, but differs in having leaves with oblanceolate blades aggregated at the top of the stem, 3–5-flowered racemes, a yellow lower sepal without reddish patches, yellowish flowers, and a dorsal petal with stalks at the base.

Type: CHINA. Yunnan, Malipo County, Tianbao, Bajiaoping, limestone forests, elev. 1250 m, 22°57′44″ N, 104°51′45″ E, 30 Oct 2012, *Yun-Hong Tan 5728* (holotype: HITBC; isotype: PE). Paratype: China. Yunnan: Malipo County, Tianbao Town, elev. 1200 m, 23° 01′02.47″ N, 104°49′34.19″ E, 20 Nov 2014, *Xin–Xin Zhu 0001* (CSH)

琴葉鳳仙花 Figs. 1, 2.

Herb perennial, 20–30 cm tall, glabrous. Stems fleshy, erect, simple or branched; inferior nodes unapparent.

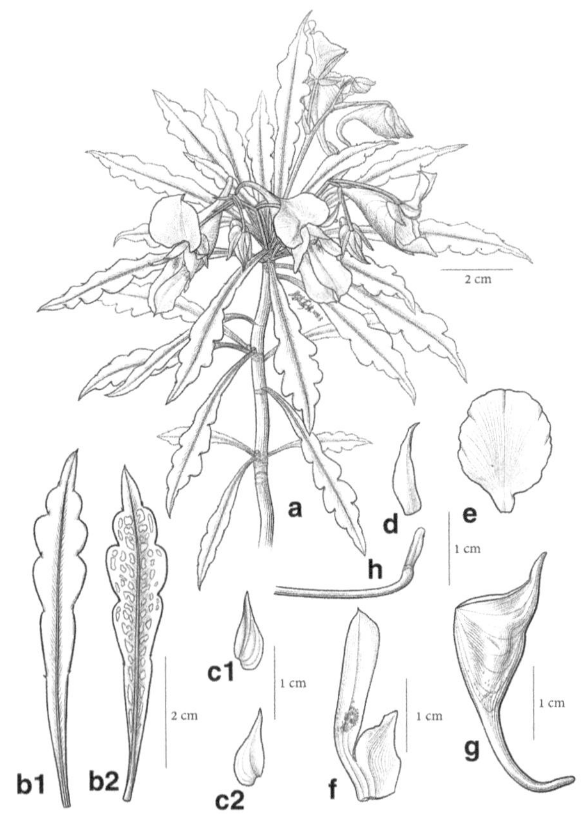

Fig. 1 *Impatiens pandurata* Y. H. Tan & S. X. Yu. **a** Habit; **b**, **b_1**, **b_2** Leaf, adaxial surface and abaxial surface; **c**, **c_1**, **c_2** Outer lateral sepal, abaxial surface and adaxial surface; **d** Inner lateral sepal; **e** dorsal petal; **f** lateral united petal; **g** lower sepal, lateral view; **h** capsule, immature. All from *Tan 5728* (HITBC) and drawn by Yun-Xi Zhu

Fig. 2 *Impatiens pandurata* Y. H. Tan & S. X. Yu. **a** Habitat; **b** habit; **c** leaf adaxial surface; **d** leaf abaxial surface; **e** flower, lateral view; **f** flower, different parts in separation, front view. All from *Tan 5728* (HITBC)

Leaves alternate, aggregated at stem apex, blades oblanceolate to linear oblanceolate, 5–7 cm long, 1–1.5 cm wide, apex acuminate, base cuneate, deep green above, pale green beneath, sometime with grey patches, margin deeply crenate, with spinose teeth. Veins unapparent. Petioles 0.8–1.2 cm. Racemes solitary in the upper axils, 5–7 cm long, 2–3 (−5)-flowered. Pedicels thin, 15–20 mm long. Bracts ovate to lanceolate, 7–9 mm long, acute. Flowers yellowish or cream. Lateral sepals 4, the outer 2 large, ovate to lanceolate, inaequilateral, 2–3 veined, yellowish-green, base rounded, apex acuminate to caudate, 7.8–9.1 × 3.3–3.7 mm; the inner 2 small, 10.7–11.4 × 1.2–1.6 mm, inaequilateral, apex acuminate. Lower sepal 2.5–3.0 cm long excluding spur, saccate, spur 5–5.6 mm. Dorsal petal 12.5–13.5 mm long, 11.6–12.3 mm wide, orbicular, apex rounded, base broadly cuneate and abruptly constricted into a stalk, midrib obvious, with a slight dorsal crest. Lateral united petals 2.1–2.4 cm long, the lower lobes 11.5–12.5 mm long, 5–5.5 mm wide, oblong, the upper ones 21–24 mm long, 4.5–5.5 wide, elliptic, apex emarginate, middle of inner margin without appendage. Stamens 5, filaments linear,

2–3 mm long, anthers obtuse. Ovary clavate, superior part inflated. Capsule hammer-shaped, seed ellipsoid.

Phenology Flowering and fruiting from September to December.

Ecology This new species grows under evergreen broad leaf forest; 1000–1200 m.

Distribution Impatiens pandurata is known from Jinping County and Malipo County, Yunnan, China.

Etymology The specific epithet 'pandurata' refers to the leaf shape of the new species.

Discussion

The phylogenetic topologies obtained with ITS and *atpB-rbcL* + *trnL-F* are congruent with those of previous studies (Yu et al. 2015). Both ITS and *atpB-rbcL* + *trnL-F* indicate that *I. pandurata* is a distinct member of the basal clade, subgenus *Clavicarpa* Yu et al. (2015) (Figs. 3, 4, Additional file 2: Figure S1, Additional file 3: Figure S2). The morphological characters, including perennial herb, racemose inflorescence, 4 lateral united petals, 4-carpellate ovary and one ovule per carpel, also support membership of *I. pandurata* in this subgenus. Although the ITS data shows that *I. pandurata* belongs to the basal clade, the relationships among the species in this clade are unclear. In the *atpB-rbcL* + *trnL-F* tree, *I. pandurata* and other species form a large polytomy, so the relationships among subgenus *Clavicarpa* are also unresolved. However, both nuclear ribosomal and plastid genes agree with the morphological evidence that *I. pandurata* is a new and distinct species.

The diagnostic morphological characters that distinguish *I. pandurata* from its allies are the oblanceolate leaves aggregated on the stem apex, with white macula beneath, the narrow lanceolate bracteole, and the subrotund dorsal petal with apparent stalk. Four species with similar morphological characters, *I. pandurata*, *I. apalophylla*, *I. clavigera*, and *I. spathulata*, are compared with each other, on the basis of their reproductive and vegetative characters in Table 1.

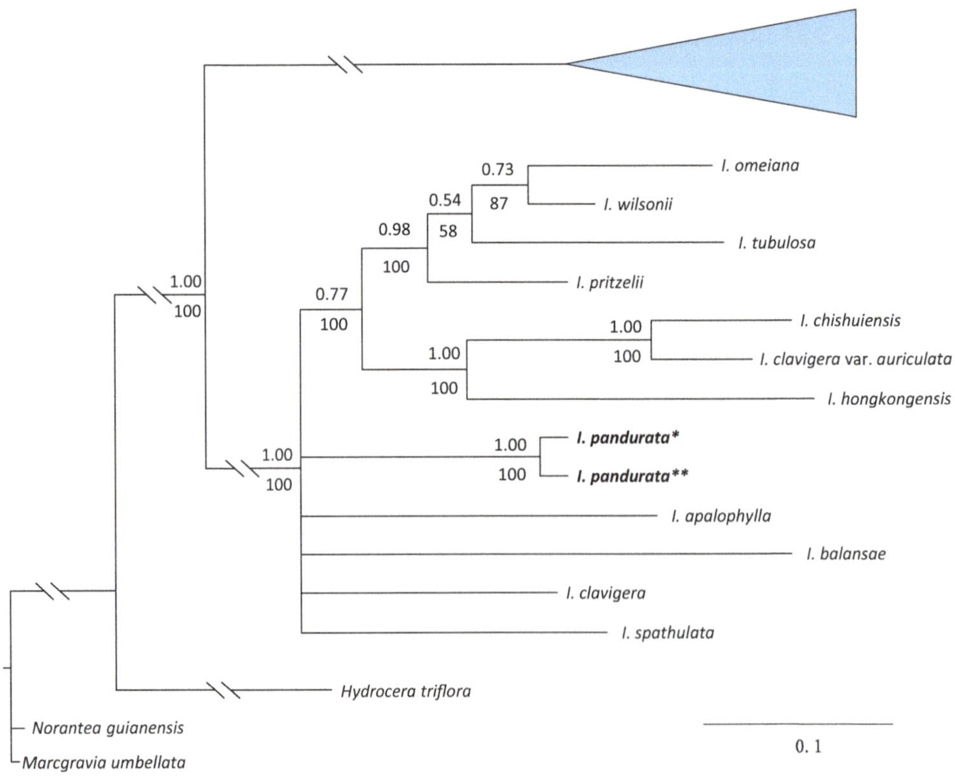

Fig. 3 Partial Bayesian consensus phylogram based on the branch lengths of the ITS data. *Numbers* above and below branches are Bayesian posterior probabilities (>0.5) and bootstrap percentages (>50 %), respectively. "—" indicates nodes not supported

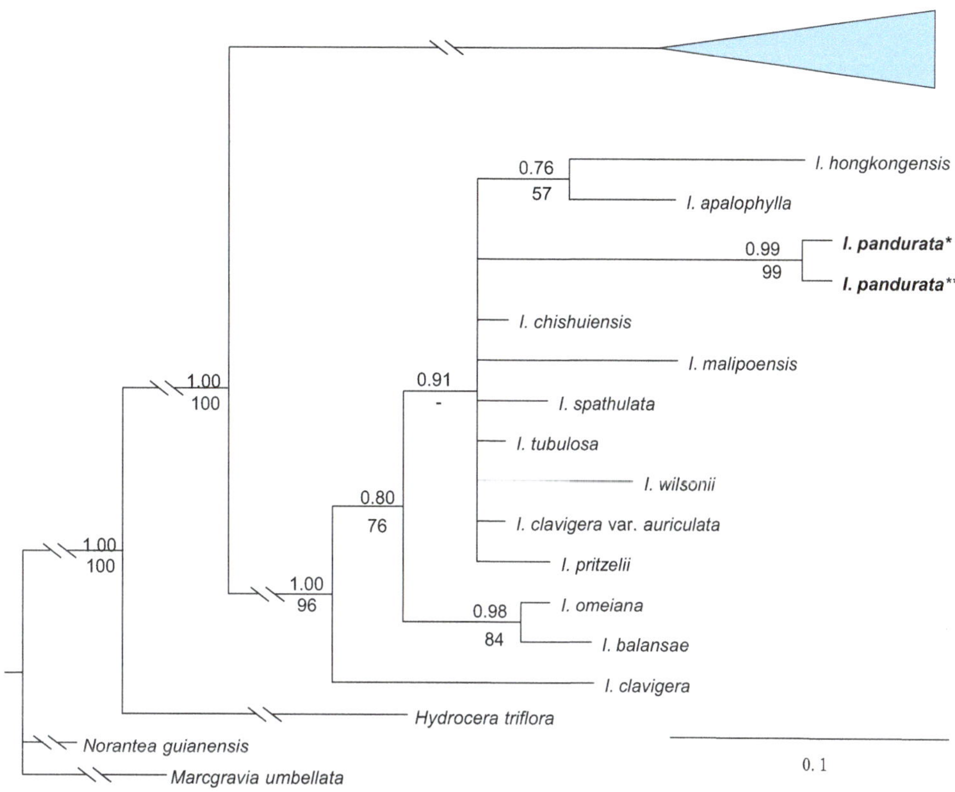

Fig. 4 Partial Bayesian consensus phylogram based on the branch lengths of the cpDNA data (*atpB-rbcL* + *trn L-F*). *Numbers* above and below branches are Bayesian posterior probabilities (>0.5) and bootstrap percentages (>50 %), respectively. "–" indicates nodes not supported

Table 1 Comparison among *Impatiens pandurata*, *I. apalophylla*, *I. clavigera* and *I. spathulata*

Characters	*I. pandurata*	*I. apalophylla*	*I. clavigera*	*I. spathulata*
Plant	Glabrous	Glabrous	Glabrous	Puberulous
Shape of leaf	Oblanceolate	Ovate to oblanceolate	Obovate to oblanceolate	Obovate
Size of leaf	5–7 cm × 1–1.5 mm	10–22 cm × 4–8 cm	5–15 cm × 3–5 cm	6–11 cm × 2.5–3.5 cm
Inflorescence	3-5-flowered	4–10-flowered	5–9-flowered	2–4-flowered
Floral color	Yellow	Yellow with reddish patches	Yellow	Pink
Outer lateral sepals	Oblique, ovate-orbiculate, aequilateral, apex acuminate	Oblique ovate, apex acuminate	Oblique ovate, apex acuminate	Oblique ovate to ovate, apex acuminate
Inner lateral sepals	Linear	Lanceolate	Linear-lanceolate	Linear-lanceolate
Lower sepal	Absent patch	With redish patches	Absent patch	Absent patch
Spur	Long, 1–1.5 cm	Long, 1.5–2 cm	Short, 5–6 mm	Long, 1.5–1.8 cm
Lower lobe of united petals	8–12 mm long, elliptic	1–1.5 cm long, oblong	1–1.2 cm long, oblong	1–1.2 cm long, oblong
Upper lobe of united petal	2–2.5 cm long, oblong, yellow, with slightly reddish patches	2.5–2.7 cm long, oblong, yellow, with apparent reddish patches	2.5–2.6 cm long, oblong, yellowish-green, absent patch	2.3–2.5 cm long, oblong, pink, absent patch
Dorsal petal	Suborbicular	Elliptic	Obovate	Elliptic to ovate

Conclusion

With the support of morphological studies and molecular phylogenetic analysis, *I. pandurata* is a species new to science. Detailed descriptions, line drawings, color plates, phylogenetic analysis and comparisons with phenetically similar species are provided to aid in identification.

Authors' contributions

YHT, XXZ and WZ collected the new species. Morphological studies were carried out by YHT, HJ and SXY. Molecular studies were carried out by YNL. YXZ prepared the line drawing. YHT, SXY and YNL prepared the manuscript. All authors read and approved the final manuscript.

Author details

[1] Center for Integrative Conservation, Xishuangbanna Tropical Botanical Garden, Chinese Academy of Sciences, Menglun, Mengla 666303, Yunnan, China. [2] State Key Laboratory of Systematic and Evolutionary Botany, Institute of Botany, Chinese Academy of Sciences, Beijing 100093, People's Republic of China. [3] Yunnan Academy of Forestry/Yunnan Laboratory for Conservation of Rare, Endangered and Endemic Forest Plants, State Forestry Administration, Kunming 650201, China. [4] Shanghai Chenshan Plant Science Research Center, Chinese Academy of Sciences, Shanghai Chenshan Botanical Garden, Shanghai 201602, China. [5] Key Laboratory of Economic Plants and Biotechnology, Kunming Institute of Botany, Chinese Academy of Sciences, Kunming 650201, Yunnan, China.

Acknowledgements

The authors wish to thank the keepers of the herbaria of Paris (P) and Royal Botanic Garden, Kew (K) for loan of type specimens, Mr Yun-Xi Zhu for making the drawing, and Professor Richard T Corlett for polishing the language. We are also grateful to Dr. Wei Wang and Xiao-Guo Xiang for checking the manuscript. This work was supported by National Natural Science Foundation of China (31170177), National Natural Science Foundation of China (Grant no. 31110103911), Main Direction Program of Knowledge Innovation of Chinese Academy of Sciences (Grant no. KSCX2-EW-Z-1) and the special program for public health of traditional Chinese medicine named "Investigation and monitoring in the raw material resources of Chinese medicine needed in the National basic medicine" (Fiscal agency [2011] No. 76) and Special program for Chinese pharmaceutical industry named "Protection and utilization of regional representative Chinese medicine resources"(201207002).

Competing interests

The authors declare that they have no competing interests.

References

Chen Y-L (2001) Balsaminaceae. In: Chen Y-L (ed) Flora Reipublicae Popularis Sinica, vol 47(2). Science Press, Beijing, pp 1–243

Chen Y-L, Akiyama S, Ohba H (2007) Balsaminaceae. In: Wu Z-Y, Raven PH (eds) Flora of China, vol 12. Science Press, Missouri Botanical Garden Press, Beijing, St. Louis, pp 43–113

Doyle JJ, Doyle JL (1987) A rapid DNA isolation procedure for small quantities of fresh leaf material. Phytochem Bull 19:11–15

Fischer E (2004) Balsaminaceae. In: Kubitzki K (ed) The Families and Genera of Vascular Plants, vol 6. Springer, Berlin, pp 20–25

Grey-Wilson C (1980) *Impatiens* of Africa. Balkema, Rotterdam, pp 1–57

Hall TA (1999) BioEdit: a user-friendly biological sequence alignment editor and analysis program for Windows 95/98/NT. Nucl Acids Symp Ser 41:95–98

Hooker JD (1908) Les espèces du genre "*Impatiens*" dans l'herbier du Museum de Paris. Nov Arch Mus Nat Hist Paris 10(4):233–272

Janssens SB, Geuten K, Yuan Y-M, Song Y, Küpfer P, Smets E (2006) Phylogenetics of *Impatiens* and *Hydrocera* (Balsaminaceae) using chloroplast *atpB-rbcL* spacer sequences. Syst Bot 31:171–180

Kuang R-P, Duan L-D, Gu J-Z, Cai X-Z, Cong Y-Y, Liu K-M (2014) *Impatiens liboensis* sp. nov (Balsaminaceae) from Guizhou, China. Nord J Bot 32:463–467

Luo Q, Wang J, Zhao H (2014) *Impatiens menghuochengensis* Q Luo, a new species of *Impatiens* (Balsaminaceae) from Sichuan, China. Nord J Bot 6:839–843

Narayanan MKR, Joseph JP, Kumar NA, Sivadasan M, Alfarhan AH (2013) *Impatiens theuerkaufiana* (Balsaminaceae), a new scapigerous species from the Western Ghats, India. Phytotaxa 83:54–60

Posada D, Crandall KA (1998) Modeltest: testing the model of DNA substitution. Bioinformatics 14:817–818

Ronquist F, Huelsenbeck JP (2003) MrBayes 3: Bayesian phylogenetic inference under mixed models. Bioinformatics 19:1572–1574

Song Y, Yuan Y-M, Küpfer P (2003) Chromosomal evolution in Balsaminaceae, with cytological observations on 45 species from Southeast Asia. Caryologia 56:463–481

Swofford DL (2003) PAUP*: Phylogenetic Analysis Using Parsimony (*and Other Methods), v4.0b10. Sinauer Associates, Sunderland

Taberlet P, Gielly L, Pautou G, Bouvet J (1991) Universal primers for amplification of three non-coding regions of chloroplast DNA. Plant Mol Biol 17:1105–1109

Thompson JD, Gibson TJ, Plewniak F, Jeanmougin F, Higgins DG (1997) The Clustal X windows interface: flexible strategies for multiple sequence alignment aided by quality analysis tools. Nucleic Acids Res 24:4876–4882

Utami N (2013) *Impatiens kerinciensis* (Balsaminaceae), a new species from Sumatra, Indonesia. Kew Bull 68:687–688

White TJ, Bruns T, Lee S, Taylor J (1990) Amplification and direct sequencing of fungal ribosomal RNA genes for phylogenetics. In: Innis MA, Gelfand DH, Sninsky JJ, White TJ (eds) PCR Protocols: a Guide to Methods and Applications. Academic Press, New York, pp 315–322

Yu S-X (2012) Balsaminaceae of China. Peking University Press, Beijing

Yu S-X, Janssens SB, Zhu X-Y, Lidén M, Gao T-G, Wang W (2015) Phylogeny of *Impatiens* (Balsaminaceae): integrating molecular and morphological evidence into a new classification. Cladistics. doi:10.1111/cla.12119

Yuan Y-M, Song Y, Geuten K, Rahelivololona E, Wohlhauser S, Fischer E, Smets E, Küpfer P (2004) Phylogeny and biogeography of Balsaminaceae inferred from ITS sequence data. Taxon 53:391–403

15

Chromosome numbers of populations of three varieties of *Bidens pilosa* in Taiwan

Ya-Lun Huang[1] and Wen-Yuan Kao[1,2*]

Abstract

Background: Hairy beggar-ticks (*Bidens pilosa* L.) is a common invasive plant in tropical and subtropical regions. The Flora of Taiwan listed three varieties of *B. pilosa* in Taiwan, var. *minor*, var. *pilosa* and var. *radiata*. Among the three varieties, var. *radiata* was the most recently, in 1970s, introduced into Taiwan. However, after its introduction into Taiwan, var. *radiata* has become dominant over the other two varieties and is considered a serious invasive plant in lowland of Taiwan. Our previous study showed that var. *radiata* is self-incompatible and the other two varieties are self-fertile. Could it be possible that different chromosome numbers contribute to the different breeding systems of these three varieties? In addition, the heterogeneities of traits of var. *radiata* were found higher than those of var. *minor* and var. *pilosa*. Is the phenomenon resulting from the hybridization between var. *radiata* with other varieties? We counted chromosome numbers of populations of these three varieties distributed in Taiwan and conducted hand pollination treatment between var. *radiata* (as pollen receiver) and var. *minor* or var. *pilosa* (as pollen donor) to provide answer for the aforementioned questions.

Results: No difference was found in chromosome numbers among populations of the same variety. Forty-eight chromosomes ($2n = 48$) were counted for var. *radiata* while 72 ($2n = 72$) chromosomes for var. *minor* and var. *pilosa*. Therefore, var. *radiata* is tetraploid and var. *minor* and var. *pilosa* are hexaploid. No successful hybridization was found between var. *radiata* and var. *minor* or between var. *radiata* and var. *pilosa*.

Conclusions: This study provided the evidence that the invasive plant (*B. pilosa* var. *radiata*) has different chromosome numbers from the other two varieties and is unlikely to hybridize with the other two varieties.

Keywords: *Bidens pilosa*, Chromosome number, Crossing, Invasive plant, Self-incompatibility

Background

Bidens pilosa L. (Hairy beggar-ticks, Asteraceae) is a cosmopolitan weed. With its diversification center in Mexico, it is widely distributed in tropical and subtropical regions (Ballard 1986). This taxon is an annual or perennial herb with square stems and opposite leaves and often occupies roadsides, disturbed sites and cultivated fields (Peng et al. 1998). Based on the morphological differences, *B. pilosa* was classified into six varieties: var. *pilosa*, var. *minor*, var. *radiata*, var. *bimucronata*, var. *calcicola*, and var. *alausensis* (Sherff 1937). Accordingly, The Flora of Taiwan listed three varieties of *B. pilosa* L. in Taiwan, i.e. var. *minor*, var. *pilosa* and var. *radiata* (Peng et al. 1998). The most obvious differences in morphology among these three varieties are their flowers (Fig. 1). The capitula of var. *pilosa* are discoid (without ray florets), and the capitula of var. *minor* and var. *radiata* are radiate (with ray florets). Although var. *minor* and var. *radiata* both have ray florets, their ray florets are quite different. *B. pilosa* var. *minor* have eight ray florets which are often shorter than 8 mm while var. *radiata* have 5–8 ray florets which are often longer than 10 mm. Thirty years ago, var. *minor* was widely distributed in Taiwan. However, after var. *radiata* being introduced into Taiwan, var. *radiata* has become a serious invasive plant and dominant over the other two varieties within 30 years (Wu et al. 2004, 2010).

*Correspondence: wykao@ntu.edu.tw
[1] Institute of Ecology and Evolutionary Biology, National Taiwan University, 1, Roosevelt Rd., Sec. 4, Taipei 106, Taiwan
Full list of author information is available at the end of the article

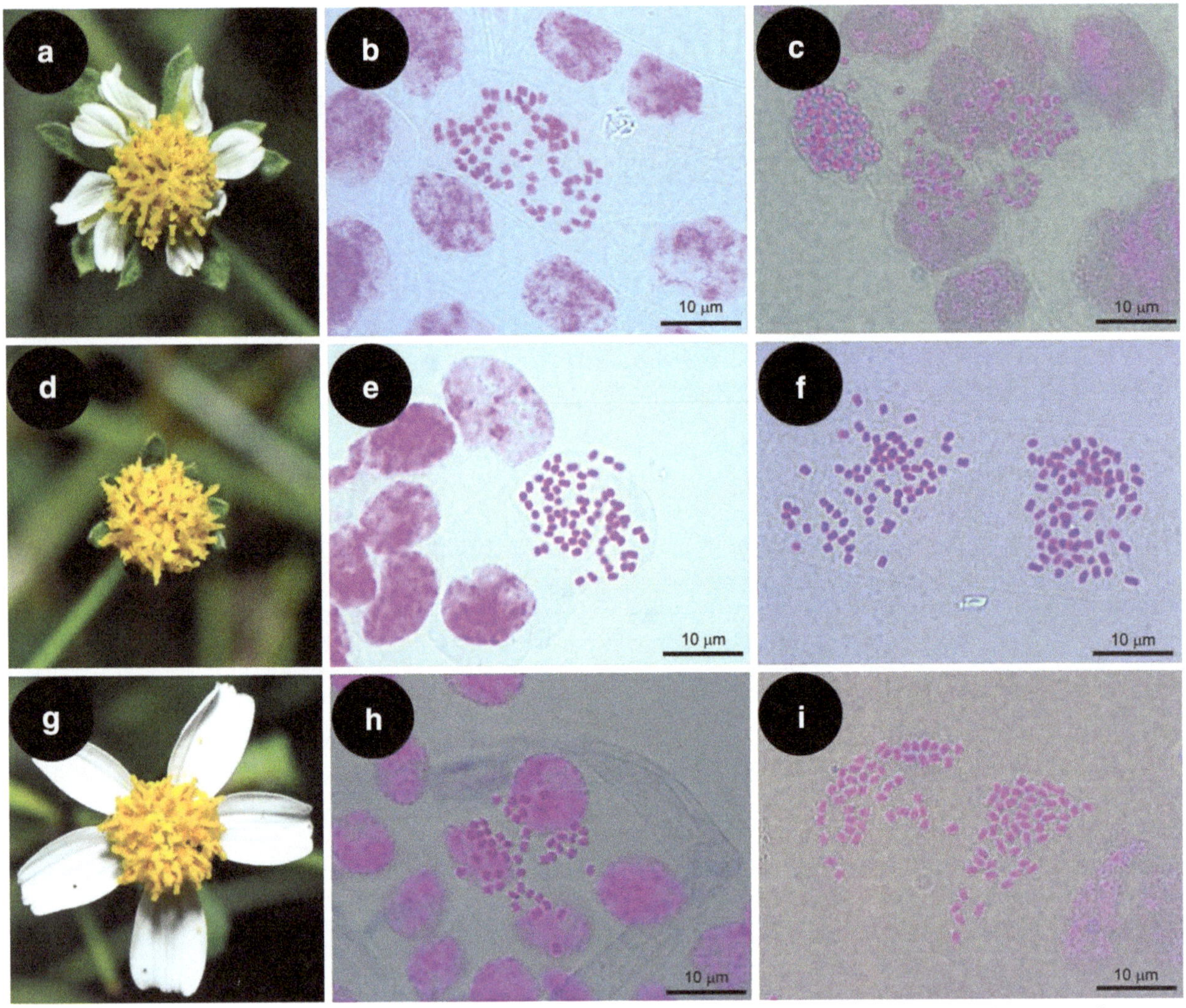

Fig. 1 Pictures of inflorescence and somatic chromosomes of three varieties of *Bidens pilosa* in Taiwan. Representatives of inflorescence of var. *minor* (**a**), var. *pilosa* (**d**) and var. *radiata* (**g**) and somatic chromosomes of roots sampled from two different populations of each variety: var. *minor* from Chiayi ($2n = 72$) (**b**) and from Yilan ($2n = 72$) (**c**), var. *pilosa* from Hualien ($2n = 72$) (**e**) and from Chiayi ($2n = 72$) (**f**) and var. *radiata* from Nantou ($2n = 48$) (**h**) and from Yilan ($2n = 48$) (**i**) county in Taiwan

A previous study on breeding systems of these three varieties demonstrated that var. *minor* and var. *pilosa* are self-compatible while var. *radiata* is self-incompatible (Huang and Kao 2014). Chromosome numbers and breeding systems of 200 populations of *B. pilosa* in southern United States, Mexico and Central America have been investigated to clarify the classification of *B. pilosa* (Ballard 1986). After the analyses, Ballard concluded that *B. pilosa* in North and Central America should be treated as a species complex containing three species with different chromosome numbers due to different ploidy: *B. odorata* ($2n = 24$), *B. alba* ($2n = 48$) and *B. pilosa* ($2n = 72$). Species with different polyploidy also had different breeding system: hexaploid populations were self-fertile while diploid and tetraploid populations were self-incompatible. According to Ballard's result, could it be possible that the chromosome number of var. *radiata* is different from that of var. *minor* and var. *pilosa*?

Furthermore, many traits of var. *radiata* were found having higher variation than those of var. *minor* and var. *pilosa*. For example, highly variation in ray floret numbers and sizes was found in var. *radiata*, while all var. *minor* had eight ray florets with similar size in each capitulum. Significant variations in pollen/ovule ratio and disc floret numbers per capitulum were only found in var. *radiata* but not in var. *minor* and var. *pilosa* (Huang and Kao 2014). Because var. *radiata* is self-incompatible and sometimes growing sympatrically with var. *minor* and var. *pilosa*, could it be possible that the var. *radiata* can hybridize with the other two varieties resulting in the highly variation in these traits?

The objectives of this study were to understand (1) whether these three varieties have different chromosome numbers, and (2) whether *B. pilosa* var. *radiata* can hybridize with var. *minor* or var. *pilosa*. To achieve the goal, we counted chromosome numbers of populations of these three varieties in Taiwan and conducted hand pollination treatments between var. *radiata* (as pollen receiver) and var. *minor* or var. *pilosa* (as pollen donors).

Methods

Chromosome count

Achenes of three varieties of *B. pilosa* were collected from 5–8 populations distributed in Taiwan (Fig. 2). The collected achenes were planted in 1 L pots for about 1 month. When plants reached 20 cm height with 4–5 paired leaves, their shoots were cut and immersed into water for growing adventitious roots (Hsu and Kao 2014).

When shoots growing adventitious roots of 3–5 cm, the emergent rootlets with root-tips (ca. 1–2 cm) were cut, pretreated with 2 mM 8-hydroxyquinoline at 14 °C

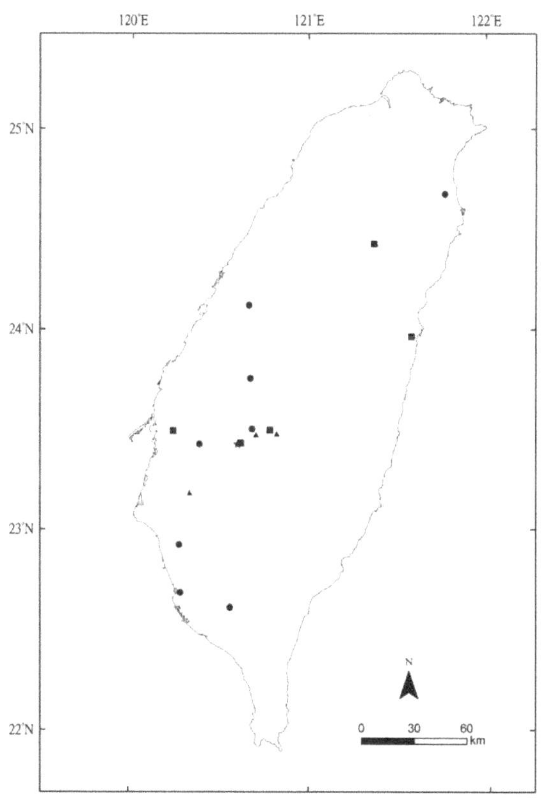

Variety	Sampling site		
	Altitude (m)	Longitude (E)	Latitude (N)
var. *minor*			
	20	120°20'	23°11'
	1420	120°41'	23°30'
	1342	120°42'	23°28'
	2274	120°49'	23°28'
	1156	121°22'	24°26'
	18	121°35'	23°58'
var. *pilosa*			
	0	120°14'	23°29'
	528	120°37'	23°26'
	2000	120°47'	23°29'
	1255	121°22'	24°25'
	26	121°34'	23°57'
var. *radiata*			
	20	120°17'	22°55'
	20	120°17'	22°41'
	20	120°23'	23°25'
	20	120°34'	22°36'
	58	120°40'	24°07'
	149	120°41'	23°45'
	1420	120°41'	23°30'
	15	121°46'	24°40'

Fig. 2 Sampling sites of the three varieties of *Bidens pilosa* in Taiwan. Achenes used in chromosome numbers experiment were collected from six populations of *B. pilosa* var. *minor* (*filled triangle*), five populations of var. *pilosa* (*filled square*), and eight populations of var. *radiata* (*filled circle*). The altitude, longitude and latitude of these 19 sampling sites were also listed. Achenes used in hand pollination experiment were collected from populations of the three varieties growing sympatrically in Chiayi (*filled star*) (120°37′E, 23°26′N, 500 m a.s.l.)

for 5 h, fixed in Carnoy's solution (95 % EtOH: acetic acid = 3:1, v/v) for 24 h and then stored in 70 % EtOH at −20 °C. After washed with distilled water twice, rootlets were hydrolyzed in 1 N HCl at 60 °C for 8 min and rinsed twice with distilled water. Following, materials were stained with leuco-basic fuchsin in the dark at room temperature for 2 h and treated with 1 % pectinase at room temperature for 1 h to separate the connected cells. Root-tips were spread on a slide and squashed in a drop of 45 % acetic acid and then observed under a microscope (Grombone-Guaratini et al. 2006). Totally, 16 cells of var. *minor*, 7 cells of var. *pilosa* and 43 cells of var. *radiata* were counted.

Hand-pollination experiment

Achenes collected from a population in Chiayi (120°37′E, 23°26′N, 500 m a.s.l.) (Fig. 2, filled star), where the three varieties grew sympatrically, were planted in pots and placed in the experimental farm of National Taiwan University. After about 2 months, when these plants bloomed, their flowers were used in crossing trials. Because var. *minor* and var. *pilosa* are highly self-compatible and capable of self-pollinating (Huang and Kao 2014), it is difficult to prevent their pollen grains depositing on their own stigma. Therefore, these two varieties were assigned as pollen donors in crossing trials. In contrast, var. *radiata* being self-incompatible (Huang and Kao 2014) was assigned as pollen receivers in crosses. In addition, crosses between individuals of var. *radiata* from different populations were also conducted as control. Plants growing naturally in the experimental farm of National Taiwan University and those plants growing from the achenes collected from the Chiayi population were used as pollen donors and receivers, respectively.

Bidens pilosa produces inflorescence of capitulum. In its capitulum, the hermaphroditic disc florets opened in centripetal sequence, and the duration from the blooming of the first disc floret to the last disc floret is about 3–4 days (Huang et al. 2012). Therefore, hand pollination was conducted once a day consecutively for 3 or 4 days during the period of anthesis. To prevent the deposition of non-target pollen grains, the capitula assigned as pollen receivers were wrapped in fine nylon-mesh netting bags (7 × 9 cm) as soon as their first disc florets opened until all disc florets on each capitulum withered. The capitula assigned as pollen donors were also wrapped into the netting bags to prevent contamination of pollen grains from other species.

At harvesting, the numbers of undeveloped ovules, immature and mature achenes were counted. Because there is only one ovule in each disc floret (Huang et al. 2012), the sum of mature achenes, immature achenes and undeveloped ovules represent the numbers of disc floret per capitulum.

Results

Chromosome count

The chromosomes of *B. pilosa* var. *radiata*, var. *minor* and var. *pilosa* shared similar size (ca. 2 μm in length) and morphology (with centromeres in the central region of the chromosomes) (Fig. 1).

For each variety, plants collected from different populations had same chromosome numbers. However, different chromosome numbers were counted among the three varieties. Chromosome numbers of var. *minor* and var. *pilosa* were counted as $2n = 72$ (Fig. 1b, c, e and f), while those of var. *radaita* was $2n = 48$ (Fig. 1h, i).

Hand-pollination experiment

In the crossing experiment, the capitulum of *B. pilosa* var. *radiata* assigned as pollen receivers had 21–57 disc florets in each capitulum (Table 1). In crossing trial, no viable achenes were produced when the stigmata of disc florets of var. *radiata* receiving either the pollen grains from var. *minor* or var. *pilosa* (Table 1) or pollen grains of their own. In contrast, 17–29 mature achenes were produced in one capitulum of var. *radiata* when their stigmata receiving pollen grains from other individuals of var. *radiata*.

Discussion

Different chromosome numbers were found between the invasive and non-invasive varieties of *B. pilosa* in Taiwan. Different ploid level is one of the most frequent reasons leading to difference in chromosome numbers in some genera of Asteraceae (Dematteis and Fernandez 2000; Keil et al. 1988; Solbrig et al. 1972). Therefore, the variation in chromosome numbers among the three varieties of *B. pilosa* might result from Polyploidy. The basic chromosome number for genus *Bidens* has been reported as

Table 1 Disc floret numbers and achene set in hand pollination experiment

Pollen donor	Pollen receiver (var. *radiata*)		
	Disc floret	Mature achene	*n*
var. *minor*	48.7 ± 3.4	0 ± 0	3
var. *pilosa*	38.8 ± 2.3	0 ± 0	5
var. *radaita* (self)	31.8 ± 4.5	0 ± 0	4
var. *radiata* (cross)	42.3 ± 3.3	23.0 ± 2.0	6

The disc floret number per capitulum, mature achene number per capitulum (mean ± SE), and the number of capitulum counted (*n*) in crossing experiment by hand pollination in which *Bidens pilosa* var. *radiata* as a pollen receiver, receiving pollen from other varieties (var. *minor* or var. *pilosa*) or same individual of var. *radiata* (self), or other individuals of var. *radiata* (cross)

$x = 12$ (Ballard 1986). Accordingly, *B. pilosa* var. *radiata* is tetraploidy, while var. *minor* and var. *pilosa* are hexaploidy. A study on the breeding system of these three varieties found that var. *radiata* was self-incompatible, while var. *minor* and var. *pilosa* were self-compatible (Huang and Kao 2014). Thus, different ploid level might result in the different breeding systems among the three varieties. Similar results were also reported for 200 populations of *B. pilosa* species complex in Central America, where the tetraploid populations were self-incompatible while the hexaploid populations were self-compatible (Ballard 1986).

Recent studies showed that there is a link between Polyploidy and invasiveness (Beest et al. 2012; Goralski et al. 2014; Lowry and Lester 2006; Treier et al. 2009). A study compared 81 invasive species and their 2356 congeners found that invasive species are generally polyploidy (Pandit et al. 2011). In this study we found that the invasive var. *radiata* is polyploidy. However, the non-invasive varieties are also polyploidy and even have higher ploid level than the invasive one. In addition, *B. bipinnata*, the congener of the three varieties, which is not invasive in Taiwan is also a polyploidy ($2n = 72$) (Peng and Hsu 1978). Furthermore, it is reported that in the native distribution range of *B. pilosa* species complex, var. *minor* and var. *pilosa* are more widespread and more dominant than var. *radiata*. Thus, Polyploidy alone cannot explain why var. *radiata* becomes more dominant than the other two varieties in Taiwan. There are some other traits, for example, the ability of vegetative reproduction (Hsu and Kao 2014) and its breeding system (Huang and Kao 2014), contributing to the invasiveness of var. *radiata* in Taiwan.

Results from hand pollination experiment and consistent chromosome numbers counted in populations of var. *radiata* indicated that var. *radiata* is unlikely to hybridize with var. *minor* or var. *pilosa*. Accordingly, highly variation in traits found in *B. pilosa* var. *radiata* is unlikely caused by hybridizing with other varieties. Instead, we have previously suggested that the highly heterogeneity in traits might result from the obligate xenogamous breeding system carried by var. *radiata* (Huang and Kao 2014). The characters of highly variation in many traits might play an important role contributing to the invasiveness of var. *radiata* in Taiwan.

Based on the morphological differences, Sherff (1937) classified *B. pilosa* into six varieties. Ballard (1986) investigated 200 populations of *B. pilosa* in central America and suggested that plants previously classified as var. *pilosa* and var. *minor* by Sherff's system still belonged to *B. pilosa* (hexaploid, $2n = 72$). But those being classified as var. *radiata* by Sherff should be treated as *B. odorata* (diploid, $2n = 24$) or *B. alba* (tetraploid, $2n = 48$), depending on their chromosome numbers. Accordingly, *B. pilosa* var. *radiata* in Taiwan might be treated as *B. alba*. Further studies on detailed morphological characters, geographic distribution and molecular phylogeny of the genus *Bidens* are suggested to confirm the systematics of the three varieties in Taiwan.

Conclusions

The invasive and non-invasive varieties of *Bidens pilosa* in Taiwan have different chromosome numbers and were different polyploid. The chromosome number of invasive var. *radiata* was $2n = 48$ (tetraploidy), while those of non-invasive var. *minor* and var. *pilosa* were $2n = 72$ (hexaploidy). The hand pollination experiment demonstrated that the invasive var. *radiata* is unlikely to hybridize with the non-invasive var. *minor* or var. *pilosa*. High variance of many traits found in invasive var. *radiata* might result from its self-incompatibility.

Authors' contributions

YLH designed the study, conducted the experiment, collected data and drafted the manuscript. WYK conceived the study, participated in its design and drafted the manuscript. Both authors read and approved the final manuscript.

Author details

[1] Institute of Ecology and Evolutionary Biology, National Taiwan University, 1, Roosevelt Rd., Sec. 4, Taipei 106, Taiwan. [2] Department of Life Science, National Taiwan University, 1, Roosevelt Rd., Sec. 4, Taipei 106, Taiwan.

Acknowledgements

We thank Cheng-En Li for plotting the figure of sampling sites. This study was partly supported by a grant from the National Science Council of Taiwan, R. O. C. (NSC 100-2621-B-002-004).

Compliance with ethical guidelines

Competing interests

The authors declare that they have no competing interests.

References

Ballard R (1986) *Bidens pilosa* complex (Asteraceae) in North and Central America. Am J Bot 73:1452–1465

Beest MTB, Le Roux JJ, Richardson DM, Brysting AK, Suda J, Kubešová M, Pyšek P (2012) The more the better? The role of polyploidy in facilitating plant invasions. Ann Bot 109:19–45

Dematteis M, Fernandez A (2000) Chromosome studies on nine South American species of *Vernonia* (Vernonieae, Asteraceae). Caryologia 53:55–61

Goralski G, Judasz A, Gacek P, Grabowska-Joachimiak A, Joachimiak AJ (2014) Polyploidy, alien species and invasiveness in Polish angiosperms. Plant Syst Evol 300:225–238

Grombone-Guaratini MT, Mansanares ME, Semir J, Solferini VN (2006) Chromosomal studies of three species of *Bidens* (L.) (Asteraceae). Caryologia 59:14–18

Hsu H-M, Kao W-Y (2014) Vegetative and reproductive growth of an invasive weed *Bidens pilosa* L.var. *radiata* and its noninvasive congener *Bidens bipinnata* in Taiwan. Taiwania 59:119–126

Huang Y-L, Kao W-Y (2014) Different breeding systems of three varieties of *Bidens pilosa* in Taiwan. Weed Res 54:162–168

Huang Y-L, Chen S-J, Kao W-Y (2012) Floral biology of *Bidens pilosa* var. *radiata*, an invasive plant in Taiwan. Bot Stud 53:501–507
Keil DJ, Luckow MA, Pinkava DJ (1988) Chromosome studies in Asteraceae from the United States, Mexico, the West Indies, and South America. Am J Bot 75:652–668
Lowry E, Lester SE (2006) The biogeography of plant reproduction: potential determinants of species' range sizes. J Biogeogr 33:1975–1982
Pandit MK, Pocock MJO, Kunin WE (2011) Ploidy influences rarity and invasiveness in plants. J Ecol 99:1108–1115
Peng C-I, Hsu C-C (1978) Chromosome numbers in Taiwan Compositae. Bot Bull Acad Sin 19:53–66
Peng C-I, Chung K-F, Li H-L (1998) Compositae. In: C HT, Taiwan ECotFo (eds) Flora of Taiwan, vol 4. 2nd edn. Department of Botany, National Taiwan University, Taipei, Taiwan, pp 807–1101
Sherff EE (1937) The genus *Bidens*. Field Mus Nat Hist Bot Ser 11:412–461
Solbrig OT, Kyhos DW, Powell M, Raven PH (1972) Chromosome numbers in Compositae VIII: Heliantheae. Am J Bot 59:869–878
Treier UA, Broennimann O, Normand S, Guisan A, Schaffner U, Steinger T, Muller-Scharer H (2009) Shift in cytotype frequency and niche space in the invasive plant *Centaurea maculosa*. Ecology 90:1366–1377
Wu S-H, Hsieh C-F, Chaw S-M, Rejmanek M (2004) Plant invasions in Taiwan: insights from the flora of casual and naturalized alien species. Divers Distrib 10:349–362
Wu S-H, Yang TYA, Teng Y-C, Chang C-Y, Yang K-C, Hsieh C-F (2010) Insights of the latest naturalized flora of Taiwan: change in the past 8 years. Taiwania 55:139–159

16

Great fraction of dissolved organic C and N in the primary per-humid *Chamaecyparis* forest soil

Chih-Wei Tsai[1], Guanglong Tian[2] and Chih-Yu Chiu[1*]

Abstract

Background: Labile organic matter plays a crucial role in a variety of forest functions, however, our understanding to its quality and quantity across various forests is limited, particularly primary forests. We investigated soil labile C and N (i.e. microbial biomass C and N, dissolved organic carbon (DOC) and nitrogen (DON), associated ammonium, and nitrate) at three topographic locations (i.e. summit, footslope and lakeshore) in a primary *Chamaecyparis* forest of Taiwan. The following hypotheses are tested in this study: (1) This undisturbed *Chamaecyparis* forest shows the great size of soil labile C and N; (2) there is an evident topographic effect on the distribution of soil labile C and N and the associated inorganic N over seasons.

Results: Fulfilling with our first hypothesis, the considerable size of labile C and N in this forest soil was quantified. Abundant C availability and the acidity of soils in this forest favoured ammonium production over nitrate. The undisturbed environment with per-humid and acidic soil was linked to the high concentrations of soil DOC and DON as the dominant form in N dynamics. In contrast to our second hypothesis, topographic effects on soil labile C and N were generally not evident, suggesting the homogeneous soil environment across various topographic locations in this *Chamaecyparis* forest.

Conclusions: This study illustrates the sustainable importance of primary montane forests for being sources of DOC and DON.

Keywords: Labile resource, DOC, DON, Ammonium, Nitrate

Background

Carbon and nitrogen play essential roles in developing, maintaining and reproducing of organisms. Transformation, movement and reuse of these nutrients are important to sustain a variety of ecosystem functions (Galloway et al. 2004). However, our understanding to the quality and quantity of soil labile C and N resources across various forests is still limited, in particular, subtropical primary montane forests. The ecological importance of soil organic matter lies in its biodegradability and availability for microbes, fauna and plant (Ghani et al. 2007). Thus, the comprehension of the size of soil labile C and N resources can help us understand the fraction of C and N immediately available to soil organisms and cycling in amount of C and N resources in forest ecosystems. Of these, microbial biomass C (C_{mic}) and N (N_{mic}) are the most labile C and N pools in soils. Although they generally constitute only 1–3 % and 1–5 % of total soil C and N (Moore et al. 2000), they are related to the size of decomposed organic material and soil physical and chemical properties.

Neff and Asner (2001) noted dissolved organic carbon (DOC) and dissolved organic nitrogen (DON) constitute the main fraction of available monomers for microorganisms. The quantity of DOC can impact the heterotrophic microbial activity and the associated N transformation (Singh and Kashyap 2006). The proportion of DON and inorganic N can be an important indicator of water quality and nutrients in hydrological cycling (Fang et al.

*Correspondence: bochiu@sinica.edu.tw
[1] Biodiversity Research Center, Academia Sinica, Taipei 11529, Taiwan
Full list of author information is available at the end of the article

2009; Lutz et al. 2011). With the shift of nitrogen cycling in forest ecosystems from organic to inorganic form, recent studies have shown that N can be lost from temperate forest ecosystems predominantly by the way of DON (Hedin et al. 1995; Perakis and Hedin 2002). So far, studies addressing the dynamics of labile C and N have mainly focused on temperate forests, but the potential contribution of subtropical primary forest soils to C and N loadings is still poorly known.

Chamaecyparis forests are valuable natural resources in eastern Asia because of its high quality timber, control of soil erosion, and storage of C and N. Much effort has been invested in investigating soil properties and fertility management of Japanese hinoki cypress (*Chamaecyparis obtusa*) plantations (e.g. Inagaki et al. 2008, 2011). By contrast, characteristics of soil organic matter under natural *Chamaecyparis* forests in subtropical montane area are not well known. Today, the *Chamaecyparis* forest in the Chi-Lan Mountain is one of the few preserved primary *Chamaecyparis* forests in northcentral Taiwan. The unique aspect of the *Chamaecyparis* forest is that it has the environment, which is rarely disturbed, high precipitation, and mild temperature.

Previous studies have shown that strongly acidic and wet environment resulted in the low diversity of soil bacterial communities in this region (Lin et al. 2010). The litter decomposition and humification in the *Chamaecyparis* forest is generally very slow (Chen et al. 2013; Chung et al. 2012), and this may result in the considerable soil organic matter and the great size of labile C and N loadings. Chen and Chiu (2000) investigated soil structure across a slope in the *Chamaecyparis* forest and found out a common phenomenon—a thick layer of organic horizon (i.e. 10–30 cm) derived from *Sphagnum* on the surface of soils. However, little is known about the size of the soil labile C and N in this primary *Chamaecyparis* forest. Additionally, Schmidt et al. (2010) demonstrated that DOC and DON were the predominant forms of C and N resources released from surface soils in a secondary *Chamaecyparis* forest of Taiwan. The objectives of the study were therefore to quantify concentrations of soil labile C and N under the *Chamaecyparis* forest and evaluate their distributions along topographic locations over seasons.

Methods

Study area

Surrounding a small and shallow lake, the primary *Chamaecyparis* forest is located in the Yuan Yang Lake Conserve area (24°35′N, 121°24′E). The Lake is small (3.6×10^4 m^2) and shallow (4.5 maximum depth). The annual precipitation is over 4000 mm and the annual mean temperature is 12 °C in this region. Daily precipitation illustrates clear pulse during summer, mainly driven by typhoons (Fig. 1). *Chamaecyparis obtusa* var. *formosana* (Hayata) Rehder is the dominant species of this Taiwan cypress forest. As a result of high annual precipitation soils are permanently moist and acidic (i.e. pH <4) (Lin et al. 2011). Although this area is geographically belonged to subtropical zone, this primary montane forest, to some extent, can be analogous to temperate rainforests due to its climatic characteristics.

Sampling locations

A slope about 28° next to the lake was divided into three sampling locations, summit, footslope and lakeshore (Fig. 2). The soil at summit is dominantly Albaquult and relatively well drained, whereas the soil at lakeshore is dominantly Histosol and poorly drained due to the frequent immersion by lake water. Footslope, between summit and lakeshore, is dominantly Dystrochrept. The thick layer of organic horizon was derived from *Sphagnum* across three locations. The depth of organic horizon decreased with an increase in the elevation of the

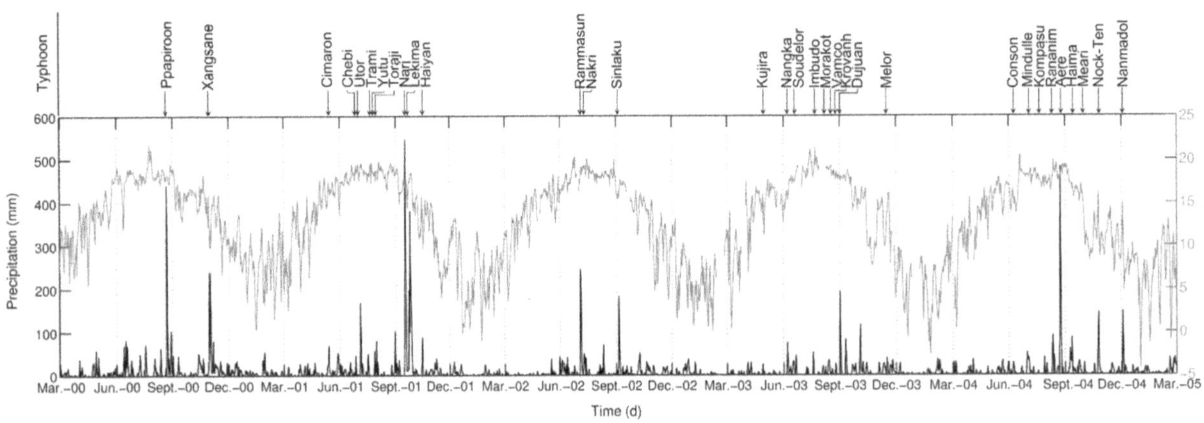

Fig. 1 Daily temperature and daily precipitation

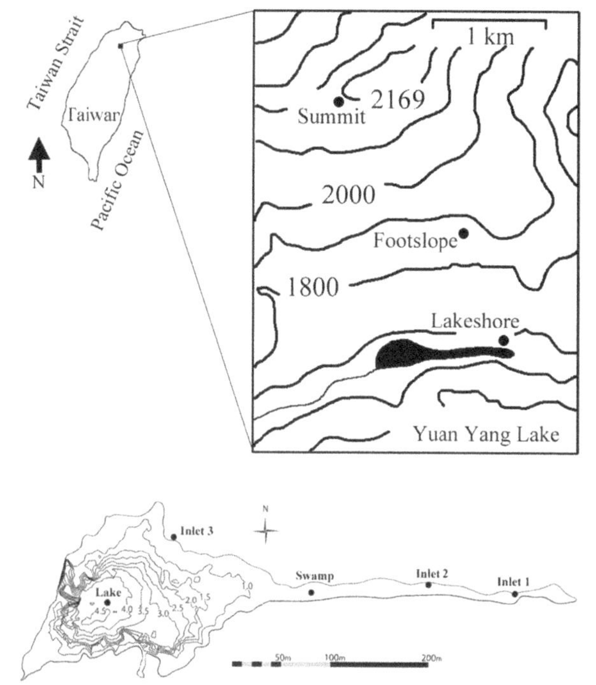

Fig. 2 Map of sampling sites at Yuan Yang Lake Nature Reserve, at the boundary point of Taoyuan and Ilan Counties, Taiwan

slope for which the depth of organic horizon for lakeshore, footslope and summit was about 30, 15 and 10 cm respectively. The depth of A horizon was small and ranged between 1 and 5 cm across three locations (Chen and Chiu 2000).

Soil sampling and processing

Five replicates of sampling plots (50 × 50 m each) at each sampling location were evenly chosen along a transect vertical to the slope. After carefully removing undecomposed litter, soil samples at 0–10 cm deep layer (O/A horizon) were collected using a soil auger (8 cm in diameter). The O/A horizon refers to the surface soil which was dominated with organic matter and humus. For lake water samples, each 100 ml water sample was collected from three inlet, one swamp (i.e. shallow) and one lake (i.e. deep) locations (Fig. 2). Three replicates were collected at each location. All samples were brought back to laboratory and kept at 4 °C before analysis. The soil subsamples were air-dried and ground to pass a 2.0 mm sieve for TC and TN analysis.

Laboratory analysis

Soil moisture was determined by oven-dried at 105 °C. The C_{mic} were determined by fumigation extraction method with a conversion factor of 2.22 (Wu et al. 1990). The N_{mic} were estimated from ninhydrin-reactive N released from the biomass and then determined using colorimetrically at 560 nm (Amato and Ladd 1988). Five sets of soil subsamples from each sampling location were incubated at 25 °C and 55 % soil moisture for 7 days for determination of net N mineralisation and net N nitrification rates (Hart and Binkley 1985). For DOC, DON and inorganic N, water extractions were made from all samples. Inorganic nitrogen was determined by using 25 g soil samples with 0.5 M K_2SO_4 extraction, followed by colorimetry flow injection analysis for ammonia and nitrate (FIA; Quikchem-Method-10-107-04-1-L, 1999 and Quikchem-Method-10-107-06-2-A, 1997, respectively). An in-line persulfate digestion followed by FIA in the same auto-analyser was applied in order to determine the concentration of the total dissolved nitrogen, TDN (Quikchem-Method-10-107-04-3-P, 2000). The concentration of the dissolved organic nitrogen (DON) was estimated as difference between TDN and inorganic N. The amount of extractable DOC was evaluated using an Aurora 1030 Wet Oxidation TOC Analyser (O.I. Analytical, USA). TC and TN in soil were analysed by high temperature combustion on a Fisons NA-1500 NCS Analyser (Italy). Water sample was analysed for DON, NH_4^+ and NO_3^- using the same procedure as soil samples but without extraction processes. All samples were analysed within 2 weeks of sample collection.

Data analysis

Linear mixed-effects models were used to assess the effect of 'Location' (i.e. summit, footslope, lakeshore) on the concentration of TC, TN, C_{mic}, N_{mic}, DOC, DON, ammonium, and nitrate for each sampling season. Additionally, linear mixed-effects models were also used to assess the effect of 'Location', 'Season', and an interactive effect of 'Location' and 'Season' on soil moisture and soil nutrients across multiple sampling years. Replicate sampling plot (Replication), and replicate sampling plot nested within each sampling year (Year/Replication) were set up as the random effect for former and latter models, respectively. Models used p value calculated based on Satterthwaite's approximations to indicate the significance of the interested fixed effects. Statistical analyses were carried out using R (R Core Team 2013).

Results

Soil total C and N and microbial biomass

Seasonal concentrations of soil TC, TN, C_{mic}, N_{mic} and the ratio of C_{mic}/TC, N_{mic} /TN, DOC/TC, DON/TN, TC/TN, C_{mic}/N_{mic}, and DOC/DON are presented in Fig. 3. The size of soil TC and TN was high, ranging between 421 and 532 g C kg^{-1} and between 19 and 26 g N kg^{-1}, respectively. Similarly, the size of soil C_{mic} and N_{mic} was also very high, ranging between 1.6 and 4.9 g C kg^{-1}

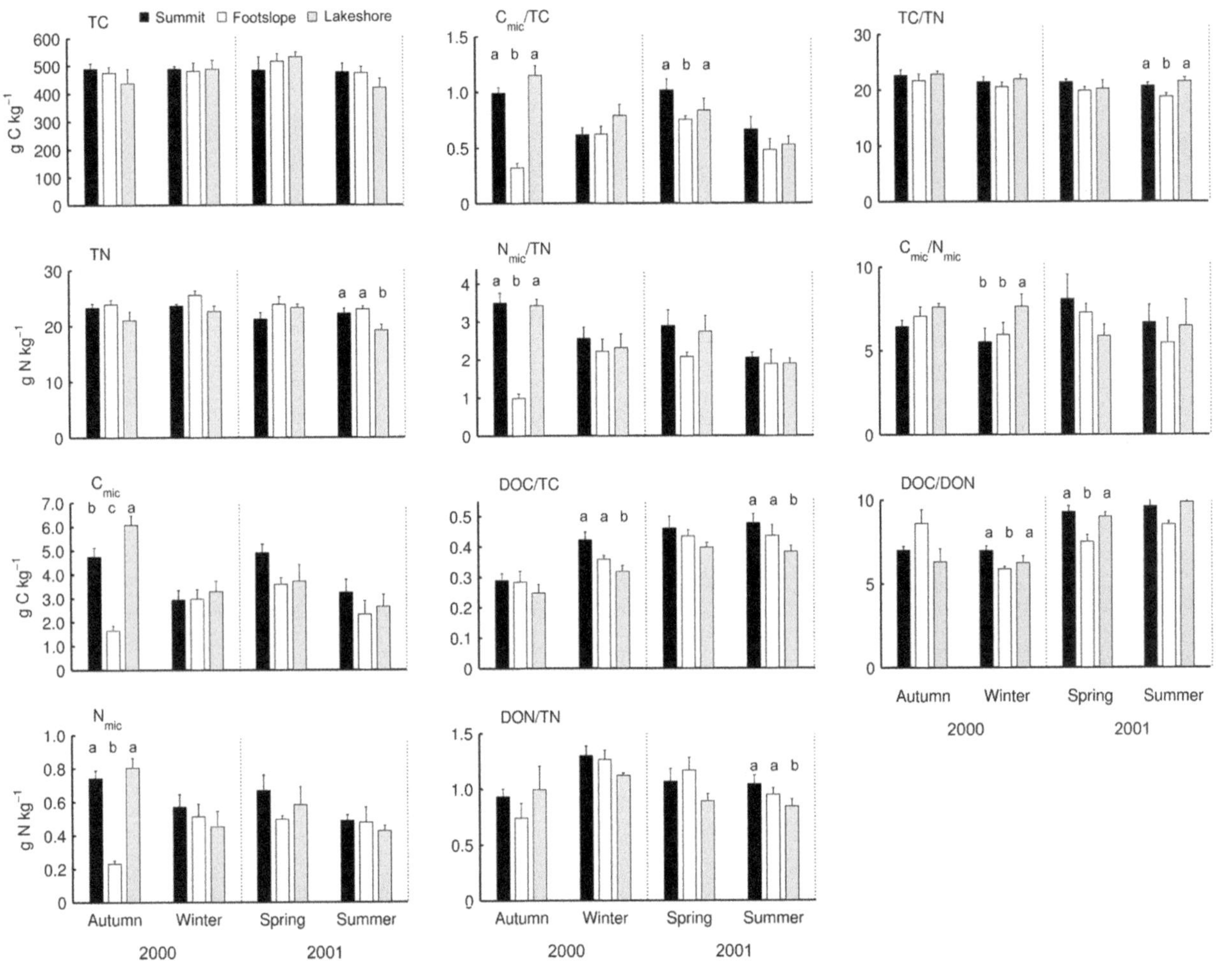

Fig. 3 The concentration of total C (TC), total N (TN), microbial biomass C (C_{mic}) and microbial biomass N (N_{mic}), dissolved organic C (DOC), dissolved organic N (DON), and associated ratios at three topographic locations in 2000 autumn and winter and in 2001 spring and summer. *Different letters* indicate significant difference among locations for a given season ($p < 0.05$)

and between 0.2 and 0.7 g N kg^{-1}, respectively. Resulting from very high amount of soil TC and TN, the ratio of soil C_{mic}/TC and N_{mic}/TN was low, between 0.3 and 1.1 and between 1.0 and 3.5. The ratio of soil DOC/TC and DON/TN was also low, between 0.2 and 0.5 and between 0.1 and 1.3. The ratio of soil TC/TN, C_{mic}/N_{mic}, and DOC/DON across three topographic locations was between 18.7 and 22.8, between 5.4 and 8.1, and between 5.9 and 9.9, respectively.

Soil moisture and the concentrations of soil DOC, DON, ammonium, and nitrate

Soil moisture levels were always high throughout the year at all topographic locations (Fig. 4). Seasonal concentrations of soil DOC, DON, ammonium and nitrate are presented in Fig. 4. The concentrations of soil DOC and DON were high, ranging from 900 to 1946 mg C kg^{-1} and 90 to 215 mg N kg^{-1}. For soil inorganic nitrogen, the concentration of soil ammonium, ranging from 16 to 80 mg N kg^{-1}, was higher than nitrate. Soil nitrate, occupying the smallest part of soil inorganic N, ranged from 0.8 to 6.8 mg N kg^{-1}.

Soil nitrogen mineralisation and nitrification

The N mineralisation in this forest soil was evident. During a week incubation, the ammonium produced by N mineralisation reached 131 mg N kg^{-1}. In contrast, the nitrification (measured as nitrate production during the incubation) was negligible (Fig. 5).

The concentration of DON, ammonium, and nitrate of lake water

The concentration of DON of lake water ranged between 0.77 and 3.37 mg N kg^{-1}, and the concentrations of ammonium and nitrate were relatively low compared to the DON (Fig. 6). Thus, organic N was the dominant

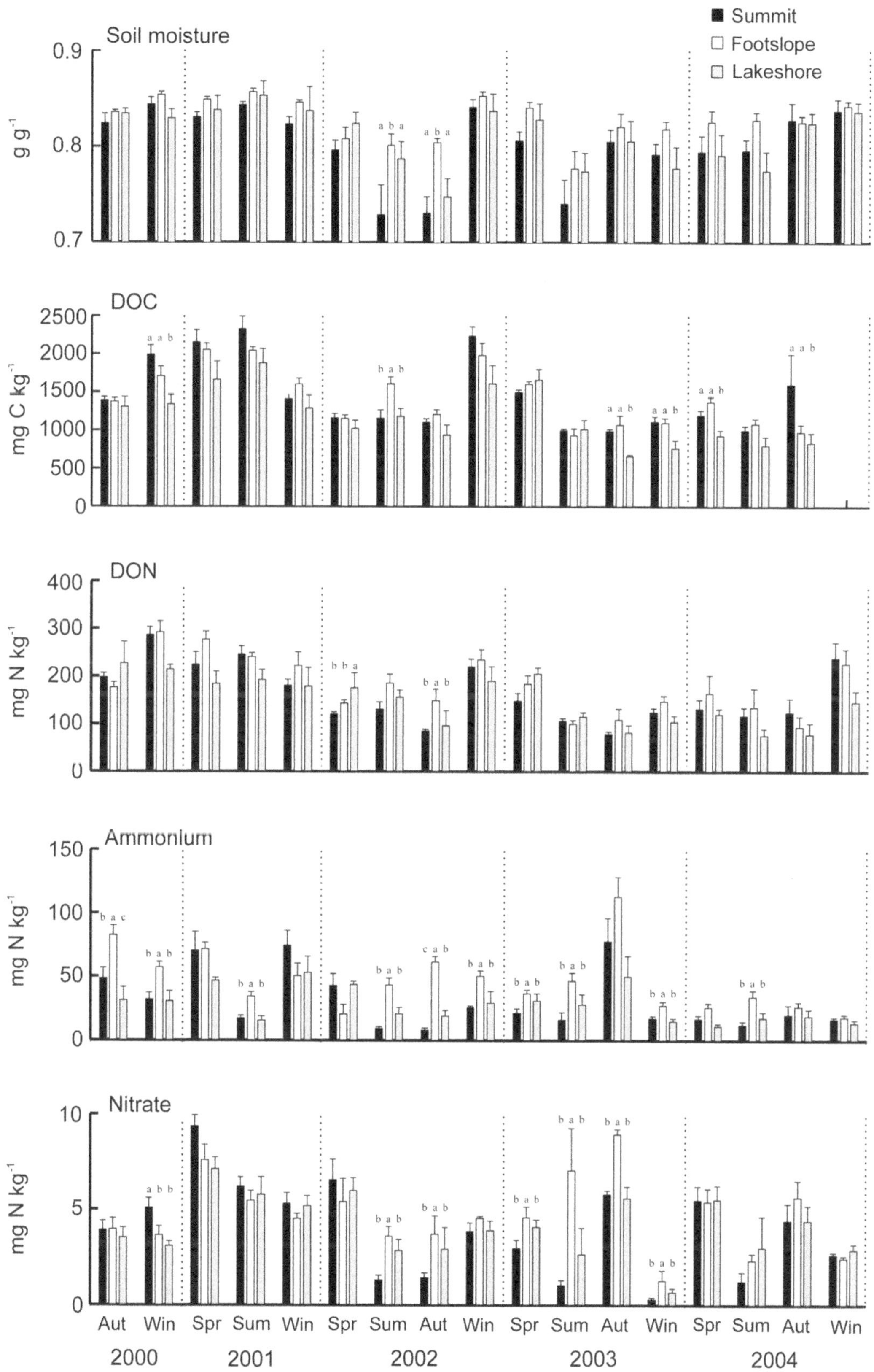

Fig. 4 Soil moisture and concentration of dissolved organic C (DOC), dissolved organic N (DON), ammonium, and nitrate of O/A horizon at three topographic locations. *Different letters* indicate significant difference among locations for a given season ($p < 0.05$)

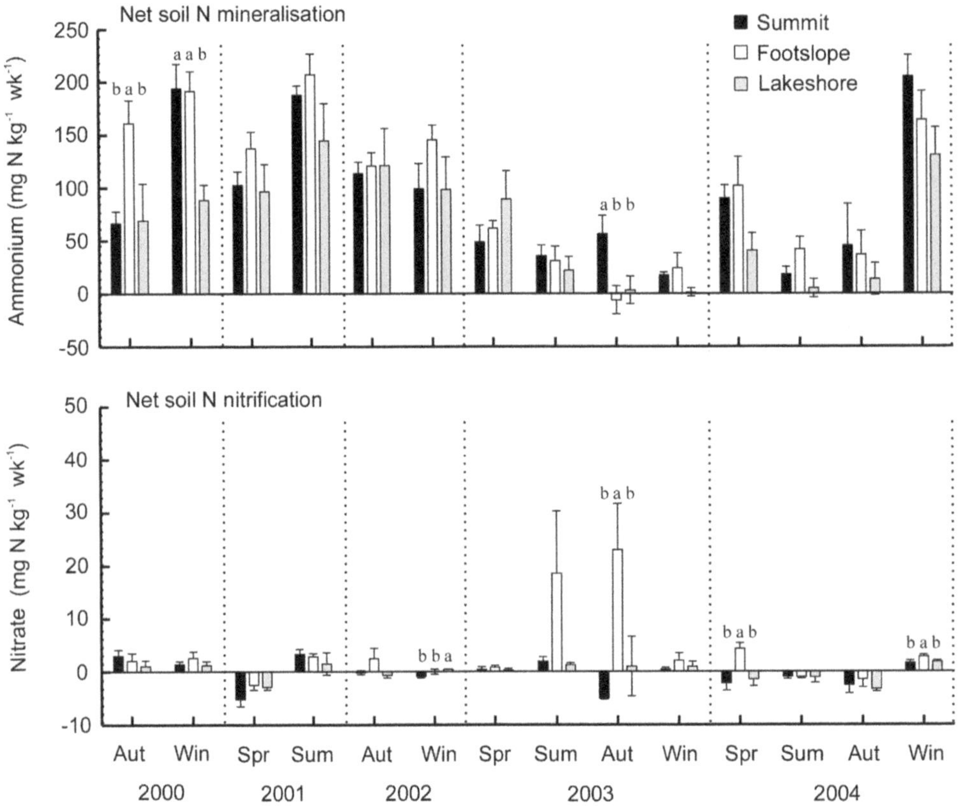

Fig. 5 Soil net N mineralisation and nitrification at three topographic locations. *Different letters* indicate significant difference among locations for a given season ($p < 0.05$)

form of N in lake water for which DON contributed to the high proportion (70–93 %) of the total N in lake water.

Topographic, seasonal, and interactive effects on soil nutrients

General speaking, the topographic effect on DOC and DON was not evident, but there were some significant cases over seasons. Surface soils at footslope had slightly higher moisture over seasons (Table 1). The soil DON concentration at footslope was highest over seasons, and that at lakeshore was lowest in winter (Table 1). Surface soils at footslope had the highest ammonium and nitrate concentrations in summer and autumn. Looking at the seasonal pattern, soil moisture was lowest in summer (Table 1). The soil DON concentration was highest in winter and lowest in autumn (Table 1). The ammonium concentration was lowest in summer and the nitrate concentration was lowest in spring, and the ammonium production through net N mineralisation processes was greatest in winter (Table 1).

Discussion

The concentration of TC, TN and labile C and N resources

The high liable C and N concentrations in soils might be related to the characteristics of this primary *Chamaecyparis* forest. This primary *Chamaecyparis* forest has accumulated abundant dead woody debris on forest floor (Jien et al. 2010), suggesting a great size of C and N pools in this forest (Gonzalez-Polo et al. 2013; Hafner et al. 2005; Wirth et al. 2009). This may partly explain high soil TC and TN concentrations in this *Chamaecyparis* forest. Long-standing fog immersion has created the suitable environment for epiphytic bryophytes to thrive (Chang et al. 2002), and the cryptogamic cover, such as algae, fungi, lichens and bryophytes, is essential for the C and N cycle in forest ecosystems (Elbert et al. 2012), suggesting that the high amount of organic C and N fluxes might be also associated with the appearance of abundant bryophytes in this primary *Chamaecyparis* forest. A relatively low ratio of C_{mic}/TC and N_{mic}/TN suggests that C and N resources used by microbes were limited in the forest, and this could be due to the low diversity of soil bacterial

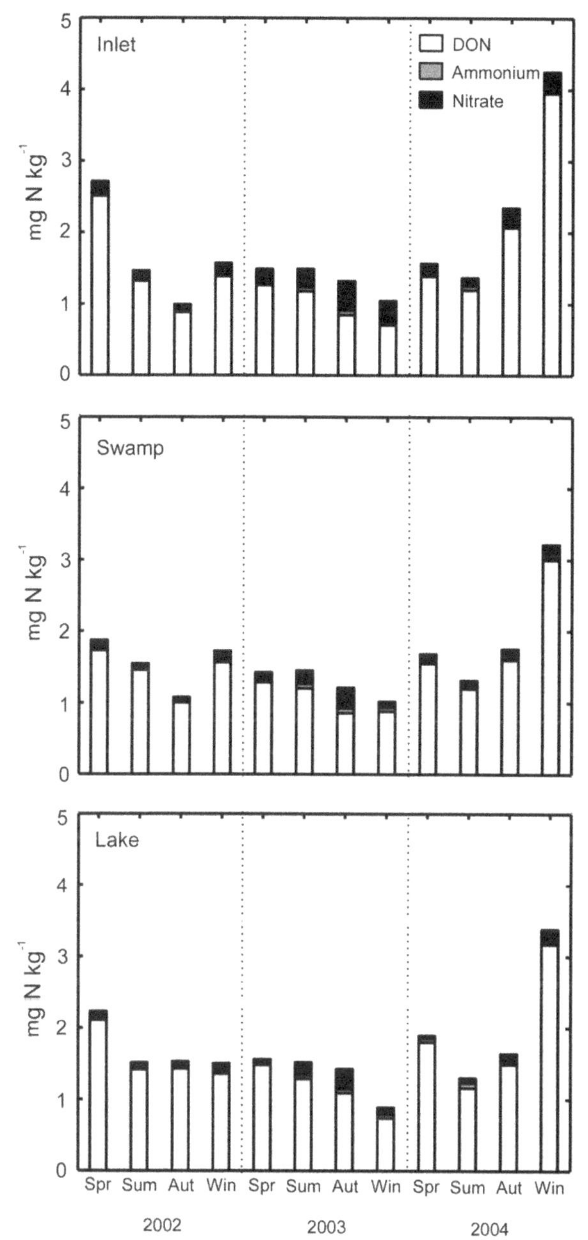

Fig. 6 Average concentrations of DON, NH_4^+ and NO_3^- of the lake water

communities under the acidic soil environments (Lin et al. 2010), and the constraint of the year-around per-humid environment might also hinder microbial activity and productivity.

The DOC and DON concentrations were profoundly high in the *Chamaecyparis* forest compared to those in the temperate deciduous forest (DOC: 150–225 mg C kg^{-1}; DON: 35 mg C kg^{-1}) in Hokkaido, Japan (Shibata et al. 2013) and the temperate coniferous (DON: 7–10 mg C kg^{-1}) and deciduous forest (DON: 6–7 mg-C kg^{-1}) in Thuringia forest, Germany (Zhong and Makeschin 2003). The fluxes of DOC and DON in forest floor leachates have been shown to be positively related to the amount of annual precipitation (Michalzik et al. 2001), suggesting that the high fluxes of dissolved organic matter in this *Chamaecyparis* forest soils might be associated with the high annual precipitation. Additionally, canopy and forest floor have been identified as the important sources of DON, in particular, in pristine ecosystems (Solinger et al. 2001). Schmidt et al. (2010) found that the DOC fluxes (478–962 kg C ha^{-1} yr^{-1}) in forest floor leachates were profoundly high and DON (8–16 kg N ha^{-1} yr^{-1}) was the major form of total N released from a secondary *Chamaecyparis* forest near our primary forest. We also observed that the O/A horizon of the *Chamaecyparis* forest soils contained the great amount of DOC (900–1946 mg C kg^{-1}), and DON was the dominant forms of N resources, suggesting the existence of high DOC and DON fluxes in this primary forest soil. In addition, we also found that DON being the predominant form of N and in lake water for which DON contributed very high proportions (70–93 %) of total dissolved N. Similarly, this phenomenon was also found in some temperate undisturbed rainforests (Hedin et al. 1995; Perakis and Hedin 2002).

Soil nitrogen mineralisation and nitrification

Compared to DON, ammonium and nitrate concentrations were relatively low at our study forest, but the magnitude was similar to the amount of other forests (Ge et al. 2014; Montaño et al. 2007; Shibata et al. 2013). With negligible amounts of external inputs of ammonium and nitrate (Chang et al. 2002), the most inorganic nitrogen in the forest was the product transformed from organic N. Montaño et al. (2007) demonstrated that the great size of DOC can stimulate heterotrophic microbes and N mineralisation, suggesting that the high ammonium production in this *Chamaecyparis* forest might be associated with its high C resources. Furthermore, nitrate may be easily lost from soils through direct absorption by plants and leaching processes due to its high mobility, and hence wet environments in such a high rainfall area may also contribute to the low nitrate concentration in surface soils. Immobilisation is one of the important processes in N dynamics. However, the ratio of C/N (i.e. TC/TN) of the O/A horizon was just 20, which was not higher than the critical level of 25 for N immobilisation, the N immobilisation should not be so intense in this forest. Acidity, in particular pH <5.5, has a detrimental effect on the nitrifying bacteria, thus reducing nitrification efficiency (Ste-Marie and Paré 1999). The soil pH was below 4, and the variation of pH was small (i.e. 3.3–3.5) in this

Table 1 Linear mixed-effects model analysis: the effects of topographic location and seasons on soil moisture, DOC, DON, NH_4^+, NO_3^- concentrations, net N mineralisation, and net N nitrification in O/A horizon

Effects	Soil moisture ($g\ g^{-1}$)		DOC ($mg\ C\ kg^{-1}$)		DON ($mg\ N\ kg^{-1}$)		NH_4^+ ($mg\ N\ kg^{-1}$)		NO_3^- ($mg\ N\ kg^{-1}$)		Net N mineralisation ($mg\ N\ kg^{-1}\ wk^{-1}$)		Net N nitrification ($mg\ N\ kg^{-1}\ wk^{-1}$)	
	Coeff.	SE	Coeff.	SE	Coeff.	SE	Coeff.	SE	Coeff.	SE	Coeff.	SE	Coeff.	SE
Location														
Summit[a]	0	0	0	0	0	0	0	0	0	0	0	0	0	0
Footslope	0.02*	0.01	NS	–	35.70*	16.78	NS	–	NS	–	NS	–	NS	–
Lakeshore	NS	–	NS	–	NS	–	NS	–	NS	–	NS	–	NS	–
Season														
Spring[a]	0	0	0	0	0	0	0	0	0	0	0	0	0	0
Summer	−0.03**	0.01	NS	–	NS	–	−24.20***	6.69	−3.62***	0.64	NS	–	NS	–
Autumn	NS	–	NS	–	−42.58*	17.11	NS	–	−1.65*	0.65	NS	–	NS	–
Winter	NS	–	NS	–	38.04*	16.07	NS	–	−2.60***	0.61	45.83*	20.45	NS	–
Interaction														
Footslope × Summer	NS	–	NS	–	NS	–	24.63**	9.46	2.50**	0.91	NS	–	NS	–
Lakeshore × Summer	NS	–	NS	–	NS	–	NS	–	NS	–	NS	–	NS	–
Footslope × Autumn	NS	–	NS	–	NS	–	31.67***	9.34	2.02*	0.90	NS	–	NS	–
Lakeshore × Autumn	NS	–	NS	–	NS	–	NS	–	NS	–	NS	–	NS	–
Footslope × Winter	NS	–	NS	–	NS	–	NS	–	NS	–	NS	–	NS	–
Lakeshore × Winter	NS	–	NS	–	−60.05**	22.86	NS	–	NS	–	NS	–	NS	–

Coeff. and SE are the estimated coefficient and standard error of each independent variable in the model

The + or − before the value of Coeff. indicates the higher or lower value than 'Summit' in location or 'Spring' in season

*** $p < 0.001$, ** $p < 0.01$, * $p < 0.05$, NS not significant ($p > 0.1$)

[a] Set to zero due to being the standard for comparison within each categorical factor

Chamaecyparis forest soil. This might have an effect on low nitrifying activity and result in low amount of nitrate production.

Topographic effect and seasonal variation

Our second hypothesis was not well supported—as the topographic effects on labile C and N and the inorganic N was not consistently evident over seasons. No apparent differences in microbial biomass among soils at various locations suggest that the soil environment of the entire forest across topography was similar. Only few cases in specific seasons could be detected. For example, DON concentrations were relatively low at lakeshore in particular in winter, and this was probably associated with the flooding events which brought organic matter away (Chung et al. 2012). We also observed that soils at footslope had highest inorganic N concentrations in rainy seasons (i.e. summer and autumn), and this might be partly due to the draining environment at footslope where soils can hold more water to increase mineralisation efficiency. The great amount of precipitation and relative warm environments in summer suggests that soils might experience intense dry-wet processes, and this probably stimulated soil microbes to greatly utilise organic N resources in soils. According to previous studies in this area (Chen and Chiu 2000), the detectable migration of humus and top soils from summit to footslope only had marginal effects on the distribution of labile nutrients, suggesting that the topographic effect of humus migration on soil nutrients might not be significant.

Conclusions

This subtropical primary montane forest has accumulated the considerable size of DOC in the O/A horizon. DON was the predominant form of total dissolved N resources in soils and lake water. Topographic effects on soil nutrients were not very significant, suggesting that the dynamics of soil nutrient components was relatively equitable across topography, and this may be due to the strong control of per-humid climate and acidic soil conditions. These findings reveal the subtropical primary montane forest soils may hold the great amount of DOC and DON resources, suggesting the importance of this primary montane forest in regional C and N biogeochemical cycles.

Authors' contributions

CWT performed statistical analyses and built statistical models. GT helped in analyzing and interpreting data. CYC originally formulated the idea and developed methodology. CWT and CYC wrote the manuscript with inputs from other authors. All authors read and approved the final manuscript.

Author details

[1] Biodiversity Research Center, Academia Sinica, Taipei 11529, Taiwan.
[2] Environmental Monitoring and Research Division, Monitoring and Research Department, Metropolitan Water Reclamation District of Greater Chicago (MWRD), Lue-Hing R&D Laboratory, 6001 W. Pershing Road, Cicero, IL 60804, USA.

Compliance with ethical guidelines

Competing interests

The authors declare that they have no competing interests.

References

Amato M, Ladd JN (1988) Assay for microbial biomass based on ninhydrin-reactive nitrogen in extracts of fumigated soils. Soil Biol Biochem 20(1):107–114

Chang S-C, Lai IL, Wu J-T (2002) Estimation of fog deposition on epiphytic bryophytes in a subtropical montane forest ecosystem in northeastern Taiwan. Atmos Res 64(1–4):159–167

Chen JS, Chiu CY (2000) Effect of topography on the composition of soil organic substances in a perhumid sub-tropical montane forest ecosystem in Taiwan. Geoderma 96:19–30

Chen JS, Chung TL, Tian G, Chiu CY (2013) Characterization of soil organic matter in perhumid natural cypress forest: comparison of humification in different particle-size fractions. Bot Stud 54(1):56

Chung TL, Chen JS, Chiu CY, Tian G (2012) 13C-NMR spectroscopy studies of humic substances in subtropical perhumid montane forest soil. J For Res 17(6):458–467

Core Team R (2013) R: a language and environment for statistical computing. R Foundation for Statistical Computing, Vienna

Elbert W, Weber B, Burrows S, Steinkamp J, Budel B, Andreae MO, Poschl U (2012) Contribution of cryptogamic covers to the global cycles of carbon and nitrogen. Nature Geosci 5(7):459–462

Fang Y, Zhu W, Gundersen P, Mo J, Zhou G, Yoh M (2009) Large loss of dissolved organic nitrogen from nitrogen-saturated forests in subtropical China. Ecosystems 12(1):33–45

Galloway JN, Dentener FJ, Capone DG, Boyer EW, Howarth RW, Seitzinger SP, Asner GP, Cleveland CC, Green PA, Holland EA, Karl DM, Michaels AF, Porter JH, Townsend AR, Vöosmarty CJ (2004) Nitrogen cycles: past, present, and future. Biogeochemistry 70(2):153–226

Ge P, Da LJ, Wang WB, Xu XN (2014) Seasonal dynamics of dissolved organic carbon, nitrogen and other nutrients in soil of Pinus massoniana stands after pine wilt disease disturbance. J Soil Sci Plant Nutr 14:75–87

Ghani A, Dexter M, Carran RA, Theobald PW (2007) Dissolved organic nitrogen and carbon in pastoral soils: the New Zealand experience. Eur J Soil Sci 58(3):832–843

Gonzalez-Polo M, Fernández-Souto A, Austin A (2013) Coarse woody debris stimulates soil enzymatic activity and litter decomposition in an old-growth temperate forest of Patagonia, Argentina. Ecosystems 16(6):1025–1038

Hafner SD, Groffman PM, Mitchell MJ (2005) Leaching of dissolved organic carbon, dissolved organic nitrogen, and other solutes from coarse woody debris and litter in a mixed forest in New York State. Biogeochemistry 74(2):257–282

Hart SC, Binkley D (1985) Correlations among indices of forest soil nutrient availability in fertilized and unfertilized loblolly pine plantations. Plant Soil 85(1):11–21

Hedin LO, Armesto JJ, Johnson AH (1995) Patterns of nutrient loss from unpolluted, old-growth temperate forests: evaluation of biogeochemical theory. Ecology 76(2):493–509

Inagaki Y, Kuramoto S, Torii A, Shinomiya Y, Fukata H (2008) Effects of thinning on leaf-fall and leaf-litter nitrogen concentration in hinoki cypress (*Chamaecyparis obtusa* Endlicher) plantation stands in Japan. For Ecol Manag 255(5–6):1859–1867

Inagaki Y, Nakanishi A, Fukata H (2011) Soil properties and nitrogen utilization of hinoki cypress as affected by strong thinning under different climatic conditions in the Shikoku and Kinki districts in Japan. J For Res 16(5):405–413

Jien SH, Hseu ZY, Iizuka Y, Chen TH, Chiu CY (2010) Geochemical characterization of placic horizons in subtropical montane forest soils, Northeastern Taiwan. Eur J Soil Sci 61:319–332

Lin YT, Huang YJ, Tang SL, Whitman WB, Coleman DC, Chiu CY (2010) Bacterial community diversity in undisturbed perhumid montane forest soils in Taiwan. Microb Ecol 59:369–378

Lin YT, Jangid K, Whitman WB, Coleman DC, Chiu CY (2011) Soil bacterial communities in native and regenerated perhumid montane forests. Appl Soil Ecol 47(2):111–118

Lutz BD, Bernhardt ES, Roberts BJ, Mulholland PJ (2011) Examining the coupling of carbon and nitrogen cycles in Appalachian streams: the role of dissolved organic nitrogen. Ecology 92(3):720–732

Michalzik B, Kalbitz K, Park JH, Solinger S, Matzner E (2001) Fluxes and concentrations of dissolved organic carbon and nitrogen—a synthesis for temperate forests. Biogeochemistry 52(2):173–205

Montaño N, García-Oliva F, Jaramillo V (2007) Dissolved organic carbon affects soil microbial activity and nitrogen dynamics in a Mexican tropical deciduous forest. Plant Soil 295(1–2):265–277

Moore JM, Klose S, Tabatabai MA (2000) Soil microbial biomass carbon and nitrogen as affected by cropping systems. Biol Fertil Soils 31(3–4):200–210

Neff JC, Asner GP (2001) Dissolved organic carbon in terrestrial ecosystems: synthesis and a model. Ecosystems 4(1):29–48

Perakis SS, Hedin LO (2002) Nitrogen loss from unpolluted South American forests mainly via dissolved organic compounds. Nature 415(6870):416–419

Schmidt BM, Wang CP, Chang SC, Matzner E (2010) High precipitation causes large fluxes of dissolved organic carbon and nitrogen in a subtropical montane Chamaecyparis forest in Taiwan. Biogeochemistry 101(1–3):243–256

Shibata H, Hasegawa Y, Watanabe T, Fukuzawa K (2013) Impact of snowpack decrease on net nitrogen mineralization and nitrification in forest soil of northern Japan. Biogeochemistry 116(1–3):69–82

Singh JS, Kashyap AK (2006) Dynamics of viable nitrifier community, N-mineralization and nitrification in seasonally dry tropical forests and savanna. Microbiol Res 161(2):169–179

Solinger S, Kalbitz K, Matzner E (2001) Controls on the dynamics of dissolved organic carbon and nitrogen in a Central European deciduous forest. Biogeochemistry 55(3):327–349

Ste-Marie C, Paré D (1999) Soil, pH and N availability effects on net nitrification in the forest floors of a range of boreal forest stands. Soil Biol Biochem 31(11):1579–1589

Wirth C, Heimann M, Gleixner G (2009) Old-growth forests: function, fate and value. Ecological Studies, vol 207. Springer, Berlin Heidelberg, Heidelberg, Germany

Wu J, Joergensen RG, Pommerening B, Chaussod R, Brookes PC (1990) Measurement of soil microbial biomass C by fumigation-extraction—an automated procedure. Soil Bio Biochem 22(8):1167–1169

Zhong Z, Makeschin F (2003) Soluble organic nitrogen in temperate forest soils. Soil Biol Biochem 35(2):333–338

17

The seasonal variations of allelopathic activity and allelopathic substances in *Brachiaria brizantha*

Ai Kobayashi and Hisashi Kato-Noguchi*

Abstract

Background: Controlling weeds through allelopathy is one strategy to reduce dependency on synthetic herbicides. The plant shoots of the grass *Brachiaria brizantha* incorporated into the field soil were found to inhibit the growth of several plant species. We investigated the variations of allelopathic activity and allelopathic substances in *B. brizantha* harvested in June, October and January.

Results: All extracts of *B. brizantha* obtained from June, October and January inhibited the root and shoot growth of cress, lettuce, *Phleum pretense* and *Lolium multiflorum* in a concentration dependent manner. However, the inhibitory activity of *B. brizantha* of June and October was greater than that of *B. brizantha* of January. Concentrations of three potent allelopathic active substances, (*6R,9S*)-3-oxo-α-ionol, (*6R,9R*)-3-oxo-α-ionol and 4-ketopinoresinol were also greater in *B. brizantha* of June and October than those in *B. brizantha* of January.

Conclusion: The results suggest that the allelopathic activity and the levels of those allelopathic active substances are greater in *B. brizantha* of June and October than those in *B. brizantha* of January. It is possible that *B. brizantha* could be useful for a weed suppressive residue or soil additive materials in the variety of agricultural settings to develop sustainable agriculture options. The effectiveness of *B. brizantha* of June and October as a weed suppressive agent may be greater than that of January.

Keywords: Allelopathy, *Brachiaria brizantha*, Growth inhibitor, Phytotoxicity, Soil additive

Background

Plant allelopathy has been investigated for long time to develop sustainable agricultural systems especially in weed control purposes (Belz 2007; Macías et al. 2007). Plants contain hundreds of secondary metabolites, and some of those compounds showed strong allelopathic activity such as germination and growth inhibitory effects on other plants (Inderjit 1996; Duke et al. 2000; Macías et al. 2007). Some plant materials and residues incorporated into field soil provided excellent weed control ability due to the allelopathic active substances in those materials. Intercropping of allelopathic plants with crop plants also inhibited weed germination and growth owing to their allelopathic property (Weston 1996; Narwal 1999). Therefore, plant allelopathy is probably useful for weed management options in the sustainable agricultural settings (Putnam 1988; Weston 1996; Narwal 1999).

Poaceae species *Brachiaria brizantha,* originating in Africa savannas, is introduced into many countries as livestock forage (McGregor et al. 1988). The extracts of *B. brizantha* shoots inhibited the seed germination of *Stylosanthes* species (Rodrigues et al. 2012) and seedling growth of cress, lettuce, *Phleum pratense* and *Lolium multiflorum* (Kato-Noguchi et al. 2014). In addition, the plant shoots of *B. brizantha* incorporated into the field soil inhibited the growth of several plant species (Martins et al. 2006; Souza et al. 2006). Three allelopathic active substances, (*6R,9S*)-3-oxo-α-ionol, (*6R,9R*)-3-oxo-α-ionol and 4-ketopinoresinol were isolated from *B. brizantha* shoots (Kato-Noguchi et al. 2014). Therefore,

*Correspondence: hisashi@ag.kagawa-u.ac.jp
Department of Applied Biological Science, Faculty of Agriculture, Kagawa University, Miki, Kagawa 761-0795, Japan

B. brizantha shoots may be also useful as soil additives for weed management options in some sustainable agricultural settings because of their allelopathic active substances.

B. brizantha is a perennial grass, 1–2 m in height, with 2 m deep roots and rhizomes. The plant shoots with long leaf blades grow from spring to autumn and die in winter (Miles et al. 1996; Cook et al. 2005). Therefore, it is important to evaluate the variations of the allelopathic activity and the levels of allelopathic active substance in *B. brizantha* for soil additive materials. However, there is no information available about those variations. In the present research, allelopathic activity of *B. brizantha* shoots and the level of allelopathic active substances in the shoots were determined in June, October and January.

Methods

Plant materials

Brachiaria brizantha (Hochst. ex A. Rich.) Stapf. (common name; palisade grass, bread grass, syn; *Urochloa brizantha* Hochst. ex A. Rich.) was grown in the research field of National Institute for Agro-Environmental Sciences, Tsukuba, Japan. Shoots of the plants were harvested in June and October 2011 and January 2012, and stored at −20 °C until extraction. Cress (*Lepidum sativum* L.) and lettuce (*Lactuca sativa* L.) were used due to their uniform establishment and sensitivity as a seedling indicator (Kato-Noguchi et al. 2010, 2014). Weed species, *Phleum pratense* L. and *Lolium multiflorum* Lam were also used for bioassay.

Extraction and bioassay

B. brizantha (100 g dry weight) shoots were extracted with 1 L of 70 % (v/v) aqueous methanol for 2 days. After filtration with filter paper (No. 2; Toyo, Tokyo, Japan), the residue was extracted again with 1 L methanol for 2 days and filtered, and the two filtrates were combined. An aliquot of the extract (final assay concentration of tested samples corresponded to the extracts obtained from 0.3, 1, 3, 10, 30, 100 and 300 mg dry weight of *B. brizantha* shoots per mL) was evaporated to dryness, dissolved in a 0.2 mL of methanol and added to a sheet of filter paper (No. 2) in a 3-cm Petri dish. The methanol was evaporated in a fume hood. Then, the filter paper in the Petri dishes was moistened with 0.8 mL of 0.05 % (v/v) polyoxyethylene sorbitan monolaurate (Tween 20). Ten seeds of cress or lettuce, or 10 seedlings of *P. pratense* or *L. multiflorum* after germination in the darkness at 25 °C for 36–48 h were placed in the Petri dishes. The length of roots and shoots of these seedlings were measured after 48 h of incubation in the darkness at 25 °C. Controls were treated exactly as described above, with the exception that 0.2 mL methanol was used instead of *B. brizantha* extracts. Inhibitory activity (%) was determined by the formula: [(control plant length − plant length incubated with extract)/control plant length] × 100. The bioassay was repeated five times using a randomized design with 10 plants for each determination. Significant differences were examined by Duncan's multiple comparison tests.

Quantification of allelopathic active substances

Brachiaria decumbens shoots was extracted as described above and the extracts were concentrated at 40 °C in vacuo to produce an aqueous residue. The aqueous residue was adjusted to pH 7.0 with 1 M phosphate buffer and partitioned three times against an equal volume of ethyl acetate as described by Kato-Noguchi et al. (2014). The ethyl acetate fraction was evaporated to dryness and separated on a column of silica gel (100 g, silica gel 60, 70–230 mesh; Merck), eluted with 20, 30, 40, 50, 60, 70 and 80 % ethyl acetate in *n*-hexane (100 mL per step) and ethyl acetate (100 mL). Allelopathic active substances, (6*R*,9*R*)-3-oxo-α-ionol and (6*R*,9*S*)-3-oxo-α-ionol, were obtained by the elution with 70 % ethyl acetate in *n*-hexane on the silica gel column, and 4-ketopinoresinol was obtained by the elution with 80 % ethyl acetate in *n*-hexane on the silica gel column (Kato-Noguchi et al. 2014).

For quantification of (6*R*,9*R*)-3-oxo-α-ionol and (6*R*,9*S*)-3-oxo-α-ionol, the fraction obtained with 70 % ethyl acetate in *n*-hexane on the silica gel column was evaporated. The residue was then dissolved in 20 % (v/v) aqueous methanol (2 mL) and loaded onto reverse-phase C_{18} cartridges (YMC Ltd., Kyoto, Japan). The cartridge was eluted with 20, 40, 60 and 80 % (v/v) aqueous methanol (15 mL per step). The active fraction was eluted by 40 % aqueous methanol and evaporated to dryness. The residue was injected into reverse-phase HPLC (4.6 mm i.d. × 250 mm, Inertsil ODS-3, GL Sciences, Osaka, Japan) eluted at a flow rate of 0.8 mL min^{-1} with 55 % aqueous methanol and detected at 220 nm. Retention time of (6*R*,9*R*)-3-oxo-α-ionol and (6*R*,9*S*)-3-oxo-α-ionol was 65 and 70 mm, respectively. Quantification of those compounds was performed by measuring their peak areas on the chromatogram of HPLC. The quantification was repeated three times independently with three assays for each determination. Significant differences were examined by Duncan's multiple comparison tests.

The ^{1}H-NMR spectrum of (6*R*,9*R*)-3-oxo-α-ionol, δ_H: 1.00 (3H, s), 1.03 (3H, s), 1.24 (3H, d, J = 6.8 Hz), 1.94 (3H, d, J = 1.0 Hz), 2.05 (1H, d, J = 16.6 Hz), 2.40 (1H, d, J = 16.6 Hz), 2.67 (1H, d, J = 9.3 Hz), 4.27 (1H, m), 5.58 (1H, dd, J = 15.1, 8.8 Hz), 5.70 (1H, dd, J = 15.2, 5.9 Hz), 5.88 (1H, s). The specific rotation of the compound $\left([\alpha]_D^{25}\right)$ was +210° (c 0.03, CH_2Cl_2). The ^{1}H-NMR spectrum of (6*R*,9*S*)-3-oxo-α-ionol, δ_H: 0.98 (3H, s), 1.03

(3H, s), 1.24 (3H, d, $J = 6.4$ Hz), 1.96 (3H, d, $J = 1.5$ Hz), 2.05 (1H, d, $J = 16.6$ Hz), 2.42 (1H, d, $J = 17.1$ Hz), 2.66 (1H, d, $J = 8.8$ Hz), 4.28 (1H, m), 5.55 (1H, dd, $J = 15.1$, 8.8 Hz), 5.69 (1H, dd, $J = 15.1$, 5.9 Hz), 5.89 (1H, s). The specific rotation of the compound $\left([\alpha]_D^{25}\right)$ was +214° (*c* 0.05, CH_2Cl_2).

For quantification of 4-ketopinoresinol obtained with 80 % ethyl acetate in *n*-hexane on the silica gel column was evaporated. The residue was then dissolved in 20 % (v/v) aqueous methanol (2 mL) and loaded onto reverse-phase C_{18} cartridges. The cartridge was eluted with 20, 40, 60 and 80 % (v/v) aqueous methanol (15 mL per step). The active fraction was eluted by 40 % aqueous methanol and evaporated to dryness. The residue was injected into reverse-phase HPLC (10 mm i.d. × 50 cm, ODS AQ-325; YMC Ltd.) eluted at a flow rate of 1.5 mL min^{-1} with 40 % aqueous methanol and detected at 220 nm. Retention time of 4-ketopinoresinol was 145 mm and quantification of the compound was performed by measuring their peak areas as described above.

The ^1H-NMR spectrum of 4-ketopinoresinol, δ_H: 3.25 (m, 1 H, H1), 3.47 (dd, $J = 9.2$, 3.8 Hz, 1 H, H5), 3.90 (s, 3 H, OCH_3), 3.92 (s, 3 H, OCH_3), 4.04 (dd, $J = 9.4$, 4.5 Hz, 1 H, H8a), 4.34 (dd, $J = 9.3$, 6.8 Hz, 1 H, H8b), 5.34 (d, $J = 4.0$ Hz, 1 H, H2), 5.35 (d, $J = 3.6$ Hz, 1 H, H6), 6.8 (m, 2H, aromatic H), 6.9 (m, 4H, aromatic H). The ^{13}C NMR spectrum of the compound, δ_C: 50.1 (C1), 53.5 (C5), 56.2 (OCH_3), 56.2 (OCH_3), 72.8 (C8), 83.5 (C2), 84.8 (C6), 107.9 (C2′), 108.2 (C2″), 114.5 (C5″), 114.8 (C5′), 118.1 (C6″), 118.5 (C6′), 131.2(C1′), 132.4 (C1″), 145.4 (C4″), 146.2 (C4′), 146.9 (C3″), 147.1 (C3′), 177.1 (C4). The specific rotation of the compound $\left([\alpha]_D^{25}\right)$ was +5.7° (*c* 0.1, CH_3OH).

Results and discussion

Three extracts obtained from 10 mg *B. brizantha* shoots harvested in June, October and January inhibited the growth of cress roots and shoots (Fig. 1). The inhibitory activity of the June extract on the cress roots and shoots was 76.1 and 79.5 %, respectively, that of the October extract on the cress roots and shoots was 55.7 and 71.5 %, respectively, and that of the January extract on the cress roots and shoots was 35.8 and 15.8 % respectively. Although *B. brizantha* is perennial plant, its shoots re-start to grow around May and die in November and only its underground parts survive after November. The shoots of the plants were growing stage in June and mature stage in October and both shoots were green. The shoots of January had already died and turned brown (Fig. 2). The results indicate that allelopathic activity of *B. brizantha* of June and October was greater than that of January. Thus, green plant materials may be better than dead plant materials for a weed suppressive residue or soil additive materials.

Inhibitory activity of *B. brizantha* shoot extracts obtained from June, October and January was determined by root and shoot growth of cress, lettuce, *P. pretense* and *L. multiflorum* (Fig. 2). All extracts inhibited the growth of the roots and shoots of those test plants with extract-concentration dependent. The concentrations required for 50 % growth inhibition on the roots and shoots of the test plants in the assay (defined as I_{50}) were determined by a logistic regression ($Y = b + a \log_{10}X$. Y; % length, X; concentration, a; slope, b; intercept) analysis (Table 1). Comparing those I_{50} values, inhibitory activity of the June extract was 1.1- to 1.8-fold and 4.4- to 27.8-fold greater than that of the October extract and January extract, respectively. These results suggest that all *B. brizantha* extracts have inhibitory effects on both dicotyledonous plants (cress and lettuce) and monocotyledonous weed plants (*P. pretense* and *L. multiflorum*) and the June extract has the greatest inhibitory activity.

(6*R*,9*S*)-3-oxo-α-ionol, (6*R*,9*R*)-3-oxo-α-ionol and 4-ketopinoresinol were isolated from *B. brizantha* shoots as three major active substances in *B. brizantha* (Kato-Noguchi et al. 2014). Thus, concentrations of those allelopathic active substances, (6*R*,9*S*)-3-oxo-α-ionol, (6*R*,9*R*)-3-oxo-α-ionol and 4-ketopinoresinol were determined in *B. brizantha* shoots of June, October and January (Fig. 3). Those concentrations in the January shoots were significantly lower than those in the June and October shoots. Concentration of (6*R*,9*S*)-3-oxo-α-ionol, (6*R*,9*R*)-3-oxo-α-ionol and 4-ketopinoresinol in the June shoots was 2.3-, 2.2- and 2.4-fold greater than that in the January shoots, respectively. This result is consistent with the result of inhibitory activity of the shoot extracts of June, October and January (Table 1) and also indicates that June and October shoots may be better than January shoots for a weed suppressive residue or soil additive materials.

(6*R*,9*S*)-3-Oxo-α-ionol, (6*R*,9*R*)-3-oxo-α-ionol and 4-ketopinoresinol inhibited the growth of cress at concentrations greater than 10, 30 and 30 μM, respectively. I_{50} values on cress roots and shoots were 223 and 156 μM for (6*R*,9*R*)-3-oxo-α-ionol, respectively, and 41.7 and 25.1 μM for (6*R*,9*S*)-3-oxo-α-ionol, respectively, and 83.5 and 141 μM for 4-ketopinoresinol, respectively (Kato-Noguchi et al. 2014). Therefore, the inhibitory activity of (6*R*,9*S*)-3-oxo-α-ionol was 5.3 to 6.2- and 2 to 5.6-fold greater than that of (6*R*,9*R*)-3-oxo-α-ionol and 4-ketopinoresinol, respectively. These results indicate that the growth inhibitory activity of (6*R*,9*S*)-3-oxo-α-ionol is the greatest, followed by 4-ketopinoresinol and (6*R*,9*R*)-3-oxo-α-ionol.

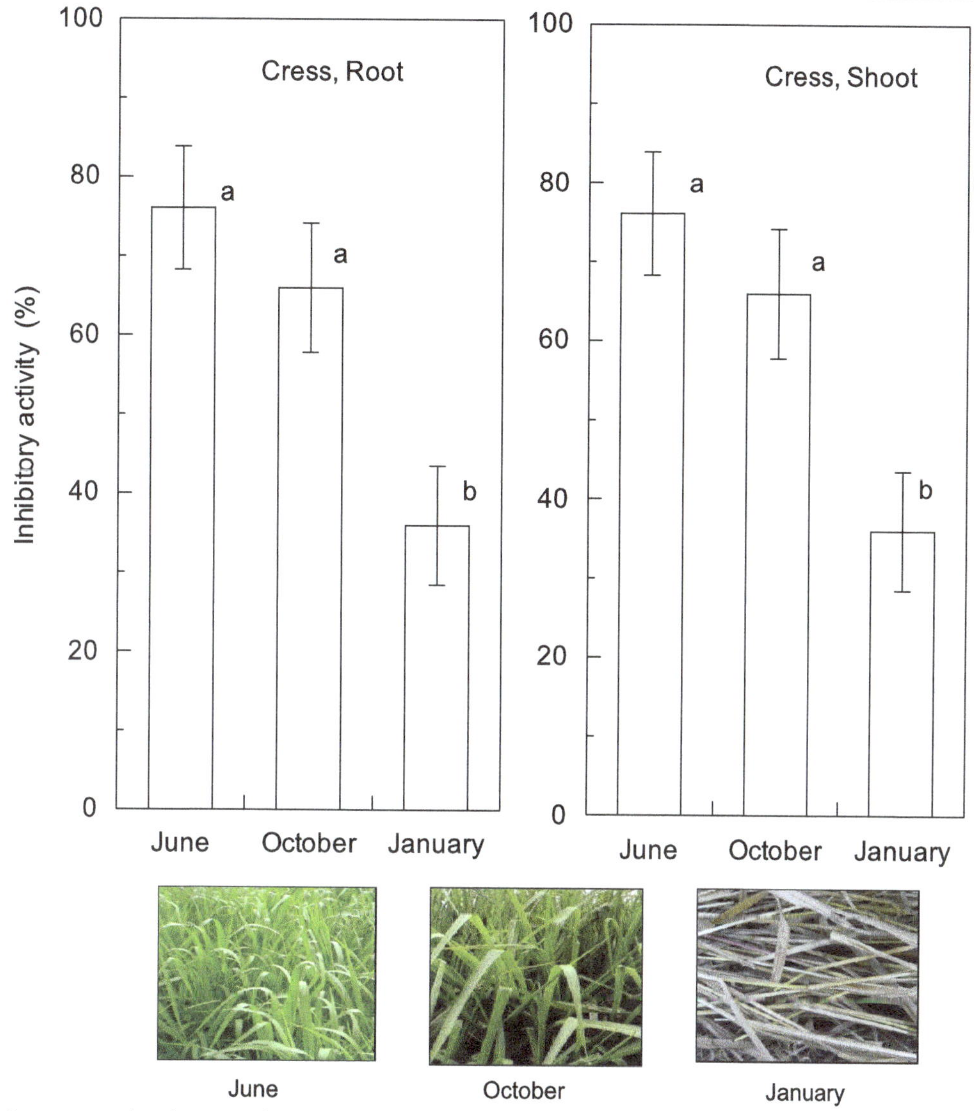

Fig. 1 Effects of aqueous methanol extracts of *B. brizantha* shoots harvested in June, October and January on cress root and shoot growth. Concentration of tested sample corresponded to the extract obtained from 10 mg dry weight of *B. brizantha* per mL. Mean ± SE from five independent experiments with ten plants for each determination are shown. The *different letters* in the same groups indicate significant difference ($P < 0.05$) according to Duncan's multiple comparison tests

3-Oxo-α-ionol was first identified as the aglycone of a glucoside isolated from *Rubus idaeus* (Pabst et al. 1992; D'Abrosca et al. 2004) and was reported to act as allelopathic active substance in *rattail fescue* (Kato-Noguchi et al. 2010). However, the specific biological activities of (*6R,9R*)-3-oxo-α-ionol and (*6R,9S*)-3-oxo-α-ionol were first reported by Kato-Noguchi et al. (2014). 4-Ketopinoresinol was first isolated from *Aegilops ovate* as germination inhibitor (Lavie et al. 1974; Cooper et al. 1977), but the information on the biological activity of the compound is very limited.

Phytotoxic active substances in plants can be released into the soil, either by exudates from living plant tissues or by decomposition of plant residues, and act as allelopathic active substances which inhibit seed germination, seedling establishment and plant growth (Bais et al. 2006; Bonanomi et al. 2006; Belz 2007). The present results showed that *B. brizantha* inhibited root and shoot growth of four test plant species including weed plants (*P. pretense* and *L. multiflorum*) in a concentration dependent manner. Thus, *B. brizantha* could be useful for a weed suppressive residue or soil additive materials in

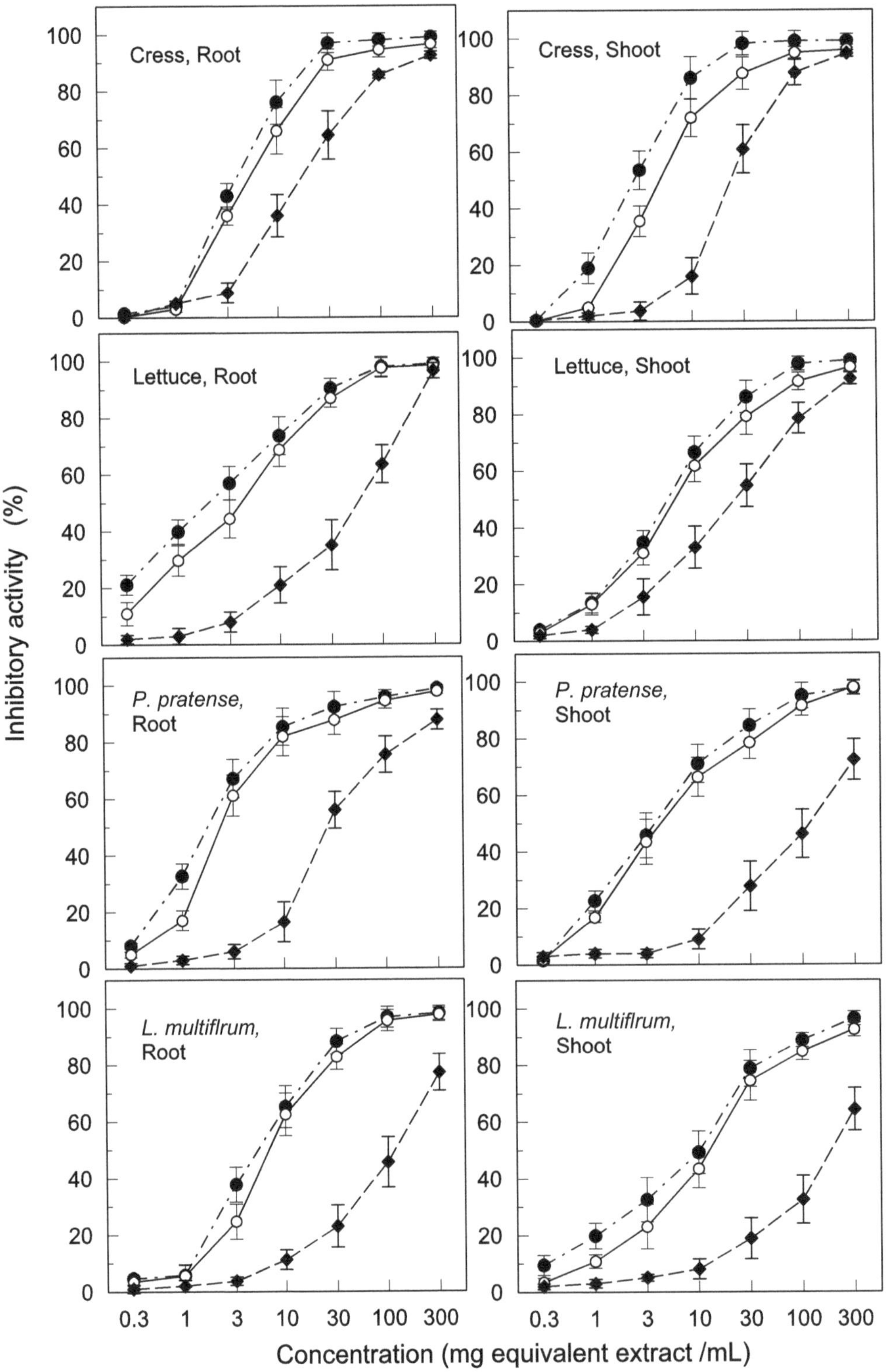

Fig. 2 Effects of aqueous methanol extracts of *B. brizantha* shoots harvested in June (*filled circle*), October (*unfilled circle*) and January (*filled diamond*) on root and shoot growth of cress, lettuce, *P. pretense* and *L. multiflorum*. Concentrations of tested samples corresponded to the extracts obtained from 0.3, 1, 3, 10, 30, 100 and 300 mg dry weight of *B. brizantha* per mL. Mean ± SE from five independent experiments with ten plants for each determination are shown

Table 1 I_{50} of aqueous methanol extracts of *B. brizantha* shoots harvested in June, October and January on root and shoot growth of cress, lettuce, *P. pretense* and *L. multiflorum*

	I_{50} (mg equivalent extract/mL)		
	June	October	January
Cress			
Root	4.5	6.3	19.8
Shoot	2.8	4.9	22.3
Lettuce			
Root	2.5	4.5	55.7
Shoot	6.3	7.9	25.3
P. pratense			
Root	1.8	2.5	23.7
Shoot	4.4	4.7	122.5
L. multiflorum			
Root	4.9	7.2	123.5
Shoot	9.8	13.8	211.3

The values were determined by a logistic regression analysis after bioassays

the variety of agricultural settings to reduce dependency on synthetic herbicides. Natural products are considered to be more environmentally benign than most synthetic herbicides. Their environmental half-lives can be expected to be shorter, and the fact that they have been part of the natural environment suggests a lesser demand for regulatory measures (Duke et al. 2000). In addition, *B. brizantha* has been reported to have a potential to enhance mineral nutrients (Souza et al. 2006). Further studies are required to determine the effectiveness of *B. brizantha* on weeds in the field conditions.

Conclusion

The extracts of *B. brizantha* shoots harvested in June, October and January inhibited root and shoot growth of cress, lettuce, *Phleum pretense* and *Lolium multiflorum* in a concentration dependent manner. However, the inhibitory activity of *B. brizantha* of June and October was greater than that of *B. brizantha* of January. Concentrations of three allelopathic active substances, (*6R,9S*)-3-oxo-α-ionol, (*6R,9R*)-3-oxo-α-ionol and 4-ketopinoresinol were also greater in *B. brizantha* shoots of June and October than those in *B. brizantha* shoots of January. Therefore, *B. brizantha* could be useful for a weed suppressive residue or soil additive materials in the variety of agricultural settings to develop sustainable agriculture options due to those allelopathic active substances. The effectiveness of *B. brizantha* of June and October as a weed suppressive agent may be greater than that of January.

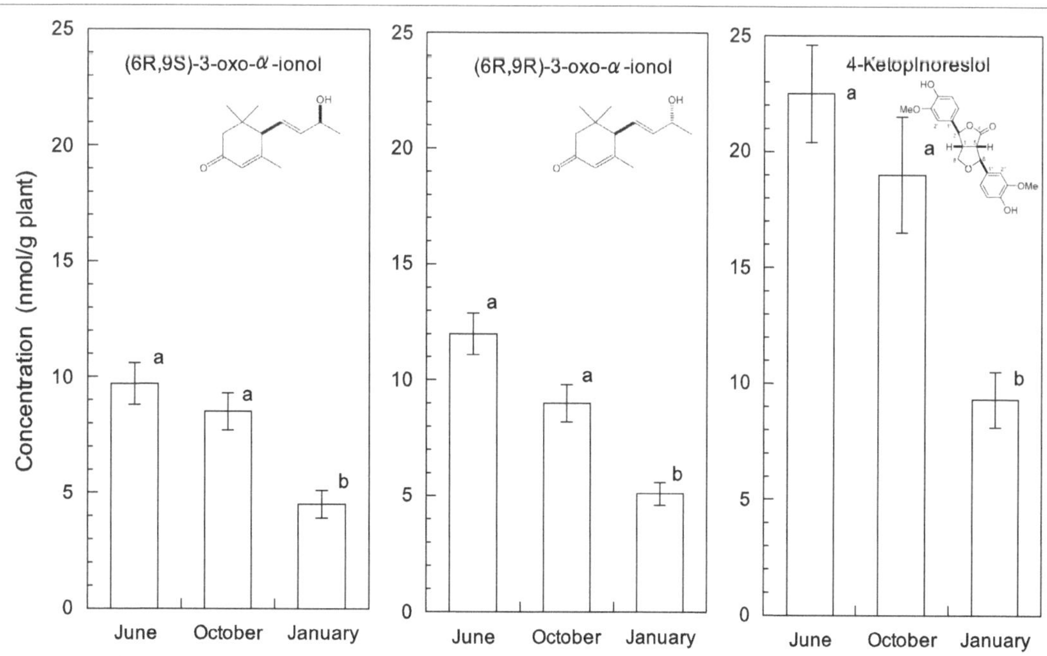

Fig. 3 Concentration of (*6R,9S*)-3-oxo-α-ionol, (*6R,9R*)-3-oxo-α-ionol and 4-ketopinoresinol in *B. brizantha* shoots harvested in June, October and January. Mean ± SE from three independent experiments with three quantifications for each determination are shown. The *different letters* in the same groups indicate significant difference ($P < 0.05$) according to Duncan's multiple comparison tests

Abbreviations

I_{50}: concentrations required for 50 % growth inhibition.

Authors' contributions

KA is responsible for the whole experiments and wrote the article. HK-N is responsible for experimental design, result interpretation, made discussion and conclusion. Both authors read and approved the final manuscript.

Acknowledgements

The authors thank Dr. Fujii Y (National Institute for Agro-Environmental Science, Tsukuba, Japan) for giving us plant materials of *B. brizantha*. The authors thank Dr. Ohno O and Suenag K (Department of Chemistry, Keio University, Yokohama, Japan) for NMR analyses and specific rotation measurements of the compounds.

Compliance with ethical guidelines

Competing interests

The authors declare that they have no competing interests.

References

Bais HP, Weir TL, Perry LG, Gilroy S, Vivanco JM (2006) The role of root exudates in rhizosphere interactions with plants and other organisms. Annu Rev Plant Biol 57:233–266

Belz RG (2007) Allelopathy in crop/weed interactions—an update. Pest Manag Sci 63:308–326

Bonanomi G, Sicurezza MG, Caporaso S, Esposito A, Mazzoleni S (2006) Phytotoxicity dynamics of decaying plant materials. New Phytol 169:571–578

Cook BG, Pengelly BC, Brown SD, Donnelly JL, Eagles DA, Franco MA, Hanson J, Mullen BF, Partridge IJ, Peters M, Schultze-Kraft R (2005) Tropical forages. CSIRO, Brisbane

Cooper R, Levy EC, Lavie D (1977) Novel germination inhibitors from *Aegilops ovata* L. J Chem Soc Chem Comm 21:794–795

D'Abrosca B, DellaGreca M, Fiorentino A, Monaco P, Oriano P, Temussi F (2004) Structure elucidation and phytotoxicity of C_{13} *nor*-isoprenoids from *Cestrum parqui*. Phytochemistry 65:497–505

Duke SO, Dayan FE, Romagni JG, Rimando AM (2000) Natural products as sources of herbicide, current status and future trends. Weed Res 40:99–111

Inderjit (1996) Plant phenolics in allelopathy. Bot Rev 62:186–202

McGregor JT Jr, Smith RJ Jr, Talbert RE (1988) Broadleaf signalgrass (*Brachiaria platyphylla*) duration of interference in rice (*Oryza sativa*). Weed Sci 36:747–750

Kato-Noguchi H, Yamamoto M, Tamura K, Teruya T, Suenaga K, Fujii Y (2010) Isolation and identification of potent allelopathic substances in *rattail fescue*. Plant Grow Regul 60:127–131

Kato-Noguchi H, Kobayashi A, Ohno O, Kimura F, Fujii Y, Suenaga K (2014) Phytotoxic substances with allelopathic activity may be central to the strong invasive potential of *Brachiaria brizantha*. J Plant Physiol 171:525–530

Lavie D, Levy EC, Choen A, Evenari M, Guttermann Y (1974) New germination inhibitor from *Aegilops ovata* L. Nature 249:388

Macías FA, Molinillo JMG, Varela RM, Galindo JGG (2007) Allelopathy—a natural alternative for weed control. Pest Manag Sci 63:327–348

Martins D, Martins CC, Costa NV (2006) Allelopathic potential of solutions of soils cultivated with *Brachiaria brizantha*: effects on some pasture grass and weeds seeds. Planta Daninha 24:61–70

Miles JW, Maas BL, do Valle CB, Kumble V (1996) Brachiaria: biology, agronomy and improvement. CIAT, Embrapa, Brazil

Narwal SS (1999) Allelopathy in weed management. In: Narwal SS (ed) Allelopathy update, vol 2., Basic and applied aspectsScience Publishers Inc, Enfield, pp 203–254

Pabst A, Barrona D, Sémonb E, Schreier P (1992) Two diastereomeric 3-oxo-α-ionol β-glucosides from raspberry fruit. Phytochemistry 31:1649–1652

Putnam AR (1988) Allelochemicals from plants as herbicides. Weed Technol 2:510–518

Rodrigues APDC, Laura VA, Pereira SR, Deiss C (2012) Allelopathy of two *Brachiaria* species in seeds of three species of *Stylosanthes*. Ciencia Rural 42:1758–1763

Souza LS, Velini ED, Martins D, Rosolem CA (2006) Allelopathic effects of *Brachiaria decumbens* on the initial development of seven crops. Planta Daninha 24:657–668

Weston LA (1996) Utilization of allelopathy for weed management in agroecosystems. Agron J 88:860–866

18

Plant bioclimatic models in climate change research

Chyi-Rong Chiou[1], Tung-Yu Hsieh[2,3,4*] and Chang-Chi Chien[5]

Abstract

Bioclimatics is an ancient science that was once neglected by many ecologists. However, as climate changes have attracted increasing attention, scientists have reevaluated the relevance of bioclimatology and it has thus become essential for exploring climate changes. Because of the rapidly growing importance of bioclimatic models in climate change studies, we evaluated factors that influence plant bioclimatology, constructed and developed bioclimatic models, and assessed the precautionary effects of the application of the models. The findings obtained by sequentially reviewing the development history and importance of bioclimatic models in climate change studies can be used to enhance the knowledge of bioclimatic models and strengthen their ability to apply them. Consequently, bioclimatic models can be used as a powerful tool and reference in decision-making responses to future climate changes. The objectives of this study were to (1) understand how climatic factors affect plants; (2) describe the sources, construction principles, and development of early plant bioclimatic models (PBMs); and (3) summarize the recent applications of PBMs in climate change research.

Keywords: Climate change, Phenological model, Theoretical model, Statistical model, Mechanistic model

Background

Bioclimatology or *bioclimatics*, which includes phenology, is an ancient science that investigates the relationship between living organisms and climates. According to historical records, China was the first country to conduct bioclimatic observation approximately 3,000 years ago. Bioclimatology is referred to as *Wuhou* (物候) in Chinese, a word that originated from the classic *Ch'un-ch'iu Tso Chuan* (春秋左傳). Western bioclimatology was established in approximately 1753 by Linnaeus, a Swedish botanist, who is known as the father of phenology. The term phenology was first introduced by the Belgian botanist Morren in 1853. One hundred years before the term was coined during Linnaeus' time, phenology was focused on the seasonal and periodic phenomena that organisms exhibit and is referred to as classic or seasonal bioclimatology. In Japan, phenology is referred to as the study of seasons and organisms. Scientists have since identified that changes in living organisms follow periodic changes in climates. Thus, the scope and definition of phenology vary constantly as new bioclimatic findings are obtained. Consequently, the early definition of phenology has become inapplicable. Although numerous scientists have attempted to redefine phenology and create linguistically specific technical terms, many people prefer to use the established term phenology, which has been used continuously since it was coined. Bioclimatology, including phenology, now involves investigations of the correlations between climates and organisms (Chu and Wan 1999; Hopkins 1938; Hsieh and Chiou 2013; Lieth 1974; Schnelle 1955; Zou 1983). To avoid confusion caused by different definitions, this article defines all types of model that have both biological and climatic variables as bioclimatic models.

Despite its ancient origin, bioclimatology has long been disregarded because of problems, such as difficulty in funding long-term research in the past. In recent years, bioclimatology has received increasing attention and has become critical for investigating the effects of climate changes on organisms (Hänninen and Tanino 2011; Hsieh and Chiou 2013; Körner and Basler 2010; Lechowicz and

*Correspondence: sdyhsieh@gmail.com
[2] Shanghai Institutes for Biological Sciences, Chinese Academy of Sciences, 320 Yue Yang Rd., Shanghai 200031, China
Full list of author information is available at the end of the article

Koike 1995). Initially, ancient people developed bioclimatology by recording the correlations between biological phenomena according to annual observations made during farming seasons and related experiences; in this way, lunar calendars and bioclimatic calendars were compiled. Thus, bioclimatic research development in ancient times was focused on agricultural phenomena and various biological indicators recorded in the bioclimatic calendars of different cultures were used as a disaster-prevention system for decision-making. Bioclimatology in the Western scientific field did not become a formal discipline until the mid-eighteenth century when Linnaeus established the first phenology observation networks in Sweden and emphasized the tasks and importance of phenological observations in his book *Philosophia Botanica* (Hsieh and Chiou 2013; Lieth 1974).

Because the threat of climate change has recently attracted increasing attention, phenology network records have been developed into two complementary research systems; one is the concept of bioclimatic fingerprints, which was developed from phenology observation networks and is used for observing and monitoring the effects of climate changes on organisms, and the other is bioclimatic modeling based on long-term bioclimatic records and variations of the phenology observation networks for clarifying the correlation between climates and organisms and predicting the possible effects of climate changes on organisms. The results can be used as references in future disaster alert systems, disaster-prevention decision-making, and the assessment of disaster effects (Peñuelas and Filella 2001).

Although bioclimatic models are essential to researching climate change effects and despite the rapid international development and application of bioclimatic models, research and reports regarding the application and exploration of bioclimatic models remain scant in many undeveloped and developing countries, which are severely threatened by climate change. To improve the capability of people to address the threat of climate changes, we reviewed the factors that influence plant bioclimatology, the construction and development of bioclimatic models, and the application of bioclimatic models in disaster prevention and impact assessment. The sequential review of the development history and importance of bioclimatic models in climate change research provided in this study can be used as references by researchers studying climate changes.

Climatic factors that affect plant growth and development

Bioclimatic models represent the phenomena, processes, or mechanisms of the effect of climate factors on organisms. Thus, before understanding the modeling principles of bioclimatic models, basic knowledge regarding the environmental factors that affect plant bioclimatology must be acquired. The effects of environmental factors on plants vary with plant species, phenological phases, geographical environments, physiological statuses, and levels and types of ecological systems, yielding complex mechanisms. Among numerous environmental factors, temperature, water availability, and air flow (i.e., wind) are more closely related to climate changes and substantially affect plants.

Temperature

The climatic conditions of different seasons and regions cause varying effects on the bioclimatology of different plants (Menzel et al. 2001). For example, the higher winter temperatures at middle latitudes cause most plants to blossom and sprout earlier (Sparkes et al. 1997). At middle and high latitudes, the end of growth periods and the beginning of dormant periods of most plants are primarily influenced by the shorter days and temperature conditions of late summer (Heide 1974; Wareing 1956). Subsequently, the low temperature of the following winter breaks plant dormancy (Fuchigami et al. 1982; Perry 1971; Vegis 1964). Fluctuating temperatures break plant dormancy more effectively than constant temperatures do (Campbell and Sugano 1975; Hänninen 1990; Murray et al. 1989). However for some plants, fluctuating and constant temperatures have the same effect (Myking 1997). Phenological variations during plant growth periods are primarily affected by accumulated temperature (Peñuelas and Filella 2001). However, selecting the initial temperature for calculating accumulated temperature has been a major difficulty in bioclimatology because it may differ substantially in plants of the same species when influenced by varying environmental factors (Heide 1993; Murray et al. 1989). This difference severely affects the precision and prediction accuracy in research regarding plant growth bioclimatology. Despite the differences, 5 °C is commonly used as the initial temperature for calculating the accumulated temperature of plants (Cannell and Smith 1986; Cannell et al. 1985; Kellomäki et al. 1995; Murray et al. 1989).

Water availability

In addition to temperature, water availability critically affects plant bioclimatology and is highly relevant to climate changes. However, the effects of water availability vary with species and other environmental conditions. In particular, the photoperiods and temperature conditions in tropical zones are relatively stable and variation in water availability is often the main factor influencing plant bioclimatology (Tissue and Wright 1995). For example, when it rains in tropical arid or semiarid climates, various plants blossom simultaneously, exhibiting high phenological synchrony (de Lampe et al. 1992). The following rainfall continuance affects the fruits of plants. A majority of tropical plants bear fruit in rainy seasons and the fruiting period is shortened or prolonged based on the precipitation of the current season (Bawa and Hadley 1991). Water shortage causes growth arrest among numerous plants, resulting in eco-dormancy (Reich and Borchert 1984). In high mountains and middle- to high-latitude areas, water availability and temperature changes resulting from thawing snow are key to plant blossom and growth (Walker et al. 1995).

Airflow

Airflow is also a critical climatic factor that affects plants. When daylight is sufficient, adequate airflow, such as a breeze or zephyr, facilitates the airflow exchange of leaves and promotes transpiration lowering the leaf and environmental temperatures. Airflow also assists the pollination of anemophilous plants; however, when the wind speed is excessively high, the photosynthesis of leaves is subdued; the stigmata of flowering plants dry up, which affects pollination and causes infertility; or soil drying and wind erosion are expedited, which results in exposed plant roots, fallen fruits, leaves, and flowers, and even severe mechanical injuries, such as broken and fallen stems. Consequently, trees are weakened because of malnutrition, diseases, pests, or infections, which cause alternate bearing, and eventually die from nutrition depletion (Campbell-Clause 1998; Duryea et al. 1996; Telewski 1995).

Climatic factors in different growth and development stages

The effects of climatic factors on plants differ according to the various growth and development stages of plants. For example, climate conditions influence germination so that the germinating of seeds of different species and in various regions differs substantially. The seeds of plants that grow in temperate latitudes require low temperatures or fluctuating temperature conditions that last for a certain amount of time to break dormancy (Hsieh et al. 2004). However, numerous studies have shown that some temperate plant species can break seed dormancy through exposure to high temperatures and long photoperiod days (Isikawa 1954; Johnson and Irgens-Moller 1964; Stearns and Olson 1958). Some temperate species can break dormancy and sprout only after exposure to a period of low temperature following exposure to high temperature, such as *Taxus sumatrana* (Miq.) deLaub. (Chien et al. 1995) and peony seeds. The germination of seeds from numerous species also varies with environmental conditions, such as those for *Tsuga canadensis* L. The seeds of *Tsuga canadensis* L. break dormancy and germinate after exposure to 10 weeks of low temperature. However, the temperature required for germination of seeds that have not been exposed to low-temperature stratification increases with the length of photoperiod. For a general photoperiod of 8–12 h, the optimal germination temperature ranges from 17 to 22 °C, Whereas if the length of the photoperiod is 16 h, the optimal germination temperature increases to 27 °C (Stearns and Olson 1958). However, *Pseudotsuga menziesii* (Mirb.) Franco seeds that have not undergone low-temperature stratification can successfully germinate after a short-photoperiod below 25 °C (Johnson and Irgens-Moller 1964). The climate requirements and resistance may differ even among the various organs of a plant species. A survey exploring the freeze injuries of *Pyrus koehnei* C.K. Schneid. showed that 90 % of 6-year-old plants were frozen to death under −14 °C and 50 % of suckers were frozen to death under −12 °C, whereas only 28 % of stem surfaces exhibited freeze injury. The median lethal temperature of the different tissues ranged from −10 to −15 °C (Nee et al. 1995).

Climatic factors in different areas

Climate changes may exert differing effects on the same species of plant in different areas with identical climatic conditions. For example, in certain areas of the former Soviet Union where the climatic conditions are identical, walnuts trees are frozen to death in autumn in certain locations but survive autumn in other places. A subsequent finding indicated that the difference is caused by varying photoperiods. In certain areas, the photoperiods shorten before the autumn frost, resulting in the early dormancy of walnuts. In other areas, the photoperiods are not short enough to induce bud dormancy. Therefore, with the same temperature during autumn frost, walnuts may be frozen to death in some areas but survive the frost in other areas (Haldane 1947). Photoperiods also influence the blossoming of strawberry flowers. Temperatures and photoperiods jointly regulate the differentiation of flower buds. Generally, long photoperiods imply that flower bud induction requires long durations at low temperatures whereas short photoperiods imply that flower bud induction requires short durations at low temperatures. Thus, in areas with the same temperature conditions, varying photoperiods may affect whether strawberry flowers blossom. The condition of the plant itself may also have an influence; for example, during flower bud induction, a decreased number of old leaves easily induces flower bud differentiation (Darnell and Hancock 1996).

Distinct microtopographies and microclimates influence precipitation; even a slight variation in rainfall may substantially affect plant growth. For example, at high altitudes, the fruiting amount of *Actinidia* is inversely proportional to the degree of overlap of flowering periods and the East Asian rainy season. A high degree of overlap implies low fruiting rates for a certain year, whereas a low degree of overlap increases the fruiting rate. Certain species lack fruit every year because the East Asian rainy season overlaps the flowering period. This severely affects the reproduction and growth of *Actinidia.* and damages economic growth related to the plants (Nee 1994). Moreover, plant species respond differently to climate changes. For example, if plants, such as *Acer saccharum* Marsh. and eastern hemlock, which originate from different regions, are planted at one location, the plants from the north areas or high altitudes stop growing early in the autumn (Nienstaedt and Olson 1961; Robak and Magnesen 1970). Altitudes also affect the temperature requirements and responses of plants. For example, the seeds and buds of *Actinidia* have different dormancy conditions at different altitudes. The higher the altitude, the higher the chilling requirement to break seed and bud dormancy (Fan and Nee 2007). By contrast, peach and cherry trees have lower chilling requirements at high altitudes (Huang 2011; Ou et al. 2000).

Based on the aforementioned research cases, we identified that understanding the physiological mechanisms through which climates affect plants is crucial to climate change research. The influence of climate changes on plants varies substantially with differences in species, region, and other influential factors. Therefore, if the physiological and ecological conditions of plants are not specifically controlled, constructing an appropriate bioclimatic model for climates with similar variable conditions and accurately evaluating and explaining the resulting influence of climate changes can be difficult.

Bioclimatic model development

The origin of plant bioclimatic modeling is earlier than the formal establishment of bioclimatology. Such models can be traced back to 1735, when Reaumur proposed that the bioclimatic events of organisms and the dates of occurrence differ with regions, species, and altitude because the temperature required for each organism to grow and develop varies and accumulates differently according to region. This is the earliest degree-day summation concept, and for hundreds of years, this concept has been a fundamental basis for constructing bioclimatic models, such as the spring index model (Schwartz 1997; Schwartz and Marotz 1986, 1988), thermal time model (Cannell and Smith 1983; Robertson 1968), and spring warming model (Hunter and Lechowicz 1992).

After Reaumur, three types of bioclimatic models were developed in response to different research needs, methods, and objectives. Scientists refer to the three model types as theoretical, statistical, and mechanistic models. The theoretical model is also called the analytical model because it emphasizes the equilibrium between the productivity and the energy and nutrition absorption of leaves. Thus, because the model focuses on growth and development, it is suitable for research regarding the evolution of the survival strategies of species. The statistical model encompasses a wide and complex research scope. The primary objective of this model is to conduct statistical modeling, such as polynomial regression and general linear models, based on bioclimatic observation to directly connect climatic factors and biological events. Therefore, this model is also referred to as the empirical model. The mechanistic model focuses on the causal relationship between bioclimatic events and environmental factors to explain the effects of environmental factors on plant physiology. Because rigorous physiological and ecological theories and experimental bases support this model, its results are accepted relatively easily by a majority of scholars. The mechanistic model has been the standard of bioclimatic model research for a long period (Zhao et al. 2013). Except for the few bioclimatic models that use simple calculations, difficulties have typically been encountered during the early development of other bioclimatic models. These models were not developed and widely used until computer software and hardware became more easily accessible and a concomitant increase in the availability of data to parameterize such models (e.g., freely available gridded climate products) resulted in a stronger emphasis on global climate changes.

Each bioclimatic model has specific application restrictions and advantages and disadvantages. Scientists use the thermal time model most often because this model considers only the accumulated temperature, threshold temperature, and mean daily temperature of bioclimatic events as the parameters, facilitating model application. The model is shown as follows:

$$S_f = \sum_{t_0}^{y} R_f(X_t) = F^* \tag{1}$$

where S_f represents the accumulated units required to promote growth that satisfies bioclimatic event occurrence; y represents the date of the bioclimatic event occurrence; t_0 represents the initial time for calculating the accumulated temperature; X_t represents the mean daily temperature; and $R_f(X_t)$ represents the calculation function of effective accumulated temperature. This function is calculated using the following equation:

$$R_f(X_t) = \begin{cases} 0 & \text{if } x_t \leq T_{b1} \\ x_t - T_{b1} & \text{if } x > T_{b1} \end{cases} \tag{2}$$

where T_{b1} represents the initial temperature for calculating the accumulated temperature. In this model, when the temperature is below the threshold growth temperature of a plant, the temperature does not influence phenological events. Only when the temperature exceeds the threshold growth temperature of a plant does the accumulated temperature affect phenological events. The higher the temperature, the greater the degree of influence is. However, this model is only applicable to the optimal temperature of plant growth. When the plant encounters extreme temperatures that exceed the optimal temperature of growth during the calculation of plant-accumulated temperature, the prediction errors of the model increase. Thus, several scientists have established the following formula to calculate the effective accumulated temperature based on the curves of plant growth development in response to temperatures.

$$R_f(x_t) = \begin{cases} 0 & \text{if } x_t < 0 \\ \frac{a}{1+e^{b(x_t-c)}} & \text{if } x_t \geq 0 \end{cases} \quad (3)$$

where c represents the optimal temperature for plant growth, b represents the parameter of plant sensitivity to variations in effective accumulated temperature, and a represents the upper limit of effective accumulated temperatures when bioclimatic events occur. This formula categorizes temperatures below 0 °C as noninfluential on bioclimatic events and involves only temperature accumulation above 0 °C.

The review of previous models shows that early thermal time models considered only the forcing units of growth, rather than the chilling requirements. In addition, during dormancy, plants are completely quiescent; thus, the phenological phase during dormancy is difficult to observe and define. However, a high number of physiological experiments in later stages have shown that low temperatures are necessary in winter for temperate plants to blossom and sprout. Bioclimatic models that neglect chilling requirements cannot effectively predict the flowering and sprouting of temperate plants. Therefore, scientists have developed numerous mechanistic models based on differing physiological plant types and have integrated chilling requirements into various models. Among these models, the most well-known are the sequential model (Hänninen 1987, 1990; Sanders 1975; Sarvas 1974), parallel model (Landsberg 1974; Sarvas 1974), alternating model (Cannell and Smith 1983; Kramer 1994; Murray et al. 1989), deepening rest model (Kobayashi et al. 1982), and four phase model (Hänninen 1990; Vegis 1964).

The differences between these bioclimatic models are as follows: The sequential model emphasizes that forcing temperature is effective only after chilling requirements are met, presenting a sequential order. Landsberg (1974) proposed the parallel model for identifying the dormancy characteristics of apple buds, indicating that regardless of temperatures, the phenological expression of plants is affected. The alternating model emphasizes that the forcing units and chilling units possess a negative indicative correlation. Thus, the two requirements alternatively influence phenological expression based on different weighting degrees with variations in the dormancy stages of plants. Kobayashi et al. (1982) proposed the deepening rest model in their study regarding the bud dormancy characteristics of *Cornus sericea* L. This model emphasizes that chilling requirements occur only during the deep rest stage, and that calculations of chilling requirements are not necessary for other dormancy stages. The four phase model emphasizes that plants have four sub-phenological phases during dormancy, which are the prerest, true-rest, postrest, and quiescence phases. The critical temperature-forced growth increases continuously during the prerest phase, but decreases during the postrest phase. In the true-rest phase, plants do not respond to any forcing growth temperature. The critical plant growth temperature decreases to the lower limit of initial temperatures for plant development in the postrest phase. When the external temperature remains below the lower limit temperature, plants enter the quiescence phase, the length of which is determined by the physiological conditions of the plant and the temperature increase in the following spring.

Regarding the measurement of the chilling requirements of plants in thermal time models, two common calculation methods exist:

$$R_c(x_t) = \begin{cases} 1 & \text{if } x_t < T_{b2} \\ 0 & \text{if } x_t \geq T_{b2} \end{cases} \quad (4)$$

where R_f becomes R_c, indicating that the growth accumulated temperature is replaced by the accumulated low temperature of chilling requirements, and T_{b2} represents the upper limit of the critical temperature of effective low temperatures. Temperatures higher than T_{b2} have no effect on the temperature accumulation of chilling requirements. Only temperatures lower than T_{b2} affect the temperature accumulation of plant chilling requirements. Binary coding is adopted to calculate the effective accumulated temperature. In other words, regardless of temperature values lower than the critical temperature, one effective chilling unit is counted. Even if the temperature is −50 °C, which freezes plants to death, an effective chilling unit is counted. This formula obviously contradicts empirical experience. Therefore, subsequent scientists have developed another formula for calculating the effective chilling unit:

$$R_c(x_t) = \begin{cases} 0 & \text{if } x_t \leq T_m \text{ or } x_t \geq T_M \\ \frac{x_t - T_m}{T_0 - T_m} & \text{if } T_0 > x_t > T_m \\ \frac{x_t - T_M}{T_0 - T_M} & \text{if } T_0 < x_t < T_M \end{cases} \quad (5)$$

where T_m and T_M represent the upper and lower limits of the effective low temperatures of plants, respectively. When the external temperature is lower or higher than the upper and lower limits, the accumulated temperatures for plant chilling requirements are not effective. The term T_0 refers to the most effective chilling requirement temperature of plants. Clearly, this formula meets the actual situation more accurately than formula (4) does.

Different plant bioclimatic models combined with various plant physiological types must be calculated using different methods. For example, when thermal time models are used to predict plant flowering on the sequential model, the plant chilling requirements must be calculated and satisfied before the growth-accumulated temperature of plants is calculated. If parallel models are used, chilling accumulated temperature and forcing accumulated temperature must also be calculated to predict bioclimatic events. Hence, dozens of model combinations for predicting plant flowering or sprouting by using the thermal time model are available. The high degree of plant bioclimatic and physiological diversity contributes to the complex development of bioclimatic models. The complexity of bioclimatic model development, to a certain degree, effectively increases the accuracy of bioclimatic prediction; however, such complexity also impedes the promotion and application of the models. To simplify the application of bioclimatic models, Chuine (2000) combined numerous major mechanistic models and developed a set of unified bioclimatic model calculation methods, which comprises two formulas to calculate the forcing and chilling requirements of plants. Through the adjustment of various parameters in the model, Chuine fitted the plant differences resulting from physiological responses, phenological phases, regions, and latitudes. Subsequently, Chuine and Beaubien (2001) further argued that the distribution of woody plants is primarily determined by the degree of fitness of the plant bioclimatology to the local climates. Thus, they integrated other models, such as those of freeze injury and fruit ripening, to develop a bioclimatic model based on biological processes, which they referred to as the PHENOFIT model. The model uses bioclimatic observation data for parameter fitting of bioclimatic models and meteorological variable map layers provided by Environment Canada, Climate Archives, the National Climatic Data Center, and the World Radiation Center to determine species distribution according to the fitting degree of the species bioclimatology to the local climates. Because the PHENOFIT model combines multiple bioclimatic models, the calculation formula is complex. Nevertheless, the PHENOFIT model requires the input of only five variables to obtain 12 variables that explain the effects of climates on species. These resulting variables altogether can determine the distribution appropriateness of species. The PHENOFIT model uses climatic data from various geographic regions to infer the distribution of numerous temperate perennial woody plants. The results indicated that the outcomes inferred using the model highly corresponded to the actual distribution of the target species.

The temperature, light, water availability, and airflow changes caused by climate changes influence the transpiration rate of leaves, which is determined by numerous factors, such as the net radiation balance of leaves, water supply conditions, leaf shapes, environmental wind speed, and the reaction of the stomata to transpiration sensitivity (Gates 1968; Raschke 1960). The model is as follows:

$$S_t(1-\alpha_l)+L_d-\varepsilon\sigma T_a^4 = \frac{\rho C_p(T_l-T_a)}{r_a}+\frac{\rho C_p}{\gamma^*}\frac{(e_o-e_a)}{r_s+r_a} \quad (6)$$

where S_t represents the incoming solar radiation (Wm^{-2}); α_l is the albedo of the leaf; L_d is the incoming longwave radiation (Wm^{-2}); $\varepsilon\sigma T^4$ is the long-wave radiation emitted by the leaf at the leaf temperature (T_l); ρ is the environmental air density around the leaf (kgm^{-3}); C_p is the specific heat of air (kPa); T_a is the air temperature (°C); r_a is the aerodynamic conductance to heat transfer (sm^{-1}); γ^* is the psychrometric constant (kPa $°C^{-1}$); e_0 is the saturated vapor pressure (kPa) at the current leaf temperature; e_a is the actual vapor pressure (kPa); and r_s represents the stomatal conductance (sm^{-1}). Formula (6) shows that a slight change in the temperature affects multiple factors simultaneously. When the air temperature increases, the long-wave radiation absorption of leaves is affected, increasing the thermal load of leaves and changing the saturated vapor pressure in the atmosphere. Consequently, the actual vapor pressure is insufficient and causes the water transpiration rate of the leaf to increase along with water consumption. Thus, the model can effectively evaluate the effects of temperature, light, water availability, and airflow changes on plants according to climate changes. Moreover, stomatal conductance differs with the sensitivity of plant species and strains to climate changes (Hofstra and Hesketh 1969).

Because of article length limitations, we introduced only three major types of plant bioclimatic models. In addition to the models introduced in this study, other bioclimatic models are of importance in separate fields of development. Basically, the diversity of relationships between organisms and climates leads to diversity among statistical (empirical) models, such as the thermal time, degree-days, heat sums, growing degree-days, physiological time, and spring warming models. The physiological and genetic diversity of organisms contributes to the diversity of mechanistic models, such as the parallel, sequential, deepening rest, four phases, Utah (Richardson et al. 1974), positive chill (Linsley-Noakes et al. 1995), and North Carolina models (Gilreath and Buchanan 1981). The diversity of biological and statistical theories contributes to the diversity of theoretical models, such as the models based on carbon equilibrium, the interaction of hormones, survival and reproductive adaptation, ecological niches, genetic behaviors, biological processes, and remote sensing. Naturally, some of the models involve a certain degree of correlation, which occasionally enables their mutual and complementary combination.

By reviewing the development of early bioclimatic models, we identified the following tendencies: (a) The number of studies regarding the bioclimatic models for perennial species substantially exceeds that of those for annual plants. (b) The number of bioclimatic model studies on temperate plants is considerably higher than that of those on tropical and subtropical plants. (c) The number of bioclimatic model studies on woody plants is substantially higher than that of those on herbal plants. (d) The number of observational bioclimatic model studies is substantially higher than that of experimental studies. (e) The number of bioclimatic model studies on plants that sprout and blossom in spring is considerably higher than that of those on plants with different growth and development stages. (f) The number of bioclimatic model studies on crops greatly exceeds that of those on forest plants. The majority of the bioclimatic model research conducted after 1753 has focused on the flowering and sprouting models of temperate plants. Regarding other bioclimatic models, only a few model studies on fruit ripening bioclimatology were found (Piper et al. 1996; Song and Ou 1997). Moreover, research on the bioclimatic model of leaf colouring periods is scant (Chuine and Beaubien 2001).

Application of plant bioclimatic models in evaluating the influence of climate changes

Plant bioclimatic models have been applied and developed in different fields, such as for predicting and evaluating the influence of climate changes on plant bioclimatology (Hänninen and Tanino 2011; Hänninen et al. 2007; Hao et al. 2001; Morin et al. 2009), improving the primary productivity of ecosystem (Kramer and Mohren 1996; Watsona et al. 2013), helping patients with pollinosis predict the time when pollen will occur in the air (Frenguelli and Bricchi 1998), assisting in crop or forest management and disaster-risk decision assessment, diagnosing the effects of climate on crop growth and development, predicting or assessing the correlations between species and their survival or adaptive strategy evolution (Chuine and Beaubien 2001; Morin et al. 2008), rebuilding regional climate environments in the past (Maurer et al. 2011; Menzel 2005; Yiou et al. 2012), forecasting the flowering time of cherry blossoms for developing the tourisy industry (Allen et al. 2014), and diagnosing the growth and development conditions of organisms as well as diseases and pests (Villalta et al. 2007). Unsurprisingly, these applications are correlated with one other to a certain degree. In recent years, plant bioclimatic models have been continuously applied to climate change research to evaluate the effects of climate changes on organisms. This implies that the importance of applying plant models in climate change-related research has constantly increased (Peñuelas and Filella 2001). Thus, this study introduced the application of bioclimatic models in assessing the influence of climate changes and in disaster prevention.

Initially, scientists focused on how plant sprouting and leaf expansion in the spring are correlated with freeze and cold injuries in the spring. Thus, statistical and mechanistic models regarding plant sprouting were the first models used to evaluate the effects of climate changes on plants. These models are often used to evaluate plants' ability to resist freezing or frost injuries (Cannell 1985; Cannell and Smith 1986; Hänninen 1991) or the competition for light that occurs among different species after climate changes (Cesaraccio et al. 2004). As bioclimatic model research progresses, theoretical models such as the DORMPHOT model, which is based on theoretical processes, are frequently used to assess the effects and risks of extremely low temperatures and freezing and cold injuries on forests. Theoretical models are also used to assess the risks of native species being affected by climate changes (Kramer 1995; Kramer et al. 1996; O'Neill et al. 2010). Based on an empirical experiment, the DORMPHOT model was more accurate than traditional models in assessing tree sprouting (Caffarra et al. 2011; Zottele et al. 2011).

Regarding the assessment of the effects of climate changes on plant bioclimatology, productivity, vegetation structures, vegetation dynamics, and forest landscapes, forest gap models that contain climate variables are often used to explain the effects of climate changes on forest succession, growth, landscapes, and the structural variations of plant communities (Bugmann 2001; Keane et al. 2001; Prentice et al. 1993). Additionally, because of the differing sensitivities of the models, the response degree of forest primary productivity models varies with the model adopted (Leinonen and Kramer 2002; Vitasse et al. 2011). Common instances are the effects of energy and carbon dioxide flows on leaf expansion and falling leaf bioclimatology, and the model for assessing the relationship between leaf area index and seasonal evolution (Chase et al. 1996). In addition, empirical (statistical) degree-day growing models are frequently used in investigating the bioclimatic changes and carbon sequestration cycles in land surface models (Arora and Boer 2005; Baldocchi et al. 2005; Delpierre et al. 2009; Vitasse et al. 2011). Similarly, regarding the effects of climate changes on the carbon sequestration ability of vegetation, the large-scale biological sphere model based on forest ecological system processes, BIOME-BGC, includes information on leaf growth and falling dates as parameters and applies the information to three types of vegetation research (Running and Hunt 1993).

The prediction results of bioclimatic modeling or the models themselves can be integrated with other models with various purposes to conduct research on the effects of climate changes (Halofsky et al. 2013). For example, Bonan (1998) used the monthly leaf area indices predicted using the land surface model of the National Center for Atmospheric Research as model parameters and applied the parameters to the grids of the Community Climate Model to facilitate global climate change research. Kaduk and Heimann (1996) determined the precautionary and mechanical structures that identify bioclimatology phases in environmental conditions and applied the structure to land carbon cyclic model research. Botta et al. (2000) used remote sensing data to estimate leaf sprouting time and developed empirical prediction formulas to predict leaf bioclimatology dynamics and propose a global bioclimatology precautionary structure. In addition, other professional bioclimatic models of climate change for large-scale structures based on biospheres or ecological systems exist, such as the Frankfurt biosphere model established based on the carbon equilibrium structure; the Lund-Potsdam-Jena dynamic global vegetation model, which assesses ecological system dynamics, plant geography, and land field carbon cycles (Sitch et al. 2003); the Canadian Centre for Climate Modeling and Analysis integrated biosphere simulator model, which predicts leaf bioclimatology based on light and temperature functions (Foley et al. 1996); and the forest carbon model based on photosynthesis and transpiration (Chiang and Brown 2007). These models have been widely applied in large-scale climate change research in recent years.

Recently, ecologists have focused on the effects of climate changes on species distribution, the resulting habitat fragmentation, and relevant species conservation arguments (Channell and Lomolino 2000; Crimmins et al. 2013; Fan et al. 2013; Gavin et al. 2014; Pauli et al. 2014; Pimm et al. 2014; Renton et al. 2013). Thus, numerous species distribution models developed on the based of the climate ecological niche theory of bioclimatic models have been applied in research on the effects of climate changes on species distribution and habitats. A major portion of these models are also referred to as climate envelope models (CEMs) (Hijmans and Graham 2006), such as the maximum entropy models (Phillips et al. 2004), machine-learning-based artificial neural network models, and integrated species distribution models (e.g., BIOMOD) (Coetzee et al. 2009; Thuiller 2003). However, not all species distribution models are categorized as CEMs. For example, although the PHENOFIT model was developed on the basis of biological processes and many physiologically based SDMs (Kearney and Porter 2009) are used to evaluate the effects of climate changes on species distribution, they are not CEMs.

The mapped atmosphere-plant-soil system model (Lenihan et al. 2003, 2008) can be used to assess the effects of climate changes on vegetation distribution, ecological system productivity, or forest fires. Remote-sensing time sequential data can be used to measure and assess land field surface phenology for assessing the vegetation responses after fires (van Leeuwen et al. 2010). In addition, regarding large-scale biological effect research, the BIOME-BGC, CLASS, Interannual Flux Tower Upscaling Sensitivity Experiment, third generation Coupled Global Climate Model, I/O buffer information specification, Lund-Potsdam-Jena, National Center for Atmospheric Research Land Surface Model, and remote-sensing-based NDVI/NDWI models can be used for assessing the effects of climate changes on large areas of vegetation (Bonan 1998; Desai 2010; Foley et al. 1996; Sitch et al. 2003). These models are convenient for use in large plain areas; thus, they have been widely adopted by studies in numerous temperate continental countries in recent years.

The types, application methods, and purposes of bioclimatic models are numerous, and the predictive accuracy of the models is determined by (a) the quality and quantity of data, (b) whether the user selects and uses the most appropriate model, and (c) the accuracy in forecasting climate changes. Because scientists mostly focus on (a) and (b), this paper does not discuss item (c), which requires the expertise of meteorologists. In particular, the situation described in (a) is inevitable when any model is used. However, because various models require different levels of data sensitivity, the requirements for data quality and quantity also differ. The requirements for data quality and accuracy are strict and are often based on bioclimatic models driven by data, such as the maximum entropy model, CEMs, and machine-learning models used for species distribution modeling. Thus, the preparation and compilation works of data are critical in these types of model. Two conditions are used to determine whether a user has selected and used the appropriate model. The first condition is the user's understanding of the target organisms' physiology, ecology, behavior, or biology. For example, if the constrain conditions of the distribution of a species is not a climatic factor, using CEMs and current species distribution data to assess the effects of climate changes on species distribution may lead to considerable errors. Therefore, to use bioclimatic models to assess the effects of climate changes on organisms, is necessary to identify the period in the target organisms' life cycle that is most sensitive to climate changes. Subsequently, based on the period, a suitable model should be selected for conducing assessment to maximize the effectiveness of the model. Choosing an inappropriate model to conduct assessment typically results in errors (Coetzee et al. 2009).

All applications of bioclimatic models in assessing the effects of climate changes have advantages and disadvantages (Elith et al. 2006; Hijmans and Graham 2006). For example, statistical models are the most widely used and are user-friendly and users are not required to consider biological processes, genetics, and physiology; however, they lack explanatory power for the research results and have a limited scopes of applications. Statistical models generally can not be applied to research on the effects on large areas of vegetation variations. Mechanistic models yield the highest explanatory power for the effects of climate changes, and thus have optimal assessment effectiveness. However, uncertainty of species' physiological mechanisms is a constraining factor of using such models. For instance, users may be uncertain regarding what model to use to assess the effects of climate changes on the dormancy of *Sassafras randaiense* Hay. Rehder because the bud dormancy and physiology of the plant species have not yet been thoroughly investigated. Regarding the research on bioclimatic models for exploring the effects of climate changes, the successful application of models is determined by the user's understanding of each model. Only by selecting suitable models can reliable assessment on the effects of climate changes be conducted and accurate results be attained.

In our previous review of climate change research on plants (Hsieh and Chiou 2013), we found that phenological gardens and phenological observation networks are used to record the effects of past climate changes on organisms in climate change research and monitor the direct influence of climate changes on organisms. Bioclimatic models are used to assess the possible effects of future climate changes and assist in making disaster-prevention decisions. Bioclimatic models and phenological observation networks are complementary in assessing the effects of climate changes; neither can be neglected. Without the historical records of phenological observation networks, bioclimatic models lack modeling data; without bioclimatic models, phenological observation networks lack the function of risk assessment and cannot assist in disaster-prevention decision-making. Thus, phenological fingerprints and models have been developed rapidly for applications in international climate change research. The use of regional phenological fingerprints, which was once a tool for small- to medium-scale spaces, has been expanded to continental and global scales through the establishment of global bioclimatic monitoring plans (Bruns et al. 2003; Parmesan and Yohe 2003; Root et al. 2003). Regarding the application of models, although the global bioclimatic models developed on the basis of remote sensing data have been widely applied in studies in temperate continental countries in Europe and North America, small- to medium-scale phenological fingerprints and models are more suitable for Taiwan because of its small terrain.

Conclusion

The effects of global climate changes have increased in recent years. Numerous cities in Europe, the United States, China, and Japan were measured to have had high temperatures exceeding 40 °C for several consecutive days throughout the summer of 2013. Torrential rain has caused disasters in numerous regions and weather stations all over the planet measured atmospheric carbon dioxide concentrations exceeding 400 ppm, the highest in millions of years. Moreover, climate changes have exerted increasingly severe effects on plants and wildlife (Anderegg et al. 2012; Harley 2011; Ibáñez et al. 2008; Inouye 2008; Kaschner et al. 2011; Moritz et al. 2008; Rode et al. 2010; van Mantgem et al. 2009). These disasters indicate that the threats of climate change are ubiquitous. Because of the global impacts of disasters, we suggest that all countries' government and relevant research units immediately establish international phenology gardens and network systems, develop phenological fingerprint observation technologies, improve the ability to monitor the effects of climate changes on global organisms, and employ long-term bioclimatic observation records to develop bioclimatic models that are suitable for local climates and disaster prevention. Consequently, the capacity for assessing the effects of climate changes and predicting and preventing disasters can be prepared, and measures and strategies can be prepared in response to disasters caused by climate changes.

Authors' contribution

CRC conceived and designed the topics. TYH collecting literature and wrote the paper. CCC edited the manuscript. All authors read and approved the final manuscript.

Author details

[1] School of Forestry and Resource Conservation, National Taiwan University, No. 1, Sec. 4, Roosevelt Rd., Taipei 10617, Taiwan (R.O.C.). [2] Shanghai Institutes for Biological Sciences, Chinese Academy of Sciences, 320 Yue Yang Rd., Shanghai 200031, China. [3] Shanghai Chenshan Plant Science Research Center,

Chinese Academy of Sciences, 3888 Chenhua Road, Songjiang, Shanghai 201602, China. [4] Shanghai Key Laboratory of Plant Functional Genomics and Resources, Shanghai Chenshan Botanical Garden, 3888 Chenhua Road, Songjiang, Shanghai 201602, China. [5] College of Business, Chung Yuan Christian University, 200, Chung Pei Rd., Chung Li 32023, Taiwan (R.O.C.).

Acknowledgements
We thank two anonymous reviewers for their comments and suggestions. This study was supported by the Chinese Academy of Sciences (Grant no.: 2013 TW2SA 0003, 2015TW1SA0001), the Forestry Bureau, Council of Agriculture, Executive Yuan (Grant no.: 101 agriculture-13.5.4-forestry-e1) and the National Science Council (Grant no.: NSC 102-2313-B-002-038).

Compliance with ethical guidelines

Competing interests
The authors declare that they have no competing interests

References

Allen JM, Terres MA, Katsuki T, Iwamoto K, Kobori H, Higuchi H, Primack RB, Wilson AM, Gelfand A, Silander JA Jr (2014) Modeling daily flowering probabilities: expected impact of climate change on Japanese cherry phenology. Glob Change Biol 20:1251–1263

Anderegg WR, Berry JA, Smith DD, Sperry JS, Anderegg LD, Field CB (2012) The roles of hydraulic and carbon stress in a widespread climate-induced forest die-off. Proc Natl Acad Sci 109:233–237

Arora VK, Boer GJ (2005) A parameterization of leaf phenology for the terrestrial ecosystem component of climate models. Glob Chang Biol 11:39–59

Baldocchi DD, Black T, Curtis P, Falge E, Fuentes J, Granier A, Gu L, Knohl A, Pilegaard K, Schmid H (2005) Predicting the onset of net carbon uptake by deciduous forests with soil temperature and climate data: a synthesis of FLUXNET data. Int J Biometeorol 49:377–387

Bawa KS, Hadley M (1991) Reproductive ecology of tropical forest plants. CRC Pres, Paris, p 422

Bonan GB (1998) The land surface climatology of the NCAR land surface model coupled to the NCAR community climate model. J. Climate 11:1307–1326

Botta A, Viovy N, Ciais P, Friedlingstein P, Monfray P (2000) A global prognostic scheme of leaf onset using satellite data. Glob Chang Biol 6:709–725

Bruns E, Chmielewski FM, VanVliet AJH (2003) The global phenological monitoring concept. Phenology 39:93–104

Bugmann H (2001) A review of forest gap models. Clim Change 51:259–305

Caffarra A, Donnelly A, Chuine I (2011) Modelling the timing of Betula pubescens budburst. II. Integrating complex effects of photoperiod into process-based models. Clim Res. 46:159–170

Campbell RK, Sugano AI (1975) Phenology of bud burst in Douglas-fir related to provenance, photoperiod, chilling, and flushing temperature. Bot Gaz 36:290–298

Campbell-Clause JM (1998) Stomatal response of grapevines to wind. Aust J Exp Agric 38:77–82

Cannell MGR (1985) Analysis of risks of frost damage to forest trees in Britain. Helsinki University Press, Helsinki, pp 153–166

Cannell MGR, Smith RI (1983) Thermal time, chill days and prediction of budburst in Picea sitchensis. J Appl Ecol 20:951–963

Cannell MGR, Smith RI (1986) Climatic warming, spring budburst and forest damage on trees. J Appl Ecol 23:177–191

Cannell MGR, Murray MB, Sheppard LJ (1985) Frost avoidance by selection for late budburst in Picea sitchensis. J Appl Ecol 22:931–941

Cesaraccio C, Spano D, Snyder RL, Duce P (2004) Chilling and forcing model to predict bud-burst of crop and forest species. Agr Forest Meteorol 126:1–13

Channell R, Lomolino MV (2000) Dynamic biogeography and conservation of endangered species. Nature 403:84–86

Chase TN, Pielke RA, Kittel TG, Nemani R, Running SW (1996) Sensitivity of a general circulation model to global changes in leaf area index. J Geophys Res 101:7393–7408

Chiang JM, Brown KJ (2007) Improving the budburst phenology subroutine in the forest carbon model PnET. Ecol Model 205:515–526

Chien CT, Yang JJ, Chung YL, Lin TP (1995) Germination promotion of taxus mairei seed by combination of warm and cold stratification. Bull Taiwan For Res Inst 10:331–336

Chu KC, Wan MW (1999) Phenology. Hunan Education Publishing House, Changsha, p 173

Chuine I (2000) A unified model for budburst of trees. J Theor Biol 207:337–347

Chuine I, Beaubien EG (2001) Phenology is a major determinant of tree species range. Ecol Lett 4:500–510

Coetzee BW, Robertson MP, Erasmus BF, Van Rensburg BJ, Thuiller W (2009) Ensemble models predict Important Bird Areas in southern Africa will become less effective for conserving endemic birds under climate change. Glob Ecol Biogeogr 18:701–710

Crimmins SM, Dobrowski SZ, Mynsberge AR (2013) Evaluating ensemble forecasts of plant species distributions under climate change. Ecol Model 266:126–130

Darnell RL, Hancock JF (1996) Balancing vegetative and reproductive growth in strawberry. In: Pritts CKC, Crocker TE (eds) Proceedings of the IV North American Strawberry Conference, pp 144–150

de Lampe MG, Bergeron Y, McNeil R, Leduc A (1992) Seasonal flowering and fruiting patterns in tropical semi-arid vegetation of northeastern Venezuela. Biotropica 24:64–76

Delpierre N, Soudani K, Francois C, Köstner B, Pontailler JY, Nikinmaa E, Misson L, Aubinet M, Bernhofer C, Granier A (2009) Exceptional carbon uptake in European forests during the warm spring of 2007: a data–model analysis. Glob Change Biol 15:1455–1474

Desai AR (2010) Climatic and phenological controls on coherent regional interannual variability of carbon dioxide flux in a heterogeneous landscape. J Geophys Res 115:G00J02

Duryea ML, Blakeslee GM, Hubbard WG, Vasquez RA (1996) Wind and trees: a survey of homeowners after Hurricane Andrew. J Arboric 22:44–50

Elith J, Graham CH, Anderson RP, Dudík M, Ferrier S, Guisan A, Hijmans RJ, Huettmann F, Leathwick JR, Lehmann A, Li J, Lohmann LG (2006) Novel methods improve prediction of species' distributions from occurrence data. Ecography 29:129–151

Fan YH, Nee CC (2007) Dormant character of different Actinidia callosa groups native in Taiwan. Seed Nurs (Taiwan) 9:49–62

Fan J, Li J, Xia R, Hu L, Wu X, Li G (2013) Assessing the impact of climate change on the habitat distribution of the giant panda in the Qinling Mountains of China. Ecol Model 274:12–20

Foley JA, Prentice IC, Ramankutty N, Levis S, Pollard D, Sitch S, Haxeltine A (1996) An integrated biosphere model of land surface processes, terrestrial carbon balance, and vegetation dynamics. Glob Biogeochem Cy 10:603–628

Frenguelli G, Bricchi E (1998) The use of the pheno-climatic model for forecasting the pollination of some arboreal taxa. Aerobiologia 14:39–44

Fuchigami LH, Weiser CJ, Kobayashi K, Timmis R, Gusta LV (1982) A degree growth stage (GS) model and cold acclimation in temperate woody plants. In: Li PH, Sakai A (eds) Plant cold hardiness and freezing stress. Academic Press, New York, pp 93–116

Gates DM (1968) Transpiration and leaf temperature. Annu Rev Plant Phys 19:211–238

Gavin DG, Fitzpatrick MC, Gugger PF, Heath KD, Rodríguez-Sánchez F, Dobrowski SZ, Hampe A, Hu FS, Ashcroft MB, Bartlein PJ, Blois JL, Carstens BC, Davis EB, Lafontaine G, Edwards ME, Fernandez M, Henne PD, Herring EM, Holden ZA, Kong W, Liu J, Magri D, Matzke NJ, McGlone MS, Saltre F, Stigall AL, Tsai YHE, Williams JW (2014) Climate refugia: joint inference from fossil records, species distribution models and phylogeography. New Phytol 204:37–54

Gilreath P, Buchanan D (1981) Rest prediction [mathematical] model for low-chilling 'Sungold' nectarine. J Am Soc Hortic Sci 106:426–429

Haldane JBS (1947) Science advances, 1st edn. Macmillan, London, p 252

Halofsky JE, Hemstrom MA, Conklin DR, Halofsky JS, Kerns BK, Bachelet D (2013) Assessing potential climate change effects on vegetation using a linked model approach. Ecol Model 266:131–143

Hänninen H (1987) Effects of temperature on dormancy release in woody plants: implications of prevailing models. Silva Fenn 21:279–299

Hänninen H (1990) Modelling bud dormancy release in trees from cool and temperate regions. Finnish Forest Research Institute, Finland, p 47

Hänninen H (1991) Does climatic warming increase the risk of frost damage in northern trees? Plant, Cell Environ 14:449–454
Hänninen H, Tanino K (2011) Tree seasonality in a warming climate. Trends Plant Sci 16:412–416
Hänninen H, Slaney M, Linder S (2007) Dormancy release of Norway spruce under climatic warming: testing ecophysiological models of bud burst with a whole-tree chamber experiment. Tree Physiol 27:291–300
Hao ZQ, Dai LM, He HS, Malandnoff DJ, Shao GF (2001) Potential response of major tree species to climate warming in Changbai Mountain, Northeast China. Chin J Appl Ecol 12:653–658
Harley CD (2011) Climate change, keystone predation, and biodiversity loss. Science 334:1124–1127
Heide OM (1974) Growth and dormancy in Norway spruce ecotypes. Physiol Plantarum 31:131–139
Heide OM (1993) Daylength and thermal time responses of budburst during dormancy release in some northern deciduous trees. Physiol Plantarum 88:531–540
Hijmans RJ, Graham CH (2006) The ability of climate envelope models to predict the effect of climate change on species distributions. Glob Chang Biol 12:2272–2281
Hofstra G, Hesketh JD (1969) The effect of temperature on stomatal aperture in different species. Can J Bot 47:1307–1310
Hopkins AD (1938) Bioclimatics: a science of life and climate relations. United States Department of Agriculture, Washington, DC, p 188
Hsieh TY, Chiou CR (2013) Phytophenology and its applications in climate change research: review and future perspectives. Q J Chin For 46:391–410
Hsieh TY, Nee CC, Chien CT (2004) Seed germination of Taiwanese Actinidia latifolia (Gardn. and Champ.) Merr. Taiwan J For Sci 19:173–176
Huang YW (2011) Establishment of cherry blossoming model in Taiwan. Department of Forestry and Resource Conservation, National Taiwan University, Taipei, p 95
Hunter AF, Lechowicz MJ (1992) Predicting the timing of budburst in temperate trees. J Appl Ecol 29:597–604
Ibáñez I, Clark JS, Dietze MC (2008) Evaluating the sources of potential migrant species: implications under climate change. Ecol Appl 18:1664–1678
Inouye DW (2008) Effects of climate change on phenology, frost damage, and floral abundance of montane wildflowers. Ecology 89:353–362
Isikawa S (1954) Light-sensitivity against the germination. I." Photoperiodism" of seeds. Bot Mag (Tokyo) 67:51–56
Johnson LRC, Irgens-Moller H (1964) Effect of photoperiod and light quality on germination of douglas-fir seed. Forest Sci. 10:200–205
Kaduk J, Heimann M (1996) A prognostic phenology scheme for global terrestrial carbon cycle models. Clim Res 6:1–19
Kaschner K, Tittensor DP, Ready J, Gerrodette T, Worm B (2011) Current and future patterns of global marine mammal biodiversity. PLoS One 6:e19653
Keane RE, Austin M, Field C, Huth A, Lexer MJ, Peters D, Solomon A, Wyckoff P (2001) Tree mortality in gap models: application to climate change. Clim Change 51:509–540
Kearney M, Porter W (2009) Mechanistic niche modelling: combining physiological and spatial data to predict species' ranges. Ecol Lett 12:334–350
Kellomäki S, Hänninen H, Kolström M (1995) Computations on frost damage to Scots pine under climatic warming in boreal conditions. Ecol Appl 5:42–52
Kobayashi KD, Fuchigami LH, English MJ (1982) Modeling temperature requirements for rest development in Cornus sericea. J Am Soc Hortic Sci 107:914–918
Körner C, Basler D (2010) Phenology under global warming. Science 327:1461–1462
Kramer K (1994) Selecting a model to predict the onset of growth of Fagus sylvatica. J Appl Ecol 172–181
Kramer K (1995) Modelling comparison to evaluate the importance of phenology for the effects of climate change on growth of temperate-zone deciduous trees. Clim Res 5:119–130
Kramer K, Mohren GMJ (1996) Sensitivity of FORGRO to climatic change scenarios: a case study on Betula pubescens, Fagus sylvatica and Quercus robur in the Netherlands. Clim Change 34:231–237
Kramer K, Friend A, Leinonen I (1996) Modelling comparison to evaluate the importance of phenology and spring frost damage for the effects of climate change on growth of mixed temperate-zone deciduous forests. Climate Res 7:31–41
Landsberg JJ (1974) Apple fruit bud development and growth; analysis and an empirical model. Ann Bot 38:1013–1023
Lechowicz MJ, Koike T (1995) Phenology and seasonality of woody plants: an unappreciated element in global change research? Can J Bot 73:147–148
Leinonen I, Kramer K (2002) Applications of phenological models to predict the future carbon sequestration potential of boreal forests. Clim Change 55:99–113
Lenihan JM, Drapek R, Bachelet D, Neilson RP (2003) Climate change effects on vegetation distribution, carbon, and fire in California. Ecol Appl 13:1667–1681
Lenihan JM, Bachelet D, Neilson RP, Drapek R (2008) Response of vegetation distribution, ecosystem productivity, and fire to climate change scenarios for California. Clim Change 87:215–230
Lieth H (1974) Phenology and seasonality modeling. Springer-Verlag, Berlin
Linsley-Noakes G, Louw M, Allan P (1995) Estimating daily positive Utah chill units using daily minimum and maximum temperatures. J SA Soc Hort Sci 5:19–23
Maurer C, Hammerl C, Koch E, Hammerl T, Pokorny E (2011) Extreme grape harvest data of Austria, Switzerland and France from AD 1523–2007 compared to corresponding instrumental/reconstructed temperature data and various documentary sources. Theor Appl Climatol 106:55–68
Menzel A (2005) A 500 year pheno-climatological view on the 2003 heatwave in Europe assessed by grape harvest dates. Meteorol Zeits 14:75–77
Menzel A, Estrella N, Fabian P (2001) Spatial and temporal variability of the phenological seasons in Germany from 1951 to 1996. Glob Change Biol 7:657–666
Morin X, Viner D, Chuine I (2008) Tree species range shifts at a continental scale: new predictive insights from a process-based model. J Ecol 96:784–794
Morin X, Lechowicz MJ, Augspurger C, O'Keefe J, Viner D, Chuine I (2009) Leaf phenology in 22 North American tree species during the 21st century. Glob Change Biol 15:961–975
Moritz C, Patton JL, Conroy CJ, Parra JL, White GC, Beissinger SR (2008) Impact of a century of climate change on small-mammal communities in Yosemite National Park, USA. Science 322:261–264
Murray MB, Cannell MGR, Smith RI (1989) Date of budburst of fifteen tree species in Britain following climatic warming. J Appl Ecol 26:693–700
Myking T (1997) Effects of constant and fluctuating temperature on time to budburst in Betula pubescens and its relation to bud respiration. Trees 12:107–112
Nee CC (1994) The effects of continuous rainings on the fruit set of kiwifruit (Actinidia deliciosa (A. Chev.) C. F. Lianf et A. R. Ferguson) in the highland of Taiwan. J Chin Soc Hort Sci 40:282–286
Nee CC, Lin WB, Hummer K (1995) The Recovery of Taiwan Native Pear (Pyrus koehnei Schneider) (III)-growth characteristics. J Chin Soc Hort Sci 41:107–115
Nienstaedt H, Olson JS (1961) Effects of photoperiod and source on seedling growth of eastern hemlock. For Sci 7:81–96
O'Neill BF, Caffarra A, Gleeson E, Semmler T, McGrath R, Donnelly A (2010) The use of climate projections in the modelling of bud burst. EGU General Assembly Conference Abstracts, 15011, vol 12
Ou SK, Hwang MF, Li CP, Lu SJ, Song CW (2000) The relationship between cultivar's chilling requirement and planting altitudes of peaches. J Agric Res China 49:46–53
Parmesan C, Yohe G (2003) A globally coherent fingerprint of climate change impacts across natural systems. Nature 421:37–42
Pauli H, Gottfried M, Grabherr G (2014) Effects of climate change on the aloine and nival vegetation. J Mount Eco 7:9–12
Peñuelas J, Filella I (2001) Responses to a warming world. Science 294:793–795
Perry TO (1971) Dormancy of trees in winter. Science 171:29–35
Phillips SJ, Dudík M, Schapire RE (2004) A maximum entropy approach to species distribution modeling. In: Proceedings of the twenty-first international conference on Machine learning, pp. 655–662
Pimm SL, Jenkins CN, Abell R, Brooks TM, Gittleman JL, Joppa LN, Raven PH, Roberts CM, Sexton JO (2014) The biodiversity of species and their rates of extinction, distribution, and protection. Science 344(6187):987–997
Piper EL, Boote KJ, Jones JW, Grimm SS (1996) Comparison of two phenology models for predicting flowering and maturity date of soybean. Crop Sci 36:1606–1614

Prentice IC, Sykes MT, Cramer W (1993) A simulation model for the transient effects of climate change on forest landscapes. Ecol Model 65:51–70
Raschke K (1960) Heat transfer between the plant and the environment. Annu Rev Plant Phys 11:111–126
Reich PB, Borchert R (1984) Water stress and tree phenology in a tropical dry forest in the lowlands of Costa Rica. J Ecol 72:61–74
Renton M, Childs S, Standish R, Shackelford N (2013) Plant migration and persistence under climate change in fragmented landscapes: does it depend on the key point of vulnerability within the lifecycle? Ecol Model 249:50–58
Richardson AJ, Seeley SD, Walker DR (1974) A model for estimating the completion of rest for 'Redhaven' and 'Elberta' peach trees. HortScience 9:331–332
Robak H, Magnesen S (1970) Contribution to the knowledge of the ecology of growth termination in spruce seedlings of Norwegian and Central European provenances. Silvae Genet 19:188–190
Robertson GW (1968) A biometeorological time scale for a cereal crop involving day and night temperatures and photoperiod. Int J Biometeorol 12:191–223
Rode KD, Amstrup SC, Regehr EV (2010) Reduced body size and cub recruitment in polar bears associated with sea ice decline. Ecol Appl 20:768–782
Root TL, Price JT, Hall KR, Schneider SH, Rosenzweig C, Pounds JA (2003) Fingerprints of global warming on wild animals and plants. Nature 421:57–60
Running SW, Hunt ER (1993) Generalization of a forest ecosystem process model for other biomes, BIOME-BGC, and an application for global-scale models. In: Ehleringer JR, Field CB (eds) Scaling physiological processes: leaf to globe. Academic Press, New York, pp 141–158
Sanders CG (1975) Comments on the model for estimating the completion of rest for "Red Haven" and "Elberta" peach trees. Hortic Sci 10:560–561
Sarvas R (1974) Investigations on the annual cycle of development of forest trees 2: autumn dormancy and winter dormancy. Communicationes Instituti Forestalis Fenniae 76:1–110
Schnelle F (1955) Pflanzen-Phänologie. Akademische Verlagsgesellschaft, Leipzig
Schwartz MD (1997) Spring index models: an approach to connecting satellite and surface phenology. In: Lieth H, Schwartz MD (eds) Phenology in seasonal climates. Backhuys, Leiden, pp 23–38
Schwartz MD, Marotz GA (1986) An approach to examining regional atmosphere-plant interactions with phenological data. J Biogeogr 13:551–560
Schwartz M, Marotz G (1988) Synoptic events and spring phenology. Phys Geogr 9:151–161
Sitch S, Smith B, Prentice IC, Arneth A, Bondeau A, Cramer W, Kaplan JO, Levis S, Lucht W, Sykes MT (2003) Evaluation of ecosystem dynamics, plant geography and terrestrial carbon cycling in the LPJ dynamic global vegetation model. Glob Change Biol 9:161–185
Song CW, Ou SK (1997) Thermal time required for fruit development in low-chill peaches. J Agric Res China 46:42–50
Sparkes TH, Carey PD, Coombes J (1997) First leafing dates of trees in Surrey between 1947 and 1996. Lond Nat 76:15–20
Stearns F, Olson J (1958) Interactions of photoperiod and temperature affecting seed germination in Tsuga canadensis. Am J Bot 45:53–58
Telewski FW (1995) Wind-induced physiological and developmental responses in trees. In: Coutts MP, Grace J (eds) Wind and trees. Cambridge University Press, Cambridge, pp 237–263
Thuiller W (2003) BIOMOD—optimizing predictions of species distributions and projecting potential future shifts under global change. Glob Change Biol 9:1353–1362
Tissue DT, Wright SJ (1995) Effect of seasonal water availability on phenology and the annual shoot carbohydrate cycle of tropical forest shrubs. Funct Ecol 9:518–527
van Leeuwen WJD, Casady GM, Neary DG, Bautista S, Alloza JA, Carmel Y, Wittenberg L, Malkinson D, Orr BJ (2010) Monitoring post-wildfire vegetation response with remotely sensed time-series data in Spain, USA and Israel. Int J Wildland Fire 19:75–93
van Mantgem PJ, Stephenson NL, Byrne JC, Daniels LD, Franklin JF, Fulé PZ, Harmon ME, Larson AJ, Smith JM, Taylor AH (2009) Widespread increase of tree mortality rates in the western United States. Science 323:521–524
Vegis A (1964) Dormancy in higher plants. Annu Rev Plant Phys 15:185–224
Villalta O, Holmes R, Kreidl S, Partington D, Hodson A, Atkins TA (2007) A weather-based model implemented in HortPlus MetWatch with potential to forecast brown rot infection risk in stone fruit. Acta Horticulturae 803:19–28
Vitasse Y, François C, Delpierre N, Dufrêne E, Kremer A, Chuine I, Delzon S (2011) Assessing the effects of climate change on the phenology of European temperate trees. Agr For Meteorol 151:969–980
Walker MD, Ingersoll RC, Webber PJ (1995) Effects of interannual climate variation on phenology and growth of two alpine forbs. Ecology 76:1067–1083
Wareing PF (1956) Photoperiodism in woody plants. Annu Rev Plant Phys 7:191–214
Watsona RA, Gabrielle BN, Traceya SR, Fultond EA, Bulmand CM, Edgara GJ, Barretta NS, Lylea JM, Frushera SD, Buxton CD (2013) Ecosystem model of Tasmanian waters explores impacts of climate-change induced changes in primary productivity. Ecol Model 264:115–129
Yiou P, de Cortázar-Atauri IG, Chuine I, Daux V, Garnier E, Viovy N, Leeuwen CV, Parker A, Boursiquot JM (2012) Continental atmospheric circulation over Europe during the Little Ice Age inferred from grape harvest dates. Clim Past 8:577–588
Zhao M, Peng C, Xiang W, Deng X, Tian D, Zhou X, Yu G, He H, Zhao Z (2013) Plant phenological modeling and its application in global climate change research: overview and future challenges. Environ Rev 21:1–14
Zottele F, Caffarra A, Gleeson E, Donnelly A (2011) Mapping future phenology of birch in Ireland. European GeoSciences Union General Assembly 2011:EGU2011-12307-3
Zou XM (1983) Agricultural phenology. Agricultural Press, Beijing

Distribution of new satellites and simple sequence repeats in annual and perennial *Glycine* species

Hsuan Chen[1,2], Mei-Chu Chung[1], Yuan-Ching Tsai[1], Fu-Jin Wei[1], Jaw-Shu Hsieh[2] and Yue-Ie C. Hsing[1*]

Abstract

The repeat sequences occupied more than 50 % of soybean genome. In order to understand where these repeat sequences distributed in soybean genome and its related *Glycine* species, we examined three new repeat sequences—soybean repeat sequence (SBRS1, SBRS2 and SBRS3), some nonspecific repeat sequences and 45S rDNA on several *Glycine* species, including annual and perennial accessions in this study. In the annual species, *G. soja*, signals for SBRS1 and ATT repeat can be found on each chromosome in GG genome, but those for SBRS2 and SBRS3 were located at three specific loci. In perennial *Glycine* species, these three SBR repeat frequently co-localized with 45S rDNA, two major 45S rDNA loci were found in all tetraploid species. However, an extra minor locus was found in one accession of the *G. pescadrensis* (Tab074), but not in another accession (Tab004). We demonstrate that some repetitive sequences are present in all *Glycine* species used in the study, but the abundancy is different in annual or perennial species. We suggest this study may provide additional information in investigations of the phylogeny in the *Glycine* species.

Keywords: *Glycine*, FISH, SSRs, 45S rDNA, Repeat sequence, Chromosome markers

Background

Soybean (*Glycine max* L. Merrill) is the most important legume crop because of its economic value and biological features for research. Since soybean seed consists of high protein and oil contents, it has become one of the most important food crops for human and livestock, as well as a valuable resource for biofuel energy. Improving soybean varieties has been slow and limited because of the lack of genetic diversity (Keim et al. 1990; Krishnan et al. 2001). Wild germplasm may be a significant breeding resource for current varieties that have undergone several genetic bottlenecks and have shown limited genetic variability (Hyten et al. 2006; Keim et al. 1990). These wild germplasm harbor many traits of interest to soybean breeders, such as biotic and abiotic disease resistance (Newell and Hymowitz 1975), high seed protein content (Sebolt et al. 2000) and increased yield (Concibido et al. 2003). Thus, investigating *Glycine* species is important for soybean genetics and breeding.

There are around 30 species in Genus *Glycine* which had been classified into two subgenera, *Soja* and *Glycine*. The subgenus *Soja* (2n = 40) includes only *G. max* and its annual close related *G. soja* Sieb. and Zucc., which may intercross freely with soybean (Palmer et al. 1987; Singh and Hymowitz 1988). The subgenus *Glycine* contains all perennial species with various genome types and ploidies (2n = 2x = 40, 2n = 4x = 80 and some aneuploids) (Doyle et al. 2004; Grant et al. 1984; Newell and Hymowitz 1983; Singh and Hymowitz 1985). Early investigations of phylogenetic relationships in *Glycine* species based on their ability to produce fertile hybrids and the degree of meiotic chromosomes pairing (reviewed by Ratnaparkhe et al. 2011). This led to the designation of seven genome groups, i.e. reproductively compatible within species, while reproductively isolated from other genome groups (groups A to G, Singh and Hymowitz 1985; Hymowitz et al. 1998). Other data, including isozymes (Doyle et al.

*Correspondence: bohsing@gate.sinica.edu.tw
[1] Institute of Plant and Microbial Biology, Academia Sinica, Taipei 115, Taiwan
Full list of author information is available at the end of the article

1986) and DNA phylogenetic analyses (Doyle et al. 1990a, b, c; Kollipara et al. 1997), suggested hybrid fertility as the primary criterion to identify species to genome groups and resulted in the designation of two additional genome groups, H and I. Besides, the situation in some species even more complicated, such as the *G. tabacina* and *G. tomentella* species complex, multiple separate 'races' are recognized that have not yet taxonomically arisen to species status, but that display enough molecular phylogenetic divergence from one another to be grouped into separate taxa (Doyle et al. 2004; Sherman-Broyles et al. 2014).

The *Glycine* species germplasm in Taiwan is abundant and important. Four species have been collected, including *G. soja* Siebold and Zucc. (GG genome, 2n = 40), *G. tomentella* Hayata (DDD_1D_1 genome, 2n = 80), *G. dolichocarpa* Tateishi and Ohashi (A_6A_6DD genome, 2n = 80) and *G. pescadrensis* Hayata ($A_6A_6B_3B_3$ genome, 2n = 80) (Hsing et al. 1995, 2001; Tateishi and Ohashi 1992; Thseng et al. 1999; Tsai et al. 2001; Tsai 2006). *G. soja* is widely distributed in Russia, Korea, Japan, China and Taiwan (Singh and Hymowitz 1987), with Taiwan being the southernmost among these areas. The three wild perennial *Glycine* species, *G. tomentella, G. dolichocarpa* and *G. pescadrensis,* are found in Taiwan, the Ryukyu Islands, the Philippines, the South Pacific islands and Australia, with Taiwan being the northernmost of these areas (Hymowitz et al. 1998). All the remaining perennial *Glycine* species are found only in Australia (Tindale and Craven 1988). Therefore, the Ryukyu Islands, Taiwan, and its adjacent islands are unusual because they have representatives of both subgenera of *Glycine.* The collection and study of *Glycine* species in Taiwan and its adjacent islands thus provide information for better understanding the evolution and relationships between these two subgenera (Hsieh et al. 2001).

The soybean genome has been well documented as paleopolyploidy and manifested as allopolyploid diploidy (Gill et al. 2009; Mudge et al. 2005; Schlueter et al. 2007; Shoemaker et al. 1996; Tek et al. 2010). In addition, whole-genome sequencing (Williams 82) and resequencing projects on *Glycine* species were available recently (Jackson et al. 2006; Katayose et al. 2012; Kim et al. 2010; Lam et al. 2010; Schmutz et al. 2010; Weidner et al. 2012). However, little information is available on the genome composition of wild relatives; that is, cytological research of wild soybean is still limited.

Fluorescent in situ hybridization (FISH) is a powerful tool for cytological studies. FISH used with specific markers can be used to label specific chromosomes and target given genomes or particular regions on chromosomes (Cheng et al. 2007; Jiang and Gill 2006; Kopecky et al. 2008; Pinkel et al. 1986). Furthermore, nonspecific repeats such as simple repeat sequences (SSRs) may also be used as genomic and chromosomal markers among many species (Cuadrado et al. 2008; Cuadrado and Jouve 2007).

FISH has been used in cytological research on soybean. Several repeat sequences such as ribosomal RNA genes (rDNAs) (Shi et al. 1996), SB92 (Kolchinsky and Gresshoff 1995; Vahedian et al. 1995) and STR120 (Morgante et al. 1997) were labeled on soybean chromosomes and used in chromosome "painting" (Shi et al. 1996). FISH with bacterial artificial chromosome (BAC) clones used as probes suggested heterochromatic blocks on the chromosome pericentromeric region (Lin et al. 2005), and some research has provided evidence of chromosome-level homeology in the paleopolyploid soybean genome (Pagel et al. 2004; Walling et al. 2006). In the study of soybean chromosome structure and genomic evolution, SB92 was previously suggested as a major repeat sequence in the *G. max* genome (Kolchinsky and Gresshoff 1995; Vahedian et al. 1995). Later, SB92 and its similar sequence SB91 were both defined as the centromere sequences of the GG genome by cytological features (Gill et al. 2009) and thus designated CentGm-1 (SB92) and CentGm-2 (SB91) (Tek et al. 2010). In addition, FISH revealed CentGm-1 or CentGm-2 present in about a half of all *G. max* chromosomes, which suggests allopolyploidy in the paleopolyploid genome (Gill et al. 2009). The *G. max* karyotyping system in mitotic metaphase was recently studied with a probe cocktail of BAC clones and partial centromere oligonucleotides (Findley et al. 2010), and this system has enabled the characterization of most of the chromosomal translocation lines (Findley et al. 2011).

The rDNAs have been widely used as reliable landmarks in chromosome karyotyping of several higher plants. In higher eukaryotes, the 45S and 5S rDNA loci are transcribed by different RNA polymerases and usually located in different positions on chromosomes (Srivastava and Schlessinger 1991). The number and location on chromosomes for both 45S and 5S may be diverse within related species (Chang et al. 2009; Chung et al. 2008; Pedrosa-Harand et al. 2006). In other words, karyotyping with rDNA loci may reflect the relationships of related species. Most of the recent cytological research on *Glycine* species has focused on the number of rDNA loci. The genome from diploid *Glycine* species, known as the paleopolyploidy, was predicted to consist of two 45S rDNA loci because it was ancient allopolyploidy; however, only one locus was observed on diploid *Glycine* species (Krishnan et al. 2001; Singh et al. 2001). Therefore, one 45S rDNA locus disappeared during soybean genome diploidization, whereby the ancient tetraploid became a new diploid paleopolyploidy.

Some studies revealed two 5S rDNA and two 45S rDNA loci in most allotetraploid genomes (Krishnan et al. 2001; Singh et al. 2001); however, an extra minor 45S rDNA locus was found in only two allotetraploid species (Krishnan et al. 2001). Therefore, the number of 45S loci among *Glycine* species or accessions was not as simple as the model of "diploid-doubled to tetraploid" of one locus for diploid and two loci for tetraploid species (Krishnan et al. 2001; Singh et al. 2001).

In the current study, we investigated the difference in repeat sequences between soybean and its wild related species. Here, we report three new soybean repeat sequences—SBRS1, SBRS2 and SBRS3—isolated from the soybean WSG data. In light of their distribution on chromosomes of GG genome, the three repeat sequences may be considered as FISH karyotyping markers for GG genome. We also labeled the three sequences in some wild *Glycine* species and found that they widely co-localized with 45S rDNA.

Methods

Plant materials

Glycine species used and locations where they were collected are in Table 1 along with their genome types and known USDA permanent plant introduction (PI) numbers or IL number designated by Dr. T. Hymowitz. All plant materials were previously described in detail (Hsieh et al. 2001). Cultivated soybean Shi-shi was kindly provided by Kaohsiung District Agricultural Research and Extension Station. *Glycine* species distributed in Taiwan and nearby islands were collected by Drs. J. S. Hsieh and Y. C. Huang, Department of Agronomy, National Taiwan University. Three diploid *Glycine tomentella* accessions originally from Papua New Guinea and Australia were kindly provided by Dr. T. Hymowitz, Department of Crop Science, University of Illinois, USA. These plants were grown in the greenhouse at Academia Sinica, Taiwan. Root tips of seedling were used for metaphase chromosome preparation.

Flow cytometry for estimating ploidy

Flow cytometric estimation of nuclear DNA content involved the MoFlo XDP Cell Sorter (Beckman Coulter, Fullerton, CA, USA), with propidium iodide (PI) as the fluorescent stain. A total of 20 mg of young leaves of each plant and standard nuclear fractions were prepared and mixed. Chicken erythrocyte nuclei (CEN; BioSure Inc., Grass Valley, CA, USA) and rainbow trout erythrocyte nuclei (TEN; NPE Systems Inc, Pembroke Pines, FL, USA) were used as internal standards and the DNA index (Vindelov et al. 1983). Leaf samples were chopped by using of a razor in a Petri dish containing 0.5 ml extraction buffer (Partec CyStain PI Absolute P Nuclei Extraction Buffer; Partec GMBH, Munster, Germany). The resulting extract was passed through two 30-μm filters before the addition of 0.5 ml Partec CyStain PI Absolute P Staining Buffer. Samples were kept in the dark for up to 30 min before flow cytometry. Six thousand cells were counted in each sample, with three replicates. The predicted genome size was calculated by assuming that 1 pg DNA = 965 Mbp (Bennett et al. 2000).

Predictions of repeat sequences in the soybean genome

Candidate repeat sequences were identified on the basis of the soybean WGS data (Schmutz et al. 2010). "eTandem" in EMBOSS 6.0 was used to scan all scaffold sequences, and the outputs were filtered by shell scripts and AWK script in Linux with repeat unit length >70 bp, homologous identity >70 % and copies >30. The longest

Table 1 The accessions of the *Glycine species* used in this study, with their nuclear DNA content and collection locations

Species	Genome types	Accession numbers		2C DNA content		Collection location
		TW[a]	IL or PI[c]	PG[b]	Mbp[b]	
G. max	GG	Shi-shi	PI416967	2.44 ± 0.01	2360	Cultivar
G. soja	GG	Soja001	PI393551	2.00 ± 0.02	1930	Shimen, Taoyuan, Taiwan
G. syndetika	A_6A_6	Tom052	PI441000	2.58 ± 0.00	2490	North Queensland, Australia
G. tomentella	DD	Tom062	PI446993	2.11 ± 0.01	2040	Papua New Guinea
G. tomentella	D_1D_1	Tom051	PI505301	2.24 ± 0.10	2170	West Australia
G. pescadrensis	$A_6A_6B_3B_3$	Tab004	IL875	4.52 ± 0.02	4370	Houliao, Penghu, Taiwan
G. pescadrensis	$A_6A_6B_3B_3$	Tab074	–	4.90 ± 0.03	4730	Kingmen, Taiwan
G. dolichocarpa	A_6A_6DD	Tom039	PI320548	5.17 ± 0.05	4990	Tungho, Taitung, Taiwan
G. tomentella	DDD_1D_1	Tom034	IL872	5.18 ± 0.21	5000	Daguang, Pintung, Taiwan

[a] Accession number in our laboratory

[b] 1 pg DNA = 965 Mbp (Bennett et al. 2000), N = 3

[c] Illinois (IL) numbers or USDA Plant introduction (PI) numbers

repeats were used for primer design (Additional file 1: Table S1).

Syntheses of probes for FISH analysis

To prepare probes for FISH analysis, the pTA71 plasmid that contains a 9.1-kb fragment of 18S-5.8S-26S rDNA from common wheat (Gerlach and Bedbrook 1979) was labeled with biotin-16-dUTP by nick translation (Roche Diagnostics, Penzberg, Germany). Other repeat sequences were subcloned into pGEM-T Easy vector (Promega) by using Shi-shi genomic DNA and were labeled with digoxigenin-11-dUTP by using the Roche PCR DIG Probe synthesis kit (Roche).

Southern blot analysis

Genomic DNA (30 μg) of each *Glycine* accession was completely digested with restriction enzymes (*Eco*RI or *Nde*I, New England Biolabs) before separation by electrophoresis overnight on a 0.8 % agarose gel in 0.5X TBE buffer, then blotting onto Hybond-N+ nylon membranes (Amersham Pharmacia Biotech). Probe preparation, membrane hybridization, and signal detection followed the instructions of the ECL direct nucleic acid labeling and detection system (Amersham Pharmacia Biotech).

Chromosome preparation and FISH analysis

The terminal 0.5 cm of young roots was collected from individual seedlings, pretreated with 2 mM 8-hydroxyquinoline for 2 h at room temperature and fixed overnight in Farmer's fixative [95 % ethanol + glacial acetic acid (3:1 v/v)]. The fixed root tips were washed with distilled water before incubation in a cell wall digestive enzyme mixture of 6 % pectinase (Sigma Chemical, St. Louis, MO, USA) and 6 % cellulase (Onozuka R-10; Yakult Honsha, Tokyo, Japan) in 75 mM KCl (pH = 4.0) buffer at 37 °C for 1.5–3 h depending on the characteristics of each accession.

After a wash with distilled water, the terminal 1 mm of softened tissues was squashed on a slide (Chang et al. 2009). Samples were pretreated with pepsin (1.25 mg/ml for *G. soja*, 5 mg/ml for other accessions, in 10 mM HCl) at 37 °C for 3 to 8 h depending on the characteristics of each accession. FISH analysis was performed as described previously (Chung et al. 2008).

Results

Estimation of genome size

The DNA contents of the *Glycine* species were analyzed by flow cytometry (Table 1). The estimated 2C DNA contents of the diploid species, cultivated soybean Shi-shi (*Glycine max*, GG genome), Soja001 (*Glycine soja*, GG genome), Tom052 (*Glycine syndetike*, A_6A_6 genome), Tom062 (*Glycine tomentella*, DD genome), and Tom051 (*Glycine tomentella*, D_1D_1 genome), were 2360, 1930, 2500, 2040 and 2170 Mb, respectively, and contents varied between different species by 29 %. The estimated 2C DNA contents of the tetraploid species, Tab 004 (*G. pescadrensis*, $A_6A_6B_3B_3$ genome), Tab 074 (*G. pescadrensis*, $A_6A_6B_3B_3$ genome), Tom039 (*Glycine dolichocarpa*, A_6A_6DD genome) and Tom034 (*Glycine tomentella*, DDD_1D_1 genome), were 4370, 4730, 4990 and 5000 Mb, respectively, and differed between different species or accessions by 15 %. The DNA for two *G. pescadrensis* accessions, Tab 004 and Tab 074, was collected from Penghu and Kingmen islands, respectively and the content differed by 9 %. When compare the tetraploid with its two diploid parental species, the DNA content of *G. dolichocarpa* (Tom039) is higher than the total DNA content from *G. syndetike* (Tom052) and *G. tomentella* (Tom062), the DNA content of tetraploid *G. tomentella* (Tom034) is higher than the total DNA content from two diploid *G. tomentella* (Tom051 and Tom062).

Repeat sequences from WGS data and the analysis

To investigate the repeat sequence component in the *Glycine* genome, we used the WGS data to find new repeat sequences. We identified 17 candidate repeat sequences by bioinformatics analysis. Several of these candidates belong to the LTR120 family (Morgante et al. 1997) and SB92 family (Kolchinsky and Gresshoff 1995). Finally, we chose 3 new repeat sequences, designated soybean repeat sequence1 (SBRS1), SBRS2 and SBRS3; their sequences and designed primers are in Additional files 1 and 2.

SBRS1

A 48 bp minisatellite sequence, represents approximately 131,000 copies in the WGS data; 123,000 copies were aligned into 13 pseudomolecules, including Gm1, 2, 3, 5, 11, 12, 14, 15, 16, 17, 18, 19 and 20. The remaining copies were aligned into contigs. SBRS1 is a head-to-tail tandem repeat with tens to hundreds of copies in each region.

SBRS2

A 124 bp minisatellite sequence, represents approximately 6000 copies in the WGS data. About 95 % of the SBRS2 repeats were aligned into pseudomolecules. The SBRS2 is a head-to-tail tandem repeat with usually less than 10 copies in each continguous region and is dispersed on all of the 20 pseudomolecules in the WSG data. The most significant region rich in SBRS2 was located at Gm9, where 1095 copies of the SBRS2 clustered in a 270-kb region.

SBRS3

A 201 bp minisatellite sequence, represents 8206 copies in the WGS data; 8016 copies were aligned into pseudomolecules. It is also a tandem repeat, with usually less than 10 copies in each region, dispersed on all 20 pseudomolecules. Three regions especially contain a high density of SBRS3 repeats, including 2 on Gm14 (1512 copies were separated into 2 groups 26,000 kb apart) and the other one on Gm20 (400 copies clustered together).

To understand the existence and the distribution of these three repeat sequences in the *Glycine* genome, we used them as probes in Southern blot analyses with genomic DNA of the cultivated soybean and the collected *Glycine* species. All three repeats, SBRS1, SBRS2 and SBRS3, showed heavy ladder signals in the GG genome species, *G. max* and *G. soja* (Lanes 1–4 of Fig. 1; Additional file 3: Figure S1, Additional file 4: Figure S2). These heavy ladder signals are similar between *G. max* and *G. soja*, but the patterns among the three repeat sequences are different. However, the 3 repeat sequences showed single band and the similar size in other genome species, including *G. tomentella*, *G. dolichocarpa* and *G. pescadrensis*, a 3.8- and a 9-kbp band were found in *Eco*RI- and *Nde*I-digested DNA samples, respectively, of all perennial *Glycine* species tested (Lanes 5–12 of Fig. 1. Additional file 3: Figure S1, Additional file 4: Figure S2). Therefore, the three repeat sequences exist in each *Glycine* species tested, with high copies in GG genome, but with very low copies in other species.

FISH analysis with SBRS1, SBRS2, SBRS3 and 45S rDNA

FISH analysis was used in further evaluated the distribution of the three repeat sequences in the genome of *Glycine* species. We used the root tips of *G. soja* to represent the GG genome on FISH analysis because the cultivated soybean and *G. soja* both represent the GG genome, and band patterns on Southern blot analysis were similar (Fig. 1, Additional file 3: Figure S1, Additional file 4: Figure S2). The 45S rDNA probe was used as a control, and it was located at the distal ends of one pair of chromosomes of *G. soja* (Fig. 2a, green).

SBRS1 signals were dispersed on every chromosome of *G. soja* as shown by metaphase (Panel A) and interphase (Panel B) chromosomes (Fig. 2a, red). Throughout the genome, the distribution of SBRS1 signals varied among chromosomes. SBRS2 signals were detected as clusters on two pairs of chromosomes of *G. soja* (Fig. 3). FISH signals for one of the SBRS2 clusters overlapped with 45S rDNA at the distal ends of one pair of chromosomes (Fig. 3a, yellowish green), and the second SBRS2 cluster was detected in the proximal regions of another pair of chromosomes (Fig. 3a, b, red). SBRS3 signals were detected on three pairs of chromosomes of *G. soja* (Fig. 4, red), and none overlappped with 45S rDNA (Fig. 4, green). One of the SBRS3 signals was detected at the subdistal ends of one pair of chromosomes, and the other 2 were detected in the proximal regions and the distal ends of another pair of chromosomes.

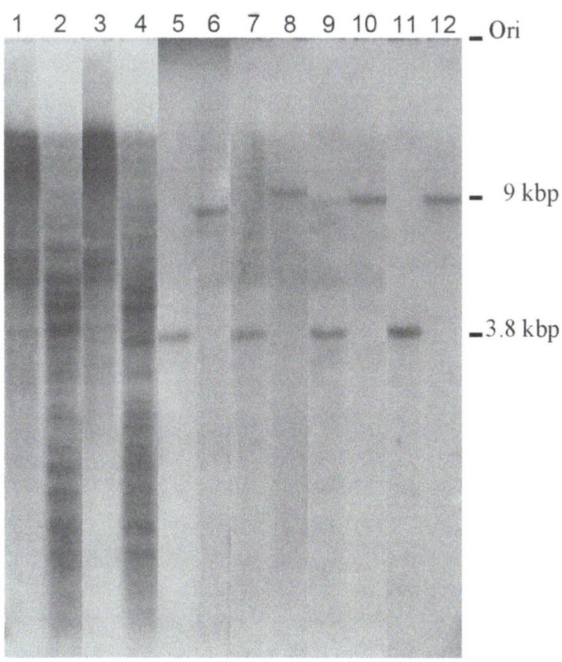

Fig. 1 Southern blot analysis of the SBRS1 on *Glycine* species

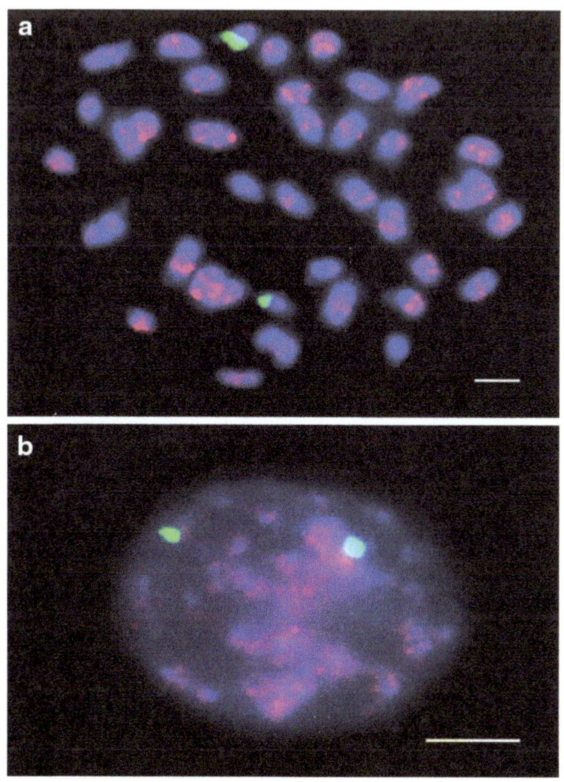

Fig. 2 FISH analysis of SBRS1 and 45S rDNA on *G. soja* chromosomes

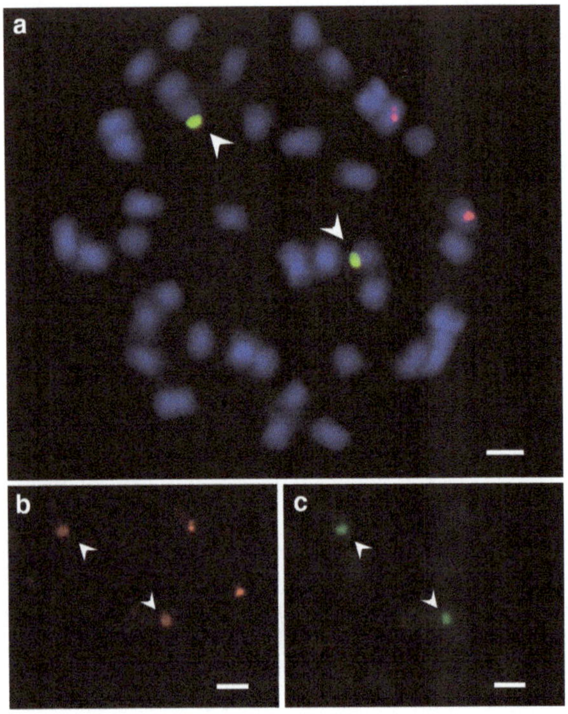

Fig. 3 FISH analysis of SBRS2 and 45S rDNA on *G. soja* chromosomes

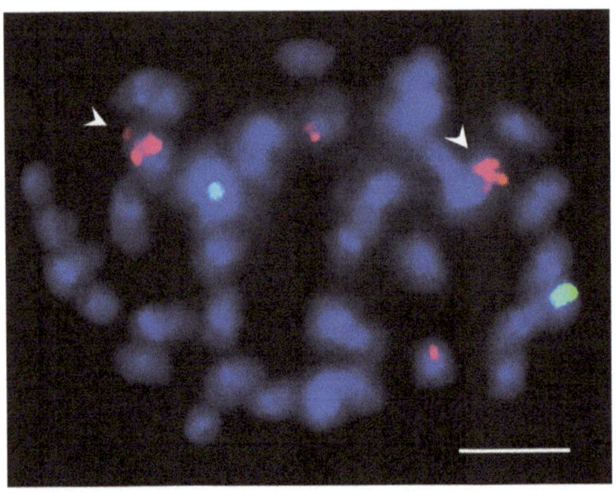

Fig. 4 Localization of SBRS3 and 45S rDNA on *G. soja* mitotic chromosomes

We labeled probes of the three repeat sequences and 45S rDNA in allotetraploid *Glycine* species (i.e., Tom034, Tom039 and Tab 004; Fig. 5), and their diploid relatives (i.e., Tom052, Tom062 and Tom051; Additional file 5: Figure S3; Additional file 6: Figure S4, Additional file 7: Figure S5, Additional file 8: Figure S6). Most of the three tandem repeat sequences signals co-localized with 45S rDNA signals among the tetraploid and diploid relatives tested (Figs. 3, 5; Additional file 5: Figure S3, Additional file 6: Figure S4, Additional file 8: Figure S6). SBRS1 and SBRS3 signals did not co-localize with 45S rDNA signals in the 2 diploid genomes, GG genome (Figs. 2, 4) or DD genome (Additional file 7: Figure S5, Additional file 8: Figure S6). In the genome of Tom062 (DD genome), very weak SBRS1 signals located on many chromosomes (Additional file 7: Figure S5), SBRS2 signals co-localized with 45S rDNA (Additional file 8: Figure S6), and no substantial SBRS3 signal was detected. In addition, all tetraploid accessions showed two 45S rDNA loci, with a minor signal detected in the proximal regions of another pair of chromosomes of Tab 074 (Fig. 6). FISH analysis results are summarized in Table 2.

FISH analysis with SSR probe in *Glycine* species

We additionally used 5 types of SSR probes (i.e., ATT, AT, CAA, CT and CTT) in FISH analysis. However, only one SSR probe, with 33 copies of the repeat ATT, on *G. soja*, gave interesting results, and, similar to SBRS1, ATT signals were detected as dispersed signals on every chromosomes of the GG genome (Fig. 7, red). There was no ATT signal on the mitotic chromosomes of most other *Glycine* species except Tom034. Few very weak ATT probe signals were detected on this accession, but the signals were unstable. In addition, the signal distribution for these SSRs varied among chromosomes.

Discussion

The predicted repeat sequences were confirmed by cytological study

The soybean WGS sequence project has set significantly new milestones for understanding the soybean genome and its evolutionary history (Jackson et al. 2006; Schmutz et al. 2010). In the current study, we identified three new repeat sequences from soybean WGS data. Using bioinformatics, we propose their characterization in the soybean genome. We used WGS as a resource in chromosome marker design for the cultivated soybean and other *Glycine* species, and found that use of the WGS data was powerful for searching for candidates of repeat sequences and designing probes for cytological research.

For the GG genome, the WGS data could be used to predict general copy numbers of the repeat sequences in each cluster and the location of repeat-sequence condensed regions. FISH analyses confirmed these bioinformatics findings (Figs. 2, 3, 4; Table 2). WGS data may be also used in cytological research for the wild related species. Combining bioinformatics and cytological strategies may help bridge the gap between cultivated species and related species. In the current study, we analysis the three new repeat sequences on many *Glycine* species by FISH. Similar studies had shown that FISH analysis would reveal heterochromatin

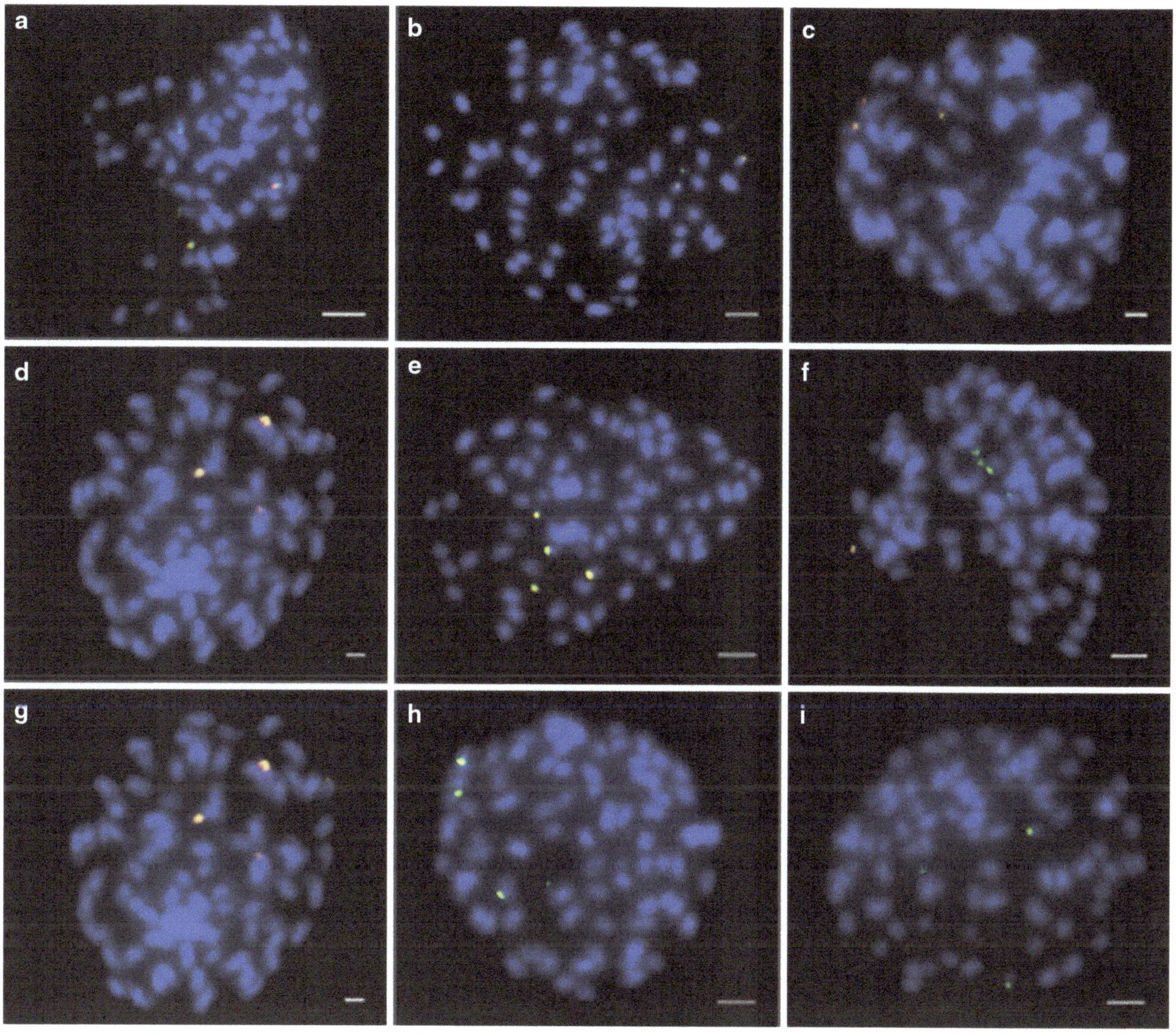

Fig. 5 Co-localization of the 3 repeat sequences and 45S rDNA on chromosomes of *Glycine* species

organization using the centromeric or satellite repeats (Ananiev et al. 1998 for maize; Badaeva et al. 1996 for *Aegilops* species; Kulikova et al. 2004 for *Medicago* species). Because *Glycine* species consist of high chromosome numbers (38, 40, 78, or 80; Hymowitz et al. 1998), as well, these chromosomes are compact and small, the utilization of repetitive sequences as probe for FISH analysis may help on the cytogenesis and breeding analysis in the post-genomics era.

Possible evolutionary process for the three new repeat sequences among *Glycine* species

Differences in the distribution of repetitive DNA sequences were shown between *Glycine* species in the current study (summarized in Table 2). The chromosomal localization of these DNA probes was different. The possible association of the these repeat sequences and rDNA loci may be due to (1) the three repeat sequences having existed in the rDNA of the genome of their common earlier ancestor, (2) all having expanded and dispersed in GG genome, and finally (3) SBRS1 and SBRS3 being reduced or disappeared in 45S rDNA loci during the GG (Soja001) and DD (Tom062) genome evolution and SBRS2 still existing in the 45S rDNA of all tested *Glycine* species (Table 2). Dolye and his colleague used global, gene-specific, and relaxed clock methods to date the polyploid history of soybean and suggested the date for the *G. max/G. tomentella* split averaged between 3.8 and 6.86 Mya across methods (Egan and Doyle 2010). It coincides well with the fact that the shared repeat sequences, but with different amount of such fragments, among the *Glycine* species.

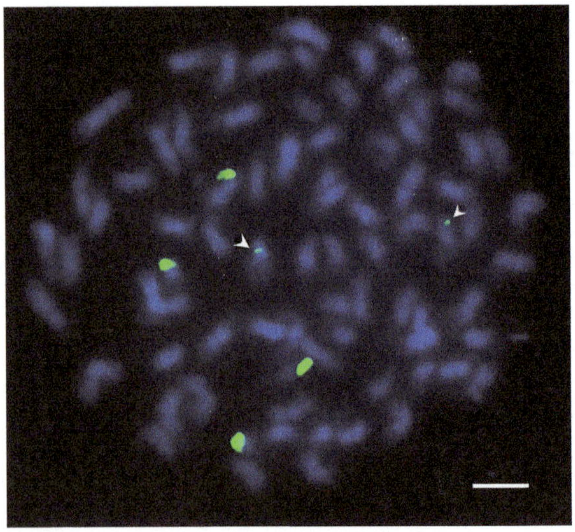

Fig. 6 FISH analysis of 45S rDNA on mitotic chromosomes of Tab074

Many satellite sequences are homologs to the IGS subrepeats of rDNA from other plants genome, such as legumes (Falquet et al. 1997), potato (Stupar et al. 2002), tobacco (Lim et al. 2004; Volkov et al. 1999) and tomato (Jo et al. 2009). In tomato study, the insertion of a retrotransposon into the rDNA gene enhanced the amplification of the nearby IGS subrepeats. Those amplified subrepeats then dispersed into other loci and became repeat sequences (Jo et al. 2009). Other studies also proposed that some transposon elements, such as *En*/*Spm*-like transposons, could be involved in rDNA movement and as one of the key steps when the rDNA subrepeat turned into repeat sequences (Raskina et al. 2004a, 2008; Schubert and Wobus 1985).

Diversity of the genome in *Glycine* species

We analyzed the DNA content of the nine *Glycine* accessions, five of them are diploid and another four are allotetraploid, by flow cytometry. The two *G. pescadrensis* accessions Tab 004 and Tab 074 collected from Penghu and Kingmen, differed in DNA content (Table 1). In addition, the DNA content of Soja001 and the cultivated soybean (GG genome) differs up to 22 % (Table 1). The DNA content of *G. dolichocarpa* (Tom039) and *G. tomentella* (Tom034) are more than the sum of their two parental diploid species (Table 1).

Research using *Helianthus* (Ungerer et al. 2006) suggested that retrotransposons may play an important role in the genome expansion process of allotetraploid evolution. The dynamics of transposon element driven genome evolution have been reported for several plant genomes (El Baidouri and Panaud 2013). In fact, retrotransposons occupy the largest part of the cultivated soybean genome (Schmutz et al. 2010) and may play a role in *Glycine* genome size variations.

The number of 45S rDNA loci among wild soybeans is dynamic. In previous studies, the number of 45S rDNA loci among the tetraploid *Glycine* species was suggested to be the same (Krishnan et al. 2001; Singh et al. 2001). However, two major and one minor loci of 45S rDNA signals were detected in Tab 074 (*G. pescadrensis*, $A_6A_6B_3B_3$ genome). The minor signals are found on the middle area of a pair of the chromosomes. Only two major loci of 45S rDNA were found for the other tetraploids, including Tab 004. These two *G. pescadrensis* accessions were collected from Penghu and Kingmen, and the distance between these two islands is about 140 km, the geographical isolation may explain the diversity of 45S rDNA loci. Thus,

Table 2 Summary of FISH analysis for soybean repeat sequence1 (SBRS1), SBRS2, and SBRS3, 45S rRNA and general ATT repeat

Accessions	Genome	SBRS1	SBRS2	SBRS3	45S	ATT
Soja001	GG	All	2 and 1[a]	3	1	All[b]
Tom052	A_6A_6	1[a]	1[a]	1[a]	1	ND
Tom062	DD	Many[c]	Many[a,d]	ND[d]	1	ND
Tom051	D_1D_1	1[a]	1[a]	1[a]	1	ND
Tab004	$A_6A_6B_3B_3$	2[a]	2[a]	2[a]	2	ND
Tab074	$A_6A_6B_3B_3$	2[a]	2[a]	2[a]	3[b]	ND
Tom039	A_6A_6DD	2[a]	2[a]	2[a]	2	ND
Tom034	DD_1D_1	2[a]	2[a]	2[a]	2	2–3

One locus has 2 signals on a set of chromosomes, and 2 loci have 4 signals

ND No significant signal detected on metaphase chromosome

[a] Signals co-localized with 45S rDNA

[b] 2 major and 1 minor signal

[c] Very weak signals

[d] Interphase data were used

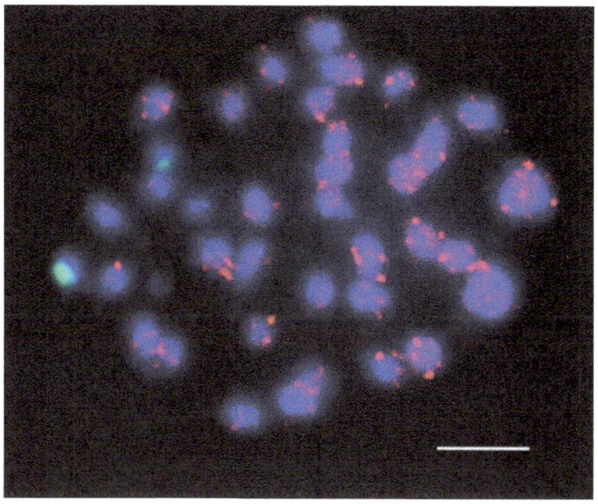
Fig. 7 Co-localization of the ATT probe and 45S rDNA on *G. soja* chromosomes

the evolution of the 45S rDNA in the *Glycine* genome is more complex than the model of "diploid-doubled to tetraploid".

Variations in the 45S rDNA loci among natural hybrids and related species were reported for *Lycoris* (Chang et al. 2009), *Oryza* (Chung et al. 2008), *Triticeae* (Kim et al. 1993), and *Arabidopsis* (Maluszynska and Heslopharrison 1993). As well, the ability to alter the number and location of rDNA loci among closely related species or even accessions was noted (Chung et al. 2008; Raskina et al. 2004b). In our study, the variation in number of rDNA loci in *Glycine* was more dramatic than in previous studies (Krishnan et al. 2001; Singh et al. 2001). Such differences in number or location of 45S rDNA may be used as chromosome markers.

Genome and chromosome markers of *Glycine* species

Specific repeat sequences, such as SBRS1-3, and nonspecific repeat sequences, such as SSR and rDNA, may be potential genome or chromosome markers. The nonspecific repeat sequences can also be used as specific markers, if the distributions of repeats are different enough among genomes. For instance, the SSRs, which are widely identified among species (Varshney et al. 2005), have been used to label the chromosomes of different genomes in hexaploidy *Triticum aestivum* (Cuadrado et al. 2000, 2008). We use nonspecific repeat sequences as probes for labeling chromosomes of several *Glycine* species. Signals of ATT were exclusively detected on the GG genome, with no difference between *G. soja* and *G. max* (data not shown). Although 2–3, weak ATT signals could be detected on the chromosome of Tom034 (data not shown), they were easy to be separated from the strong ones of GG genome.

The study of the *Glycine* genome is important for soybean improvement; combing bioinformatics and cytology techniques can advance investigations. Our study is just a beginning to show that the soybean WSG data and cytological methods can be used to find new repeat sequences, reveal the genome evolution process, and provide addition information in investigation of the phylogeny in the *Glycine* species.

Additional files

Additional file 1. Table S1. Sequence of the primers used in the study.

Additional file 2. Table S2. Sequence of repeat sequences and probes used in the study.

Additional file 3. Figure S1. Southern blot analysis of SBRS2 on *Glycine* species.

Additional file 4. Figure S2. Southern blot analysis of SBRS3 on *Glycine* species.

Additional file 5. Figure S3. FISH analysis of SBRS and 45S rDNA on Tom051 chromosomes.

Additional file 6. Figure S4. FISH analysis of SBRS and 45S rDNA by FISH on Tom052 chromosomes.

Additional file 7. Figure S5. FISH analysis of SBRS1 and 45S rDNA on Tom062 chromosomes.

Additional file 8. Figure S6. FISH analysis of SBRS2 and 45S rDNA on Tom062 chromosomes.

Authors' contributions

HC and MCC performed the FISH analysis, YCT and JSH collected the Glycine species and performed the genome size analysis, FJW performed the bioinformatics analysis, YIC wrote the manuscript. All authors read and approved the final manuscript.

Author details

[1] Institute of Plant and Microbial Biology, Academia Sinica, Taipei 115, Taiwan.
[2] Department of Agronomy, National Taiwan University, Taipei 106, Taiwan.

Acknowledgements

We deeply thank the Kaohsiung District Agricultural Research and Extension Station for the Shi-shi seeds, Professor Hymowitz at University of Illinois for the three diploid *G. tomentella* seeds, and the Plant Cell Biology Core Lab in the Institute of Plant and Microbial Biology, Academia Sinica, for flow cytometry analysis. We also appreciate the technical support from Ms. Yu-Chu Chang and English editing from Ms. Laura Smales (BioMedEditing, Toronto, Canada). This research was supported by the National Science Council and Academia Sinica to JSH and YICH.

Competing interests

The authors declare that they have no competing interests.

References

Ananiev EV, Phillips RL, Rines HW (1998) Chromosome-specific molecular organization of maize (*Zea mays* L.) centromeric regions. Proc Nat Acad Sci USA 95:13073–13078

Badaeva ED, Friebe B, Gill BS (1996) Genome differentiation in *Aegilops*. 1. Distribution of highly repetitive DNA sequences on chromosomes of diploid species. Genome 39:293–306

Bennett, MD, Bhandol, P, Leitch IJ (2000) Nuclear DNA amounts in Angiosperms and their modern uses new estimates. Ann Bot 86:859–909
Chang Y-C, Shii C-T, Chung M-C (2009) Variations in ribosomal RNA gene loci in spider lily (*Lycoris* spp.). J Amer Soc Hort Sci 134:567–573
Cheng YY, Fang SA, Lin YC, Chung MC (2007) A repetitive sequence specific to *Oryza* species with BB genome and abundant in *Oryza punctata* Kotschy ex Steud. Bot Stud 48:263–272
Chung MC, Lee YI, Cheng YY, Chou YJ, Lu CF (2008) Chromosomal polymorphism of ribosomal genes in the genus *Oryza*. Theor Appl Genet 116:745–753
Concibido VC, La Vallee B, Mclaird P, Pineda N, Meyer J, Hummel L, Yang J, Wu K, Delannay X (2003) Introgression of a quantitative trait locus for yield from *Glycine soja* into commercial soybean cultivars. Theor Appl Genet 106:575–582
Cuadrado A, Jouve N (2007) Similarities in the chromosomal distribution of AG and AC repeats within and between *Drosophila*, human and barley chromosomes. Cytogenet Genome Res 119:91–99
Cuadrado A, Schwarzacher T, Jouve N (2000) Identification of different chromatin classes in wheat using in situ hybridization with simple sequence repeat oligonucleotides. Theor Appl Genet 101:711–717
Cuadrado A, Cardoso M, Jouve N (2008) Physical organisation of simple sequence repeats (SSRs) in *Triticeae*: structural, functional and evolutionary implications. Cytogenet Genome Res 120:210–219
Doyle MJ, Grant JE, Brown AHD (1986) Reproductive isolation between isozyme groups of *Glycine tomentella* (Leguminosae), and spontaneous doubling of their hybrids. Aust J Bot 34:523–535
Doyle JJ, Doyle JL, Brown AH, Grace J (1990a) Multiple origins of polyploids in the *Glycine tabacina* complex inferred from chloroplast DNA polymorphism. Proc Nat Acad Sci USA 87:714–717
Doyle JJ, Doyle JL, Brown AHD (1990b) A chloroplast-DNA phylogeny of the wild perennial relatives of soybean (*Glycine* subgenus *Glycine*): congruence with morphological and crossing groups. Evolution 44:371–389
Doyle JJ, Doyle JL, Brown AHD (1990c) Chloroplast DNA polymorphism and phylogeny in the B genome of *Glycine* subgenus *Glycine* (Leguminosae). Amer J Bot 77:772–782
Doyle JJ, Doyle JL, Rauscher JT, Brown AHD (2004) Evolution of the perennial soybean polyploid complex (*Glycine* subgenus *Glycine*): a study of contrasts. Biol J Linn Soc 82:583–597
Egan AN, Doyle J (2010) A comparison of global, gene-specific, and relaxed clock methods in a comparative genomics framework: dating the polyploid history of soybean (*Glycine max*). Syst Biol 59:534–547
El Baidouri M, Panaud O (2013) Comparative genomic paleontology across plant kingdom reveals the dynamics of TE-driven genome evolution. Gen Biol Evol 5:954–965
Falquet J, Creusot F, Dron M (1997) Molecular analysis of *Phaseolus vulgaris* rDNA unit and characterization of a satellite DNA homologous to IGS subrepeats. Plant Physiol Bioch 35:611–622
Findley SD, Cannon S, Varala K, Du J, Ma J, Hudson ME, Birchler J, Stacey G (2010) A fluorescence in situ hybridization system for karyotyping soybean. Genetics 185:727–744
Findley SD, Pappas AL, Cui Y, Birchler JA, Palmer RG, Stacey G (2011) Fluorescence in situ hybridization—based karyotyping of soybean translocation lines. G3 (Genes/Genomes/Genetics 1:117–129
Gerlach WL, Bedbrook JR (1979) Cloning and characterization of ribosomal RNA genes from wheat and barley. Nucl Ac Res 7:1869–1885
Gill N, Findley S, Walling JG, Hans C, Ma J, Doyle J, Stacey G, Jackson SA (2009) Molecular and chromosomal evidence for allopolyploidy in soybean. Plant Physiol 151:1167–1174
Grant JE, Brown AHD, Grace JP (1984) Cytological and isozyme diversity in *Glycine tomentella* Hayata (Leguminosae). Aust J Bot 32:665–677
Hsieh JS, Hsieh KL, Tsai YC, Hsing YI (2001) Each species of *Glycine* collected in Taiwan has a unique seed protein pattern. Euphytica 118:67–73
Hsing YIC, Hsieh KL, Huang YC, Hsieh JS (1995) The relationships of cultivated soybeans and their wild relatives collected from Taiwan—revealed by seed proteins. Bot Bull Acad Sinica 36:65–72
Hsing YLC, Hsieh JS, Peng CI, Chou CH, Chiang TY (2001) Systematic status of the *Glycine tomentella* and *G. tabacina* species complexes (Fabaceae) based on ITS sequences of nuclear ribosomal DNA. J Plant Res 114:435–442
Hymowitz T, Singh RJ, Kollipara KP (1998) The genomes of the *Glycine*. Plant Breed Rev 16:289–317
Hyten DL, Song Q, Zhu Y, Choi I-Y, Nelson RL, Costa JM, Specht JE, Shoemaker RC, Cregan PB (2006) Impacts of genetic bottlenecks on soybean genome diversity. Proc Nat Acad Sci USA 103:16666–16671
Jackson SA, Rokhsar D, Stacey G, Shoemaker RC, Schmutz J, Grimwood J (2006) Toward a reference sequencing of the soybean genome: a multiagency effort. Crop Sci 46:S55–S61
Jiang J, Gill BS (2006) Current status and the future of fluorescence in situ hybridization (FISH) in plant genome research. Genome 49:1057–1068
Jo S-H, Koo D-H, Kim J, Hur C-G, Lee S, Yang T-j, Kwon S-Y, Choi D (2009) Evolution of ribosomal DNA-derived satellite repeat in tomato genome. BMC Plant Biol 9:42
Katayose Y, Kanamori H, Shimomura M, Ohyanagi H, Ikawa H, Minami H, Shibata M, Ito T, Kurita K, Ito K, Tsubokura Y, Kaga A, Wu JZ, Matsumoto T, Harada K, Sasaki T (2012) DaizuBase, an integrated soybean genome database including BAC-based physical maps. Breed Sci 61:661–664
Keim P, Diers BW, Olson TC, Shoemaker RC (1990) RFLP mapping in soybean: association between marker loci and variation in quantitative traits. Genetics 126:735–742
Kim NS, Kuspira J, Armstrong K, Bhambhani R (1993) Genetic and cytogenetic analyses of the genome of *Triticum-Monococcum*.VIII. Localization of rDNAs and characterization of 5 s ribosomal-DNA genes. Genome 36:77–86
Kim MY, Lee S, Van K, Kim TH, Jeong SC, Choi IY, Kim DS, Lee YS, Park D, Ma J, Kim WY, Kim BC, Park S, Lee KA, Kim DH, Kim KH, Shin JH, Jang YE, Do Kim K, Liu WX, Chaisan T, Kang YJ, Lee YH, Kim KH, Moon JK, Schmutz J, Jackson SA, Bhak J, Lee SH (2010) Whole-genome sequencing and intensive analysis of the undomesticated soybean (*Glycine soja* Sieb. and Zucc.) genome. Proc Nat Acad Sci USA 107:22032–22037
Kolchinsky A, Gresshoff PM (1995) A major satellite DNA of soybean is a 92-base-pairs tandem repeat. Theor Appl Genet 90:621–626
Kollipara KP, Singh RJ, Hymowitz T (1997) Phylogenetic and genomic relationships in the genus *Glycine* Willd. based on sequences from the ITS region of nuclear rDNA. Genome 40:57–68
Kopecky D, Lukaszewski AJ, Dolezel J (2008) Cytogenetics of festulolium (*Festuca* X *Lolium hybrids*). Cytogenet Genome Res 120:370–383
Krishnan P, Sapra VT, Soliman KM, Zipf A (2001) FISH mapping of the 5S and 18S-28S rDNA loci in different species of *Glycine*. J Hered 92:295–300
Kulikova O, Geurts R, Lamine M, Kim DJ, Cook DR, Leunissen J, de Jong H, Roe BA, Bisseling T (2004) Satellite repeats in the functional centromere and pericentromeric heterochromatin of *Medicago truncatula*. Chromosoma 113:276–283
Lam HM, Xu X, Liu X, Chen W, Yang G, Wong FL, Li MW, He W, Qin N, Wang B, Min J, Guo X, Murat F, Ham BK, Zhang Z, Gao S, Huang M, Xu Y, Zhong S, Bombarely A, Mueller LA, Zhao H, He H, Zhang Y, Zhang Z, Huang S, Tan T, Pang E, Lin K, Hu Q (2010) Resequencing of 31 wild and cultivated soybean genomes identifies patterns of genetic diversity and selection. Nat Genet 42:1053–1059
Lim KY, Skalicka K, Koukalova B, Volkov RA, Matyasek R, Hemleben V, Leitch AR, Kovarik A (2004) Dynamic changes in the distribution of a satellite homologous to intergenic 26-18S rDNA spacer in the evolution of *Nicotiana*. Genetics 166:1935–1946
Lin JY, Jacobus BH, SanMiguel P, Walling JG, Yuan Y, Shoemaker RC, Young ND, Jackson SA (2005) Pericentromeric regions of soybean (*Glycine max* L. Merr.) chromosomes consist of retroelements and tandemly repeated DNA and are structurally and evolutionarily labile. Genetics 170:1221–1230
Maluszynska J, Heslopharrison JS (1993) Molecular cytogenetics of the genus *Arabidopsis* : *in situ* localization of rDNA sites, chromosome numbers and diversity in centromeric heterochromatin. Annal Bot 71:479–484
Morgante M, Jurman I, Shi L, Zhu T, Keim P, Rafalski JA (1997) The STR120 satellite DNA of soybean: organization, evolution and chromosomal specificity. Chromosome Res 5:363–373
Mudge J, Cannon S, Kalo P, Oldroyd G, Roe B, Town C, Young N (2005) Highly syntenic regions in the genomes of soybean, *Medicago truncatula* and *Arabidopsis thaliana*. BMC Plant Bio 5:15
Newell CA, Hymowitz T (1975) *Glycine canescens* F. J. Herm. A wild relative of soybean Crop Sci 15:879–881
Newell CA, Hymowitz T (1983) Hybridization in the Genus *Glycine* Subgenus *Glycine* Willd (Leguminosae, Papilionoideae). Am J Bot 70:334–348
Pagel J, Walling JG, Young ND, Shoemaker RC, Jackson SA (2004) Segmental duplications within the *Glycine max* genome revealed by fluorescence in situ hybridization of bacterial artificial chromosomes. Genome 47:764–768

Palmer RG, Newhouse KE, Graybosch RA, Delannay X (1987) Chromosome structure of the wild soybean—Accessions from China and the Soviet Union of *Glycine soja* Sieb. & Zucc. J Hered 78:243–247

Pedrosa-Harand A, de Almeida CCS, Mosiolek M, Blair M, Schweizer D, Guerra M (2006) Extensive ribosomal DNA amplification during Andean common bean (*Phaseolus vulgaris* L.) evolution. Theor Appl Genet 112:924–933

Pinkel D, Straume T, Gray JW (1986) Cytogenetic analysis using quantitative, high-sensitivity, fluorescence hybridization. Proc Nat Acad Sci USA 83:2934–2938

Raskina O, Belyayev A, Nevo E (2004a) Activity of the *En/Spm*-like transposons in meiosis as a base for chromosome repatterning in a small, isolated, peripheral population of *Aegilops speltoides* Tausch. Chromosome Res 12:153–161

Raskina O, Belyayev A, Nevo E (2004b) Quantum speciation in *Aegilops*: Molecular cytogenetic evidence from rDNA cluster variability in natural populations. Proc Nat Acad Sci USA 101:14818–14823

Raskina O, Barber JC, Nevo E, Belyayev A (2008) Repetitive DNA and chromosomal rearrangements: speciation-related events in plant genomes. Cytogenet Genome Res 120:351–357

Ratnaparkhe MB, Singh RJ, Doyle JJ (2011) *Glycine*. In: Kole C (ed) Wild crop relatives: genomic and breeding resources: legume crops and forages. Springer, New York, pp 83–116

Schlueter J, Lin J-Y, Schlueter S, Vasylenko-Sanders I, Deshpande S, Yi J, O'Bleness M, Roe B, Nelson R, Scheffler B, Jackson S, Shoemaker R (2007) Gene duplication and paleopolyploidy in soybean and the implications for whole genome sequencing. BMC Genom 8:330

Schmutz J, Cannon SB, Schlueter J, Ma J, Mitros T, Nelson W, Hyten DL, Song Q, Thelen JJ, Cheng J, Xu D, Hellsten U, May GD, Yu Y, Sakurai T, Umezawa T, Bhattacharyya MK, Sandhu D, Valliyodan B, Lindquist E, Peto M, Grant D, Shu S, Goodstein D, Barry K, Futrell-Griggs M, Abernathy B, Du J, Tian Z, Zhu L, Gill N, Joshi T, Libault M, Sethuraman A, Zhang X-C, Shinozaki K, Nguyen HT, Wing RA, Cregan P, Specht J, Grimwood J, Rokhsar D, Stacey G, Shoemaker RC, Jackson SA (2010) Genome sequence of the palaeopolyploid soybean. Nature 463:178–183

Schubert I, Wobus U (1985) In situ hybridization confirms jumping nucleolus organizing regions in *Allium*. Chromosoma 92:143–148

Sebolt AM, Shoemaker RC, Diers BW (2000) Analysis of a quantitative trait locus allele from wild soybean that increases seed protein concentration in soybean. Crop Sci 40:1438–1444

Sherman-Brotles S, Bombarely A, Powell AF, Doyle JL, Egan AN, Coate JE, Doyle JF (2014) The wild side of a major crop—Soybean's perennial cousins from down under. Amer. J. Bot. 101:1651–1665

Shi L, Zhu T, Morgante M, Rafalski JA, Keim P (1996) Soybean chromosome painting: A strategy for somatic cytogenetics. J Hered 87:308–313

Shoemaker RC, Polzin K, Labate J, Specht J, Brummer EC, Olson T, Young N, Concibido V, Wilcox J, Tamulonis JP, Kochert G, Boerma HR (1996) Genome duplication in soybean (*Glycine* subgenus *soja*). Genetics 144:329–338

Singh RJ, Hymowitz T (1985) The genomic relationships among six wild perennial species of the genus *Glycine* subgenus *Glycine* Willd. Theor Appl Genet 71:211–230

Singh RJ, Hymowitz T (1987) Taxonomy and speciation. Wilcox, Soybeans: Improvements, Production, and Uses. Agronomy Monograph, pp 23–48

Singh RJ, Hymowitz T (1988) The genomic relationship between *Glycine max* (L.) Merr. and *G. soja* Sieb. and Zucc. as revealed by pachytene chromosome analysis. Theor Appl Genet 76:705–711

Singh RJ, Kim HH, Hymowitz T (2001) Distribution of rDNA loci in the genus *Glycine* Willd. Theor Appl Genet 103:212–218

Srivastava AK, Schlessinger D (1991) Structure and organization of ribosomal DNA. Biochimie 73:631–638

Stupar RM, Song JQ, Tek AL, Cheng ZK, Dong FG, Jiang JM (2002) Highly condensed potato pericentromeric heterochromatin contains rDNA-related tandem repeats. Genetics 162:1435–1444

Tateishi Y, Ohashi H (1992) Taxonomic studies on *Glycine* of Taiwan. J Jap Bot 67:127–147

Tek AL, Kashihara K, Murata M, Nagaki K (2010) Functional centromeres in soybean include two distinct tandem repeats and a retrotransposon. Chromosome Res 18:337–347

Thseng FS, Tsai SJ, Abe J, Wu ST (1999) *Glycine formosana* Hosokawa in Taiwan: pod morphology, allozyme, and DNA polymorphism. Bot Bull Acad Sin 40:251–257

Tindale M, Craven L (1988) Three new species of *Glycine* (Fabaceae: Phaseolae) from north-western Australia, with notes on amphicarpy in the genus. Aust Syst Bot 1:399–410

Tsai YC (2006) Study on phylogenetic relationship of *Glycine species* collected in Taiwan. Doctoral Dissertation, National Taiwan University

Tsai YC, Hsing YIC, Chung CI, Hsieh JS (2001) Wild soybean and its relatives collected in Taiwan. Chinese Agro J 11:217–230

Ungerer MC, Strakosh SC, Zhen Y (2006) Genome expansion in three hybrid sunflower species is associated with retrotransposon proliferation. Curr Biol 16:R872–R873

Vahedian M, Shi L, Zhu T, Okimoto R, Danna K, Keim P (1995) Genomic organization and evolution of the soybean SB92 satellite sequence. Plant Mol Biol 29:857–862

Varshney RK, Sigmund R, Börner A, Korzun V, Stein N, Sorrells ME, Langridge P, Graner A (2005) Interspecific transferability and comparative mapping of barley EST-SSR markers in wheat, rye and rice. Plant Sci 168:195–202

Vindelov LL, Christensen IJ, Nissen NI (1983) Standardization of high-resolution flow cytometric DNA analysis by the simultaneous use of chicken and trout red blood cells as internal reference standards. Cytometry 3:328–331

Volkov RA, Borisjuk NV, Panchuk II, Schweizer D, Hemleben V (1999) Elimination and rearrangement of parental rDNA in the allotetraploid *Nicotiana tabacum*. Mol Biol Evol 16:311–320

Walling JG, Shoemaker R, Young N, Mudge J, Jackson S (2006) Chromosome-level homeology in paleopolyploid soybean (*Glycine max*) revealed through integration of genetic and chromosome maps. Genetics 172:1893–1900

Weidner S, Becker A, Bonilla I, Jaenicke S, Lloret J, Margaret I, Puhler A, Ruiz-Sainz JE, Schneiker-Bekel S, Szczepanowski R, Vinardell JM, Zehner S, Gottfert M (2012) Genome sequence of the soybean symbiont *Sinorhizobium fredii* HH103. J Bacteri 194:1617–1618

In vitro culture and production of syringin and rutin in *Saussurea involucrata* (Kar. et Kir.) – an endangered medicinal plant

Chao-Lin Kuo[1], Dinesh-Chandra Agrawal[2], Hung-Chi Chang[2,3], Ya-Ting Chiu[2], Chu-Peng Huang[2], Yi-Lin Chen[4], Shih-Hung Huang[2] and Hsin-Sheng Tsay[2,5*]

Abstract

Background: *Saussurea involucrata* (Kar. et Kir.) commonly known as 'snow lotus' or 'Xue Lian' is an important plant in the traditional Chinese system of medicine. The plant contains flavonoids such as syringin and rutin. These compounds have been reported to be anti-rheumatic, anti-inflammatory and dilate blood vessels, lower blood pressure, prevent cardiovascular diseases, enhance immunity, and act as anti-aging, anti-cancer, and anti-fatigue agents. The species has become endangered due to the excessive collection of *S. involucrata* plants in the wild, slower plant growth and ecological destruction of natural habitats. There is a severe shortage of plant material, while the market demand is ever increasing. Hence, it is very important to apply tissue culture technique for plant propagation and production of the bioactive compounds of this species.

Results: Multiple shoot induction and proliferation in shoot base explants derived from *in vitro* raised seedlings of *S. involucrata* was achieved on 3/4 strength of Murashige and Skoog's (MS) basal medium (MSBM) supplemented with 1.0 mg/L^{-1} BA and 1.5 mg/L^{-1} NAA. Rooting was induced in 100 % shoots cultured on 1/2X MSBM supplemented with 1.0 mg/L^{-1} IBA for one week and then transfer to auxin free medium. The plantlets could be acclimatized successfully by sachet technique and established in the greenhouse. Maximum callus induction and proliferation in leaf segments was achieved on 1/2X MSBM supplemented with 0.5 mg/L^{-1} BA, 0.5 mg/L^{-1} NAA, 0.4 % gelrite and on incubation at 20 °C. Container closures had an influence on the quality and quantity of callus and production of the active compounds. The HPLC analysis showed much higher syringin content in *in vitro* shoots and callus as compared to commercially available market crude drug.

Conclusion: The present study describes an *in vitro* culture protocol of *Saussurea involucrata*. The bioactive compounds, syringin and rutin could be produced through tissue culture technique without sacrificing the endangered *Saussurea involucrata* plants in the wild.

Keywords: *In vitro* culture; Medicinal plant; Rutin; *Saussurea involucrata*; Snow lotus; Syringin

Background

Saussurea involucrata (Kar. et Kir.) belonging to the family Asteraceae and commonly known as snow lotus or 'Xue Lian' in Chinese mainly grows in the high rocky mountains at 4000 to 4300 m in the Tianshan and A'er Tai areas in Xinjiang province of China (Fu 1992). The plant has been used in the traditional Chinese system of medicine for over 2000 years for the treatment of many diseases such as rheumatoid arthritis, cough and cold, stomachache, dysmenorrhea, altitude sickness, and has been found to have properties of anti-inflammatory, cardiotonic, abortifacient, anticancer, and antifatigue (Jia et al. 2005). Considering the importance of the species, recently several researchers carried out investigations with regard to the beneficial effects of *S. involucrata*. In pharmacological studies on *S. involucrata*, anti-inflammatory and anti-nociceptive properties have been reported (Yi et al.

* Correspondence: hstsay@cyut.edu.tw
[2]Department of Applied Chemistry, Chaoyang University of Technology, Taichung, Taiwan
[5]Department of Agronomy, National Chung Hsing University, Taichung, Taiwan
Full list of author information is available at the end of the article

2010). In another investigation, petroleum ether extract of *S. involucrata* showed high anti-hypoxic activity effective in preventing acute mountain sickness (Ma et al. 2011). While studying to identify a potential agent for androgen-independent prostate cancer patients and to investigate its biological mechanism as an antineoplastic agent, Way et al. (2010) demonstrated that *S. involucrata* effectively inhibited EGFR signaling in human hormone-resistant prostate cancer PC-3 cells. A study on antimetastatic effects on SK-Hep1 human HCC cells found that *S. involucrata* could inhibit cell growth of metastatic cells in dose- and time-dependent manner demonstrating its potential as an anti-tumor agent (Byambaragchaa et al. 2013). *S. involucrata* produces several bioactive flavonoids that are derived from the phenylpropanoid pathway. A method validation study showed that the LC/MS technique was a powerful analytical tool for detecting trace amounts of the flavonoid compounds in extracts of *S. involucrata* (Xu et al. 2009). In a separate study, ethyl acetate extract from the aerial parts of *S. involucrata* led to the isolation of three new sesquiterpene lactones having significant anti-inflammatory and cytotoxic activities against A549 cells (Xiao et al. 2011). Pharmacological properties of syringin and rutin have been studied by several research groups. Syringin has been reported to possess anti-inflammatory and antinociceptive effects (Choi et al. 2004), hypoglycemic effect (Niu et al. 2008), antidepressant activity (Kurkin et al. 2006) and antitumor activity (Zhang et al. 2007). Rutin has several pharmacological properties including antioxidant (Katsube et al. 2006), anticarcinogenic, cytoprotective (Janbaz et al. 2002), antiplatelets (Sheu et al. 2004), and vasoprotective activities (La Casa et al. 2000). Moreover, rutin was found to be a neuroprotective agent, and can ameliorate ischemic-reperfusion injury in the heart (Lebeau et al. 2001), brain (Gupta et al. 2003) and skeletal muscle (Neumayer et al. 2006).

Owing to slow plant growth, over exploitation of wild plants, and ecological destruction of the natural habits in recent years, there is a severe shortage of plant material of this species, while the market demand is ever increasing. The wild plant population has diminished to the extent that *S. involucrata* is now considered as an endangered species and has been listed as a national protected wild plant in China (Fu 1992). Due to these reasons, the isolation of active constituents from the wild plant is a serious constraint. Hence, there is a need to investigate alternative method for plant propagation and isolation of the active compounds. Plant tissue culture system can be an effective means to achieve these objectives. In the present study, Response Surface Methodology (RSM) was used to optimize a suitable medium for *in vitro* culture and callus induction in *S. involucrata.* HPLC analysis of *in vitro* cultures and market crude drug was carried out to estimate the active compounds syringin and rutin.

Methods

Plant material

Seeds of *Saussurea involucrata* were provided by Professor C.T. Wei of Department of Biotechnology, China Medical University, Taichung, Taiwan. Seeds were surface sterilized by submerging them into 70 % ethyl alcohol for 30 s followed by 10 % sodium hypochloride for 10 min, and then rinsing them 3 times with sterile double distilled water. Disinfected seeds were inoculated in glass test tubes (measuring 120 mm long, 22 mm dia) containing Murashige and Skoog's (Murashige and Skoog 1962) salts and vitamins, hereinafter referred as MS basal medium (MSBM) supplemented with 0.9 % agar and 3 % sucrose. The pH of the medium was adjusted to 5.7 ± 0.1, prior addition of agar and before autoclaving for 15 min under 1.05 kg/cm at 121 °C. All the cultures were incubated at 25 ± 2 °C, under 14/10 h light and dark period and light intensity of 100 μE/m^{2}s provided by cool-white fluorescent lights. *In vitro* raised seedlings were used as source material for different experiments in the present study.

Initiation and proliferation of shoots

In an initial trial, five different basal media (salts and vitamins) viz. MS (Murashige and Skoog, 1962), WPM - Woody plant medium (Lloyd and McCown 1981); B5 (Gamborg et al. 1968); N6 (Chu et al. 1975) and White (1963) without plant growth regulators (PGRs) were tested for initiation of shoots. Since, the best response was observed with MS basal medium, hence all further experiments were carried out with MS alone. Shoot base (Fig. 1a) excised from *in vitro* raised seedlings of *S. involucrata* was used as an explant for shoot initiation. Using RSM, a central composite factorial design was used for media optimization with 3 components viz. MS strength (1X, 3/4X, 1/2X and 1/4X), four concentrations of BA (1.0, 1.5, 2 and 2.5 mg/L^{-1}) and NAA (0.5, 1.0 and 1.5 mg/L^{-1}). All media were supplemented with 3 % sucrose and pH adjusted to 5.7 ± 0.1 prior to autoclaving. Glass test tube was used as a culture vessel closed with 2 layers of aluminum foil (2AF). After first subculture, explants with initiated shoots were transferred to 650 ml capacity glass orchid flasks for further proliferation. Each culture flask containing 100 ml medium was closed with 4 layers of dispense paper (DP4) and an additional layer of parafilm which was removed after 2 weeks. Both vitrified and normal shoots were cultured as separate sets. Cultures were incubated for total 7 weeks at 25 ± 2 °C, under 14/10 h light and dark period and light intensity of 100 μE/m^{2}s provided by cool-white fluorescent lights.

Fig. 1 *In vitro* culture of *Saussurea involucrata* Kar. et Kir. **a**: Shoot base explant (shown as in circle); **b**: Multiple shoots; **c**: Proliferation of shoots; **d**: Rooted shoot; **e** and **f**: Tissue culture plants in green house. Scale bar for **a**–**d** = 1 cm; **e** & **f** = 2 cm

Rooting of *in vitro* shoots

For induction of rooting in *in vitro* shoots, an initial trial was carried out with 3 auxins (IAA, IBA and NAA) individually, at a fixed concentration of 0.5 mgL^{-1}. Based on the higher response of rooting on MSBM containing IBA, further experiments were carried out with IBA alone. To optimize the rooting medium, 3 concentrations of IBA (0.1, 0.5 and 1 mgL^{-1}), 1/2 strength of MSBM and two concentrations of sucrose (1.5 and 3 %) were tested in a three factorial design. *In vitro* shoots were inoculated on medium with IBA for 1, 2, 3, and 4 weeks and then transferred to medium without IBA for remaining 6, 5,4 and 3 weeks, respectively. All media were gelled with 0.9 % agar. Orchid culture flasks were used as culture vessel and closed with two layers of non-permeable aluminum foil for two weeks, then replaced by 4 layers of dispense paper with an additional layer of parafilm which was removed after 2 weeks. All the cultures were incubated at 25 ± 2 °C, light and dark period of 14/10 h and light intensity of 100 $\mu E/m^2s$ provided by cool-white fluorescent lights. Observations like the number of rooted shoots, number of roots and length of roots were recorded after 50 days of incubation.

Transfer to soil

Rooted shoots were removed from the culture vessel and agar adhering to the roots was removed by gently rinsing with running tap water. Then, these rooted shoots were transferred to plastic pots (15 cm dia. each) containing a mixture of peat soil: perlite: vermiculite (1:1:1) and kept in a greenhouse. Hardening of the plantlets was achieved by following the sachet technique described in our earlier report (Huang et al. 2014). The plants were irrigated once daily and survival was recorded after two months of transfer to pots.

Induction and proliferation of callus

Petiole, leaf and root parts of in vitro raised seedlings of *S. involucrata* were used as explants for callus induction. Explants were cultured on 1/2X MSBM supplemented with 2,4-D (0.5, 1.0 mgL – 1), BA (0.5, 1.0, 2.0 mgL – 1), and NAA (0.5 mgL – 1), 3 % sucrose and 0.4 % gelrite. The pH of all media was adjusted to 5.7 ± 0.1. Petri-dishes (9 cm dia), each containing 25 ml of medium were used for callus induction experiment. While, for proliferation experiment, glass test tubes, each containing 10 ml medium were used. Callus (200 mg) was inoculated in each test tube and tubes were closed either with two layers of non-permeable aluminum foil (2AF) or with 3 dispense papers (3DP). Dispense papers had an additional layer of parafilm which was removed after 2 weeks. Observations on callus fresh weight under each treatment was recorded after 4 weeks of incubation. Experiments were carried out to optimize the callus proliferation using factors like temperature (10, 20, 25 °C), and 5 chemical elicitors viz. salicylic acid, abscisic acid, phenylalanine, sodium acetate and jasmonic acid (at a fixed concentration 3 mgL – 1) supplemented individually to 1/2X MSBM. All the cultures were incubated at 25 ± 2 °C (except for temperature experiment), under dark.

HPLC analysis

HPLC analysis was carried out to quantify syringin and rutin contents in *in vitro* shoots, callus and market crude drug. A reversed-phase symmetry-1 column (250 mm, 4.6X; 5 μm particle size; Waters, USA) was used. The mobile phase was made up from solvent A (methanol) and solvent B (aqueous 0.3 % formic acid in HPLC grade water). The solvent system used was a gradient of aqueous 0.3 % formic acid and methanol. The gradient was as follows: 0 min, A; 3 min, A; 5 min, 15 %, B; 15 min, 30 %, B; 20 min, 30 %, B. The flow rate was 1.0 ml/min and the injection volume was 10 μL. The chromatogram was monitored at 270 nm with a Waters DAD detector. Pure syringin and rutin samples used in the study as standards were procured from Formosa Kingstone Bioproducts International Corporation, Taiwan. Quantification was based on the peak area of the original sample injected. The recovery rate of extraction was evaluated by known quantities of syringin and rutin added to the biomass samples. The accuracy and reproducibility of syringin and rutin measurements were confirmed by analyzing different quantities of the samples.

Statistical analysis

Software SAS 9.1 was used for statistical analysis. Data were subjected to the least significant difference (LSD) tested at 5 % probability level ($p \geq 0.05$). Each treatment had minimum 30 replicates and all experiments were repeated minimum three times.

Results and discussion

Initiation and proliferation of shoots

Shoot initiation response in shoot base explants derived from *in vitro* raised seedlings of *S. involucrata* was very poor. Out of the five basal media tested, the maximum response in 49 % explants with an average 0.85 shoot/explant was observed on 1/2X MSBM after 50 days of incubation. Non-responding shoots turned brown. In the first sub-culture, number of shoots per explant increased to 3.89 in 93.3 % explants however shoots became vitrified (Fig. 1b). In the second subculture, a marked difference was observed between vitrified and normal (non-vitrified) explants and media composition with respect to the number of proliferating shoots. The maximum shoots (13 shoots/per explant) were observed on 3/4X MSBM supplemented with BA (1.0 mg/L^{-1}) and NAA (1.5 mg/L^{-1}) in case of normal (non-vitrified) explants, while vitrified explants resulted in 4.2 shoots/explant (Fig. 1c). The next best shoot proliferation (11.1 shoots/explant) was achieved on 1/4X MSBM supplemented with 2.0 mg/L^{-1} BA and 0.5 mg/L^{-1} NAA in case of non-vitrified explants. Since, media combinations in the 3-factorial experiment are too many to tabulate, hence, here we have described only the best responses observed with regard to the number of shoots.

In an earlier study, a supplement of BA and NAA was reported to be the most effective growth regulator combination for induction of shoots in leaf explants of *S. involucrata* via organogenesis route (Guo et al. 2007). They initially recorded 5.2 shoots/explant on BA and NAA containing medium, but this shoot number could be increased to 9.3 by pre-treatment of leaf explants to 4 °C for 5 days. In the present study, a dramatic increase in shoot proliferation in second subculture could be due to culture of shoots in the larger vessels (orchid flasks having 650 ml volume) compared to glass tubes (45 ml volume) used in the first two cultures. Also, culture tube and orchid flask had differential medium content of 10 ml and 100 ml, respectively. Thus, orchid flask had more volume and more medium content compared to the culture tube. Akin to our observation, a positive influence of larger culture vessel on shoot proliferation has been earlier been reported (McClelland and Smith 1990; Agrawal et al. 1997; Hazra et al. 2000). The positive influence of the larger vessel on increased shoot proliferation could be due to the availability of higher amounts of nutrients and a larger area for gaseous exchange.

Rooting of *in vitro* shoots

In an initial trial, out of three auxins (IAA, IBA and NAA) tested at a fixed concentration of 0.5 mg/L^{-1}, the best rooting response was observed with IBA. Therefore, in order to further optimize the rooting medium, 3 concentrations of IBA (0.1, 0.5 and 1.0 mgL^{-1}) and two concentrations of sucrose (1.5 and 3 %) in the 1/2X MSBM were investigated. The maximum rooting response in 100 % shoots, with an average of 6 roots/explant, and 6.5 cm root length was observed on 1/2X MSBM supplemented with 1.0 mg/L^{-1} IBA and 3 % sucrose (Table 1). This treatment induced healthier roots and shoots compared to all other treatments (Fig. 1d). Medium with 0.1 mg/L^{-1} IBA, though induced rooting in 100 % shoots, however average number of roots per shoot were limited to 3.57 and 2.71 with 1.5 % and 3 % sucrose in the medium, respectively (Table 1). Out of 4 IBA treatments for 1,2,3 and 4 weeks, the maximum rooting response was observed with IBA for 1 week and other three treatments did not show significant differences. Hence, Table 1 contains data pertaining to treatment of IBA for 1 week only. It was observed that shoots incubated in medium with auxin continuously for 50 days induced excessive callusing at the base and roots were shorter in length.

Efficient rooting of *in vitro* grown shoots is a prerequisite for the success of micropropagation. It is a common practice nowadays to employ lower salt concentrations in the culture medium for rooting of *in vitro* shoots. Higher salt levels in the medium have often been

Table 1 The influence of IBA and sucrose concentrations on rooting of *in vitro* shoots in *Saussurea involucrata* Kar. et Kir

IBA (mg/L^{-1})*	Sucrose %	% of shoots rooted**	Av. no. of roots/explant**	Av. root length (cm)**	Av. shoot height (cm)**
0.1	1.5	100^{a}	3.57^{abc}	5.8^{a}	4.50^{bc}
	3.0	100^{a}	2.71^{a}	5.6^{a}	3.57^{ab}
0.5	1.5	79^{c}	3.00^{ab}	4.3^{ab}	4.87^{a}
	3.0	93^{a}	3.92^{abc}	5.7^{ab}	4.14^{abc}
1.0	1.5	86^{ab}	3.33^{abc}	4.5^{a}	5.91^{abc}
	3.0	100^{a}	6.00^{abc}	6.5^{a}	5.07^{a}

*1/2X MSBM: Murashige and Skoog (1962) basal medium (salts and vitamins), supplemented with 3 % sucrose, 0.9 % agar, pH 5.7 ± 0.1. Data were recorded after 7 weeks of culture. Shoots were cultured on IBA containing medim for 1 week and then transferred to PGR free MSBM containing 1.5 % or 3 % sucrose and 0.9 % agar and further incubation for 6 weeks, respectively

**Each treatment had thirty explants with three replicates. Means followed by the same letter (a,b,c) of a column are not significantly different at 5 % level by least significant difference (LSD) test

found inhibitory to root initiation (George et al. 2008). In a study on *in vitro* propagation of grapes, it was found that an appropriate salt concentration (1/2X or 1/4X MS) was more important than the sucrose level for induction of rooting in *in vitro* cuttings of grapevine (Harris and Stevenson 1979). The benefit of low salt medium for root initiation may be due to the need for a low nitrogen level, than for an increased osmotic potential (Harris and Stevenson, 1979). In an another study, it was demonstrated that microcuttings of several tree fruit rootstocks rooted best on MS salts, and the sucrose concentration in the medium was not critical, though its presence in the medium was essential (Dunstan 1981). Besides salts and sucrose, plant growth regulators (PGRs), especially, auxins are known to affect both root and shoot growth parameters and play an important role in root development. The success of auxin supply depends on its type, concentration and duration in the culture medium. Different plant species respond differently to auxins for the induction of rooting, hence, it is

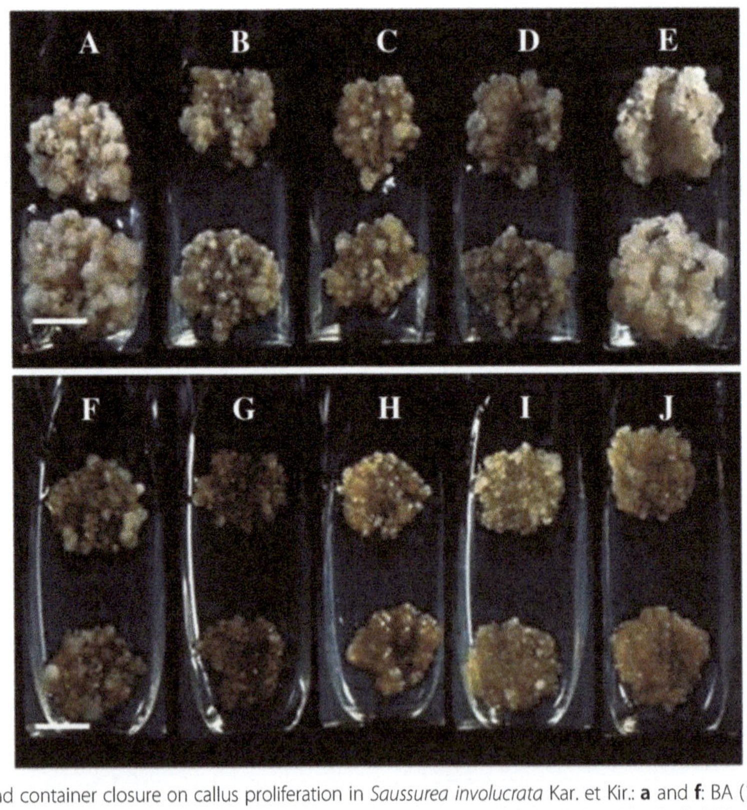

Fig. 2 Influence of PGRs and container closure on callus proliferation in *Saussurea involucrata* Kar. et Kir.: **a** and **f**: BA (1.0) + 2,4-D (1.0); **b** and **g**: BA (1.0) + 2,4-D (0.5); **c** and **h**: BA (2.0) + 2,4-D (0.5); **d** and **i**: BA (0.5) + 2,4-D (0.5); **e** and **j**: BA (0.5) + NAA (0.5). PGR conc. - mg/L^{-1}; Scale bar for **a**–**j** = 1 cm. Culture medium: 1/2X MSBM, 3 % sucrose, 04 % gelrite. Culture tubes in treatments **a** to **e** were closed with 2 layers of aluminum foil (2AF), while in treatments **f** to **j** were closed with. 3 layers of dispense papers (3DP). Incubation at 20 °C for 4 weeks

important to optimize the type, concentration of an auxin and duration for which shoots need to be cultured in the auxin medium. In several cases, a pulse treatment with auxins for a short period has also been sufficient for root induction. While shoots of some species require longer exposure. In the present study, exposure of shoots to IBA containing medium for 1 week was sufficient. Continuous exposure of shoots to auxin medium for 7 weeks induced excessive callus, shorter and unhealthy shoots, an observation similar to our earlier report on *G. davidii* (Chueh et al. 2001). However, in contrast to our results, rooting in shoots of *S. involucrata* was achieved with a continuous exposure to IAA containing medium (Guo et al. 2007).

Transfer of plantlets to soil and their survival in the greenhouse

The sachet technique used in our previous report (Huang et al. 2014) was quite successful in hardening the tissue culture plantlets. Plantlets (100 %) could be hardened and established in the soil (Fig. 1e and f).

Callus induction and proliferation

Induction of callus in 3 explants viz. leaf, petiole and root of *S. involucrata* varied among the different media compositions. The highest response (80 %) was observed in leaf explants cultured on 1/2X MSBM supplemented with 0.5 or 1.0 mg/L^{-1} BA, 0.5 mg/L^{-1} NAA and 0.4 % Gelrite. In the experiment with the incubation temperature, the maximum callus proliferation was recorded at 20 °C. Sealing of culture vessels with aluminium foil or dispense papers had an influence on the quality and quantity of callus induced (Fig. 2). The maximum callus proliferation (1.036 g) was recorded on medium supplemented with 0.5 mg/L^{-1} BA and 0.5 mg/L^{-1} NAA and culture tubes closed with 3 layers of dispense papers (Table 2). Sealing of culture tubes with 2 layers of aluminum foil resulted in callii compact in texture and darker in color. While, in case of sealing of culture tubes with 3 layers of dispense papers, the callii were friable and fairer. Overall, the quality and quantity of callus proliferated in culture tubes sealed with 3 layers of dispense papers were superior and higher under all treatments compared to culture tubes sealed with 2 layers of aluminum foil (Table 3; Fig. 2). Inclusion of chemical elicitors had a mixed influence on the fresh weight of callus and syringin content (Table 2) All the five elicitors marginally enhanced the callus fresh weight, however, had a negetive influence on syringin content with an exception of jasmonic acid, which mariginally enhanced syringin content in callus (14.788 mg/g dw) compared to control treatment (13.524 mg/g dw). We did not observe any significant effect of any of the five elicitors on rutin content in callus which remained more or less same in all the treatments (Table 2).

Table 2 Influence of different elicitors on syringin and rutin content in callus *of Saussurea involucrata* Kar. et Kir

Elicitor*	Fresh weight (g)**	Syringin (mg/g dw)***	Rutin (mg/g dw)
Control	1.31^{b}	13.52^{a}	0.29
Salicylic acid	1.65^{a}	11.30^{bc}	0.29
Abscisic acid	1.74^{a}	8.75^{d}	0.31
Phenylalanine	1.73^{a}	10.73^{c}	0.37
Sodium acetate	1.58^{a}	12.85^{ab}	0.32
Jasmonic acid	1.59^{a}	14.79^{a}	0.32

*Basal medium:1/2X MSBM, BA (0.5 mg/L^{-1}), NAA 0.5 (mg/L^{-1}), elicitor (3.0 mg/L^{-1}), sucrose (3 %), Gelrite (0.4 %) and pH 5.7 ± 0.1. Cultures were incubated at 20 °C. Data were recorded after 4 weeks of culture

**Callus (0.2 g) was taken as inoculum. Each treatment had 50 explants. Four random samples were analyzed in HPLC

***dw: Freeze dry weight; Means followed by the same letter (a,b,c,d) were not significantly different at 5 % level

Table 3 Influence of PGRs and container closure on callus proliferation in *Saussurea involucrata* Kar. et Kir

PGR (mg/L^{-1})*			Callus (g)	
2,4-D	BA	NAA	2 Aluminum foils**	3 Dispense papers**
0	0.5	0.5	0.57^{b}	1.04^{a}
0.5	0.5	0	0.32^{e}	0.64^{bc}
1.0	1.0	0	0.46^{c}	0.74^{b}
0.5	1.0	0	0.44^{dc}	0.63^{bc}
0.5	2.0	0	0.38^{de}	0.52^{c}

*Basal medium: 1/2X MSBM, 3 % sucrose, 0.4 % Gelrite and pH 5.7 ± 0.1. Data were recorded after 4 weeks of culture

Each treatment had 30 explants in three replicates. Cultures were incubated at 20 °C. Callus (0.2 g) was taken as the initial inoculum

**Means followed by the same letter (a,b,c,d,e) in a column were not significantly different at the % level by LSD test

The physical and chemical micro-environments of culture containers can influence growth rate and other physiological and morphological characteristics of plants developed under *in vitro* conditions (Walker et al. 1988). It has been reported that the type of closure affects gaseous exchange, availability of water, micronutrients, and balance of hormones in the culture container (Kataeva et al. 1991; Lai et al. 2005; Chen et al. 2006a; Tsay et al. 2006). The head space of culture vessels with low ventilation, accumulates various gaseous compounds like ethylene and carbon dioxide (Zobayed et al. 2001; Lai

Table 4 HPLC analysis: Syringin and rutin conents in different materials of *Saussurea involucrata* Kar. et Kir

Plant material	Syringin (mg/g dw)*	Rutin (mg/g dw)*
Tissue culture shoots	8.37	0.34
Callus	13.52	0.29
Market crude drug	0.95	0.24

*dw: Freeze dry weight

et al. 2005). These undesirable compounds can alter biochemical responses and leaf development of *in vitro* cultured plants (Pierik et al. 2007) and also affect enzymes involved in oxidative activities (Synková and Pospíšilová 2002). Some closures cause restriction of gaseous exchange between the container atmosphere and the outside environment (Buddendorf-Joosten and Woltering 1994), which can result in poor aeration and hyperhydric condition of cultures. Also, Different species show different requirement with respect to container closures. Hence, it is important to optimize a closure type in a micropropagation protocol of a particular plant species. In our laboratory, aluminum foil and dispense paper have been used successfully to improve ventilation for cultures of *Scrophularia* (Chen et al. 2006a), and *Bupleurum* (Chen et al. 2006b). It was observed that aluminum foil, a less air permeable material for first 4 weeks, followed by more air permeable dispense papers for the next 4 weeks was an adequate ventilation treatment for optimum root/shoot growth and subsequent survival of *G. scabra* plantlets (Huang et al. 2014). In the present study also a combination of foil and dispense paper was found to be a favorable combination for callus proliferation in *S. involucrata.* Similar to our results, a selective influence of chemical elicitors on the enhancement of secondary plant metabolites in callus has been reported earlier (Bulgakov et al. 2002). It was found that salicylic acid and methyl jasmonate had a positive influence on anthraquinone accumulation in both transgenic and non-transgenic calluses of *Rubia cordifolia* but ethephon did not (Bulgakov et al. 2002).

HPLC analysis

HPLC analysis of *in vitro* shoots, callus and market crude drug revealed marked differences in syringin content in these materials (Table 4). The highest syringin content (13.52 mg/g dw) was observed in callus followed by *in vitro* shoots (8.57 mg/g dw) compared to market crude drug (0.95 mg/g dw). There was a marginal difference in rutin content in all the three materials. However, the positive outcome of the present study is that production of both syringin and rutin was achieved in an *in vitro* culture system.

Conclusions

An *in vitro* culture protocol for *S. involucrata* has been developed using Response Surface Methodology (RSM). Callus induction and proliferation could be achieved by *in vitro* leaf culture method. HPLC analysis results showed that syringin content was much higher both in callus and *in vitro* shoots compared to crude drug available in the market. Between the *in vitro* shoots and callus, the latter showed a higher syringin accumulation and could be a method of choice for the production of syringin and rutin. The present study demonstrates the usefulness of a tissue culture system with regard to production of the active compounds (syringin and rutin) round the year without the destruction of wild *S. involucrata* plants which are already under endangered category.

Abbreviations

AF: Aluminum foil; BA: 6-Benzylaminopurine; DP: Dispense paper; HPLC: High- performance liquid chromatography; IAA: Indole-3-acetic acid; IBA: Indole-3-butyric acid; MS: Murashige and Skoog; MSBM: Muruashige & Skoog's basal medium (salts and vitamins); NAA: 1-Naphthalene acetic acid; PGR: Plant growth regulator; RSM: Response surface methodology; WPM: Woody plant medium.

Competing interests

The authors declare that they have no competing interests.

Authors' contributions

YTC and CPH carried out the experimental work under the guidance/ supervision of CLK, HCC, YLC, SHH and HST. DCA wrote the manuscript. All authors have read and approved the final manuscript.

Acknowledgement

A research grant (NSC 102-2313-B-324-001-MY3) from the Ministry of Science and Technology (MOST), Taiwan is gratefully acknowledged.

Author details

[1]Department of Chinese Pharmaceutical Sciences and Chinese Medicine Resources, China Medical University, Taichung, Taiwan. [2]Department of Applied Chemistry, Chaoyang University of Technology, Taichung, Taiwan. [3]Department of Golden-Ager Industry Management, Chaoyang University of Technology, Taichung, Taiwan. [4]Department of Biotechnology and Animal Science, National Ilan University, Yilan City, Taiwan. [5]Department of Agronomy, National Chung Hsing University, Taichung, Taiwan.

References

Agrawal DC, Banerjee AK, Kolala RR, Dhage AB, Nalawade SM, Kulkarni AV, Hazra S, Krishnamurthy KV (1997) *In vitro* induction of multiple shoots and plant regeneration in cotton (*Gossypium hirsutum* L.). Plant Cell Rep 16:647–652. doi:10.1007/BF01275508

Buddendorf-Joosten JMC, Woltering EJ (1994) Components of the gaseous environment and their effects on plant growth and development *in vitro*. Plant Growth Regul 15:1–16. doi:10.1007/BF00024671

Bulgakov VP, Tchernoded GK, Mischenko NP, Khodakovskaya MV, Glazunov VP, Radchenko SV, Zvereva EV, Fedoreyev SA, Zhuravlev YN (2002) Effect of salicylic acid, methyl jasmonate, ethephon and cantharidin on anthraquinone production by *Rubia cordifolia* callus cultures transformed with the *rolB* and *rolC* genes. J Biotechnol 97:213–221. doi:10.1016/S0168-1656(02)00067-6

Byambaragchaa M, de la Cruz J, Yang SH, Hwang SG (2013) Anti-metastatic potential of ethanol extract of *Saussurea involucrata* against hepatic cancer *in vitro*. Asian Pac J Cancer Prev 14(9):5397–5402, http://dx.doi.org/10.7314/APJCP.2013.14.9.5397

Chen UC, Hsia CN, Agrawal DC, Tsay HS (2006a) Influence of ventilation closures on plant growth parameters, acclimation and anatomy of leaf surface in *Scrophularia yoshimurae* Yamazaki - a medicinal plant native to Taiwan. Bot Stud 47:259–266

Chen UC, Hsia CN, Yeh MS, Agrawal DC, Tsay HS (2006b) *In vitro* micropropagation and *ex vitro* acclimation of *Bupleurum kaoi* - an endangered medicinal plant native to Taiwan. In Vitro Cell Dev Biol–Pl 42:128–133. doi:10.1079/IVP2005744

Choi J, Shin KM, Park HJ, Jung HJ, Kim HJ, Lee YS, Rew JH, Lee KT (2004) Anti-inflammatory and antinociceptive effects of sinapyl alcohol and its glucoside syringing. Planta Med 70:1027–1032. doi:10.1055/s-2004-832642

Chu CC, Wang CC, Sun CS, Hsu C, Yin KC (1975) Establishment of an efficient medium for anther culture of rice through comparative experiments on the nitrogen sources. Sci Sin 18:659–668

Chueh FS, Chen CC, Sagare AP, Tsay HS (2001) Quantitative determination of secoiridoid glucosides in *in vitro* propagated plants of *Gentiana davidii* var.

formosana by high performance liquid chromatography. Planta Med 67:70–73. doi:10.1055/s-2001-10622

Dunstan DI (1981) Transplantation and post-transplantation of micropropagated tree-fruit rootstocks. Comb Proc Int Plant Prop Soc 31:39–44

Fu LG (1992) China plant red data book - rare and endangered plants, vol 1. Chinese Science Press, Beijing

Gamborg OL, Miller RA, Ojima K (1968) Nutrient requirements of suspension cultures of soybean root cells. Exp Cell Res 50:151–158. doi:10.1016/0014-4827(68)90403-5

George EF, Hall MA, De Klerk GJ (eds) (2008) Plant propagation by tissue culture. Vol. 1. The background. Springer, The Netherlands

Gupta R, Singh M, Sharma A (2003) Neuroprotective effect of antioxidants on ischaemia and reperfusion-induced cerebral injury. Pharmacol Res 48:209–215. doi:10.1016/S1043-6618(03)00102-6

Guo B, Gao M, Liu CZ (2007) *In vitro* propagation of an endangered medicinal plant *Saussurea involucrata* Kar. et Kir. Plant Cell Rep 26:261–265. doi:10.1007/s00299-006-0230-6

Harris RE, Stevenson JH (1979) Virus elimination and rapid propagation of grapes *in vitro*. Comb Proc Int Plant Prop Soc 29:95–108

Hazra S, Kulkarni AV, Nalawade SM, Banerjee AK, Agrawal DC, Krishnamurthy KV (2000) Influence of explants, genotypes and culture vessels on sprouting and proliferation of pre-existing meristems of cotton (*Gossypium hirsutum* L. and *Gossypium arboreum* L.). In Vitro Cell Dev Biol-Pl 36:505–510. doi:10.1007/s11627-000-0090-7

Huang SH, Agrawal DC, Wu FS, Tsay HS (2014) *In vitro* propagation of *Gentiana scabra* Bunge – an important medicinal plant in the Chinese system of medicines. Bot Stud 55:56. doi:10.1186/s40529-014-0056-4

Janbaz KH, Saeed SA, Gilani AH (2002) Protective effect of rutin on paracetamol- and CCl_4-induced hepatotoxicity in rodents. Fitoterapia 73:557–563. doi:10.1016/S0367-326X(02)00217-4

Jia JM, Wu CF, Liu W, Yu H, Hao Y, Zheng JH, Ji YR (2005) Anti-inflammatory and analgesic activities of the tissue culture of *Saussurea involucrata*. Biol Pharma Bull 28(9):1612–1614. doi:org/10.1248/bpb.28.1612

Kataeva NV, Alexandrova IG, Butenko RG, Dragavtceva EV (1991) Effect of applied and internal hormones on vitrification and apical necrosis of different plants cultured *in vitro*. Plant Cell Tiss Org 14:31–40. doi:10.1007/BF00041283

Katsube T, Imawaka N, Kawano Y, Yamazaki Y, Shiwaku K, Yamane Y (2006) Antioxidant flavonol glycosides in mulberry (*Morus alba* L.) leaves isolated based on LDL antioxidant activity. Food Chem 97:25–31. doi:10.1016/j.foodchem.2005.03.019

Kurkin VA, Dubishchev AV, Ezhkov VN, Titova IN, Avdeeva EV (2006) Antidepressant activity of some phytopharmaceuticals and phenylpropanoids. Pharm Chem J 40:614–619. doi:10.1007/s11094-006-0205-5

La Casa C, Villegas I, de la Lastra CA, Motilva V, Martin Calero MJ (2000) Evidence for protective and antioxidant properties of rutin, a natural flavone, against ethanol induced gastric lesions. J Ethnopharmacol 71:45–53. doi:10.1016/S0378-8741(99)00174-9

Lai CC, Lin HM, Nalawade SM, Fang W, Tsay HS (2005) Hyperhydricity in shoot cultures of *Scrophularia yoshimurae* can be effectively reduced by ventilation of culture vessels. J Plant Physiol 162(3):355–61. doi:10.1016/j.jplph.2004.07.015

Lebeau J, Neviere R, Cotelle N (2001) Beneficial effects of different flavonoids on functional recovery after ischemia and reperfusion in isolated rat heart. Bioorg Med Chem Lett 11:23–27. doi:10.1016/S0960-894X(00)00589-8

Lloyd G, McCown B (1981) Commercially feasible micropropagation of mountain laurel, *Kalmia latifolia* by shoot tip culture. Intl Plant Prop Soc Proc 30:421–427

Ma HP, Fan PC, Jing LL, Yao J, He XR, Yang Y, Chen KM, Jia ZP (2011) Anti-hypoxic activity at simulated high altitude was isolated in petroleum ether extract of *Saussurea involucrata*. J Ethnopharmacol 137(3):1510–5. doi:10.1016/j.jep.2011.08.037

McClelland MT, Smith MAL (1990) Vessel type, closure and explant orientation influence on *in vitro* performance of five woody species. HortSci 25:797–800

Murashige T, Skoog F (1962) A revised medium for rapid growth and bioassays with tobacco tissue cultures. Physiol Plant 15:473–497. doi:10.1111/J.1399-3054.1962.TB08052.X

Neumayer C, Fugl A, Nanobashvili J, Blumer R, Punz A, Gruber H, Polterauer P, Huk I (2006) Combined enzymatic and antioxidative treatment reduces ischemia-reperfusion injury in rabbit skeletal muscle. J Surg Res 133:150–158. doi:10.1016/j.jss.2005.12.005

Niu HS, Liu IM, Cheng JT, Lin CL, Hsu FL (2008) Hypoglycemic effect of syringin from *Eleutherococcus senticosus* in streptozotocin-induced diabetic rats. Planta Med 74:109–113. doi:10.1055/s-2008-1034275

Pierik R, Sasidharan R, Voesenek LCJ (2007) Growth control by ethylene: adjusting phenotypes to the environment. J Plant Growth Regul 26:188–200. doi:10.1007/s00344-006-0124-4

Sheu JR, Hsiao G, Chou PH, Shen MY, Chou DS (2004) Mechanisms involved in the antiplatelet activity of rutin, a glycoside of the flavonol quercetin, in human platelets. J Agric Food Chem 52:4414–4418. doi:10.1021/jf040059f

Synková H, Pospíšilová J (2002) *In vitro* precultivation of tobacco affects the response of antioxidative enzymes to *ex vitro* acclimation. J Plant Physiol 159:781–789. doi:10.1078/0176-1617-0511

Tsay HS, Lee CY, Agrawal DC, Basker S (2006) Influence of ventilation closure, gelling agent and explant type on shoot bud proliferation and hyperhydricity in *Scrophularia yoshimurae* – a medicinal plant. In Vitro Cell Dev Biol-Pl 42:445–449. doi:10.1079/IVP2006791

Walker PN, Heuser CW, Heinemann PH (1988) Micropropagation: studies of gaseous environments. Acta Hort 230:145–151

Way TD, Lee JC, Kuo DH, Fan LL, Huang CH, Lin HY, Shieh PC, Kuo PT, Liao CF, Liu H, Kao JY (2010) Inhibition of epidermal growth factor receptor signaling by *Saussurea involucrata*, a rare traditional Chinese medicinal herb, in human hormone-resistant prostate cancer PC-3 cells. J Agric Food Chem 58(6):3356–65. doi:10.1021/jf903793p

White PR (1963) The cultivation of animal & plant cells, 2nd edn. Ronald Press, New York

Xiao W, Li X, Li N, Bolati M, Wang X, Jia X, Zhao Y (2011) Sesquiterpene lactones from Saussurea involucrata. Fitoterapia 82(7):983–987. doi:10.1016/j.fitote.2011.05.015

Xu YJ, Zhao DX, Fu CX, Cheng LQ, Wang NF, Han LJ, Ma FS (2009) Determination of flavonoid compounds from *Saussurea involucrata* by liquid chromatography electrospray ionisation mass spectrometry. Nat Prod Res 23:1689–98. doi:10.1080/14786410802187742

Yi T, Zhao ZZ, Yu ZL, Chen HB (2010) Comparison of the anti-inflammatory and anti-nociceptive effects of three medicinal plants known as "Snow Lotus" herb in traditional Uighur and Tibetan medicines. J Ethnopharmacol 128(2):405–11. doi:10.1016/j.jep.2010.01.037

Zhang W, Zhang WD, Zhang C, Liu RH, Li TZ, Fu P, Shan L (2007) Antitumor activities of extracts and compounds from the roots of *Daphne tangutica* Maxim. Phytother Res 21:1113–1115. doi:10.1002/ptr.2227

Zobayed SMA, Armstrong J, Armstrong W (2001) Leaf anatomy of *in vitro* tobacco and cauliflower plantlets as affected by different types of ventilation. Plant Sci 161:537–548. doi:10.1016/S0168-9452(01)00438-1

21

Distyly and floral morphology of *Psychotria cephalophora* (Rubiaceae) on the oceanic Lanyu (Orchid) Island, Taiwan

Kenta Watanabe[1*], T Y Aleck Yang[2,3], Chihiro Nishihara[4], Tai-Liang Huang[2], Koh Nakamura[5], Ching-I Peng[6] and Takashi Sugawara[7]

Abstract

Background: *Psychotria cephalophora* Merr. (Rubiaceae), a shrub in oceanic islands of Taiwan and the Philippines, appears to be distylous, but distyly is usually rare on oceanic islands. To elucidate the functional breeding system of *P. cephalophora* can improve our understanding of plant reproductive ecology on oceanic islands.

Results: Field investigations on Lanyu (Orchid Island) off the coast of southeastern Taiwan revealed the flowers to be distylous with short (S)- and long (L)-styled morphs, with only one morph per individual. Laboratory observations revealed that both morphs had stainable pollen grains and indicated dimorphism in stigmatic papillae and pollen size. In hand pollination experiments, the pollen tubes reached the base of the style in intermorph crossing, whereas they rarely penetrated stylar tissue in intramorph crossing and selfing. Open pollinated S- and L-styled flowers produced fruit.

Conclusions: The results indicate that the breeding system of *P. cephalophora* is morphologically and functionally distylous.

Keywords: Breeding system; Distyly; Heterostyly; Lanyu; Oceanic island; Orchid Island; Pollen dimorphism; *Psychotria cephalophora*; Rubiaceae

Background

Psychotria L. (Rubiaceae), one of the largest genera in flowering plants, with more than 1800 species (Davis et al. 2009), occurs mainly in tropical and subtropical regions (Hamilton 1989; Nepokroeff et al. 1999; Davis et al. 2001; Sakai and Wright 2008). Within *Psychotria*, distyly is regarded as the ancestral breeding system (Hamilton 1990). Distyly is a genetically controlled dimorphism reported from 26 angiosperm families (Naiki 2012). In distylous species, populations are composed of two distinct floral morphs: long (L) -styled morph with stigmas exceeding the anthers and short (S)-styled morph with anthers exceeding the stigmas. Between the two morphs male and female organs are usually at the reciprocal height (Ganders 1979; Barrett et al. 2000). This is a mechanism to promote animal-mediated cross-pollination.

Four species of *Psychotria*: three shrubs *P. cephalophora*, *P. manillensis*, and *P. rubra*, and a woody climber *P. serpens* are in Taiwan (Yang 1998b). *Psychotria cephalophora* occurs only on Lanyu (Orchid Island) (Yang 1998b) and Lutao (Green Island) off the coast of southeastern Taiwan (Chen and Lu 2008); *P. manillensis* occurs only on Lanyu, Lutao (Yang 1998b) and Kuei-Shan Tao (Turtle Island) off the coast of northeastern Taiwan (Tsai 2009); *P. rubra* occurs on Taiwan Island and adjacent islands, including Lutao, but not on Lanyu; *P. serpens* occurs on Taiwan Island and adjacent islands including Lanyu. *Psychotria serpens* is distylous (Yang 1998b; Sugawara et al. 2013); *P. rubra* is dioecious (Yang 1998b; Watanabe et al. 2014b); and *P. manillensis* is

* Correspondence: kenta-w@okinawa-ct.ac.jp
[1]Okinawa National College of Technology, 905 Henoko, Nago, Okinawa 905-2192, Japan
Full list of author information is available at the end of the article

monoecious (Watanabe et al. in preparation). There has been no study of the breeding system in *P. cephalophora*, except that Yang et al. (1999) mentioned it to be morphologically heterostylous without detailed description. The reproductive nature of *P. cephalophora* is left to be explored.

The circumscription of *Psychotria cephalophora* has been controversial. *Psychotria cephalophora* was described by Merrill (1908) from the Philippines, while *P. kotoensis* was described by Hayata (1920) from Lanyu. *Psychotria kotoensis* was later merged with *P. cephalophora* (Yang 1998b). Recently, Sohmer and Davis (2007) stated, based on a study of specimens, that *P. cephalophora* from the Philippines and *P. kotoensis* from Taiwan are not conspecific, but they did not describe differences between the two species. Tao and Taylor (2011) provisionally applied the name *P. cephalophora* to plants from Lanyu because of the lack of data in Sohmer and Davis (2007). In this study, we tentatively use the name *Psychotria cephalophora* Merr. for the plants from Lanyu Island. Until now, details of the floral morphology of *P. cephalophora* have been insufficiently examined (Yang 1998b; Sohmer and Davis 2007; Tao and Taylor 2011). Therefore, more information on the floral morphology of *P. cephalophora* is important in considering the taxonomy of these plants.

In this paper, we present the results of a detailed study of the floral morphology and breeding system, focusing specifically on distyly, of *Psychotria cephalophora* on Lanyu . We also briefly discuss the evolutionary significance of distyly on Lanyu Island.

Methods

Study site and sample collection

Psychotria cephalophora on Lanyu grows to 2.5 m tall. It is terrestrial in relatively dark, wet evergreen forests (Figure 1) and blooms from February to April, according to specimen records. The fruit matures from September to January.

Observations and collections of *Psychotria cephalophora* were carried out in one natural population at Tienchi Pond (N22° 01', E121° 34', 337 m a.s.l.). We collected non-flowering branch cuttings on 12 September 2011 and 13 September 2012, and planted them in the greenhouse of Okinawa National College of Technology, Okinawa, Japan. To study sexual and morphological differentiation of flowers, more than two inflorescences per plant were collected from 40 plants from the Tienchi population on 30 March 2013 and preserved in 70% ethanol.

Measurements and observations of floral morphology

Ethanol-preserved flowers from the 40 plants were used. We measured the diameter of the inflorescence (Figure 1Cx), anther height (Figure 2a), corolla height (Figure 2b), and stigma height (Figure 2c) using two flowers from different inflorescences on each plant, and the number of flowers per inflorescence was determined. The mean values of two flowers were calculated to represent each plant.

In addition to the measurements of the floral traits, we observed floral morphology of *P. cephalophora* both in the field and in the greenhouse.

Figure 1 Flowers and fruit of *Psychotria cephalophora* in the Tienchi population, Lanyu Island, Taiwan. **(A)** Flowering branches. **(B)** Young fruit and inflorescences. **(C)** Short-styled morphs. **(D)** Long-styled morphs. X, diameter of inflorescence. All scale bars **(B–D)** = 1 cm.

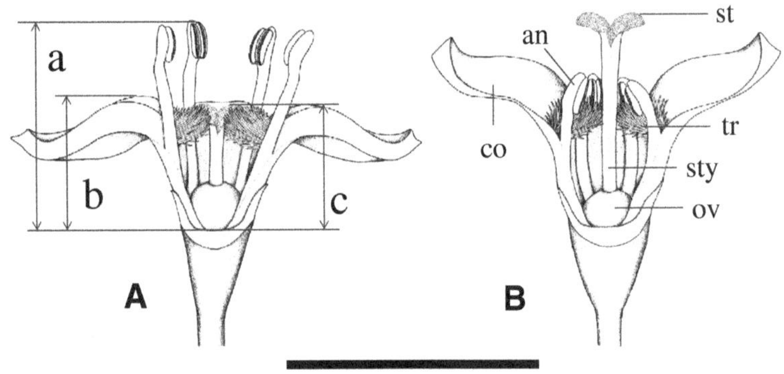

Figure 2 Drawings of short **(A)** and long-styled **(B)** flowers from the Tienchi population of *Psychotria cephalophora*, Lanyu Island, Taiwan. a, anther height; b, corolla height; c, stigma height. an, anther; co, corolla lobe; ov, ovary; st, stigma; sty, style; tr, oblique-facing long hairs. Scale bar = 5 mm.

Pollen size and stainability

Equatorial and polar diameters of 30 pollen grains from a single flower were measured using a light microscope in 10 plants for each morph. We excluded pollen grains that were extremely small or had irregular shapes. Pollen stainability with aniline blue was determined using 200 pollen grains per flower for 10 plants for each morph to examine pollen viability (Kearns and Inouye 1993). A single anther within a flower before anthesis was crushed on a glass slide with 0.05% aniline blue (Nakarai esque, Kyoto, Japan) in lactophenol, and counted the number of stained and non-stained pollen grains under light microscope. Two flowers per plant were examined to calculate the mean value.

SEM observations of stigma and pollen

Fresh stigmas from the two morphs, collected in the greenhouse in April 2013, were observed using environmental scanning electron microscopy (SEM, S-3000 N; Hitachi, Tokyo, Japan). Pollen grains from two morphs were treated by the standard acetolysis method (Erdtman 1960), dehydrated in an ethanol: t-butanol series, freeze-dried using a freeze-drying device (FDU-1000; Eyela, Tokyo, Japan), mounted onto SEM stubs on double sided carbon tape, coated with Pt using an ion sputter coater (E-1010; Hitachi) and observed using SEM (S-3000 N) and Field Emission SEM (S-4800; Hitachi).

Pollination experiments

To elucidate incompatibility systems by observing pollen tube penetration in the style, intermorphic, intramorphic and self- pollinations were carried out in the greenhouse in April 2013. Approximately 24 h after hand pollination the pistils were collected, fixed in formalin acetic alcohol (FAA) solution (5 : 5 : 90 = solution of formalin : glacial acetic acid : 70% ethanol) or 70% ethanol, and transferred to new 70% ethanol for storage. The fixed styles were softened in 8 N NaOH for 24 h at 4°C, rinsed with water and then stained with 0.005% aniline blue in Na_2HPO_4 (pH 11) for 3 h at 4°C according to (Kearns and Inouye 1993). The styles were mounted on a slide in glycerol beneath a cover glass and gently pressed to spread tissue. Pollen grains on the stigmas and pollen tubes in the styles were observed using a fluorescence light microscope (Axio Imager Z2; Zeiss, Oberkochen, Germany) with U-filter set 49 (Excitation 365 nm, Emission 445 nm; Zeiss).

Statistical analyses

All statistical tests were performed using R ver. 3.0.0 software (The R Project for statistical computing; http://www.r-project.org/).

Results

Floral morphology and distyly

The terminal inflorescences were composed of 14 – 60 pedicellate flowers in a head-like cluster. The corolla tube, whitish and funnel shaped, averaging about 3 mm long, was glabrous outside and has white hairs in the throat. The five or six, triangular to oblong corolla lobes were about 3.5 mm long. All flowers examined had a single style, two ovules per ovary, and five or six anthers. The flowers opened in the morning and lasted until the next morning/noon. They produced a perceptible scent.

Psychotria cephalophora in wild was distylous, with L- and S-styled flowers (Figure 3). Each plant had either S- or L-styled flowers. Measurements of the stigmas and anthers of the two morphs are presented in Table 1 and Figure 4. Anther height was significantly greater in the S-styled morph than in the L-styled morph, while stigma height was significantly less in the S-styled morph than in the L-styled morph (Table 1). No significant difference was found between the morphs (Table 1) in corolla length. No significant difference was found between stigma height of the L-styled morph and anther height of the S-styled morph (Figure 4). Anther height of the L-styled morph was significantly greater than stigma height of the S-styled morph but the difference in mean values was only 0.6 mm (Table 1). Anthers and

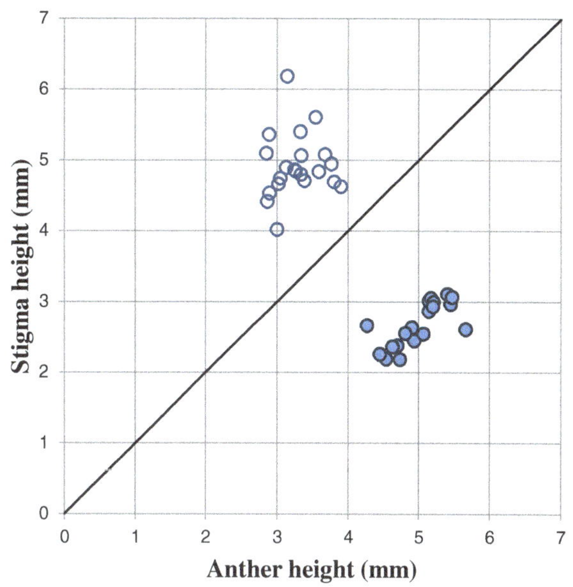

Figure 3 Scatter diagram showing the relationship between stigma and anther heights of *Psychotria cephalophora* in the Tienchi population. The bold line in a figure is y = x, meaning that anther and stigma heights are equal. ●, a flower of short-styled morph; ○, a flower of long-styled morph.

stigmas of the different morphs were roughly at the same height. The stigmas were bifurcated and produced well-developed stigmatic papillae on both morphs. The fine structures of the stigmatic surface differed between the two morphs. In the S-styled morph the papillae are longer and less crowded than in the L-styled morph (Figure 5).

The inflorescence was 12.9 – 23.6 mm in diameter in the S-styled morph and 11.32 – 19.99 mm in the L-styled morphs. Flowers per inflorescence ranged from 22 – 57 in the S-morph and 15 – 37 in the L-styled morph. The size of the inflorescence and the number of flowers per inflorescence were greater in the S-styled morph, but the differences between the two morphs were insignificant (Table 1).

In the Tienchi population, we found 21 L-styled morph individuals and 21 S-styled morph individuals. There was therefore no significant deviation from a 1:1 morph ratio (Binomial test: $p > 0.05$).

Pollen stainability, size and exine sculpturing

The pollen grains of *Psychotria cephalophora* were spherical, had 3 or 4 colpi, and had reticulate exine. Polar axis/equatorial axis ratio (E/P ratio) of pollen grains was about 1.08 in both morphs. There was no significant difference between the two morphs (Mann Whitney *U*-test; $p > 0.05$). In addition, we did not notice major differences between the two morphs in exine sculpturing (Figure 6). Pollen size of the two morphs, however, was clearly different. The pollen of the S-styled morph was more than 50 μm in diameter and about 10 μm larger than in the L-styled morph (Mann Whitney *U*-test; $p < 0.001$). Pollen stainability was over 90% in both the morphs (Table 2).

Pollination experiments

In intermorph pollinations, more than five pollen tubes reached the base of the style in all the plants (Figure 7; Table 3). In intramorph and self-pollinated flowers, the pollen tubes did not reach the base of the style except for three cases in intramorph pollinations of the L-styled morph and two cases of selfing in the L-styled morph.

Discussion

Distyly of *Psychotria cephalophora* on Lanyu Island

Our observations revealed *Psychotria cephalophora* to be morphologically and functionally distylous with S- and L- styled morphs. Generally, reciprocal positioning of stamens and anthers of the two morphs is characteristic of distylous species (Barrett 2002), and reciprocal placement of the organs in distylous flowers is considered to promote disassortative pollination (Keller et al. 2014). In *P. cephalophora*, although stigma height of the S-styled morph is significantly lower than anther height of the L-styled morph (Figure 4), the difference was only 0.6 mm (Table 1). Thus, probably this small gap does not prevent disassortative pollination between the two morphs. Although we did not observe pollinators in natural populations, moths

Table 1 A comparison of several floral traits between Short- and Long-styled morphs of *Psychotria cephalophora*

	Short-styled morph		Long-styled morph		Mann Whitney's
	N†	Mean ± S. D.	N†	Mean ± S. D.	*U* test
Inflorescence size (mm)	20§	18.2 ± 2.36	20	16.8 ± 1.81	ns‡
Number of flowers/inflorescence	20§	32.5 ± 7.53	20	28.1 ± 5.65	ns‡
Stigma height (mm)	20¶	2.7 ± 0.32	20	4.9 ± 0.48	p < 0.001
Anther height (mm)	20¶	5.0 ± 0.38	20	3.3 ± 0.32	p < 0.001
Corolla height (mm)	20¶	3.1 ± 0.36	20	3.2 ± 0.32	ns‡

†N: Number of individuals examined.
§:Two inflorescences on each individual were examined.
¶:Two flowers from different inflorescences on each individual were examined.
‡ns: Not significantly differnet between two floral morphs (p > 0.05).

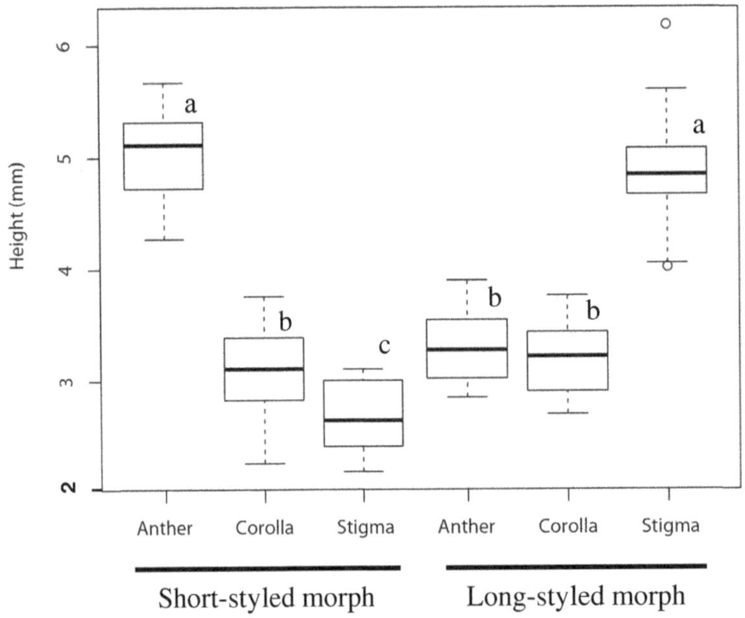

Figure 4 Box plots of three floral traits (anther, corolla and stigma heights) of the short- and long-styled morphs of *Psychotria cephalophora*. Small letters **(a, b, c)** indicate significant differences at $p < 0.001$ determined by Tukey-type multiple comparison after one-way ANOVA.

may be one of the important pollinators because the flowers have a strong scent and white corolla (Van der Pijl 1961). We need to study pollination in natural populations to understand the ecological significance of the positioning of the style and stamens.

The pollen of the S-styled morph is significantly larger than in the L-styled morph (Table 2), a commonly observed phenomenon in distylous species (Ganders 1979). Darwin (1877) suggested that the larger pollen of the S-styled morphs is related to the need for longer

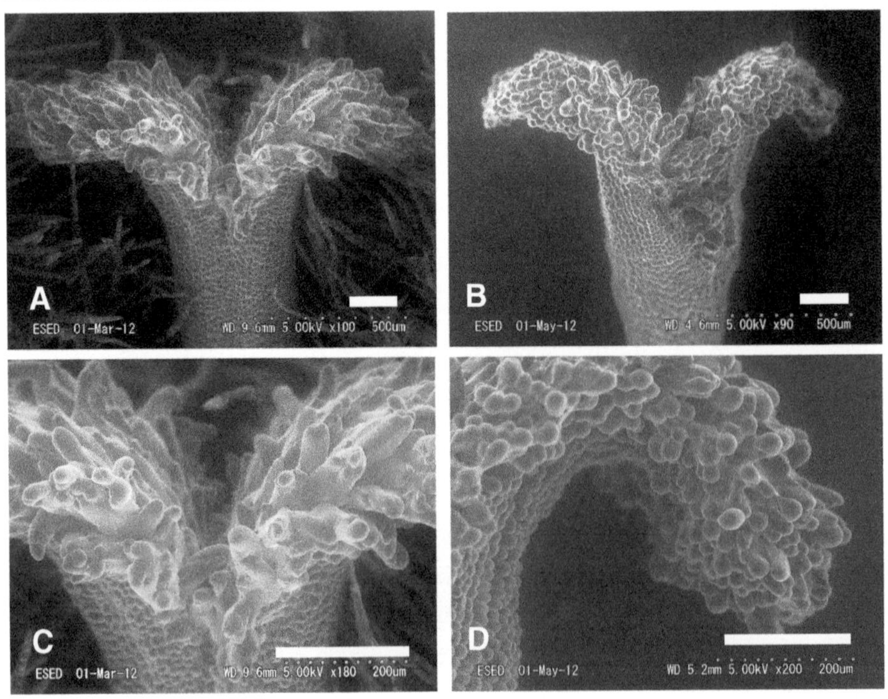

Figure 5 Scanning electron micrographs of stigma of short **(A, C)** and long-styled **(B, D)** morphs of *Psychotria cephalophora*. Scale bars = 200 μm.

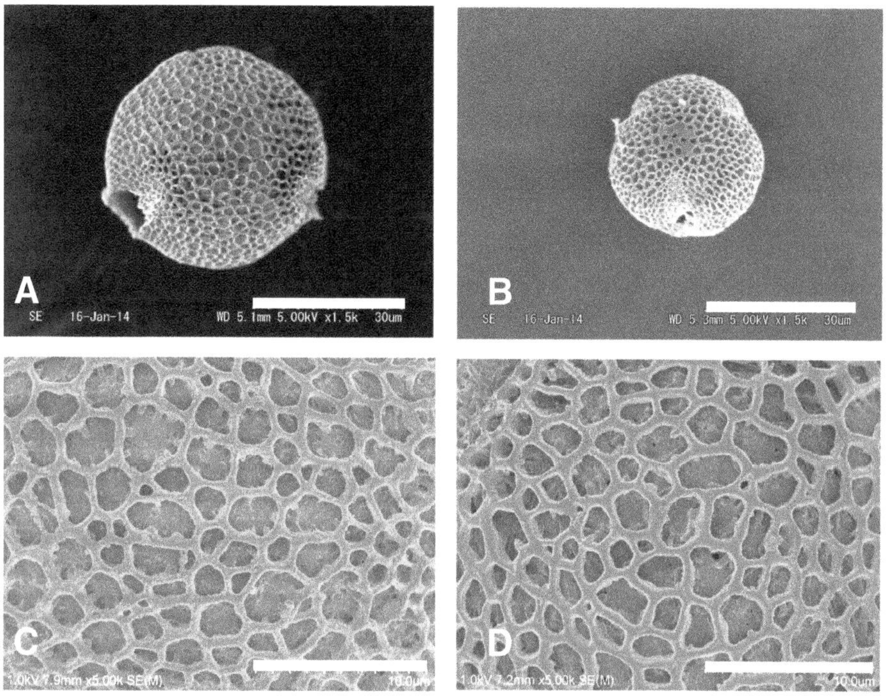

Figure 6 Scanning electron micrographs of pollen grains in short- **(A, C)** and long-styled **(B, D)** morphs in *Psychotria cephalophora*. Scale bars = 30 μm **(A, B)** and 10 μm **(C, D)**.

pollen tubes to reach the ovules through the longer styles of the L-styled morph. However, Ganders (1979) found no correlation between L / S style length ratio and S / L pollen volume ratio. The adaptive significance of pollen size dimorphism in distylous plants is left to be explored. In *P. cephalophora*, the pollen exine is reticulated with no distinguishable differences between the two morphs. These features are similar in *P. serpens* (Sugawara et al. 2013), *P. boninensis* (Kondo et al. 2007) and *P. homalosperma* (Watanabe et al. 2014a). Since the stamens of the two morphs produce stainable pollens they are considered to be functional. Meanwhile, both morphs have well developed stigmatic papillae, two ovules per ovary, and frequently set fruit at the study site, indicating that the flowers of both morphs of *P. cephalophora* are hermaphroditic.

In distylous plants, self- and intramorphic incompatibility is commonly observed (Klein et al. 2009). Although we did not test fruit set, hand pollination experiments indicated that *P. cephalophora* is basically self- and intramorphic incompatible (Table 3). However, we also observed a few cases of successful self- and intramorphic pollination in the L-styled morph. This relaxed self- and intramorphic incompatibility in the L-styled morph was also reported for two distylous species of *Coussarea* and *Rudgea* (Rubiaceae) by Bawa and Beach (1983), although they reported that those species did not set fruit in the field. To confirm the

Table 2 Pollen size and stainability of *Psychotria cephalophora*

	Short-styled morph		Long-styled morph		Mann Whitney's
	N†	Mean ± S. D.	N†	Mean ± S. D.	*U* test
Pollen size (μm)§					
Equatral	10	54.7 ± 2.5	10	44.2 ± 2.4	p < 0.001
polar	10	50.5 ± 2.7	10	41.1 ± 2.5	p < 0.001
E/P	10	1.08 ± 0.03	10	1.08 ± 0.02	ns‡
Pollen stainability (%)¶	10	93.1 ± 13.4	10	95.9 ± 2.6	ns‡

†N: Number of individuals examined.
§: 30 pollen grains from single flowers on each individual were measured.
¶: 200 pollen grains from single flowers on each individual were examined.
‡ns: Not significantly differnet between two floral morphs (p > 0.05).

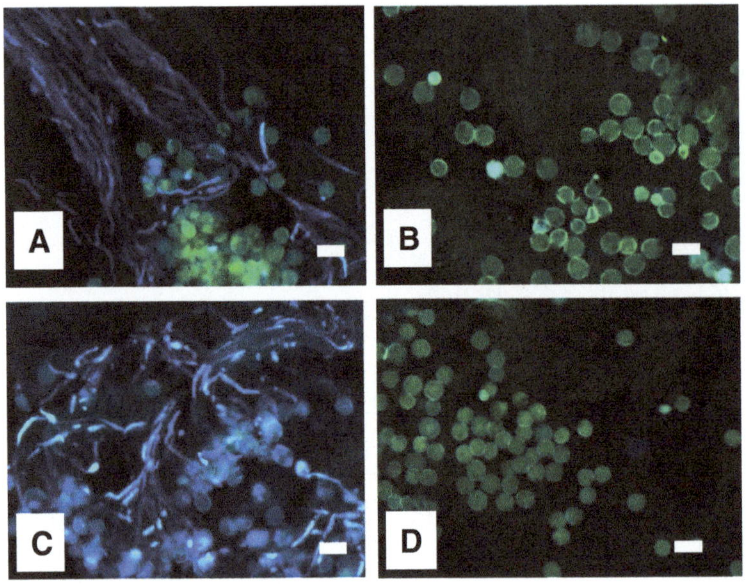

Figure 7 Pollen tubes in pistils approximately 24 h after intermorphic **(A, C)** and self- **(B, D)** pollinations in short- **(A, B)** and long-styled **(C, D)** morphs in *Psychotria cephalophora* observed with fluorescence microscopy. All scale bars = 100 μm.

compatibility system in *P. cephalophora* more closely, we also need to examine fruit set after hand pollination experiments. The morph ratio was approximately 1:1 in the studied population. This is also typical for distylous plant species, because distyly is a genetically controlled polymorphism following Mendelian inheritance (Barrett and Shore 2008). We did not find any influence of partial self- and intramorphic compatibility in morph ratio.

Recently, many reproductive studies of the genus *Psychotria* have been performed in neotropics (e. g., Faivre and McDade 2001; Castro et al. 2004; Sakai and Wright 2008). Among them, most species are functionally distylous, but some are intramorphically compatible, and others monomorphic. It is reported that *Psychotria carthagenesis* in Brazil shows different sexuality among different populations: typical distyly, pin-monomorphism and homostyly (Consolaro et al. 2011; Faria et al. 2012). Since our investigation is restricted only one population of *P. cephalophora* in Lanyu, we need to study more populations to confirm stability of its breeding system.

We described detailed floral morphology of *P. cephalophora* for the first time. To clarify whether *P. cephalophora* on Lanyu Island is endemic to the island (as *P. kotoensis*) or conspecific with the species in the Philippines,

Table 3 Pollen tube growth after hand pollination treatments in *Psychotria cephalophora*

Experiments	n‡	N‡	Status of pollen tube growth			Percentages of the pollen tube reached the base of styles
			A§	B§	C§	
Intermorphic pollination						
S→L cross†	50	7	50	0	0	100%
L→S cross†	51	8	51	0	0	100%
Intramorphic pollination						
SS cross†	30	8	0	0	30	0%
LL cross†	60	13	3	8	49	5.0%
Self pollination						
S†	77	14	0	2	75	0%
L†	61	15	2	4	45	3.3%

†S: Short-styled morph; L: Long-styled morph.
‡n: Number of flowers examined; N: Number of individuals examined.
§A: More than five pollen tubes reached at the base of the style.
B: 1 - 10 pollen tubes reached at one third of the style.
C: No pollen tube penetrate into the stigma.

we need to examine floral traits of *P. cephalophora* in the Philippines. Molecular based analyses would also be useful to solve this issue.

Distyly on the oceanic Lanyu Island

In this study, we reported distyly in *P. cephalophora* on the oceanic island Lanyu. Although we are unaware of the exact number of distylous species on Lanyu, there are at least several (e.g., *Ophiorrhiza japonica*, Liu 1998; *Guettarda speciosa*, Yang 1998a). In our preliminary observations, *P. serpens* also maintains distyly on Lanyu (Watanabe unpublished data). Distyly is, both theoretically and actually, rare or absent in oceanic islands (Carlquist 1974; Pailler et al. 1998; Meeus et al. 2011; McMullen 2012; Sugawara et al. 2014; Watanabe et al. 2014a). Thus it is of great interest to study how *P. cephalophora* colonized and reproduces on the oceanic island. Self-incompatible plants are thought to be less successful in colonization to oceanic islands than self-fertilizers (Baker 1955; Stebbins 1957; Barrett et al. 1996); self-fertilizers can establish populations from single individual and reproduce successfully without suitable pollinators (Baker 1955; Baker and Cox 1984; Barrett et al. 1996). In general, island florae have higher ratio of hermaphroditic self-compatible plants than in mainland (Crawford et al. 2011). Lanyu and northern islands of the Philippines, where *P. cephalophora* occurs, have not been land-connected but only a few hundred km away from each other and these islands share many floristic elements (Nakamura and Kokubugata 2014). Colonization of *P. cephalophora* into Lanyu may have occurred repeatedly unlike more isolated oceanic islands. Also Lanyu is only 60 km away from Taiwan Island and this is probably why the island has entomofauna rich enough to support self-incompatible distylous plants. There are contradictory cases where oceanic islands have a high proportion of dioecism (Carlquist 1974; Baker and Cox 1984; Sakai et al. 1995; Abe 2006). It is possibly because outcrossing has a merit to avoid inbreeding depressions in small populations on oceanic islands (Sakai et al. 1995; Barrett et al. 1996). Tseng et al. (2008) reported the proportion of dioecism on Lanyu to be very high (11.9%) compared to the global average (6%; Renner and Ricklefs 1995) and Taiwan proper (7.9%). The outcrossing-merit hypothesis may also be applicable to distyly on Lanyu. To understand how *P. cephalophora* colonized and reproduces in natural populations, phylogeographic analyses and pollinator observations and testing open fruit sets are required.

Conclusions

We described detailed floral morphology and incompatibility of *Psychotria cephalophora* on Lanyu for the first time. Floral dimorphism and intramorphic incompatibility indicate that the breeding system of *P. cephalophora* is morphologically and functionally distylous. Because distyly is usually rare on oceanic islands, it is important to clarify how this species colonized and reproduces on islands. Further reproductive studies of *P. cephalophora* in the natural populations on Lanyu and the Philippines are needed.

Competing interests

The authors declare that they have no competing interests.

Authors' contributions

Collection of flower samples: KW, AY, CN, TLH, KN, TS, CIP. Flower Drawing: CN, KW. Morphological study: KW, CN, TS. Draft preparation: KW, TS. Paper writing: KW, TS, KN. All authors read and approved the final manuscript.

Acknowledgements

We thank Mr. CF. Chen for helping in the field study, Dr. D. Boufford for improving the English, and Drs. H. X. Chiura, M. Hara, N. Ogawa, K. Kogure and Y. Watanabe for their help in using the FE-SEM at the Atmosphere and Ocean Research Institute of the University of Tokyo. This study was partly supported by a Grant –in –Aid for Scientific Research (no. 23923008; 24923009; 26840130 to KW and no. 22570096 to TS).

Author details

[1]Okinawa National College of Technology, 905 Henoko, Nago, Okinawa 905-2192, Japan. [2]Department of Botany, National Museum of Natural Science, Kuan Chien Rd, Taichung 404, Taiwan. [3]Department of Life Science, National Chung Hsin University, Taichung 402, Taiwan. [4]774-1-302, Umusa, Nago, Okinawa 905-0006, Japan. [5]Botanic Garden, Field Science Center for Northern Biosphere, Hokkaido University, North 3, West 8, Chuo-ku, Sapporo 060-0003, Japan. [6]Biodiversity Research Center, Academia Sinica, Taipei, Nangang 115, Taiwan. [7]Makino Herbarium, Graduate School of Science, Tokyo Metropolitan University, 1-1 Minami-Ohsawa, Hachioji, Tokyo 192-0397, Japan.

References

Abe T (2006) Threatened pollination systems in native flora of the Ogasawara (Bonin) Islands. Ann Bot 98:317–334, doi:10.1093/aob/mcl117

Baker HG (1955) Self-compatibility and establishment after long-distance'dispersal. Evolution 9:347–349

Baker HG, Cox PA (1984) Further thoughts on dioecism and islands. Ann Missouri Bot Gard 71:244–253

Barrett SCH (2002) The evolution of plant sexual diversity. Nat Rev Genet 3:274–284, doi:10.1038/nrg776

Barrett SCH, Shore JS (2008) New insights on heterostyly: comparative biology, ecology and genetics. In: Self-incompatibility in Flowering Plants. Springer, Berlin Heidelberg, pp 3–32

Barrett SCH, Emerson B, Mallet J (1996) The reproductive biology and genetics of island plants. Philos Trans R Soc Lond B Biol Sci 351:725–753, doi:10.1098/rstb.1996.0067

Barrett SCH, Jesson LK, Baker AM (2000) The evolution and function of stylar polymorphisms in flowering plants. Ann Bot 85:253–265, doi:l0.1006/anbo.1999.1067

Bawa KS, Beach JH (1983) Self-incompatibility systems in the Rubiaceae of a tropical lowland wet forest. Am J Bot 70:1281–1288

Carlquist S (1974) Island biology. Columbia Univ Press, New York & London

Castro CC, Oliveira PE AM, Alves MC (2004) Breeding system and floral morphometry of distylous *Psychotria* L. species in the Atlantic rain forest, SE Brazil. Plant Biol 6:755–760, doi:10.1055/s-2004-830349

Chen TY, Lu DH (2008) The Inventory of Vascular Plant of Green Island. Construction and Planning Agency, Ministry of the Interior Research Project Report, Taipei

Consolaro H, Silva SCS, Oliveira PE (2011) Breakdown of distyly and pin-monomorphism in *Psychotria carthagenensis* Jacq. (Rubiaceae). Plant Species Biol 26:24–32, doi:10.1111/j.1442-1984.2010.00300.x

Crawford DJ, Anderson GJ, Bernardello G (2011) The reproductive biology of island plants. In: Barmwell D, Caujape-Castells J (eds) The Biology of Island floras. Cambridge University Press, Cambridge, pp 11–36

Darwin C (1877) The different forms of flowers on plants of the same species. John Murray, London
Davis AP, Bridson D, Jarvis C, Govaerts R (2001) The typification and characterization of the genus *Psychotria* L. (Rubiaceae). Bot J Linn Soc 135:35–42, doi:10.1111/j.1095-8339.2001.tb02366.x
Davis AP, Govaerts R, Bridson DM, Ruhsam M, Moat J, Brummitt N (2009) A global assessment of distribution, diversity, endemism, and taxonomic effort in the Rubiaceae. Ann Missouri Bot Gard 96:68–78, doi:10.3417/2006205
Erdtman G (1960) The acetolysis method. Sven Bot Tidskr 54:561–564
Faivre AE, McDade LA (2001) Population-level variation in the expression of heterostyly in three species of Rubiaceae: does reciprocal placement of anthers and stigmas characterize heterostyly? Am J Bot 88:841–853
Faria RR, Ferrero V, Navarro L, Araujo AC (2012) Flexible mating system in distylous populations of *Psychotria carthagenensis* Jacq. (Rubiaceae) in Brazilian Cerrado. Plant Syst Evol 298:619–627, doi:10.1007/s00606-011-0571-7
Ganders FR (1979) The biology of heterostyly. New Zeal J Bot 17:607–635
Hamilton CW (1989) A revision of mesoamerican *Psychotria* subgenus *Psychotria* (Rubiaceae), part I: introduction and species 1-16. Ann Missouri Bot Gard 76:67–111
Hamilton CW (1990) Variations on a distylous theme in Mesoamerican *Psychotria* subgenus *Psychotria* (Rubiaceae). Mem NY Bot Gard 55:62–75
Hayata B (1920) *Psychotria kotoensis*. In: Icones Plant. Formosanarum nec non Contrib. ad Floram Formosanum. 9. Bureau of productive industry, Government of Formosa, Taipei, p 62
Kearns CA, Inouye DW (1993) Techniques for pollination biologists. University Press of Colorado, Colorado
Keller B, Thomson J, Conti E (2014) Heterostyly promotes disassortative pollination and reduces sexual interference in Darwin's primroses: evidence from experimental studies. Funct Ecol 28:1413–1425, doi:10.1111/1365-2435.12274
Klein DE, Freitas L, Da Cunha M (2009) Self-incompatibility in a distylous species of Rubiaceae: is there a single incompatibility response of the morphs? Sex Plant Reprod 22:121–131, doi:10.1007/s00497-009-0097-0
Kondo Y, Nishide M, Watanabe K, Sugawara T (2007) Floral dimorphism in *Psychotria boninensis* Nakai (Rubiaceae) endemic to the Bonin (Ogasawara) Islands. J Jap Bot 82:251–258
Liu HY (1998) Ophiorrhiza. In: Huang TC (ed) Flora of Taiwan 4, 2nd edn. Editorial Committee, Dept. Bot., NTU, Taipei, pp 306–311
McMullen CK (2012) Pollination of the heterostylous Galápagos native, *Cordia lutea* (Boraginaceae). Plant Syst Evol 298:569–579, doi:10.1007/s00606-011-0567-3
Meeus S, Jacquemyn H, Honnay O, Pailler T (2011) Self-incompatibility and pollen limitation in the rare tristylous endemic *Hugonia serrata* on La Réunion Island. Plant Syst Evol 292:143–151, doi:10.1007/s00606-010-0400-4
Merrill ED (1908) Psychotria cephalophora. Philipp J Sci 3:438
Naiki A (2012) Heterostyly and the possibility of its breakdown by polyploidization. Plant Species Biol 27:3–29, doi:10.1111/j.1442-1984.2011.00363.x
Nakamura K, Kokubugata G (2014) In situ glacial survival at the northern limit of tropical insular Asia by a lowland herb *Begonia fenicis* (Begoniaceae). Bot J Linn Soc 174:305–325, doi:10.1111/boj.12138
Nepokroeff M, Bremer B, Sytsma KJ (1999) Reorganization of the genus *Psychotria* and tribe Psychotrieae (Rubiaceae) inferred from ITS and rbcL sequence data. Syst Bot 24:5–27
Pailler T, Humeau L, Thompson JD (1998) Distyly and heteromorphic incompatibility in oceanic island species of *Erythroxylum* (Erythroxylaceae). Plant Syst Evol 213:187–198, doi:10.1007/BF00985199
Renner SS, Ricklefs RE (1995) Dioecy and its correlates in the flowering plants. Am J Bot 82:596–606
Sakai S, Wright J (2008) Reproductive ecology of 21 coexisting *Psychotria* species (Rubiaceae): when is heterostyly lost? Biol J Linn Soc 93:125–134, doi:10.1111/j.1095-8312.2007.00890.x
Sakai AK, Wagner WL, Ferguson DM, Herbst DR (1995) Origins of dioecy in the Hawaiian flora. Ecology 76:2517–2529
Sohmer SH, Davis AP (2007) The genus *Psychotria* (Rubiaceae) in the Philippine Archipelago. Sida, Botanical Miscellany, vol 27. Botanical Research Institute of Texas, Texas
Stebbins GL (1957) Self fertilization and population variability in the higher plants. Am Nat 91:337–354
Sugawara T, Watanabe K, Tabata M (2013) Distyly in *Psychotria serpens* (Rubiaceae) in the Ryukyu Islands, Japan. Acta Phytotax Geobot 64:113–122
Sugawara T, Yumoto T, Tsuneki S, Watanabe K (2014) Incompatibility and reproductive output in distylous *Psychotria boninensis* (Rubiaceae), endemic to the Bonin (Ogasawara) Islands, Japan. J Jap Bot 89:22–26
Tao C, Taylor CM (2011) *Psychotria*. In: Chen T, Taylor CM, Swartz C (eds) Flora of China, vol 19. Science Press and Missouri Botanical Garden Press, Beijing and St. Louis, pp 294–301
Tsai C-C (2009) A study on flora of Kushan Island. Master's thesis. Providence University, Taichung
Tseng YH, Hsieh CF, Hu JM (2008) Incidences and ecological correlates of dioecious angiosperms in Taiwan and its outlying Orchid Island. Bot Stud 49:261–276
Van der Pijl L (1961) Ecological aspects of flower evolution. II. Zoophilous flower classes. Evolution (N Y) 15:44–59
Watanabe K, Kato H, Sugawara T (2014a) Distyly and incompatibility in *Psychotria homalosperma* (Rubiaceae), an endemic plant of the oceanic Bonin (Ogasawara) Islands. Flora 209:641–648, doi:10.1016/j.flora.2014.09.006
Watanabe K, Shimizu A, Sugawara T (2014b) Dioecy derived from distyly and pollination in *Psychotria rubra* (Rubiaceae) occurring in the Ryukyu Islands, Japan. Plant Species Biol 29:181–191, doi:10.1111/1442-1984.12013
Yang TYA (1998a) Guettarda. In: Huang TC (ed) Flora of Taiwan 4, 2nd edn. Editorial Committee, Dept. Bot., NTU, Taipei, p 263
Yang TYA (1998b) Psychotria. In: Huang TC (ed) Flora of Taiwan 4, 2nd edn. Editorial Committee, Dept. Bot., NTU, Taipei, pp 315–318
Yang YP, Liu HY, Shih BL, Lu SY (1999) Rubiaceae. In: Yang YP, Liu HY, Shih BL, Lu SY (eds) Man. Taiwan Vasc, Pl, vol 4. The Council of Agriculture, Taipei, pp 74–99

22

Six new species of *Begonia* (Begoniaceae) from limestone areas in Northern Vietnam

Ching-I Peng[1*], Che-Wei Lin[2], Hsun-An Yang[1], Yoshiko Kono[1] and Hieu Quang Nguyen[3]

Abstract

Background: Species of *Begonia* are richly represented in limestone karst areas across the Sino-Vietnamese border. More than one hundred species were known, many of which were documented recently.

Results: In continuation of our systematic studies of Asian *Begonia*, we report six species of *Begonia* that are unknown to science, namely *B. caobangensis* [sect. *Platycentrum*], *B. circularis, B. melanobullata, B. langsonensis, B.locii* and *B. montaniformis* [sect. *Coelocentrum*] from Northern Vietnam. Diagnostic features that separate them from morphologically allied species are provided. Somatic chromosome numbers were determined, which supports their placement in the respective sections. Foliar SEM microphotographs were taken and described.

Conclusion: A careful study of the literature, herbarium specimens and living plants, both in the wild and in cultivation in the experimental greenhouse, supports the recognition of the six new species, which are described and illustrated.

Keywords: *Begonia caobangensis*; *Begonia circularis*; *Begonia melanobullata*; *Begonia langsonensis*; *Begonia locii*; *Begonia montaniformis*; Limestone; New species; sect. *Coelocentrum*; sect. *Platycentrum*,Vietnam

Background

Begonia sect. *Coelocentrum* is richly represented in limestone karst areas across the Sino-Vietnamese border region, comprising more than 60 species (Peng et al. 2014a, Chung et al. 2014). Nearly half of the species in sect. *Coelocentrum* were discovered in the past decade (Chung et al., 2014; Peng et al. 2013, 2014a), mostly from southern China. The area from South China to North Vietnam harbors very high levels of biodiversity (Sodhi et al. 2004). Recently, Averyanov and Nguyen (2012) documented ten new species of *Begonia* from Vietnam and one new species from Laos. They also estimated the total actual number of *Begonia* species in eastern Indochina to be 180–200 species. It seems reasonable to speculate that there are more species of *Begonia* sect. *Coelocentrum* to be discovered in the future. In this study we report five new species of sect. *Coelocentrum* from Vietnam, some with very handsome maculation pattern and/or attractive leaf texture and an unusual new species of sect. *Platycentrum* with lanceolate and symmetric leaves. Detailed morphological descriptions, line drawings, color plates and foliar SEM microphotographs are provided, and chromosome numbers are reported.

* Correspondence: bopeng@sinica.edu.tw
[1]Herbarium (HAST), Biodiversity Researach Center, Academia Sinica, Taipei, Nangang 115, Taiwan
Full list of author information is available at the end of the article

Methods

Chromosome preparations

Somatic chromosomes were studied for plants of *Begonia caobangensis* (*Peng 23895*, HAST), *B. circularis* (*Peng 22610*, HAST), *B. melanobullata* (*Peng 22609*, HAST), *B. langsonensis* (*Peng 21946*, HAST), *B. locii* (*Peng 21943*, HAST) and *B. montaniformis* (*Peng 24609, 24610, 24613*, HAST) using root tips. The procedures of pretreatment, fixation, and staining for chromosome observations were described in our previous paper (Peng et al. 2014b). Classification of the chromosome complements based on centromere position at mitotic metaphase followed Levan et al. (1964). Voucher specimens of all new species were deposited in the Herbarium of the Biodiversity Research Center, Academia Sinica (HAST).

Cryo scanning electron microscopy

Fresh leaves of *Begonia caobangensis*, *B. circularis*, *B. melanobullata*, *B. langsonensis*, *B. locii* and *B. montaniformis* were dissected and attached to a stub. The samples were frozen with liquid nitrogen slush, then transferred to a sample preparation chamber at −160°C and etched for 15 min at −85°C. After etching, the temperature reached −130°C for sample fracturing and coating. After coating, the samples were transferred to the SEM chamber and observed at −190°C with a cryo scanning electron microscope (FEI Quanta 200 SEM/Quorum Cryo System PP2000TR FEI).

Results and discussion

Species description

1. *Begonia caobangensis* C.-I Peng & C. W. Lin, sp. nov. (sect. *Platycentrum*) **—TYPE:** VIETNAM. Cao Bang Province, Tra Linh District, Thang Hen Lake, elev. ca. 1000 m. Growing as solitary plants on soil slope of 30° in forest with light to medium shade, plant not situated in proximity to the water but in drier area up the

Figure 1 *Begonia caobangensis* C.-I Peng & C. W. Lin. **A**. Habit; **B**. Stipule, abaxial view; **C**. Bract; **D**, **D'**, **D"**. Staminate flowers; **E**. Androecium; **F**, **G**, **H**. Stamens; **I**, **I'**. Pistillate flowers; **J**, **J'**. Style and stigmatic band; **K**. Capsule; **L**. Cross section of an immature capsule. All from *Peng 23895* (HAST).

bank. Originally collected by Ms. Mary Sizemore on 19 May 2005 and distributed to American Begonia Society as *Begonia* U555. Type specimens pressed from plants cultivated in the experimental greenhouse in Academia Sinica, Taiwan, 9 June 2014, *Peng 23895* (holotype: HAST; isotypes: A, E, HN) (Figures 1 and 2).

Herbs, monoecious, perennial, rhizomatous. **Rhizome** creeping, 1–2 cm thick, internodes 0.2–0.5 cm long, minutely appressed tomentose. **Stipules** deciduous, reddish, ovate, 2–2.5 cm long, 0.7–1 cm wide, keeled, abaxially minutely appressed tomentose, sometimes puberulous along midrib, margin denticulate, apex aristate, arista ca. 1 mm long. **Petioles** reddish to crimson, flat front, rounded back (D-shaped in cross section), 3.5–8 (–10) cm long, 0.5–0.7 cm thick, minutely appressed tomentose to glabrous, narrowly winged on both sides, wings 1–2.5 mm wide. **Leaves** alternate, adaxially lime green, abaxially reddish to crimson, juvenile blade folding inward, mature blade flat, symmetric, narrowly elliptic to elliptic or slightly rhomboid, (8–) 11–21.5 cm long, 3.9–7.3 cm wide, apex acuminate, base attenuate, margin repand to sparsely denticulate, chartaceous to thinly coriaceous, glabrous

Figure 2 *Begonia caobangensis* C.-I Peng & C. W. Lin. **A**. Habit; **B**. Blade, abaxial view with an inflorescence; **C**. Winged petiole; **D**. Cross section of petiole; **E**. Infolded juvenile leaves; **F, G**. Portion of leaf, adaxial and abaxial surface; **H**. Stipules on rhizome; **I**. Inflorescence, showing pistillate flowers; **J**. Bracts; **K, L**. Staminate flowers; **M, N**. Pistillate flowers; **O**. Cross section of an immature capsule; **P**. Capsule. All from *Peng 23895* (HAST).

(minutely appressed tomentose on young leaves), venation 6–8-pinnate on each side of the midrib, secondary veins branching dichotomously or nearly so, tertiary veins weakly percurrent. **Inflorescences** terminal or axillary on upper nodes, cymes dichasial, branched 3–4 times; peduncle 8–12 cm long, tomentose or subglabrous; bracts and bracteoles caducous, reddish, bracts ovate, ca. 1.8 cm long, 0.6–0.9 cm wide, boat-shaped, veins crimson, margin denticulate or ciliate, bracteoles similar but smaller. **Staminate flower:** pedicel 1.3–2.1 cm long, tepals 4, outer 2 broadly ovate, 12–18 mm long, 10–17 mm wide, abaxially reddish, sparsely setulose, inner 2 pinkish, obovate, 13–16 mm long, 6–8 mm wide; androecium actinomorphic, spheric, 6–8 mm across; stamens 85–120; anthers yellow, slightly compressed, lanceolate to oblanceolate, apex obtuse to retuse, ca. 2.5 mm long; filaments shorter than anthers, slightly fused at base. **Pistillate flower:** pedicel 1.2–1.8 cm long, tepals 6, pinkish-white, outer 3 obovate to elliptic, 8–14 mm long, 6–9 mm wide, inner 3 oblong to narrowly elliptic,9–18 mm long, 4–6.5 mm wide; ovary reddish, trigonous-ellipsoid, 12–15 mm long, 4–5 mm thick (wings excluded), 3-winged, wings unequal, reddish-yellow, each tipped with several succulent bristles 2–4 mm long; the 2 lateral wings 5.5–8 mm wide, shallowly triangular, verrucose, abaxial wing crescent-shaped, 10–13 mm wide; styles 2, fused at base, yellow, ca. 5 mm long, stigma strongly undulate and spirally twisted. **Capsule** trigonous-ellipsoid, 18–25 mm long, 4–7 mm thick (wings excluded), greenish or reddish when fresh; wings unequal, lateral wings 7–15 mm wide, abaxial wing crescent-shaped, 20–25 mm wide. Somatic chromosome number, $2n = 22$.

Distribution and ecology

Northern Vietnam, around Thang Hen Lake, Tra Linh District, Cao Bang Province (Figure 3). Situated among limestone mountains, near the border of Vietnam and China. Growing on soil slope in evergreen broad-leaved forest, in light shade, at ca.1,000 m in elevation (Mary Sizemore, pers. comm.). *Begonia caobangensis* occurs also in Tuyen Quang province (Trinh Ngoc Bon, pers. comm.), which is adjacent to the southwest border of Cao Bang.

Etymology

The specific epithet refers to the locality of the type collection.

Notes

Begonia caobangensis resembles *B. aequilateralis* Irmsch., also a member of sect. *Platycentrum*, in having symmetric and elliptic leaves. However, *B. caobangensis* is markedly distinct in many other features such as blade base decurrent (vs. obtuse to cuneate), narrowly winged (vs. not winged) petiole, presence (vs. lacking) of the succulent bristles at tips of ovary wings and pistillate tepals 6 (vs. 5). A comparison of the two species is presented in Table 1. The combination of symmetric leaves, 6-tepalled pistillate flowers and the succulent bristles at tips of ovary wings make it a unique species in sect. *Platycentrum*.

Figure 3 Distribution map of *Begonia caobangensis* (open circle), *B. circularis* (Inverted solid triangle), *B. langsonensis* (open triangle), *B. locii* (solid circle), *B. melanobullata* (square) and *B. montaniformis* (star) in Northern Vietnam.

Table 1 Comparison of *Begonia caobangensis* and *B. aequilateralis*

	B. caobangensis	*B. aequilateralis*
Stipule		
Size (cm)	2–2.5 × 0.7–1	1 × 0.3
Petiole	Narrowly winged	Not winged
Leaf		
Base	Decurrent	Obtuse to cuneate
Staminate flower		
Tepal size (mm)	Outer 10–12 × 4–8, inner 7–15 × 2–6	Outer 12–18 × 10–17, inner 13–16 × 6–8
Pistillate flower		
Tepal number	6	5
Tepal size (mm)	Outer 8–14 × 6–9, inner 9–18 × 4–6.5	Outer 13–17 × 5–7.5, inner 10–14 × 5–7
Ovary size, wings excluded (mm)	12–15 × 4–5	*Ca.* 8 × 3
Wings on ovary body	Lateral wings 5.5–8 mm wide, abaxial wing 10–13 mm wide	Lateral wings ca. 4 mm wide, abaxial wing ca. 7 mm wide
Tip of wings	With succulent bristles	Glabrous
Distribution	Northern Vietnam	Malay peninsula

2. *Begonia circularis* C.-I Peng & C. W. Lin, sp. nov. (sect. *Coelocentrum*) —**TYPE:** VIETNAM. Cao Bang Province, Thach An District, N 22°30'49", E 106°22'40". Seeds collected and presented by Mr. Etsuo Kobayashi. Type collection made from plants raised from seeds, 7 July 2014, *Peng 22610* (holotype: HAST) (Figures 4 and 5).

Herb, monoecious, rhizomatous. **Rhizome** stout, creeping, 1.2–2 cm thick, internodes congested, villous near the petiole insertion. **Stipules** deciduous, reddish, ovate-triangular, ca.1 cm long, 0.8 cm wide, strongly keeled, abaxially villous along midrib, apex aristate, arista ca. 0.5 cm long. **Petioles** red to reddish-green, terete, 5–9.5 (–14) cm long, 0.6–0.8 cm thick, densely pinkish villous. **Leaves** alternate; blade asymmetric, widely ovate to widely elliptic, 13.5–21 cm long, 9.5–17 cm wide, broad side 6–10 cm wide, basal lobes cordate, 4.8–7.5 cm, apex caudate,

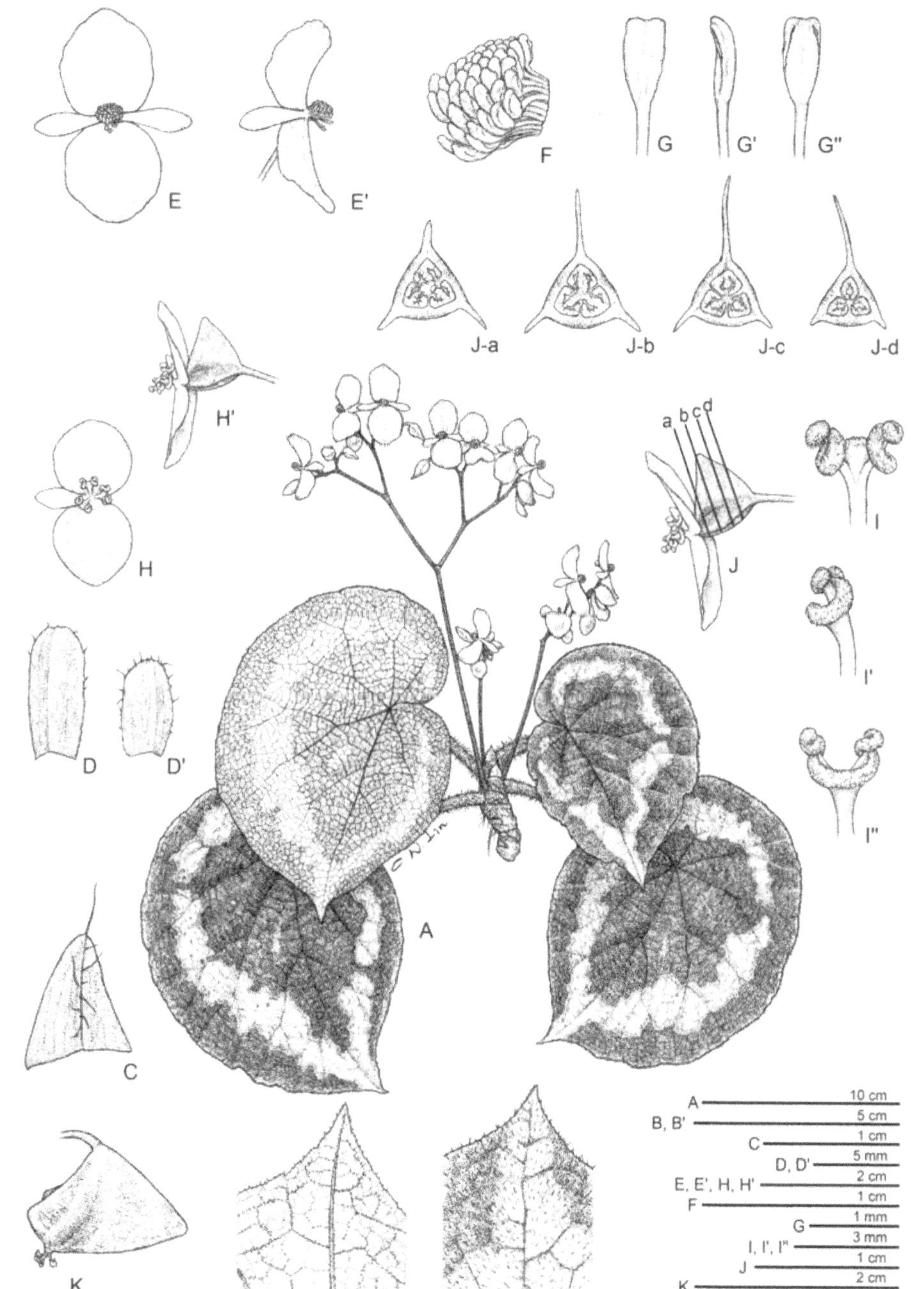

Figure 4 *Begonia circularis* C.-I Peng & C. W. Lin. **A**. Habit; **B**, **B'**. Portion of leaf, abaxial and adaxial surface; **C**. Stipule, abaxial view; **D**, **D'**. Bract and bracteole; **E**, **E'**. Staminate flowers; **F**. Androecium; **G**, **G'**, **G"**. Stamen; **H**, **H'**. Pistillate flowers; **I**, **I'**, **I"**. Style and stigmatic band; **J**. Serial cross sections of an immature capsule; **K**. Capsule. All from *Peng 22610* (HAST).

Figure 5 *Begonia circularis* C.-I Peng & C. W. Lin. **A**, **B**. Habitat; **C**. Habit; **D**, Leaf; **E**, **F**. Portion of leaf, adaxial and abaxial surface; **G**. Petiole; **H**. Bract; **I**. Inflorescence; **J**. Staminate flowers; **K**. Pistillate flowers; **L**. Capsule; **M**. Serial cross section of immature capsule, **N**. Same, closer view. All from *Peng 22610* (HAST).

base strongly oblique-cordate, margin repand to denticulate, thickly chartaceous, densely magenta hispid, surface slightly rugose between veins, adaxially olive-green to dark green, embellished with a wide silvery green to lemon green ring and adorned with small similar-colored patches between primary palmate veins; abaxially reddish to pale magenta, with a ring of pale green corresponding to the pattern on adaxial surface; venation palmate with 6–8 primary veins, midrib distinct, with 2 or 3 secondary veins on each side, secondary veins branching dichotomously or nearly so, tertiary veins weakly percurrent or reticulate; all venation prominently raised and densely magenta hispid abaxially. **Inflorescences** axillary, dichasial cymes branched 3–5 times; peduncle 9.5–16 cm long, subglabrous; bracts and bracteoles caducous, reddish, oblong to ovate, ca. 0.8 cm long, 0.25–0.4 cm wide, boat-shaped, margin sparsely shortly fimbriate, bracteoles similar but smaller. **Staminate flower:** pedicel 0.9–2 cm long, tepals 4, outer 2 broadly ovate, 14–19 mm long, 13–16 mm wide, abaxially reddish, sparsely red setulose to glabrous, inner 2 oblanceolate, pinkish to white, 10–13 mm long, 2–4.5 mm wide; androecium zygomorphic, golfclubshaped, ca. 5 mm across; stamens 45–65; anthers yellow, slightly compressed, obovate, ca. 1.3 mm long, apex retuse, filaments longer than anthers, slightly fused at base. **Pistillate flower:** pedicel 0.7–1 cm long, tepals 3, pinkish-white, outer 2 widely obovate to orbicular, 7–17 mm long, 8–13 mm wide, inner 1 narrowly elliptic, 5–12 mm long, 3–4 mm wide; ovary reddish, trigonous-ellipsoid, 6–10 mm long, 3–4 mm thick (wings excluded), with sparse sessile or subsessile glands, 3-winged, wings unequal, reddish-yellow, shallowly triangular, sparse subsessile-glandular, lateral wings 2–4 mm wide; abaxial wing crescent-shaped, entire, 6–10 mm wide; styles 3, fused at base, yellow, ca. 4 mm long, stigma spirally twisted. **Capsule** trigonous-ellipsoid, 10–17 mm long, 4–9 mm thick (wings excluded), reddish when fresh; wings unequal, shallowly triangular, lateral wings ca. 5 mm wide, abaxial wing 7–17 mm wide. Somatic chromosome number, $2n = 30$.

Distribution and ecology

Endemic to Thach An District, Cao Bang Province, Vietnam (Figure 3), near the border between Vietnam and China. Growing in cracks of mossy rocks on semi-shaded limestone cliffs in evergreen broad-leaved forest, elevation ca. 600 m.

Etymology

The specific epithet refers to the circular foliar variegation of the new species.

Notes

Begonia circularis somewhat resembles *B. lanternaria* Irmsch. and *B. picturata* Yan Liu, S. M. Ku & C.-I Peng (SW Guangxi, China), all members of sect. *Coelocentrum*, in the variegated foliage. A comparison of salient features of the three species is shown in Table 2.

3. *Begonia melanobullata* C.-I Peng & C. W. Lin, sp. nov. (sect. *Coelocentrum*) **—TYPE:** VIETNAM. Cao Bang Province, Thach An District, N 22°31'23", E 106°20'41". Living collection made on June 29, 2010; type specimens pressed from plants cultivated in the experimental greenhouse in Academia Sinica, Taiwan, 22 April 2014, *Ching-I Peng 22609* (holotype: HAST; isotypes: E, HN) (Figures 6 and 7).

Herbs, monoecious, perennial, rhizomatous. **Rhizome** stout, creeping, to 50 cm long,1–2 (–2.5) cm thick, internodes 1–2 cm long, villous near petiole insertion. **Stipules** yellowish to pale green, ovate-triangular, 1.4–2.5 cm long, 1–1.8 cm wide, strongly keeled, abaxially hairy along midrib, apex aristate, arista *ca.* 0.2 cm long. **Petiole** red to olive-green, terete, 8–21 cm long, 0.4–0.8 cm thick, densely white villous when young, turning brownish tomentose, sometimes subglabrous with age. **Leaves** alternate, blade asymmetric, widely ovate to widely elliptic, 13–21 cm long, 9–15 cm wide, broad side 6.5–12 cm wide, basal lobes cordate, 5–8.5 cm, apex caudate, base strongly oblique-cordate, margin repand, thick chartaceous, villous

Table 2 Camparison of *Begonia circularis* with *B. lanternaria* and *B. picturata*

	B. circularis	*B. lanternaria*	*B. picturata* (Liu et al. 2005: figures one and two)
Leaf			
Maculation	A pale ring ca. 1.5–2 cm from the margin against a dark green background	With digitate, green or dark green bands along major veins and leaf margin	With digitate, dark brown bands along major veins and leaf margin
Adaxial surface	Hispid	Puberulous	Villous-setose or tomentose-setose, bullate at base
Abaxial surface	Hispid on all veins	Villous on major veins	Densely shortly villous, particularly on veins
Bract			
Shape	Oblong to ovate, boat-shaped, margin sparsely shortly fimbriate	Elliptic to broadly-ovate, margin glandulose-ciliate	Ovate, oblong to rounded, margin ciliate
Length (mm)	6–10 × 2.5–4	5–9 × 4–6.5	4.5–15 × 4.5–8
Male flower			
Outer tepal size (mm)	14–19 × 13–16	7–9 × 6–8	14–22 × 14–17
Inner tepal size (mm)	10–13 × 2–4	5.5–6 × 2–3	10–19 × 4.–5.5
Vestiture on outer tepals	Sparsely setulose to glabrous	Glandulose-pilose	Villous-setose
Female flower			
Ovary	With minute sessile glands	Glandulose-pilose	Villous-setose or hispid-setose
Wings on ovary	Deltate	Crescent-shaped	Crescent-shaped

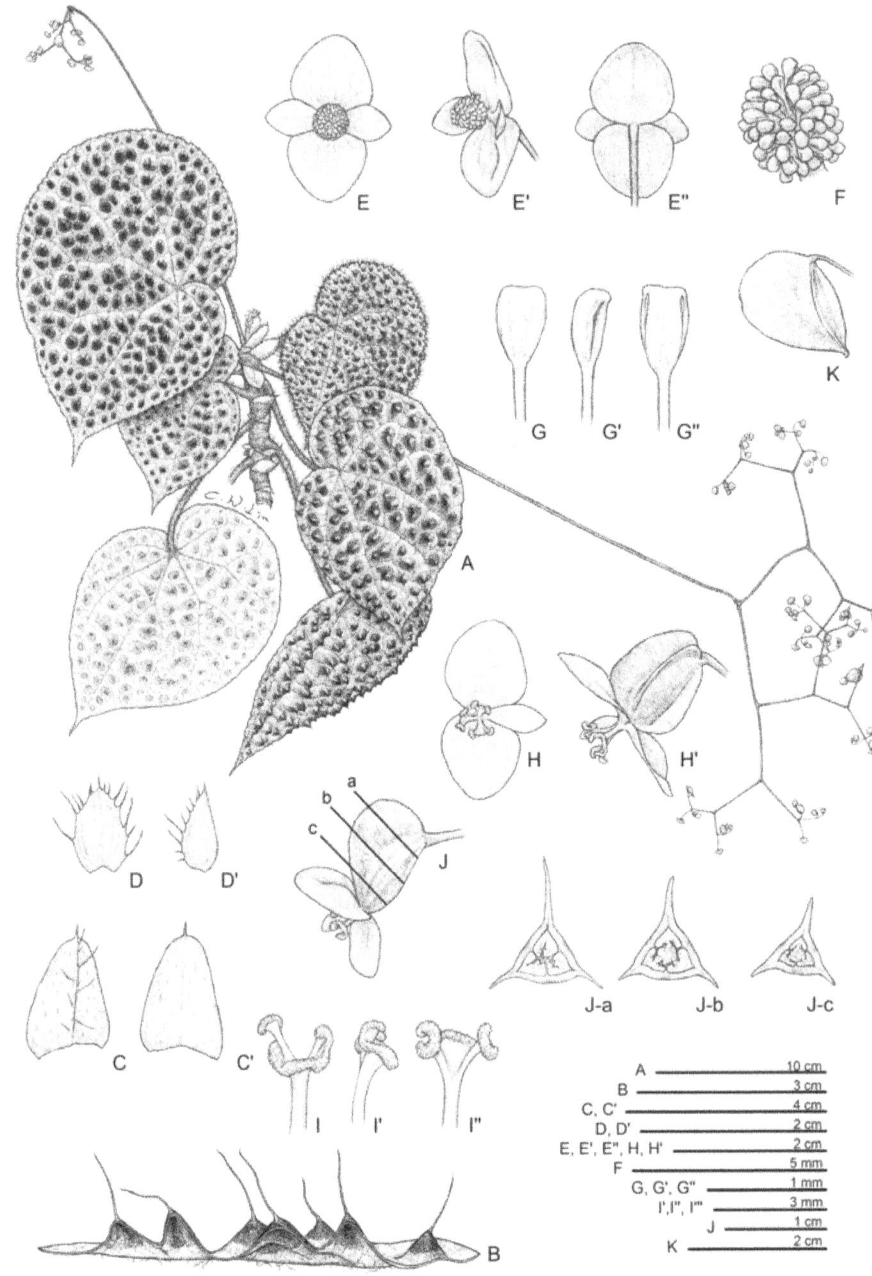

Figure 6 *Begonia melanobullata* C.-I Peng & C. W. Lin. **A**. Habit; **B**. Cross section of leaf, showing bullae; **C**, **C'**. Stipule, both surface views; **D**, **D'**. Bract, abaxial and side views; **E**, **E'**, **E"**. Staminate flowers; **F**. Androecium; **G**, **G'**, **G"**. Stamen; **H**, **H'**. Pistillate flowers; **I**, **I'**, **I"**. Style and stigmatic band; **J**. Cross section of an immature capsule; **K**. Capsule. All from *Peng 22609* (HAST).

when juvenile, adaxially emerald green to yellowish green, surface densely bullate, bulla dark maroon, conical, 2–8 mm high, 3–8 mm across, tipped with a velutinous hair 6–10 mm long, abaxially pale green, reddish on veins and bullae, brownish tomentose on veins; venation palmate with 7–9 primary veins, midrib distinct, with 2–4 secondary veins on each side, tertiary veins reticulate or percurrent, minor veins reticulate. **Inflorescences** arising directly from rhizome, dichasial cymes branched 4–6 times; peduncle 15–38 cm long, tomentose; bracts at cyme base pale green, ovate, ca. 1 cm long, 0. 5 cm wide, boat-shaped, veins reddish, margin fimbriate, bracteoles on upper nodes similar but gradually smaller. **Staminate flower:** pedicel ca. 1.5 cm long, tepals 4, outer 2 broadly ovate, 6–9 mm long, 7–9 mm wide, abaxially greenish-reddish, sparsely setulose, adaxially yellowish-green, inner 2 greenish, elliptic, 5–7 mm long, 3–4 mm wide; androecium actinomorphic, spheroid, 2.5–3.5 mm across; stamens 65–80;

Figure 7 *Begonia melanobullata* C.-I Peng & C. W. Lin. **A**, Habitat; **B**. Habit; **C**, **D**. Portion of leaf, adaxial and abaxial surface, **E**. Same, leaf side view, showing conical bullae, each tipped by an erect hair; **F**. Leaf cross section; **G**. Inflorescence; **H**. Young inflorescence, showing bracts; **I**, **I'**, **I''**. Staminate flowers; **J**, **J'**, **J''**. Pistillate flowers; **K**. Capsule; **L**. Cross section of an immature capsule. All from *Peng 22609* (HAST).

anthers yellowish green to gold yellow, slightly compressed, obovate, 0.5–0.8 mm long, apex retuse to obtuse; filaments fused at base into a column ca. 1.5 mm long; **Pistillate flower:** pedicel ca. 1.5 cm long, tepals 3, outer 2 suborbicular or broadly ovate, abaxially reddish, adaxially yellowish-pinkish, 7–8 mm long, 7–8 mm wide, inner 1 yellowish-pinkish, elliptic, 4–6 mm long, 0.2–0.3 cm wide; ovary trigonous-ellipsoid, 8–10 mm long, 4 mm thick (wings excluded), reddish, sparsely dotted with sessile glands, 3-winged, wings unequal, reddish-green, lateral wings 2–3 mm wide, abaxial wing crescent-shaped, 8–10 mm long, 4–6 mm wide; styles 3, fused at base, yellow or greenish, ca. 3 mm long, stigma spirally twisted. **Capsule** trigonous-ellipsoid, 12–19 mm long, 5–8 mm thick (wings excluded), greenish or reddish when fresh; wings unequal, lateral wings 2–4 mm wide, abaxial wing crescent-shaped, 5–7 mm wide. Somatic chromosome number, $2n = 30$.

Distribution and ecology

Endemic to Thach An District, Cao Bang Province, Vietnam (Figure 3), near the border between Vietnam and China. On semi-shaded wet cliffs or steep limestone slopes in evergreen broad-leaved forest, elevation at 300–400 m.

Etymology

The specific epithet refers to the color of the bullae on leaf surface.

Notes

Begonia melanobullata resembles *B. ferox* C.-I Peng & Yan Liu (Guangxi, China) and *B. nahangensis* Aver. & H. Q. Nguyen (Northern Vietnam) in having blackish bullae on leaf upper surface. However, *Begonia melanobullata* is sharply distinct in other aspects. The new species is similar to *B. ferox*, differing by the leaves widely ovate to widely elliptic (vs. ovate), peduncle 15–38 cm (vs. 5–13 cm) long, inflorescence branched 4–6 (vs. 3–4) times, staminate flower tepals greenish (vs. yellowish-reddish) and smaller anthers. *Begonia melanobullata* also resembles *B. nahangensis*, differing in the leaf apex acuminate (vs. obtuse), staminate flower greenish (vs. white to pinkish); tepals of pistillate flowers yellowish-pinkish (vs. light olive-green), suborbicular or broadly ovate (vs. broadly reniform). A comparison of salient features of the three species is shown in Table 3. *Begonia melanobullata* is known only from limestone areas in Cao Bang Province.

4. *Begonia langsonensis* C.-I Peng & C. W. Lin, sp. nov. (sect. *Coelocentrum*) —**TYPE:** VIETNAM. Lang Son Province, Huu Lung District, Huu Lien Commune, Tam Lai Village. N 21°39'60", E 106°22'53", elev. ca. 140 m. Living collection made on 20 November 2008; type specimens pressed from plants cultivated in the experimental greenhouse in Academia Sinica, Taiwan, 15 November 2012, *Peng 21946-A* (holotype: HAST; isotypes: A, E, HN, MO) (Figures 8 and 9).

Herb, monoecious, rhizomatous. **Rhizome** creeping, to 10 cm or longer, 0.6–1.7 cm thick, internodes congested, glabrous. **Stipules** deciduous, reddish pale green, widely ovate-triangular, 0.7–1 cm long, 0.7–1.1 cm wide, herbaceous, strongly keeled, glabrous, apex aristate, arista ca. 0.3 cm long. **Petioles** brownish green, terete, 13–25 cm long, 0.3–0.7 cm thick, glabrous. **Leaves** alternate, blade asymmetric, obliquely ovate, 11–22.5 cm long, 6–13.5 cm wide, broad side 3.3–6.8 cm wide, basal lobes cordate, 2.7–5.5 cm long, apex attenuate, caudate, base strongly oblique-cordate, margin serrulate and puberulous, thickly chartaceous, adaxially uniformly emerald green, very sparsely

Table 3 Camparison of *Begonia melanobullata* with *B. ferox* and *B. nahangensis*

	B. melanobullata	*B. ferox* (Peng et al. 2013: figures five and six)	*B. nahangensis* (Averyanov and Nguyen 2012: figure seven)
Stipule			
Length	1.4–2.5 cm	1–1.7 cm	0.4–0.6 cm
Leaf			
Apex	Acuminate	Acuminate	Obtuse to nearly rounded
Shape	Widely ovate to widely elliptic	Ovate	Broadly oblique-ovate or oblique-reniform
Conical bullae on upper surface	Present on all leaves; erect hairs tipping bullae persistent	Progressively developed on leaves as the plant matures; hairs on bullae soon deciduous	Unknown
Inflorescence			
Peduncle length (cm)	15–38	5–13	(8–) 10–12 (–14)
Male flower			
Tepal color	Greenish	Pale pinkish-yellow	White to light pink
Tepal size (mm)	Outer 6–9 × 7–9, inner 5–7 × 3–4	Outer 9–11 × 6–11, inner 7–11 × 4	Outer 8–9 in diam., inner 8–9 × 2.5–3.5
Another length (mm)	0.5–0.8	Ca. 1	Ca. 0.5
Female flower			
Tepal color	Yellowish-pinkish	Pinkish-white	Light olive-green
Tepal size (mm)	Outer 7–8 × 7–8, inner 4–6 × 2–3	Outer 8–11 × 7–11 mm, inner 8–9 × 3–4	Outer (4.5) 5–6 (7) × (8) 9–11 (12) mm, inner 5–6 × 3–3.5
Capsule			
Size	1.2–1.9 cm long; abaxial wing crescent-shaped, 0.5–0.7 cm wide	1–1.5 cm long; abaxial wing crescent-shaped, 0.6–0.9 cm wide	0.8–1 cm long; abaxial wing oblique-triangular, 0.4 cm wide
Distribution	Northern Vietnam	Guangxi, China	Northern Vietnam

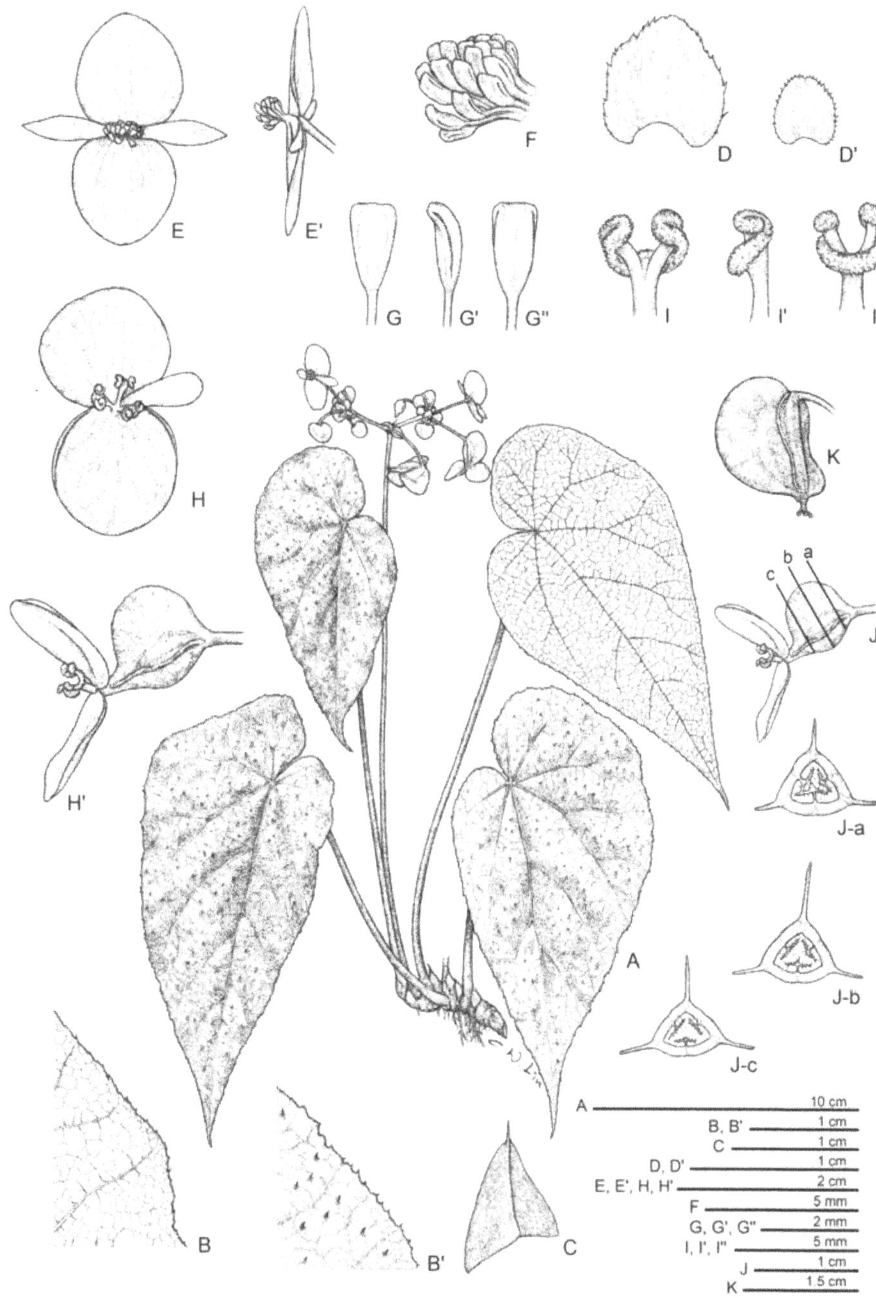

Figure 8 *Begonia langsonensis* C.-I Peng & C. W. Lin. **A**. Habit; **B**, **B'**. Portion of leaf, abaxial and adaxial surface; **C**. Stipule, abaxial view; **D**, **D'**. Bract and bracteole; **E**, **E'**. Staminate flowers; **F**. Androecium; **G**, **G'**, **G"**. Stamen; **H**, **H'**. Pistillate flowers; **I**, **I'**, **I"**. Style and stigmatic band; **J**. Serial cross section of immature capsule; **K**. Capsule. All from *Peng 21946* (HAST).

minutely scabrescent, abaxially pale green, sparingly puberulous or subglabrous; venation palmate with 6–8 primary veins, midrib distinct, secondary veins 2–4 on each side, minor veins reticulate, all venation prominently raised on upper surface, major and secondary veins prominent on lower surface. **Inflorescences** axillary, arising directly from rhizome, dichasial cymes branching 4–6 times; peduncle 12–26 cm long, subequal to leaves, glabrous; bracts and bracteoles caducous, white, pale green to pinkish, bracts widely ovate to very widely ovate, 0.6–0.9 cm long, 0.6–0.8 cm wide, flat, margin denticulate to biserrate, puberulous or glandular hairy, bracteoles widely ovate to orbicular, margin denticulate, glandular hairy, gradually smaller toward upper part of the inflorescence. **Staminate flower:** pedicel ca. 2.2 cm long, tepals 4, outer 2 broadly ovate to orbicular, 10–15 mm long, 11–14 mm wide, abaxially pink, glabrous, adaxially pinkish, inner 2 pinkish, lanceolate, 6–9 mm long, 2.5–3.5 mm wide; androecium zygomorphic,

Figure 9 *Begonia langsonensis* C.-I Peng & C. W. Lin. **A**. Habitat; **B**. Habit; **C**, **D**. Portion of leaf, adaxial and abaxial surface; **E**. Stipule; **F**. Inflorescence; **G**. bract and bracteole; **H**. Staminate flowers; **I**. Pistillate flowers; **J**. Cross section of immature capsule; **K**. Capsule. All from *Peng 21946* (HAST).

ca. 5 mm across; stamens 25–40; filaments coalescent, not forming a stalk at base; anthers yellow, slightly compressed, obovate, ca. 1.5 mm long, apex retuse, shorter than filaments. **Pistillate flower:** pedicel 1.8–2.6 cm long, tepals 3, outer 2 pink or pinkish on both surfaces, orbicular, 11–15 mm long, 13–18 mm wide, inner 1 pinkish, oblanceolate, 7–11 mm long, 3–4 mm wide; ovary whitish green to pinkish, trigonous-ellipsoid, 12–15 mm long, ca. 5 mm thick (wings excluded), with sparse sessile glands, 3-winged, wings unequal, pinkish, lateral wings narrower, 6–4.5 mm wide, abaxial wing crescent-shaped, 5–7 mm wide; styles 3, fused at base, yellow, 4–5.5 mm long, stigma spirally twisted. **Capsule** trigonous-ellipsoid, 13–16 mm long, ca. 5 mm thick (wings excluded), greenish or reddish when fresh; wings unequal, lateral wings 3–5 mm wide, abaxial wing crescent-shaped, 7–9 mm wide. Somatic chromosome number, $2n = 30$.

Distribution and ecology

Endemic to Huu Lung Protected Area, Huu Lien District, Lang Son Province, Vietnam (Figure 3). On semishady cliff face or steep limestone slopes in evergreen broad-leaved forest, elevation ca. 125–165 m.

Etymology

The specific epithet refers to Lang Son Province, where the new species was discovered.

Notes

Begonia langsonensis somewhat resembles *B. ornithophylla* Irmsch. (southwest Guangxi, China) in leaf shape and aspect. However, *B. langsonensis* is nearly glabrous throughout, whereas *B. ornithophylla* has sericeous petiole, puberulous abaxial blade, as well as velutinous to

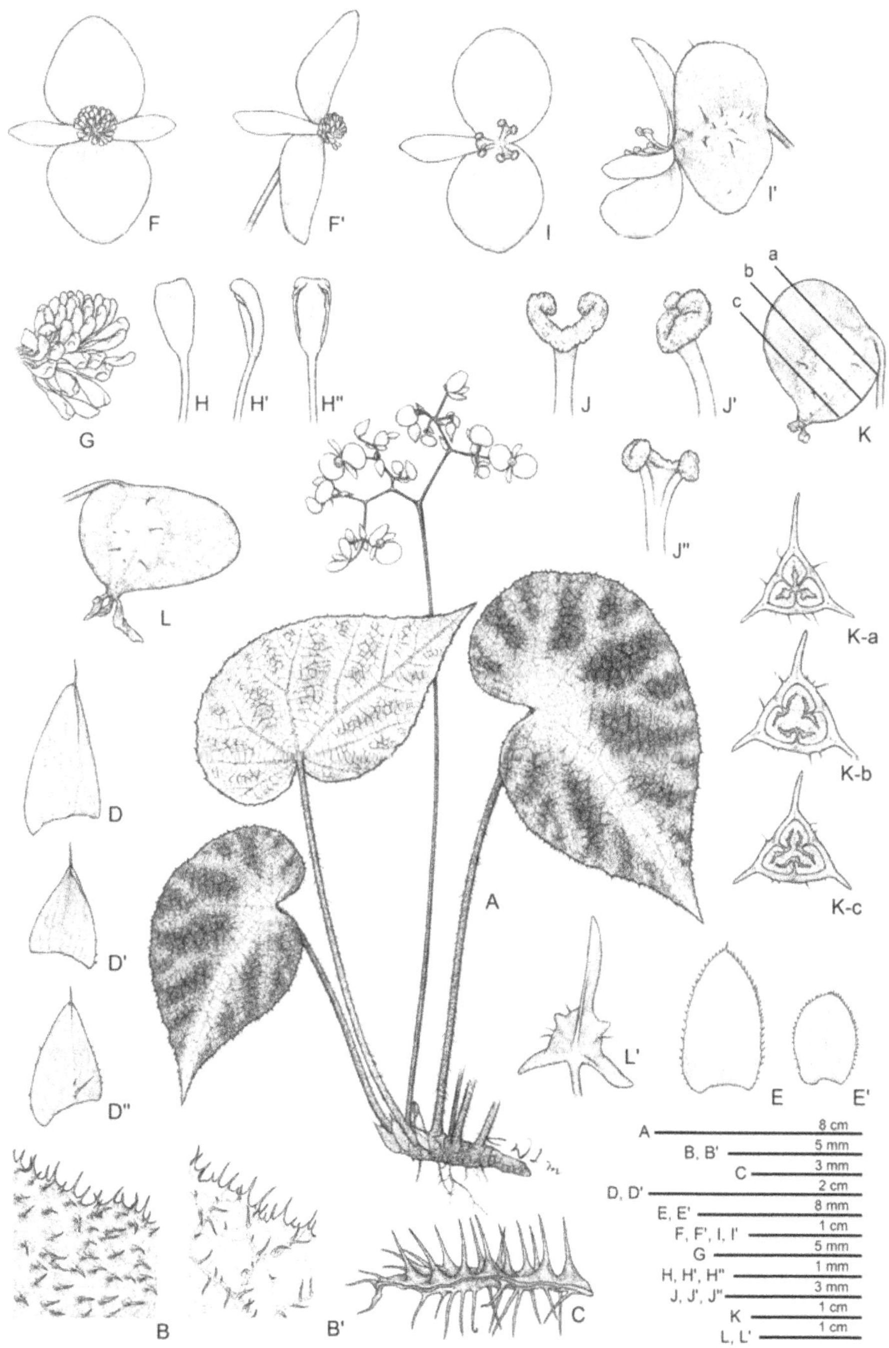

Figure 10 *Begonia locii* C.-I Peng, C. W. Lin & H. Q. Nguyen. **A**. Habit; **B**, **B'**. Portion of leaf, adaxial and abaxial surface; **C**. Cross section of leaf; **D**, **D"**. Stipule, abaxial view, **D'**. adaxial view; E. Bract and bracteole; **F**, **F'**. Staminate flowers; G. Androecium; **H**, **H'**, **H"**. Stamen; **I**, **I'**. Pistillate flowers; **J**, **J'**, **J"**. Style and stigmatic band; **K**. Serial cross section of an immature capsule; **L**. Capsule, **L'**. Capsule, back view, showing the occasional presence of 2 additional, reduced wings. All from *Peng 21943* (HAST).

villous pedicel and peduncle. In addition, *B. langsonensis* differs from the latter in the bracts widely to very widely ovate (vs. ovate-triangular) as well as the primary and secondary veins on leaf adaxial surface elevated (vs. sunken).

5. ***Begonia locii*** C.-I Peng, C. W. Lin & H. Q. Nguyen, sp. nov. (sect. *Coelocentrum*) **—TYPE:** VIETNAM. Lang Son Province, Huu Lung District, Huu Lien Commune, Tam Lai Village, N 21°39'60", E 106°22'53', elev. ca. 185 m. Living collection made on 20 November 2008; type specimens pressed from plants cultivated in the experimental greenhouse in Academia Sinica, Taiwan, 8 April 2010, *Peng 21943-A* (holotype, HAST; isotype, A, E, HN, KEP, MO) (Figures 10 and 11).

Herb, monoecious, rhizomatous. **Rhizome** stout, creeping, to 30 cm long, 0.8 – 1.6 (–2) cm thick, internodes 0.5–1.3 cm long, subglabrous. **Stipules** deciduous, reddish to olive with red veins, ovate-triangular, 0.6–1.4 cm long, 0.5–0.85 cm wide, strongly keeled, abaxially sparsely velutinous along midrib, margin entire or sparsely ciliate-dentate, apex aristate, arista ca. 0.25 cm long. **Petioles** pale brownish-green to rwddish, terete, (11–) 20–25 (–33) cm long, 0.3–

Figure 11 *Begonia locii* C.-I Peng, C. W. Lin & H. Q. Nguyen. **A, B.** Habit and habitats; **C.** Habit; **D, E.** Portion of leaf, adaxial and abaxial surface; **F.** Stipule; **G.** Inflorescence; **H.** Staminate flowers; **I.** Pistillate flowers; **J.** Capsule, side view, showing the additional reduced wing (arrows); **K.** Capsule, back view, showing 3 wings; **L.** Capsule; **M.** Cross section of an immature capsule. All from *Peng 21943* (HAST).

0.6 cm thick, densely villous. **Leaves** alternate, blade asymmetric, obliquely ovate, (9–) 11–22 cm long, 6.5–12.5 cm wide, broad side 3.2–7.4 cm wide, basal lobes cordate, 3.3–6.6 cm long, chartaceous, apex acuminate, base strongly obliquely cordate, margin denticulate and densely puberulous with pale magenta hairs, adaxially deep olive-green to maroon with light green zones along primary and secondary veins, sometimes embellished with small bits of silvery white spots, midrib veins forming a wide silvery white zone; upper surface densely covered by small raised cones each topped by a magenta hair, giving a wrinkled texture; abaxially pale green, purplish red between primary and secondary veins, reddish-white tomentose on all veins; venation palmate with 7–8 primary veins, midrib distinct, with 3–5 secondary veins on each side, tertiary veins red, percurrent or reticulate. **Inflorescences** axillary, arising directly from rhizome, dichasial cymes branched 3–4 times; peduncle 13–35 cm long, tomentose or subglabrous; bracts and bracteoles caducous, pale yellow-green, ovate to elliptic, orbicular, bracts ca. 0.8 cm long, 0.5 cm wide, margin gland-tipped denticulate, bracteoles similar but smaller. **Staminate flower:** pedicel ca. 1 cm long, tepals 4, outer 2 broadly ovate to orbicular, 9–12 mm long, 8–11 mm wide, abaxially white to pinkish, adaxially whitish, glabrous, inner 2 elliptic, white, 7–11 mm long, 3–4.5 mm wide; androecium zygomorphic, subglobose, ca. 4 mm across; stamens 35–60; filaments fused at base; anthers yellow, slightly compressed, obovate, ca. 1 mm long, apex retuse, shorter than filaments. **Pistillate flower:** pedicel ca. 1 cm long, tepals 3, outer 2 pinkish to white, suborbicular or broadly ovate, 8–11 mm long, 8–12 mm wide, inner 1 white, oblanceolate to obovate, 6–10 mm long, 3–5 mm wide; ovary trigonous-ellipsoid, sometimes strongly keeled at middle, 7–12 mm long, 3–4.5 mm thick (wings excluded), whitish, sparsely hirsute and glandular, 3-winged, wings whitish green to pinkish, unequal, lateral wings narrower, narrowly crescent-shaped to trapezium, 2.5–6 mm wide, abaxial wing crescent-shaped, 4.5–7 mm wide, sometimes with 2 very reduced wings running parallel along the side of the ovary, sparsely hirsute to glabrous, margin entire to hirsute; styles 3, fused at base, yellow to greenish, 3–4.5 mm long, stigma spirally twisted. **Capsule** trigonous-ellipsoid, 10–15 mm long, 4–5 mm thick (wings excluded), greenish or reddish when fresh, glabrous to hirsute; wings unequal, lateral wings 4–6 mm wide, abaxial wing crescent-shaped, 6–7 mm wide. Somatic chromosome number, $2n = 30$.

Additional specimens examined

Vietnam. Lang Son, Huu Lung District, Huu Lien Commune, Tam Lai Village, E106°22'53", N21°39'60", elevation ca. 185 m, *Peng 21938* (HAST).

Distribution and ecology

Begonia locii is endemic to Huu Lung Protected Area, Huu Lien District, Lang Son Province, Vietnam (Figure 3), occurring on semi-shaded limestone rock face, elevation 125–165 m.

Etymology

Begonia locii is named in honor of Professor Phan Ke Loc for his contribution to Vietnamese botany.

Notes

Begonia locii resembles *B. luochengensis* S.M. Ku, C.-I Peng & Yan Liu from Guangxi, China in leaf shape and maculation pattern, but is clearly distinct from the latter in the percurrent and spiderweb-like leaf venation (vs. pinnate) near the attachment of the petioles; upper leaf surface densely covered with small raised cones (vs. nearly flat); and hirsute (vs. glabrous) ovaries. The new species is also similar to *B. pengii* S. M. Ku & Yan Liu in leaf maculation pattern, vestiture, and hirsute ovaries, differing by the leaf basifixed (vs. peltate); bract margin denticulate and glandular (vs. denticulate and ciliate). A comparison of salient characters of the three species, all of sect. *Coelocentrum*, is shown in Table 4.

6. *Begonia montaniformis* C.-I Peng, C. W. Lin & H. Q. Nguyen, sp. nov. (sect. *Coelocentrum*) —**TYPE:** North VIETNAM. Thanh Hoa Prov., Ba Thuoc Distr., Co Lung Municipality, Eo Dieu village. Primary evergreen seasonal broad-leaved closed lowland forest on crystalline marble-like highly eroded limestone, very steep rocky slopes and cliffs of S exposition at elev. 450–600 m. Lithophytic herb on shady cliff. Introduced from home garden plants in Hanoi for cultivation in Academia Sinica. Type specimen pressed from the greenhouse plant on 5 June 2006. *Ching-I Peng 24609* (holotype: HAST; isotype: HN) (Figures 12 and 13).

Herbs, monoecious, perennial, rhizomatous. **Rhizome** stout, creeping, to 30 cm long, 1–2.3 cm thick, internodes 0.5–1.7 cm long, villous near the petiole insertion. **Stipules** pale green to pinkish, ovate-triangular, 0.7–1.4 cm long, 0.6–1.3 cm wide, strongly keeled, abaxially hairy along midrib, margin velutinous, apex aristate, arista ca. 0.2 cm long. **Petiole** red to greenish-purple, terete, 7–20 cm long, 0.4–0.7 cm thick, densely white villous. **Leaves** alternate, blade asymmetric, ovate to widely ovate, 11.5–19 cm long, 8–13.5 cm wide, broad side 6–8 cm wide, apex acuminate to acute, base strongly oblique-cordate, margin repand to shallowly denticulate and ciliate, thickly chartaceous to succulent, adaxially blackish-malachite green, purplish-olive or dark bluish-brown, with silvery green zone along primary and secondary veins, surface bullate, bulla conical, 0.4–0.8 (–1) cm high, 0.5–0.8 cm across, tipped by 2–4 (–6) peak-like hispidulous protrusions, abaxially pale green, reddish

Table 4 Camparison of *Begonia locii* with *B. luochengensis* and *B. pengii*

	B. locii	*B. luochengensis* (Ku et al. 2004: figures four and five)	*B. pengii* (Ku et al. 2008: figures one and two)
Leaf			
Attachment	Basifixed	Basifixed	Peltate
Venation near petiole	Spider-web like	Pinnate	Spider-web like
Adaxial vestiture	Scabrous, bullate at base	Villous	Scabrous, bullate at base
Stipule	Triangular-ovate, abaxially sparsely velutinous along midrib, margin entire or sparsely ciliate-dentate	Widely triangular-ovate, abaxially sparsely villous to glabrous, margin ciliate-dentate or long-ciliate	Narrowly triangular-ovate, glabrous, margin eciliate
Bract			
Margin	Gland-tipped denticulate	Sparsely ciliate-serrate.	Denticulate and ciliate
Inflorescence			
Peduncle	Tomentose to glabrous	Glabrous	Pilose or hispid-villous
Female flower			
Ovary	Hirsute	Glabrous	Pilose or villous-pilose
Capsule			
Body size (mm)	10–15 × 4–5	11–17 × 6–7	18–25 × 6–13
Width of lateral wings (mm)	4–6	3–4	2–3
Width of abaxial wing (mm)	6–7	10–12.5	7–11

on hollow bullate region, densely brownish-white tomentose on veins; venation palmate with 7–9 primary veins, midrib distinct, 2–4 secondary veins on each side, tertiary veins percurrent or reticulate, sometimes spiderweb-like near petiole attachment, minor veins reticulate. **Inflorescences** axillary, arising directly from rhizome, dichasial cymes branching 3–8 times; peduncle 11–28 cm long, tomentose to subglabrous; bracts and bracteoles caducous, yellowish-green to reddish, bracts ovate, ca. 1 cm long, 0. 5 cm wide, boat-shaped, abaxially sparsely velutinous along midrib, margin densely tomentose, bracteoles similar to bracts but gradually smaller on upper part of the inflorescence. **Staminate flower:** pedicel 1.1–2 cm long, tepals 4, outer 2 broadly ovate to obovate-orbicular, 7–12 mm long, 6–10 mm wide, abaxially yellowish-green to reddish, sparsely velutinous, adaxially yellowish to pinkish-green, inner 2 greenish, elliptic, 6–8 mm long, 3–5 mm wide; androecium sub-actinomorphic, spherical, ca. 5 mm across; stamens 55–90; anthers yellow, slightly compressed, obovate, ca. 1 mm long, apex obtuse, filaments fused at base into a column. **Pistillate flower:** pedicel 1.4–2 cm long, tepals 3, outer 2 suborbicular or broadly ovate, abaxially reddish-green, adaxially yellowish-greenish, 8–12 mm long, 9–13 mm wide, inner 1 yellowish-greenish, elliptic, 5–8 mm long, 2.5–4.5 mm wide; ovary reddish-green, trigonous-ellipsoid, 5–9 mm long, ca. 3 mm thick (wings excluded), sparsely sessile-glandular, 3-winged, wings unequal, reddish-green, lateral wings 2–3 mm wide, abaxial wing crescent-shaped, 3–8 mm wide; styles 3, fused at base, yellow or greenish, *ca.* 4 mm long, stigma spirally twisted. **Capsule** trigonous-ellipsoid, 7–11 mm long, 3–4 mm thick (wings excluded), greenish or reddish when fresh; wings unequal, lateral wings 2–4 mm wide, abaxial wing crescent-shaped, 4–9 mm wide. Somatic chromosome number, $2n = 30, 31, 32, 33$.

Additional specimens examined

VIETNAM, Thanh Hoa Prov., Ba Thuoc Distr., Co Lung Municipality, Eo Dieu village. Primary evergreen seasonal broadleaved and coniferous submontane forest on crystalline marble-like highly eroded limestone, very steep upper slopes and tops of limestone rocky ridge at elev. 800–850 m. Lithophyte. Introduced from home garden plants in Hanoi for cultivation in Academia Sinica. Specimens pressed from greenhouse plants on 1 April 2013, *Ching-I Peng 24610* (HAST); Khuyn village. Primary evergreen seasonal broadleaved lowland forest on crystalline marble-like highly eroded limestone, N slopes of limestone ridge at elev. 400–460 m. Lithophyte on vertical cliff. Occasional. Introduced from home garden plants in Hanoi for cultivation in Academia Sinica. Specimens pressed from greenhouse plants on 22 April 2014, *Ching-I Peng 24613* (HAST).

Distribution and ecology

Lithophytic herb on shady cliff. Endemic to Ba Thuoc Dist, Thanh Hoa Province, Northern Vietnam (Figure 3), known only from the type locality. In primary evergreen seasonal broad-leaved closed lowland forest on crystalline

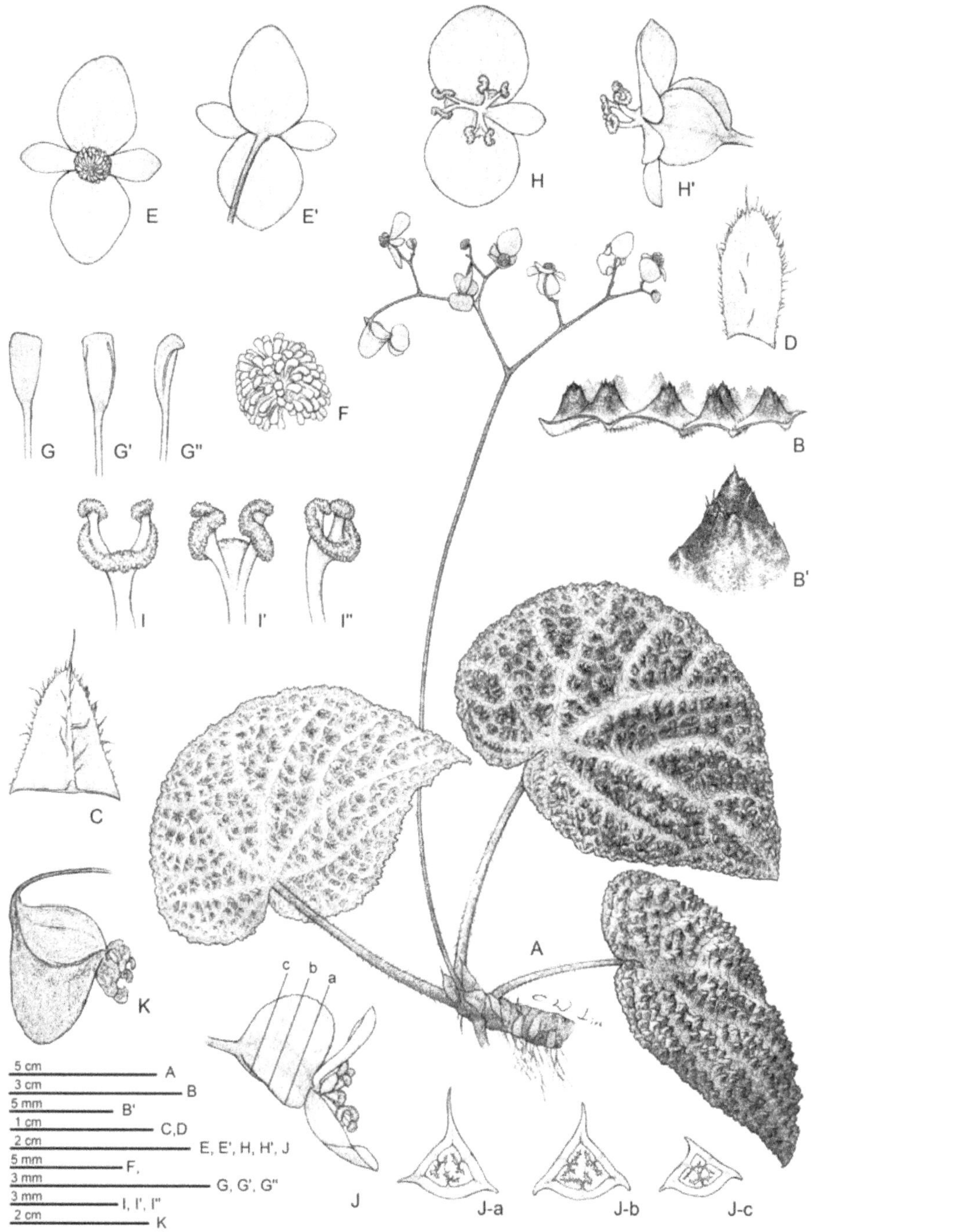

Figure 12 *Begonia montaniformis* C.-I Peng, C. W. Lin & H. Q. Nguyen. **A**. Habit; **B**. Cross section of leaf, **B'**. 2–4 tipped conic bulla; **C**. Stipule, abaxial view; **D**. Bract; **E**, **E'**. Staminate flowers; **F**. Androecium; **G**, **G'**, **G"**. Stamen; **H**, **H'**. Pistillate flowers; **I**, **I'**, **I"**. Style and stigmatic band; **J**. Cross section of an immature capsule; **K**. Capsule. All from *Peng 24609* (HAST).

marble-like highly eroded limestone, very steep rocky slopes and cliffs at 450–600 m elevation.

Etymology

The specific epithet refers to the leaf surface with dense conic bullae, which is reminiscent of the peak clusters in Sino-Vietnamese limestone karst landform.

Notes

Begonia montaniformis resembles *B. ferox* C.-I Peng & Yan Liu, differing in the stipule margin tomentose (vs. entire), veins silvery (vs. dark green), each bullae 2–4-tipped (vs. 1-tipped) and shortly hispidulous (vs. long velutinous), staminate flower yellowish-green (vs. whitish). The new species is also similar to *B. nahangensis* Aver. & H. Q.

Figure 13 *Begonia montaniformis* C.-I Peng, C. W. Lin & H. Q. Nguyen. **A**, **B**. Habit; **C**. Leaf, abaxial surface; **D**, **E**, **F**. Portion of leaf, showing conic bullae on adaxial surface; **G**. Inflorescence; **H**, **I**. Staminate flowers; **J**, **K**. Pistillate flowers; **L**. Serial cross sections of an immature capsule. All from *Peng 24609* (HAST).

Nguyen, another bullate-leaved species recently documented from north Vietnam (Averyanov & Nguyen 2012), differing in the leaves widely ovate to widely elliptic (vs. broadly oblique-ovate or oblique-reniform), apex caudate (vs. obtuse) and each conic bullae 2–4-tipped (vs. 1-tipped). A detailed comparison of the new species with *B. ferox* and *B. nahangensis* are provided in Table 5.

Chromosome cytology

1. *Begonia caobangensis* (*Peng 23895*), $2n = 22$, Figure 14A

The 22 chromosomes, ranging from ca. 1.1 to 1.8 μm long, show a gradual change in length. Most chromosomes had centromeres at the median/submedian positions. No satellite chromosomes were observed.

2. *Begonia circularis* (*Peng 22610*), $2n = 30$, Figure 14B

Among the 30 chromosomes, the first pair of median chromosomes (ca. 2.1 μm long) was relatively longer than the rest and the remaining 28 gradually varied from ca. 1.2 to 1.9 μm long. Except for the shorter chromosomes, most

Table 5 Camparison of *Begonia montaniformis* with *B. ferox* and *B. nahangensis*

	B. montaniformis	*B. ferox*	*B. nahangensis*
Leaf			
Bullae	2–4-tipped	Uni-tipped	Uni-tipped
Color	Silvery green along primary and secondary veins against blackish-malachite green background	Embellished dark maroon against green background	Embellished dark maroon against green background
Inflorescence			
Peduncle length (cm)	11–26	5–13	(8) 10–12 (14)
Male flower			
Tepal color	Yellowish to pinkish-green	Pale pinkish yellow	White to light pink
Female flower			
Tepal color	Yellowish-greenish	Pinkish-white	Light olive-green
Tepal duration	Persistent	Deciduous	Unknown
Capsule			
Abaxial wing	Oblique-triangular	Crescent-shaped	Oblique-triangular

chromosomes had centromeres at the median/submedian positions. No satellite chromosomes were observed.

3. *Begonia melanobullata* (*Peng 22609*), $2n = 30$, Figure 14C

Among the 30 chromosomes, the first pair of median chromosomes (ca. 1.7 μm long) was longer than the remaining 28 chromosomes (ca. 0.9–1.5 μm). Most chromosomes had centromeres at median/submedian positions. Satellites were not observed.

4. *Begonia langsonensis* (*Peng 21946*), $2n = 30$, Figure 14D

The 30 chromosomes gradually varied from ca. 0.9 to 1.2 μm long. Because of the small chromosome size, centromere positions of chromosomes were difficult to ascertain. Satellites were not observed.

5. *Begonia locii* (*Peng 21943*), $2n = 30$, Figure 14E

The 30 chromosomes gradually varied from ca. 0.9 to 1.7 μm long. Almost all chromosomes had centromeres at median/submedian positions. Satellites were observed at distal regions of short arms in the second pair of the longest median chromosomes.

6. *Begonia montaniformis* (*Peng 24609, Peng24610, Peng24613*), $2n = 30, 31, 32, 33$, Figure 15

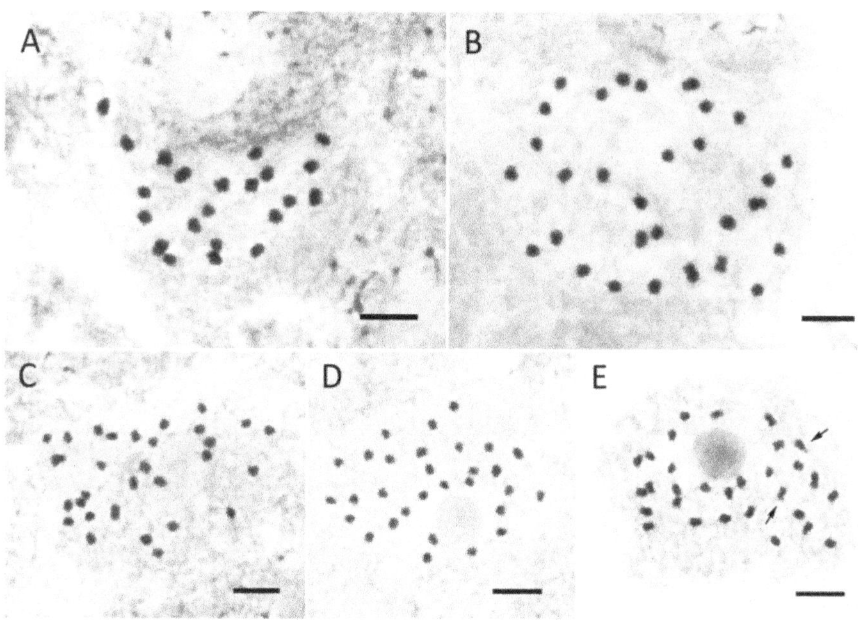

Figure 14 Somatic chromosomes at metaphase of *Begonia*. **A**. *B. caobangensis* (2*n* = 22, *Peng 23895*); **B**. *B. circularis* (2*n* = 30, *Peng et al. 22610*); **C**, *B. melanobullata* (2*n* = 30, *Peng 22610*); **D**. *B. langsonensis* (2*n* = 30, *Peng et al. 21946*); **E**, *B. locii* (2*n* = 30, *Peng et al. 21943*), arrows indicate median chromosomes with satellites. Scale bars = 5 μm.

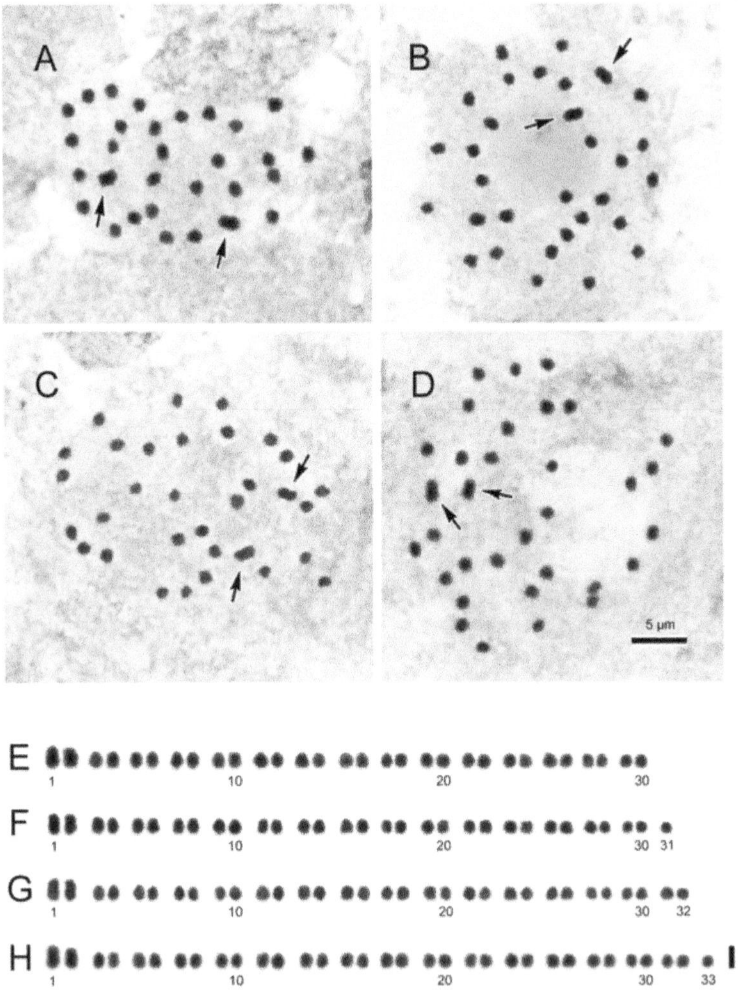

Figure 15 Somatic chromosomes at metaphase of *Begonia montaniformis* (*Peng 24610*, HAST). **A–D**, micrographs; arrows indicate a pair of longer metacentric chromosomes. **E–H**, somatic chromosomes serially arranged by chromosome length and centromere position. **A**, **E**, $2n = 30$; **B**, **F**, $2n = 31$; **C**, **G**, $2n = 32$; **D**, **H**, $2n = 33$. Scale bars = 5 μm **(A–D)**; 2 μm **(E–H)**.

The holotype (*Peng 24609*) and one of the paratypes (*Peng 24613*) had $2n = 30$, which is the prevalent chromosome number for members of sect. *Coelocentrum*. However, another collection (*Peng 24610*) showed aneuploidy with $2n = 30$, 31, 32, 33 in different individuals. The aneuploids shared chromosomal features in common in the respective chromosome complements: The first two chromosomes were relatively longer, about 1.8–2.0 μm (arrows in Figures A-D; Nos. 1, 2 in Figures E-H), and the rest were shorter, about 1.0-1.5 μm long. The two longest chromosomes were clearly metacentric, however, the centromere positions of most other chromosomes could not be determined. Satellites were not observed.

To sum up, the chromosome number $2n = 22$ for *B. caobangensis* agrees with those of the majority species delimited in Asian section *Platycentrum* (Y. Kono and C.-I Peng, unpublished data). The chromosome count of $2n = 30$ in four species (*Begonia circularis*, *B. melanobullata*, *B. langsonensis*, *B. locii*) of sect. *Coelocentrum* is consistent with previous studies on members in this section (Peng et al. 2013, 2014a). Unexpected aneuploid chromosome numbers ($2n = 30$, 31, 32, 33) however were observed in *B. montaniformis*. A review of literature revealed that aneuploidy was documented in several species of *Begonia*, e.g. *B. formosana* ($2n = 60$, 64: Oginuma and Peng, 2002); *B. longipetiolata* ($2n = 34$, 36, 38, 71–73: Arends, 1992).

Leaf anatomy and vestiture

1. *Begonia caobangensis*

Cross section ca. 550 μm thick; epidermis single-layered on both surfaces, hypodermis absent; palisade parenchyma cells 1-layered; spongy parenchyma cells ca. 10-layered

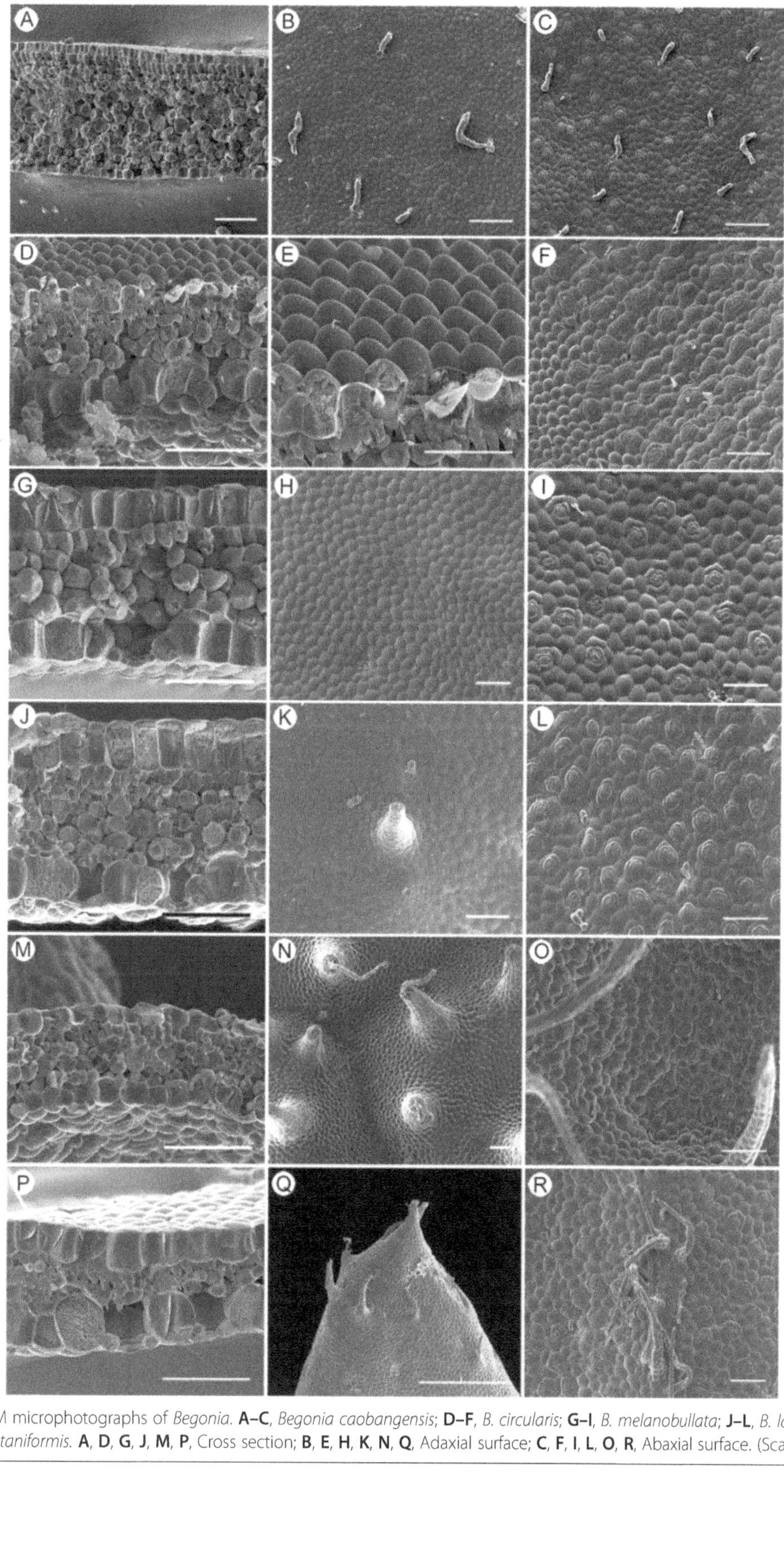

Figure 16 Leaf SEM microphotographs of *Begonia*. **A–C**, *Begonia caobangensis*; **D–F**, *B. circularis*; **G–I**, *B. melanobullata*; **J–L**, *B. langsonensis*; **M–O**, *B. locii*; **P–R**, *B. montaniformis*. **A**, **D**, **G**, **J**, **M**, **P**, Cross section; **B**, **E**, **H**, **K**, **N**, **Q**, Adaxial surface; **C**, **F**, **I**, **L**, **O**, **R**, Abaxial surface. (Scales = 200 μm).

(Figure 16A); adaxial and abaxial surfaces with glandular trichomes (Figures 16B,C); stomata complex single, helicocytic, flat, subsidiary cells 3 (Figure 16C).

2. *Begonia circularis*

Cross section ca. 250 µm thick; epidermis single-layered on both surfaces, hypodermis absent; palisade parenchyma cells 1-layered; spongy parenchyma cells ca. 3-layered (Figure 16D); adaxial surface conoidal (Figure 16E) like those of *B. pengii* (Ku et al. 2008: figure six C); abaxial surface with sparse glandular trichomes (Figure 16F); stomata complex single, helicocytic, slightly elevated, subsidiary cell 5–6 (Figure 16F).

3. *Begonia melanobullata*

Cross section ca. 390 µm thick, epidermis single-layered, hypodermis absent; palisade parenchyma cells 1-layered; spongy parenchyma cells ca. 3-layered (Figure 16G); both surfaces with sparse glandular trichomes (Figures 16H,I); stomata complex single, helicocytic, nearly flat, subsidiary cell 3–7 (Figure 16I).

4. *Begonia langsonensis*

Cross section ca. 420 µm thick, epidermis single-layered, hypodermis absent; palisade parenchyma cells 1-layered; spongy parenchyma cells ca. 5-layered (Figure 16J); both surfaces with glandular trichomes (Figures 16K,L); stomata complex single, helicocytic, nearly flat, subsidiary cell 5 (Figure 16L).

5. *Begonia locii*

Cross section ca. 200 µm thick; epidermis single-layered, hypodermis absent; palisade parenchyma cells 1-layered; spongy parenchyma cells ca. 4-layered (Figure 16M); adaxial surface slightly conoidal and with multiseriate trichomes (Figure 16N), both surfaces with sparse glandular trichomes (Figures 16N,O); stomata complex single, helicocytic, slightly elevated, subsidiary cell 4–5 (Figure 16O).

6. *Begonia montaniformis*

Cross section ca. 280 µm thick, epidermis single-layered, hypodermis absent; palisade parenchyma cells 1-layered; spongy parenchyma cells ca. 2-layered (Figure 16P); adaxial surface conical-bullate, each bullate with several multiseriate trichomes near the tip, abaxial surface with multiseriate trichomes on veins, both surfaces with sparse glandular trichomes (Figure 16Q,R); stomata complex single, helicocytic, nearly flat, subsidiary cell 5 (Figure 16R).

Conclusion

A careful study of the literature, herbarium specimens and living plants, both in the wild and in cultivation in the experimental greenhouse, supports the recognition of the six new species, which are described and illustrated. Our study also suggests that there are more new species in limestone karst areas across the Sino-Vietnamese border region to be discovered in the future.

Competing interests

The authors declare that they have no competing interests.

Authors' contributions

CIP collected type specimens, studied literature and herbarium materials and prepared the final version of the manuscript; CWL drafted the earlier version of the manuscript and prepared color plates and line drawings; HAY carried out the Cryo-SEM studies and helped with various versions of the manuscript; YK carried out the cytological studies. HQN accompanied a field trip and provided discussion. All authors read and approved the final manuscript.

Acknowledgments

We thank Prof. Tien Hiep Nguyen for applying for the collect permits, logistic supports and accompanying the field trips; Mr. Etsuo Kobayashi for providing plant materials and information on *B. circularis* and *B. melanobullata*; Mr. Dean Turney for providing plant material, Ms. Mary Sizemore and Mr. Trinh Ngoc Bon for information on the distribution and habitat of *B. caobangensis*; Mr. Shin-Ming Ku, Wai-Chao Leong and Chih-Kai Yang (HAST) for accompanying and assisting the fieldwork. This study was supported in part by grants from Academia Sinica to CIP.

Author details

[1]Herbarium (HAST), Biodiversity Researach Center, Academia Sinica, Taipei, Nangang 115, Taiwan. [2]Division of Botanical Garden, Taiwan Forestry Research Institute, No. 53, Nan-Hai Road, Taipei 10066, Taiwan. [3]Center for Plant Conservation (CPC), Vietnam Union of Science and Technology Associations, No. 25/32, Lane 191, Lac Long Quan Road, Nghia DoCau Giay District, Hanoi, Vietnam.

References

Arends JC (1992) Studies in Begoniaceae IV: biosystematics of *Begonia squamulosa* Hook. f. and affiliated species in section *Tetraphila* A. DC. Wageningen Agric Univ Pap 91–6:1–222

Averyanov LV, Nguyen HQ (2012) Eleven new species of *Begonia* L. (Begoniaceae) from Laos and Vietnam. Turczaninowia 15(2):5–32

Chung K-F, Leong W-C, Rubite RR, Repin R, Kiew R, Liu Y, Peng C-I (2014) Phylogenetic analyses of *Begonia* sect. *Coelocentrum* and allied limestone species of China shed light on the evolution of Sino-Vietnamese karst flora. Bot Stud 55:e1. doi:10.1186/1999-3110-55-1

Ku S-M, Peng C-I, Liu Y (2004) Notes on *Begonia* (sect. *Coelocentrum*, Begoniaceae) from Guangxi, China, with the report of two new species. Bot Bull Acad Sin 45:353–367

Ku S-M, Kono Y, Liu Y (2008) *Begonia pengii* (sect. *Coelocentrum*, Begoniaceae), a new species from limestone areas in Guangxi, China. Bot Stud 49:167–175

Levan A, Fredga K, Sandberg AA (1964) Nomenclature for centromeric position on chromosomes. Hereditas 52:201–220

Liu Y, Ku S-M, Peng C-I (2005) *Begonia picturata* (sect. *Coelocentrum*, Begoniaceae), a new species from limestone areas in Guangxi, China. Bot Bull Acad Sin 46:367–376

Oginuma K, Peng C-I (2002) Karyomorphology of Taiwanese *Begonia* (Begoniaceae): taxonomic implications. J Pl Res 115:225–235

Peng C-I, Yang H-A, Kono Y, Chung K-F, Huang Y-S, Wu W-H, Liu Y (2013) Novelties in *Begonia* sect. *Coelocentrum*: *B. longgangensis* and *B. ferox* from limestone areas in Guangxi, China. Bot Stud 54:e44. doi:10.1186/1999-3110-54-44

Peng C-I, Ku S-M, Yang H-A, Leong W-C, Liu Y, Nguyen TH, Kono Y, Chung K-F (2014a) Two new species of *Begonia* sect. *Coelocentrum*, *B. guixiensis* and *B.*

longa, from Sino-Vietnamese limestone karsts. Bot Stud 55:e52. doi:10.1186/s40529-014-0052-8
Peng C-I, Wang H, Kono Y, Yang H-A (2014b) *Begonia wui-senioris* (sect. *Platycentrum*, Begoniaceae), a new species from Myanmar. Bot Stud 55:e13. doi:10.1186/1999-3110-55-13
Sodhi NS, Koh LP, Brook BW, Ng PKL (2004) Southeast Asian biodiversity: an impending disaster. Trends Ecol Evol 19:654–660

Chemical characterization and antioxidative activity of four 3-hydroxyl-3-methylglutaroyl (HMG)-substituted flavonoid glycosides from *Graptopetalum paraguayense* E. Walther

Hsin-Yi Liu[1], Hsin-Yi Peng[2], Shih-Lan Hsu[3], Ting-Ting Jong[1*] and Su-Tze Chou[2*]

Abstract

Background: *Graptopetalum paraguayense* E. Walther is a popular traditional Chinese herb and possesses several health benefits. In earlier studies, we demonstrated that *G. paraguayense* showed no genotoxicity and showed several biological activities. However, the constituents of *G. paraguayense* have not been studied yet. In this present study, we isolated and identified the constituents of the leaves of *G. paraguayense* E. Walther.

Results: A total of seven flavonoid compounds were isolated from the methanolic extract of *G. paraguayense*. The four major compounds isolated were flavonoid glucoside derivatives of quercetin (**1**, **3**) and kampferol (**2**, **4**), each presenting a 3-hydroxyl-3-methylglutaroyl (HMG) substituent; compounds **3** and **4**—the 2′′-acetyl derivatives of **1** and **2**, respectively—are novel compounds isolated from nature for the first time. High-performance liquid chromatography for the quantitative analyses of the four major HMG-substituted flavonoid glycosides in *G. paraguayense* E. Walther were accomplished to acquire the high yields of **1**–**4** in the methanolic extract (4.8, 5.7, 4.3, and 2.5 mg/g, respectively). Furthermore, the antioxidant activities, including radical-scavenging, reducing power and lipid peroxidation inhibitory effects of these isolated flavonoids were also evaluated. All seven of the isolated flavonoid compounds possessed antioxdative activity.

Conclusions: In this study of the constituents of the leaves of *G. paraguayense* E. Walther, we isolated four major components from its methanolic extract and determined their structures to be (acetylated) HMG-substituted flavonol glycosides, which are rare in nature. All seven of the isolated compounds possessed antioxdative activity, and those flavonoid compounds may be responsible for the functional ingredients in *G. paraguayense*. Further investigation of their bioactivities or pharmacological activities will be continued.

Keywords: Antioxidative activity; Flavonoid glycoside; *Graptopetalum paraguayense*; 3-hydroxyl-3-methylglutaroyl (HMG)

Background

Graptopetalum paraguayense E. Walther is a popular traditional Chinese herb and possesses several health benefits. According to an archaic Chinese prescription, *G. paraguayense*, belonging to the *Crassulaceae* family, is used for alleviating hepatic disorders, lowering blood pressure, relieving pains and infections. We had shown that *G. paraguayense* showed no genotoxicity by using Ames test, *in vitro* chromosome aberration assay and *in vivo* erythrocyte micronucleus assay. Additionally, the acute oral toxicity and 28-day repeated feeding toxicity tests showed that the aqueous extract of *G. paraguayense* is generally considered to be safe, with an no observed adverse effect level (NOAEL) value of 1.0 g/kg BW in rats (Chou et al. 2005; Chung et al. 2012). In earlier studies, we demonstrated several biological activities of *G. paraguayense*, including its dose-dependent inhibition

* Correspondence: ttjong@mail.nchu.edu.tw; stchou@pu.edu.tw
[1]Department of Chemistry, National Chung Hsing University, 250 Kuo-kuang Road, Taichung 402, Taiwan
[2]Department of Food and Nutrition, Providence University, 200, Sec. 7, Taiwan Boulevard, Shalu Dist., Taichung 43301, Taiwan
Full list of author information is available at the end of the article

of mushroom tyrosinase-catalyzed oxidation of L-dopa (Huang et al. 2005), radical-scavenging and lipid peroxidation-preventive effects (Chung et al. 2005), and antihepatoma potential (Chen et al. 2008). Further, a 50% aqueous ethanolic extract of *G. paraguayense* (GE50) acts as a mixed-type inhibitor of angiotensin-converting enzyme (ACE) (Chen et al. 2007), and our *in vivo* study confirmed that daily oral administration of GE 50 (2.5 g/kg BW) for 4 weeks reduces blood pressure in spontaneously hypertensive rats (Chung et al. 2009) and provide hepatoprotection against *tert*-butylhydroperoxide- and CCl_4-induced oxidative liver damage in rats (Chou et al. 2008). In an 8-week clinical study, Lin et al. noted that oxidative stress in hypercholesterolemic subjects decreases on daily consumption of 100 g *G. paraguayense* (Lin et al. 2011). Duh et al. observed that an aqueous extract of *G. paraguayense* has hepatoprotective effects by promoting antioxidant and antiinflammatory activities in an animal model of carbon tetrachloride (CCl_4)-induced liver damage (Duh et al. 2011). A reported patent depicted that *G. paraguayense* is potent for the protection of animal liver diseases and medical conditions, such as inflammation, steatosis, and fibrosis. In particular, *G. paraguayense* inhibits proliferation of activated hepatic stellate cells, which play a pivotal role in liver fibrosis (Hsu 2006). Furthermore, *G. paraguayense* is useful against fibrosis or inflammation of tissues or organs other than the liver, in particular lung, kidney, and bladder. They mentioned in the patent that clinical trial results indicate almost all patients (14 persons) completely recovered from hepatic steatosis, thus, *G. paraguayense* is effective in patients with fatty liver.

However, the constituents of *G. paraguayense* have not been studied yet. In this present study, we isolated and identified several components of the herb and by using high-performance liquid chromatography (HPLC), we performed quantitative analysis for the four major compounds **1–4** (structures see Figure 1) which appeared to be the main components in the crude methanol extract. In addition, we used several assays to test the antioxidant potentials of these isolated compounds, including their radical-scavenging activities, reducing power, and antioxidative effects on Fe/ascorbate–induced lipid peroxidation in a liposome model system. Herein, we compare these results with those of quercetin, a reference antioxidant.

Methods

General experimental procedures

Optical rotations were measured using a Perkin–Elmer 241 polarimeter. IR spectra were recorded using a Perkin–Elmer Spectrum 100 FT-IR spectrometer. UV spectra were recorded using a Cintra UV 101 spectrometer. ^{1}H NMR spectra were recorded using Varian Unity Inova 400 and 600 MHz FT-NMR spectrometers; chemical shifts (δ) are expressed herein in units of ppm, with coupling constants (*J*) in Hz. HR-FAB/MS spectra were recorded using a Finnigan Thermo Quest MAT 95XL mass spectrometer. Column chromatography (CC) was performed using silica gel 60 F_{254} (70–230 and 230–400 mesh, Merck), an ODS column [Lichroperp RP-18 (15–25 μm, Merck)], and Sephadex LH-20 (Amersham Biosciences, Sweden). Thin layer chromatography (TLC) and preparative TLCs were performed using precoated silica gel plates; spots were visualized at 254 and 365 nm. Preparative HPLC was performed using an Agilent 1100 series system coupled with a UV detector.

Plant material

The plant, *G. paraguayense*, was identified by an herb expert, Mr. Chiu Nien-Yuan, Department of Pharmacology, China Medical University (Taiwan). The leaves of *G.*

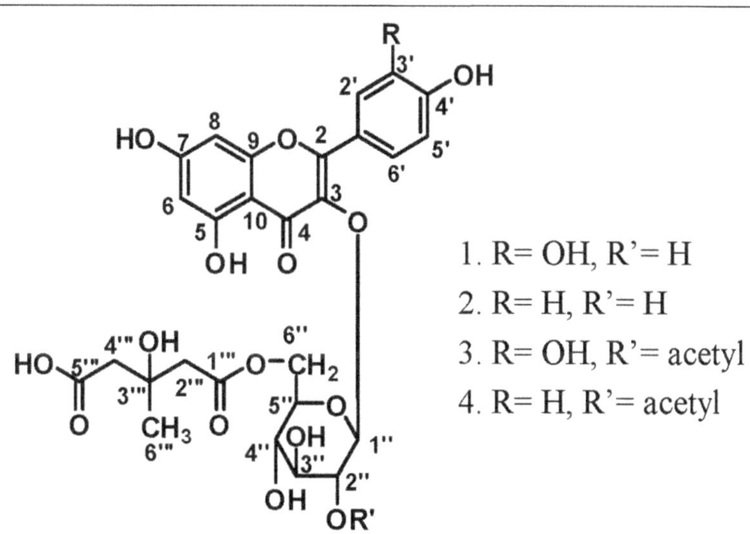

Figure 1 Structures of **1**–**4**.

paraguayense were collected from a farm in Yunlin County in Taiwan. A few batches of leaves samples were bought and found consistent by their HPLC profiles.

Extraction and isolation

The freeze-dried leaves of *G. paraguayense* (GP, 530 g) were extracted with n-hexane; the residue was extracted with methanol and then this extract was concentrated. The methanolic extract (89.6 g) of *G. paraguayense* was partitioned between EtOAc and water, the aqueous layer discarded, and the organic layer dried over Na_2SO_4. After evaporation of the solvent under reduced pressure, we obtained 10.0 g of EtOAc extract (GP-EtOAc, yield from methanolic extract). The GP-EtOAc extract was coated with silica gel and subjected to CC (SiO_2, 230–400 mesh, 9.5 × 22 cm), eluting with a gradient of hexane, EtOAc, and MeOH (hexane/EtOAc, 9:1, 2:8, 4:6, 8:2, 0:1; then EtOAc/MeOH, 3:1, 0:1) to yield seven fractions. Fraction 6 was subjected to ODS CC (Lichroperp RP-C18, 15–25 μm, Merck), eluting with a gradient of MeOH and H_2O (MeOH/H_2O, 8:2, 7:3, 1:1) to yield three fractions. Fraction 3 (MeOH/H_2O, 1:1) was further purified six times through ODS CC (gradient from MeOH/H_2O to MeOH) to obtain compounds **1** (30 mg), **2** (46 mg), **3** (51 mg), and **4** (35 mg).

Experimental data of the novel flavonoid glycosides (3) and (4)

Quercetin 3-*O*-[6′′-(3-hydroxyl-3-methylglutaroyl)]-[2′′-acetyl]-β-D-glucopyranoside (**3**): $C_{29}H_{30}O_{17}$; $[\alpha]D^{24} = -46.6°$ (c = 0.16, MeOH); UV (MeOH) λ_{max}(log ε): 208 (3.17), 257 (2.81), 354 (2.76) nm; IR ν_{max} (MeOH) cm^{-1}: 1764; (+) ESI-MS m/z: 673 $[M+Na]^+$, 651 $[M+H]^+$; (+) ESI-MS^2 m/z: 349 $[M+H - quercetin]^+$, 303 $[quercetin+H]^+$; HR(+) ESI/MS $[M+H]^+$m/z, 651.1587 (calcd for $C_{29}H_{30}O_{17}$, 651.1556). ^1H NMR (d_6-acetone, 600 MHz) and ^{13}C NMR (d_6-acetone, 150 MHz) spectral data of **3** are provided in Table 1. Kampferol 3-*O*-[6′′-(3-hydroxyl-3-methylglutaroyl)]-[2′′-acetyl]-β-D- glucopyranoside (**4**): $C_{29}H_{30}O_{16}$; $[\alpha]D^{24} = -43.0°$ (c = 0.40, MeOH); UV (MeOH) λ_{max} (logε): 209 (3.18), 266 (3.00), 350 (2.94) nm; IR ν_{max} (MeOH) cm^{-1}: 1764; (+)ESI-MS m/z: 657 $[M+Na]^+$, 635 $[M+H]^+$; (+)ESI-MS2 m/z: 349 $[M+H - kaempferol]^+$, 287 $[kaempferol+H]^+$; HR(+)ESI/MS: $[M+H]^+$ m/z 635.1632 (calcd for $C_{29}H_30O_{17}$, 635.1607). ^1H NMR (d_6-acetone, 600 MHz) and ^{13}C NMR (d_6-acetone, 150 MHz) spectral data of 4 are provided in Table 1.

Quantitative analysis through HPLC

Calibration solutions of standards **1–4** were prepared at concentrations of 2–500 μg/mL. The analysis was performed using a reversed-phase Lichrospher 100 RP-18 column (250 mm × 4.6 mm, 5 μm). The extracts were separated using aqueous solutions containing 0.1% phosphoric acid (A) and acetonitrile (B). The eluent flow rate was maintained at 1 mL/min; the injection volume was 20 μL; the detection wavelength was 270 nm; the column temperature was set at ambient temperature. The elution program with a linear gradient was optimized as follows: 0 min, 6% B; 30 min, 45% B; 33 min, 100% B; 40 min, 6% B, followed by a 5-min conditioning period. The identities of the analytes were confirmed by comparing their HPLC retention times with that of the analytical standard. Calibration was achieved through linear regression analysis performed at eight different concentrations (2, 10, 25, 50, 100, 200, 400, 500 μg/mL). Quantitative analysis was performed after integration of the peaks. All experiments in this study were carried out in triplicate.

Antioxidative activity assay: DPPH radical-scavenging activity

The DPPH radical-scavenging activities of the compounds isolated from *G. paraguayens* were measured using a method described previously (Chen et al. 2008). An aliquot of an extracted compound or quercetin (0.1 mL, 0.005–600.0 μM) was mixed with 100 mM Tris-HCl buffer (pH 7.4, 0.4 mL) and then added to 500 μM DPPH in EtOH (0.5 mL; to a final concentration of 250 μM). The mixture was shaken vigorously and left in the dark at room temperature for 20 min. The absorbance of the mixture was measured spectrophotometrically at 517 nm. The ability to scavenge DPPH radicals was calculated using the equation: scavenging effect (%) = [1 − (absorbance of sample at 517 nm/absorbance of control at 517 nm)] × 100%. The value of IC_{50}, the effective concentration at which the antioxidant activity was 50%, was obtained by interpolation from linear regression analysis. All assays were conducted at least three times with three different sample preparations. All data are expressed as the mean ± standard deviation (SD).

Antioxidative activity assay: $ABTS^+$ radical-scavenging activity

The $ABTS^+$ radical-scavenging activities of the isolated compounds were determined through ABTS decolorization assays (Chen et al. 2008). The $ABTS^+$ radical generation system, comprising 2.45 mM potassium persulfate ($K_2S_2O_8$) and 7 mM ABTS salt in 0.01 M phosphate-buffered saline (pH 7.4), was maintained for 16 h at room temperature in the dark. Each tested compound (0.01–600.0 μM) was diluted (100X) with the $ABTS^+$ solution to a total volume of 1 mL and then left to react for 10 min. The control (without sample) was used as a blank; 990 μL of PBS was added to these control samples instead. All assays were conducted at least three times with three different sample preparations. All data are expressed as the mean ± standard deviation (SD).

Table 1 Summarizes pertinent ^{1}H and ^{13}C NMR spectral data and assignments

Position	Compound 1 (in Acetone-d_6)		Compound 2 (in Acetone-d_6)		Compound 3 (in Acetone-d_6)		Compound 4 (in Acetone-d_6)	
	δ_H	δ_C	δ_H	δ_C	δ_H	δ_C	δ_H	δ_C
2		158.4		158.6		158.0		158.1
3		135.2		133.3		134.4		134.2
4		178.9		177.4		178.5		178.6
5		162.6		161.2		162.8		162.8
6	6.28 (1H, d, 1.8)	99.5	6.28 (1H, d, 1.8)	98.8	6.26 (1H, d, 1.8)	99.4	6.26 (1H, d, 1.8)	98.2
7		165.1		164.2		164.8		164.8
8	6.52 (1H, d, 1.8)	94.6	6.52 (1H, d, 1.8)	93.7	6.49 (1H, d, 1.8)	94.4	6.50 (1H, d, 1.8)	93.5
9		157.8		156.5		157.8		157.8
10		105.4		104.1		105.5		103.1
1′		122.7		120.9		122.8		121.7
2′	7.77(1H, d, 1.8)	117.2	8.12 (1H, d, 9.0)	130.7	7.76(1H, d, 2.4)	117.1	8.08 (1H, d, 9.0)	132.0
3′		145.2	6.96 (1H, d, 8.4)	115.1		145.2	6.97 (1H, d, 9.0)	115.4
4′		149.1		160.0		148.9		160.8
5′	6.94 (1H, d, 8.4)	115.6	6.96 (1H, d, 8.4	115.1	6.95 (1H, d, 8.4)	115.7	6.97 (1H, d, 9.0)	115.4
6′	7.68 (1H, dd, 1.8,8.4)	123.1	8.12 (1H, d, 9.0)	130.7	7.63 (1H, dd, 2.4,8.4)	123.0	8.08 (1H, d, 9.0)	132.0
1″	5.30 (1H, d, 7.2)	104.2	5.28 (1H, d, 7.8)	101.3	5.61 (1H, d, 7.8)	100.2	5.62 (1H, d, 7.8)	100.1
2″	3.40-3.52 (1H, m)	75.2	3.45 (1H, dd, 7.8, 9.0)	74.1	5.01 (1H, dd, 7.8, 9.6)	75.0	4.97 (1H, dd, 8.4, 9.6)	74.8
3″	3.40-3.52 (1H, m)	77.7	3.40 (1H, t, 9.0)	76.4	3.70 (1H, t, 9.0)	75.3	3.68 (1H, t, 9.0)	75.2
4″	3.40-3.52 (1H, m)	70.6	3.48-3.54 (1H, m)	70.1	3.42 (1H, dd, 9.0, 9.6)	71.0	3.46 (1H, dd, 9.0, 9.6)	71.2
5″	3.40-3.52 (1H, m)	75.1	3.48-3.54 (1H, m)	74.1	3.55 (1H, dd, 1.8, 7.8)	74.9	3.54 (1H, dd, 1.8, 7.8)	75.0
6″	4.11 (1H, dd, 5.4, 12.0) 4.22 (1H, dd, 1.8, 12.0)	64.0	4.10 (1H, dd, 6.0, 12.0) 4.22 (1H, dd, 1.8, 12.0)	63.2	4.12 (1H, dd, 5.4, 12.0) 4.24 (1H, dd, 1.8, 12.0)	63.7	4.10 (1H, dd, 5.4, 12.0) 4.22 (1H, dd, 1.8, 12.0)	63.8
1‴		171.6		170.0		171.5		171.4
2‴	2.52 (1H, d, 14.4, H-2‴A or H-2‴B) 2.56 (1H, d, 14.4, H-2‴A or H-2‴B)	45.7	2.51 (1H, d, 14.4, H-2‴A or H-2‴B) 2.56 (1H, d, 14.4, H-2‴A or H-2‴B)	45.7	2.51 (1H, d, 13.8, H-2‴A or H-2‴B) 2.54 (1H, d, 13.8, H-2‴A or H-2‴B)	45.7	2.48 (1H, d, 13.8, H-2‴A or H-2‴B) 2.51 (1H, d, 13.8, H-2‴A or H-2‴B)	45.7
3‴		69.9		68.7		69.9		69.9
4‴	2.50 (1H, d, 15.6, H-4‴A or H-4‴B) 2.59 (1H, d, 15.6, H-4‴A or H-4‴B)	45.1	2.52 (1H, d, 15.6, H-4‴A or H-4‴B) 2.59 (1H, d, 15.6, H-4‴A or H-4‴B)	44.9	2.48 (1H, d,15.6, H-4‴A or H-4‴B) 2.56 (1H, d, 15.6, H-4‴A or H-4‴B)	45.1	2.47 (1H, d, 15.6, H-4‴A or H-4‴B) 2.54 (1H, d, 15.6, H-4‴A or H-4‴B)	45.0
5‴		173.6		172.2		173.5		173.4
6‴	1.23 (3H, s)	27.5	1.24(3H, s)	27.0	1.20 (3H, s)	27.5	1.19 (3H, s)	27.4
1⁗						170.5		170.3
2⁗					2.10 (3H, s)	21.1	2.10 (3H, s)	21.0

Antioxidative activity assay: antioxidative effect on liposome peroxidation

Liposome peroxidation was induced by Fe^{2+}-ascorbate and aquatinted by the malondialdehyde-thiobarbituric acid (MDA-TBA) adduct, according to the method described by Liao and Yin but with slight modification (Liao and Yin 2000). Liposomes (multi-lamellar vesicles) were prepared from phosphatidyl choline (PC) (30 mg), cholesterol (12 mg), and dicetyl phosphate (3 mg) at 4°C. The solvents ($CHCl_3$, MeOH) used for liposome preparation were removed and a lipid film was formed through rotary evaporation under a N_2 flush. The liposomes were then suspended in 50 mM sodium phosphate buffer (pH 7.2, 10 mL). A mixture of the liposome suspension (1 mL), sodium phosphate buffer (0.6 mL), 25 mM $FeCl_3$ (0.05 mL), 25 mM ascorbic acid

(0.05 mL), and the isolated compounds (0.08–600.0 μM, 0.25 mL) was incubated for 1 h at 37°C. After incubation, the solution was mixed with TBA (0.4% in 0.2 M HCl) and BHT (0.2% in 95% EtOH) at a ratio of 1:2:0.3 and then heated at 100°C for 20 min. After cooling, an equal volume of n-BuOH was added to the mixture to extract the chromogen. The absorbance of the n-BuOH layer was measured spectrophotometrically at 532 nm. The ability to inhibit MDA formation was calculated using the equation: inhibition effect (%) = [1 – (absorbance of sample at 532 nm/absorbance of control at 532 nm)] × 100%. The MDA level for no compound added was 18.04 μM in the control. The standard reference compound for all antioxidative tests was quercetin. All assays were conducted at least three times with three different sample preparations. All data are expressed as the mean ± standard deviation (SD).

Antioxidative activity assay: reducing power

The reducing powers of the *G. paraguayense* extracts and the isolated compounds were determined according to the method of Yen and Chen (1995). An isolated compound (0.001–1.0 μM) was mixed with an equal volume of 0.2 M phosphate buffer (pH 6.6) and 1% potassium ferricyanide and then incubated at 50°C for 20 min. An equal volume of 1% trichloroacetic acid was added to stop the reaction and then the mixture was centrifuged (2790 g, 10 min). The supernatant was mixed with distilled water and 0.1% $FeCl_3$ at a ratio of 1:1:0.2 and then the absorbance was measured at 700 nm. All assays were conducted at least three times with three different sample preparations. All data are expressed as the mean ± standard deviation (SD).

Results and discussion

Figure 2 displayed the HPLC traces of the MeOH extract and the three partitioned fractions (GP-Hex, GP-EtOAc, GP-H_2O) of the MeOH extract from the leaves of *G. paraguayense*. Since the GP-EtOAc fraction contained the same major four compounds of the herb as found in the crude MeOH extract, we subjected the EtOAc fraction to further purification. After successive normal and reversed-phase chromatography, we obtained a total of seven flavonoid compounds. The four major components isolated were quercetin 3-*O*-[6′′-(3-hydroxyl-3-methylglutaroyl)]-β-d- glucopyranoside (**1**), kampferol 3-*O*-

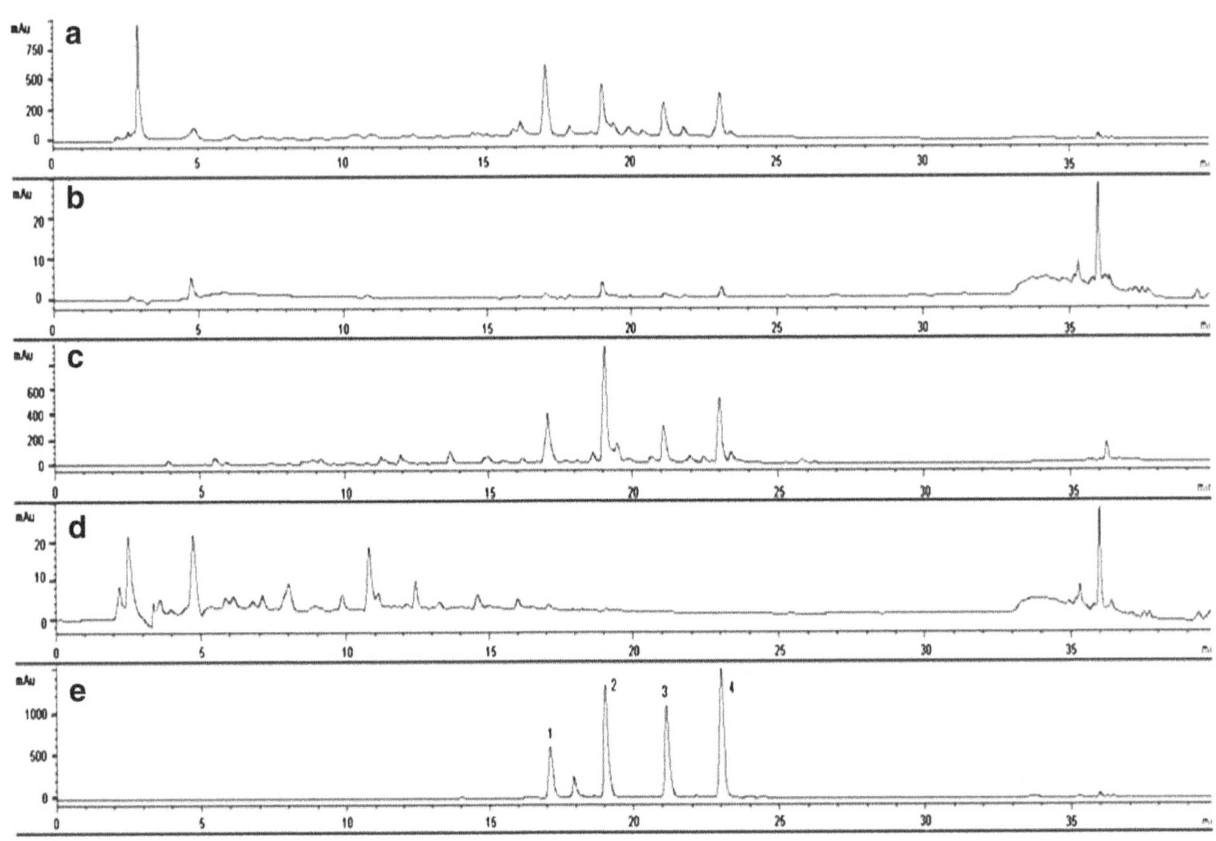

Figure 2 HPLC chromatograms (270 nm) of the **(a)** crude MeOH extract of *G. paraguayense*, **(b)** *n*-hexane extract, **(c)** EtOAc extract, **(d)** water extract, and **(e)** isolated flavonoid glycosides **1**–**4** (each 250 ppm). (Note: The impurity appearing at R_f 18 min on **(e)** is due to the contamination of compound **6**.

[6′′-(3-hydroxyl-3-methylglutaroyl)]-β-d- glucopyranoside (**2**), quercetin 3-*O*-[6′′-(3-hydroxyl-3-methylglutaroyl)-2′′-acetyl]- β-d-glucopyranoside (**3**), and kampferol 3-*O*-[6′′-(3-hydroxyl-3-methylglutaroyl)- 2′′-acetyl]-β-d-glucopyranoside (**4**); the three minor components were isoquercetin (**5**), kaempferol 3-*O*-β-d-glucopyranoside (**6**), and kaempferol (**7**). Among these components, compounds **1**, **2**, and **5–7** were known; compounds **3** and **4**–derivatives of **1** and **2**, respectively—are novel compounds isolated from nature for the first time in this study. Compounds **3** and **4** are new acylated derivatives of **1** and **2**, respectively. The characteristic feature of these major flavonoid glycosides is the 3-hydroxyl-3- methylglutaroyl (HMG) substituent on their sugar moieties. Compound 1 had been isolated previously as a UV shield from Himalayan Rheum species and from another Euphorbia species (Iwashina et al. 2004; Liu et al. 2004), compound **2** had been found solely in 10-year-old callus cultures of Mexican lime, with the authors of that study suggesting that the compound may have been stress-induced and found only in the wounded tissues from which the callus was derived (Berhow et al. 1994). We used ESI and FAB mass spectrometry and 1D and 2D NMR spectroscopy to elucidate the structures of the four HMG substituted flavonoid glycosides **1–4**.

Structural elucidation of 1–4

The characteristic UV absorptions at 256 and 357 nm of compounds **1–4** and the presence of signals for 15 aromatic carbon atoms in the downfield region in their ^{13}C NMR spectra indicated their flavonoid skeleton structures. We deduced that the A-ring of each of the flavonoids **1–4** featured a 5,7-disubsituted structure, due to the appearance of a small coupling constant (2 Hz) between H-6 and H-8 in the ^{1}H NMR spectrum; the B-ring featured either the ABX (i.e., 1 and 3) or A_2X_2 (i.e., **2** and **4**) system of protons that is characteristic of quercetin or kaempferol, respectively. For each of these compounds, we confirmed that the sugar moiety was a glucose unit, as revealed by the relationships among the coupled protons in the ^{1}H-2D COSY spectra and by the chemical shifts of the corresponding signals of six carbon atoms-for example, at 104.2 (C-1′′), 75.2 (C-2′′), 77.7 (C-3′′), 70.6 (C-4′′), 75.0 (C-5′′), and 64.0 (C-6′′) ppm for compound **1** - in ^{13}C NMR spectra. The β-glycoside linkage was established on the large coupling constant (7.8 Hz) of the glucose proton H-1′′. We established the site of attachment of the glucose moiety to the flavonoid as being the C-3 position, due to the correlation found between proton H-1′′ and C-3 in the HMBC spectra.

In addition to the signals in the ^{1}H and ^{13}C NMR spectra for the glucose moieties in these compounds, the remaining signals including an upfield (ca. 2.45-2.55 ppm) multiplet corresponding to four protons, a signal at ~1.2 ppm for methyl protons, and peaks for six carbon atoms-two C = O carbon atoms (171.6, 173.6 ppm); one oxygenated quaternary carbon atom (69.9 ppm), two upfield methylene carbon atoms (45.1, 45.7 ppm), and one methyl carbon (27.5 ppm)-that characterized the HMG substituent on these flavonoid glycosides. We confirmed that the attachment of the HMG units to the sugar moieties occurred at the C-6 position of the glucose unit, through HMBC correlation of the H-6′′ protons of the glucose with the carbonyl carbon (C = O) of the HMG unit.

For compounds **3** and **4**, spectroscopic analysis revealed one extra substituent, which corresponded to an acetyl group (i.e. a methyl proton signal at 2.1 ppm and a carbonyl carbon signal at 170.5 ppm for compound **3**). From the correlations between the glucose H-2′′ protons and the acetyl C = O carbon atoms in the HMBC spectra, we deduced that the site of attachment was the C-2′′ hydroxyl group.

Quantitative analyses of the amounts of the flavonoids in the leaves

Figure 2a reveals that The MeOH extract of the freeze-dried leaves of *G. paraguayense* powder yielded the four major compounds **1**–**4**. According to HPLC analysis, with UV detection at 270 nm, the retention times of compounds **1–4** were 17.0, 19.0, 21.1, and 23.0 min, respectively. Figure 2e displays the chromatogram of the crude extract sample spiked with the four isolated standards; the signals of the major components were coherent.

Through quantitative analysis based on triplicate injections into the HPLC system, we found that the linearity of detection of the four compounds ranged from 2 to 500 ppm (μg/mL), with the squared correlation coefficient exceeding 0.999 in each case (Table 2). These experiments provided the necessary information for estimating the limits of detection (LODs) from the slopes of the calibration curves and the threefold standard deviation (3S) of the lowest detectable concentration. We estimated the LODs for compounds **1–4** to be 0.6, 0.8, 0.5, and 0.4 μg/mL, respectively.

The overall yields of the four flavonoids **1–4** from the crude MeOH extract of the leaves of G. *paraguayense*

Table 2 Linear ranges, calibration curves, correlation coefficients, and detection limits for 1–4 analyzed using HPLC

Compound	Linear range (μg /mL)	Linear equation	Squared correlation coefficient (r^2)	LOD* (μg/mL)
1	2–500	$y = 20.930x–20.053$	0.9999	0.6
2	4–500	$y = 14.028x–0.100$	0.9998	0.8
3	2–500	$y = 25.918x–24.031$	0.9998	0.5
4	2–500	$y = 29.500x–16.592$	0.9999	0.4

*LOD = 3 × SD/slope.

were 4.8, 5.7, 4.3, and 2.5 mg/g, respectively, totaling 17.3 mg/g of the extract. This content corresponds to 5.0 mg/g of the freeze-dried sample and further to 0.2 mg/g of the fresh leaves, after taking into account a measured water content of 97%. It indicated that a high abundance of these special flavonoids in the leaves of this plant.

Antioxidative activities of the seven compounds isolated from *G. Paraguayense*

In this present study, we evaluated the antioxidative activities-namely the radical-scavenging activities, reducing powers and anti-lipid peroxidation activities-of the flavonoids isolated from *G. paraguayense*. Many food materials possess free radical-scavenging ability and their effect on radical scavenging was thought to be due to their hydrogen-donating ability (Yamaguchi et al. 1998). Figures 3A - B show the dose–response curves of DPPH and ABTS radical-scavenging activities of the flavonoids isolated from *G. paraguayense*. It was found that the radical-scavenging abilities of all isolated compounds increased with the increase in their concentrations. Figure 3C revealed the reducing powers-that is, the abilities of the isolated compounds to reduce Fe^{3+} to Fe^{2+}. The reducing powers of isolated compounds also increased with the increase of concentrations. The results revealed that the isolated flavonoids displayed reductive potential and could serve as an electron donor and terminate the radical chain reaction (Yen and Chen 1995). Lipid peroxidation is a typical free radical oxidation and proceeds via a cyclic chain reaction. Liposomes, artifical biomembanes, have been used extensively as a model system for *in vitro* lipid peroxidation studies (Duh et al. 1999; Yin and Cheng 1998; Tsuda et al. 1996). The antioxidant activities of the

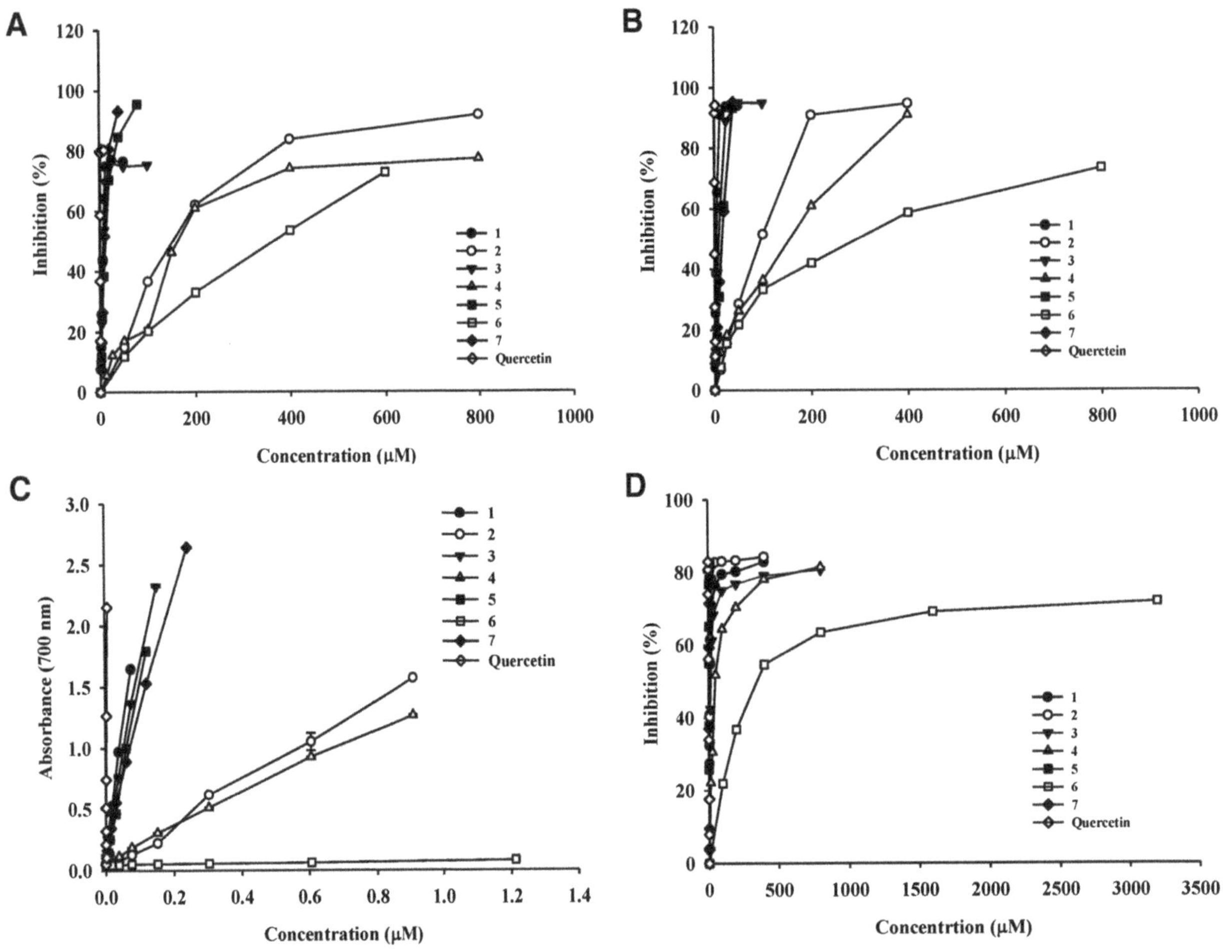

Figure 3 Scavenging effects of the compounds **1**–**7** isolated from *G. paraguayens* on **(A)** DPPH, **(B)** ABTS radical, **(C)** reducing power, and **(D)** Fe^{2+}/ascorbate-induced lipid peroxidation. Each value represents mean ± SD (n = 3). **1**, quercetin 3-*O*-(6′′-HMG) -β-D-glucopyranoside; **2**, kaempferol 3-*O*-(6′′-HMG) -β-D-glucopyranoside; **3**,quercetin 3-*O*-(2′′-acetyl),(6′′-HMG) -β-D-glucopyranoside; **4**, kaempferol 3-*O*-(2′′-acetyl),(6′′-HMG)-β-D-glucopyranoside; **5**, quercetin 3-*O*-β-glucoside; **6**, kaempferol 3-*O*-β-glucoside; **7**, kaempferol.

Table 3 50% Inhibition concentrations (IC_{50}) for radical-scavenging activity and lipid peroxidation inhibition of compounds 1–7 isolated from *G. paraguayense*

Compound	*DPPH radical-scavenging activity* (μM)	*$ABTS^+$radical-scavenging activity* (μM)	*Lipid peroxidation inhibition* (μM)
1	7.88 ± 0.57	5.69 ± 0.31	10.76 ± 0.10
2	144.93 ± 3.56	96.81 ± 2.40	9.10 ± 0.17
3	8.40 ± 0.06	9.22 ± 0.67	46.54 ± 1.01
4	108.47 ± 6.94	152.21 ± 12.16	17.45 ± 0.18
5	11.06 ± 0.34	14.50 ± 1.24	1.67 ± 0.01
6	351.89 ± 17.53	277.44 ± 10.01	330.64 ± 12.63
7	19.18 ± 0.91	15.28 ± 0.06	2.96 ± 0.06
Quercetin	0.475 ± 0.00	0.35 ± 0.01	0.07 ± 0.00

Values are given as mean ± S.D. (n = 3).

isolated compounds on MDA formation in Fe^{2+}/ascorbate-mediated lipid peroxidation in the liposome system were shown in Figure 3D. It was also found that the inhibitory effect on the lipid peroxidation of isolated compounds was in a concentration-dependent manner. The mechanism of the inhibitory effect through which the isolated compounds protect against lipid peroxidation may involve radical scavenging and reducing capability. Many studies have showed that flavonoids not only have anti-radicals activities but also have the transition metal ions-chelating properties which may afford protection against lipid peroxidation (Cai et al. 2006; Malešev and Kuntić 2007; Olennikov et al. 2014). Lipid peroxidation is an oxidative alteration of polyunsaturated fatty acids in cell membranes that leads to cellular damage, which is strongly associated with, for example, aging and carcinogenesis (Yagi 1987). In addition, lipid peroxidation is a major deteriorative reaction in the processing and storage of lipid-containing foods (Duthie 1993). Hence, our findings confirm the health-promoting potential of *G. paraguayense*.

Table 3 lists the half-inhibition concentrations (IC_{50}) for the radical-scavenging capacities and anti-lipid peroxidation activities. In terms of their IC_{50} values, the radical-scavenging activities and reducing powers of the isolated compounds decreased in the order **1** > **3** > **5** > **7** > **2** > **4** > **6**; in contrast, the anti-lipid peroxidation capacities followed the order **5** > **7** > **2** > **1** > **4** > **3** > **6**. Based on these results, the radical-scavenging capacities of the glycosides are lower than that of their aglycones. In addition, the acetylated HMG-substituted flavonoid glycosides **3** and **4** possessed weaker antioxidative activities than those of **1** and **2** (Table 3 and Figure 3). Flavonoids comprise a group of polyphenolic substances that are present in most plants. The structural components common to these molecules include two benzene rings on either side of a heterocycle containing three additional carbon atoms. In general, the greater number of hydroxyl substituents on the backbone structure, the stronger the antioxidant activity of the flavonoid. Glycosylation does, however, decrease the antioxidative potential (Miller 1996). Furthermore, the antioxidative activity in the membrane system depends not only on the number of hydroxyl groups but also on the polarity and hydrophobicity of the tested compounds (Wu et al. 2007). This study demonstrated that all seven of the isolated compounds from *G. Paraguayense* possessed antioxdative activity. Although those flavonoid compounds showed less antioxidant properties than that of quercetin, our standard reference, it was evident that those compounds may be responsible for the functional ingredients in *G. paraguayense*.

Conclusions

In this study of the constituents of the leaves of *G. paraguayense* E. Walther, we isolated four major components from its MeOH extract and determined their structures to be (acetylated) HMG-substituted flavonol glycosides, which are rare in nature. Quantitative analysis results indicated that the four major flavonol glycosides totaled 17.3 mg/g of the extract (equivalent to 0.2 mg/g of fresh leaves), a notably high abundance of these special flavonoids in the leaves of this plant. All seven of the isolated compounds possessed antioxdative activity, including radical scavenging, reducing capability and lipid peroxidation inhibitory effect. In future work, due to the similarities of the structures of the four main constituents isolated herein to those commercialized inhibitors as HMG-CoA reductase, further investigation of their bioactivities or pharmacological activities will be continued.

Competing interests

The authors declare that they have no competing interests.

Authors' contributions

J-TT and C-ST conceived and designed the experiments and drafted the manuscript. L-HY and P-HY carried out most of analyses and finished the figures and tables presented in the paper. H-SL performed the statistical analysis. All authors read and approved the final manuscript.

Acknowledgements

We thank the National Science Council of the Republic of China, Taiwan, for supporting this research financially under contracts NSC 98-2313-B-126-004-MY3, NSC 99-2113-M-005-002 and NSC 102-2313-B-126-004-MY3.

Author details

[1]Department of Chemistry, National Chung Hsing University, 250 Kuo-kuang Road, Taichung 402, Taiwan. [2]Department of Food and Nutrition, Providence University, 200, Sec. 7, Taiwan Boulevard, Shalu Dist., Taichung 43301, Taiwan. [3]Department of Education and Research, Taichung Veterans General Hospital, 1650, Sec. 4, Taiwan Boulevard, Taichung City 40705, Taiwan.

References

Berhow MA, Bennet RD, Poling SM, Vannier S, Hidaka T, Omura M (1994) Acylated flavonoids in callus cultures of *Citrus aurantifolia*. Phytochemistry 36:1225–1227

Cai YZ, Sun M, Xing J, Luo Q, Corke H (2006) Structure-radical scavenging activity relationships of phenolic compounds from traditional Chinese medicinal plants. Life Sci 78:2872–2888

Chen SJ, Chang CT, Chung YC, Chou ST (2007) Studies on the inhibitory effect of *Graptopetalum paraguayense* E. Walther extracts on the angiotensin converting enzyme. Food Chem 100:1032–1036

Chen SJ, Chung JG, Chung YC, Chou ST (2008) *In vitro* antioxidant and antiproliferative activity of the stem extracts from *Graptopetalum paraguayense*. Am J Chin Med 36:369–383

Chou ST, Chen SJ, Yeh JY, Chung YC (2005) Antimutagenic potential of *Graptopetalum paraguayense* E. Walther extracts. Taiwan Food Sci Agric Chem 43:349–357 (in Chinese)

Chou ST, Chung YC, Teng KY, Yeh JY (2008) Effects of *Graptopetalum paraguayense* extract on *tert*- butylhydroperoxide-induced oxidative stress. J Sci Food Agric 88:429–434

Chung YC, Chen SJ, Hsu CK, Chang CT, Chou ST (2005) Studies on the antioxidative activity of *Graptopetalum paraguayense* E. Walther. Food Chem 91:419–424

Chung YC, Chen SJ, Peng HY, Chou ST (2009) Antihypertensive and antioxidant effects of the *Graptopetalum paraguayense* E. Walther extract in spontaneously hypertensive rats. J Sci Food Agric 89:2678–2686

Chung YC, Chou ST, Jhan JK, Liao JW, Chen SJ (2012) *In vitro* and *in vivo* safety of aqueous extracts of *Graptopetalum paraguayense* E. Walther. J Ethnopharmacol 140:91–97

Duh PD, Du PC, Yen GC (1999) Action of methanolic extract of mung bean hulls as inhibitors of lipid peroxidation and non-lipid oxidative damage. Food Chem Toxicol 37:1055–1061

Duh PD, Lin SL, Wu SC (2011) Hepatoprotection of *Graptopetalum paraguayense* E. Walther on CCl_4-induced liver damage and inflammation. J Ethnopharmacol 134:379–385

Duthie GG (1993) Lipid peroxidation. Eur J Clin Nutr 47:759–764

Hsu SL (2006) Pharmaceutical use of *Graptopetalum* and related plants., U.S. Patent US 2006/045925 A1, 2 Mar 2006

Huang KF, Chen YW, Chang CT, Chou ST (2005) Studies on the inhibitory effect of *Graptopetalum paraguayense* E. Walther extracts on mushroom tyrosinase. Food Chem 89:583–587

Iwashina T, Omori Y, Kitajima J, Akiyama S, Suzuki T, Ohba H (2004) Flavonoids in translucent bracts of the Himalayan *Rheum nobile* (Polygonaceae) as ultraviolet shields. J Plant Res 117:101–107

Liao KL, Yin MC (2000) Individual and combined antioxidant effects of seven phenolic agents in human erythrocyte membrane ghosts and phosphatidylcholine liposome systems: importance of the partition coefficient. J Agric Food Chem 48:2266–2270

Lin YL, Peng HY, Hsieh HM, Lin CH, Chou ST (2011) Effects of *Graptopetalum paraguayense* consumption on serum lipid profiles and antioxidative status in hypercholesteremic subjects. J Sci Food Agric 91:1230–1235

Liu X, Ye W, Yu B, Zhao S, Wu H, Che C (2004) Two new flavonol glycosides from *Gymnema sylvestre* and *Euphorbia ebracteolata*. Carbohydr Res 339:891–895

Malešev D, Kuntić V (2007) Investigation of metal-flavonoid chelates and the determination of flavonoids via metal-flavonoid complexing reactions. J. Serbian Chem Soc 72:921–939

Miller AL (1996) Antioxidant flavonoids: structure, function and clinical usage. Altern Med Rev 1:103–111

Olennikov DN, Kashchenko NI, Chirikova NK (2014) A novel HPLC-assisted method for investigation of the Fe^{2+}-chelating activity of flavonoids and plant extracts. Molecules 19:18296–18316

Tsuda T, Shiga K, Ohshima K, Kawakishi S, Osawa T (1996) Inhibition of lipid peroxidation and the active oxygen radical scavenging effect of anthocyanin pigments isolated from *Phaseolus vulgaris* L. Biochem Pharmacol 52:1033–1039

Wu WM, Lu L, Long Y, Wang T, Liu L, Chen Q, Wang R (2007) Free radical scavenging and antioxidative activities of caffeic acid phenethyl ester (CAPE) and its related compounds in solution and membranes: a structure–activity insight. Food Chem 105:107–115

Yagi K (1987) Lipid peroxides and human diseases. Chem Phys Lipids 45:337–341

Yamaguchi T, Takamura H, Matoba T, Terao J (1998) HPLC method for evaluation of the free radical-scavenging activity of foods by using 1,1,-diphenyl-2-picrylhydrazyl. Biosci Biotechnol Biochem 62:1201–1204

Yen GC, Chen HY (1995) Antioxidant activity of various tea extracts in relation to their antimutagenicity. J Agric Food Chem 43:27–32

Yin MC, Cheng WS (1998) Antioxidant activity of several *Allium* members. J Agric Food Chem 46:4097–4101

Two new species of *Begonia*, *B. moneta* and *B. peridoticola* (Begoniaceae) from Sabah, Malaysia

Ching-I Peng[1*], Che-Wei Lin[2], Rimi Repin[3], Yoshiko Kono[1], Wai-Chao Leong[1,4] and Kuo-Fang Chung[4]

Abstract

Background: Mount Kinabalu, reknowned for its high biodiversity and endemism, is a National Park in the State of Sabah on the northern end of the island of Borneo. Every year many visit the higher part of the Kinabalu National Park, while most lowland forests in the Park are under-explored. Two unknown species of *Begonia* were collected from a peridotic (ultramafic) cliff in the Kinabalu National Park at ca. 400 m elevation.

Results: The two species are named *B. moneta* C.-I Peng, Rimi & C. W. Lin and *B. peridoticola* Rimi, C.-I Peng & C. W. Lin. *Begonia moneta* (sect. *Baryandra*) is similar to *B. gueritziana* Gibbs, a widespread species of the same section in Borneo, differing in the peltate (vs. basifixed) leaves and the smaller flower parts. Also, their chromosome numbers are different (*B. moneta*, $2n = 30$; *B. gueritziana*, $2n = 28$). The peltate and succulent foliage of *B. moneta* is also reminiscent of *B. burttii* Kiew & S. Julia and *B. payung* S. Julia & Kiew, both of sect. *Reichenheimia*, from Sarawak. *Begonia moneta* is distinct from the two species in having branched (vs. entire) placental lamellae. Additionally, *B. moneta* differs from *B. burttii* in having 4 (vs. 5) tepals in pistillate flowers and markedly unequal (vs. equal) fruit wings. *Begonia moneta* differs from *B. payung* in the smaller leaves and conspicuously winged (vs. wingless) capsules. *Begonia peridoticola* (sect. *Petermannia*) resembles *B. punchak* Kiew & S. Julia from limestone areas in Kuching Division, Sarawak, differing in the entire leaf margin (vs. distantly dentate), much larger capsular wings (8–11 mm vs. 2–3 mm wide) and yellow, spiral (vs. crimson, U-shaped) styles.

Conclusion: A careful study of the herbarium materials and literature supports the recognition of the two new species. Detailed descriptions, line drawings, color plates, chromsome data, foliar SEM observations and comparisons with phenetically similar species are provided to aid in identification.

Keywords: *Begonia moneta*; *Begonia peridoticola*; Borneo; Chromosome number; Malaysia; New species; Sabah; Sect. *Baryandra*; Sect. *Petermannia*

Background

Mount Kinabalu, reknowned for its high biodiversity and endemism, is a National Park in the State of Sabah on the northern end of the island of Borneo (Somaweer 2007). Every year many visit the higher part of the Kinabalu National Park, while most lowland forests in the Park are under-explored. Several new species of *Begonia* were documented from Sabah in recent years (Reza & Kiew 1998; Beaman et al. 2001; Kiew 1998; 2001; Kiew & Tan 2004). In continuation of our recent taxonomic studies on Asian *Begonia* (e.g. Chung et al. 2014; Ding et al. 2014; Lin & Peng 2014, Lin et al. 2014a, b; Lin et al. 2015; Nakamura et al. 2013; Peng et al. 2013, 2014a, b, c; Rubite et al. 2013, 2014), we report the discovery of two new species of *Begonia, B. moneta* and *B. peridoticola*, that co-occur on a peridotic (ultramafic) cliff in the Kinabalu National Park at ca. 400 m elevation. The two species belong to different sections, and no sign of natural hybridization was apparent locally.

* Correspondence: bopeng@sinica.edu.tw
[1]Herbarium (HAST), Biodiversity Research Center, Academia Sinica, Nangang, Taipei 115, Taiwan
Full list of author information is available at the end of the article

Methods

Chromosome preparations

Somatic chromosomes of *Begonia moneta* were examined using root tips from plants of the type collection. The methods of pretreatment, fixation and staining for chromosome observations followed Peng et al. (2014a). Classification of the chromosome complements based on centromere position at mitotic metaphase follows

Levan et al. (1964). Voucher specimen (*Peng 22343*) is deposited in HAST.

Cryo scanning electron microscopy

Fresh leaves of *Begonia moneta and B. peridoticola* were dissected and attached to a stub. The samples were frozen with liquid nitrogen slush, then transferred to a sample preparation chamber at −160°C and etched for 15 min at −85°C. After etching, the temperature droped to −130°C for sample fracturing and coating. After coating, the samples were transferred to the SEM chamber and observed at −160°C with a cryo scanning electron microscope (FEI Quanta 200 SEM/Quorum Cryo System PP2000TR FEI). Voucher specimens (*Begonia moneta*: *Peng 22344*; *B. peridoticola*: *Peng 22343*) are deposited at HAST.

Results and discussion

Species description

1. *Begonia moneta* C.-I Peng, Rimi & C. W. Lin, *sp. nov.* -**TYPE:** MALAYSIA. Borneo. Sabah. Kota Marudu

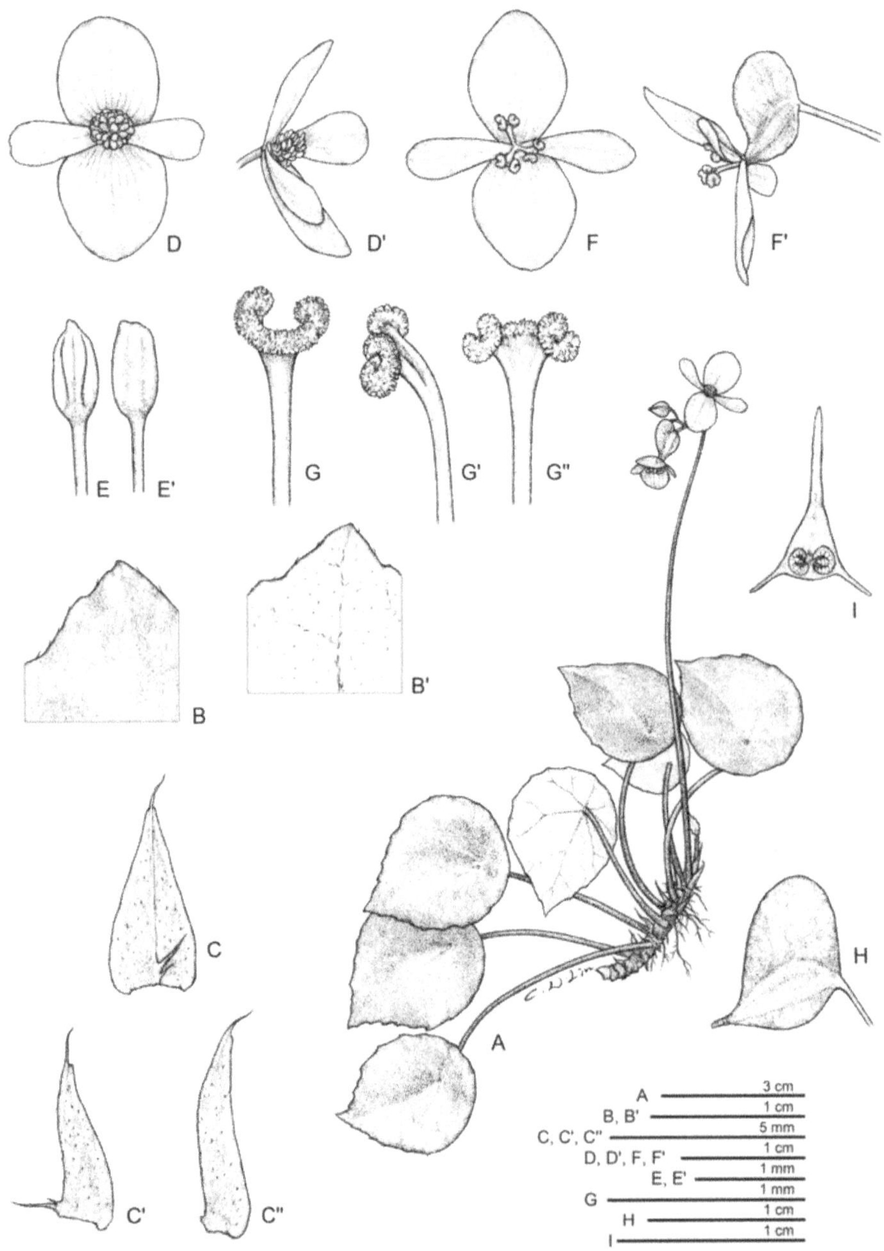

Figure 1 Begonia moneta C.-I Peng, Rimi & C. W. Lin. A. Habit; B. Portion of leaf adaxial surface; **B'**. Portion of leaf abaxial surface; **C**. Stipule, abaxial view, **C'**, **C"**. Stipules, side views; **D**, **D'**. Staminate flower; **E**, **E'**. Stamens, adaxial and abaxial views; **F**, **F'**. Pistillate flower; **G**, **G'**, **G"**. Style and stigmatic band, dorsal, side, and ventral views; **H**. Fruit; **I**. Cross sections of immature fruit. All from *Peng 22344* (HAST).

District: Kinabalu Park – Serinsim substation, Bat Cave Trail, elev. *ca.* 300–350 m. Collected on 13 November 2009; type specimens pressed from plants brought back from the field and cultivated in the experimental greenhouse. *Ching-I Peng 22344-A*, with *Kuo-Fang Chung, Wai-Chao Leong & Rimi Repin* (holotype: SNP; isotypes: A, E, HAST, KEP, SAN, TAIF).

銀幣秋海棠Figures 1 and 2.

Plant monoecious, epipetric, creeping, perennial. **Rhizome** succulent, 3.5–9 cm long, 0.3–0.4 cm across, internodes 0.2–0.6 cm, greenish to reddish brown, subglabrous. **Stipules** pale greenish, triangular-ovate, up to *ca.* 0.7 cm long, 0.3 cm wide, herbaceous, abaxially minutely appressed brown-hairy, entire, strongly keeled on abaxial midrib, apex aristate, arista *ca.* 0.15 cm long. **Leaves** 3–8, simple, peltate, petiole attachment displaced to one side, broadly ovate to orbicular, occasionally angular, base oblique, margin slightly undulate to sparsely obscurely denticulate, apex acuminate to shortly acute, 2.5–4.1 cm long, 1.9–3.8 cm wide, broad side to 2.1 cm wide, adaxially lime green, distinctly fleshy, glossy, glabrous (minutely puberulous on young leaves, hairs caducous), margin sparsely

Figure 2 Begonia moneta C.-I Peng, Rimi & C. W. Lin. A. Habit and habitat; **B.** Habit (in cultivation); **C.** Leaf, adaxial surface; **D.** Leaf, abaxial surface; **E.** Stipule on rhizome; **F.** Inflorescence; **G, H.** Staminate flower; **I, J.** Pistillate flowers; **K.** Young inflorescence with buds; **L, L'.** Cross sections of an immature fruit; **M.** Fruit. All from *Peng 22344* (HAST).

pilosulous, abaxially pale green, sparsely pilosulous on veins; venation basally 7–8 palmate, veins pinnate along primary veins, with 1–3 secondary veins on each side, these branching dichotomously or nearly so, tertiary veins weakly percurrent or reticulate. **Petioles** greenish to reddish, terete, 1.5–2.5 mm thick, 2.2–4 cm long in upper leaves, to 6.5 cm in lower leaves, pilose. **Inflorescence** reddish to crimson, compound cymose, bisexual, axillary, 1–2 arising directly from rhizome, sparsely pilose, erect, with up to 4 orders of branching, exceeding leaves; peduncles 6.5–10 cm long, 0.1–0.2 cm thick; flowers 3–6 per cyme. **Bracts** reddish, herbaceous, ovate to broadly ovate, apex obtuse to acute, at first node of inflorescence ca. 2.5 mm long, 2 mm wide, upper bracts 1.5–2 mm long, 1.2–1.8 mm wide, caducous. **Staminate flower:** tepals 4, margin entire, glabrous, outer 2 oblong to broadly obovate, base rounded, apex obtuse or rounded, 7–10 mm long, 6.5–8 mm wide, abaxially pink, adaxially pinkish to white;

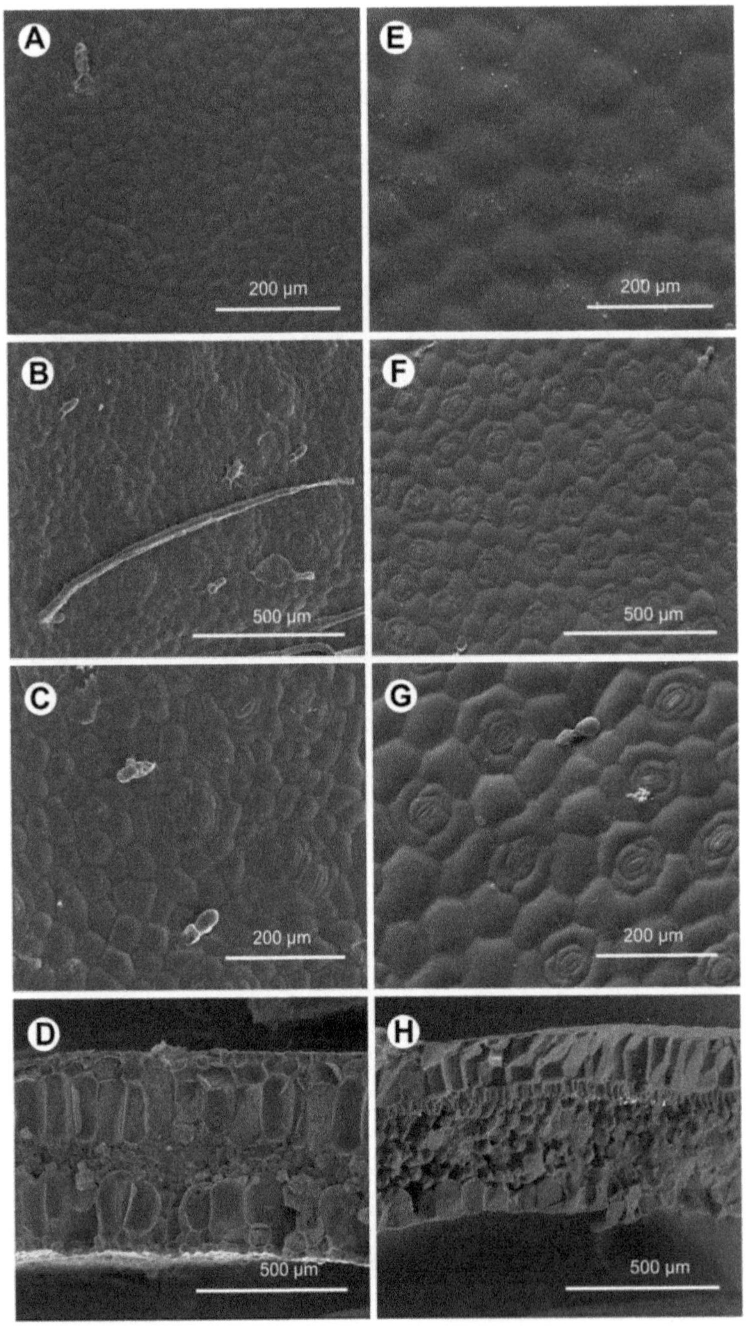

Figure 3 Cryo SEM microphotographs of Begonia leaves. **A**, **E**: Adaxial surface; **B**, **F**: Abaxaial surface; **C**, **G**: stomata complex; **D**, **H**: Cross section. **A**-**D**: *B. moneta* (*Peng et al. 22344*, HAST); **E**-**H**: *B. peridoticola* (*Peng 22343*, HAST).

inner 2 narrowly ovate to narrowly obovate, base cuneate, apex obtuse to retuse, 6–8.5 mm long, 3–4.5 mm wide, pinkish or whitish; androecium actinomorphic, stamens 13–15, golf club-shaped, slightly compressed, anthers 2-locular, 0.7–1 mm long, yellow, filaments *ca.*1 mm long, shortly fused at base. **Pistillate flower:** tepals 4, margin entire, glabrous, outer 2 adaxially white to pink, abaxially magenta to white, oblong to widely ovate, apex obtuse or rounded, 7–11 mm long, 6–8 mm wide, inner 2 pinkish or whitish, obovate, base cuneate, apex obtuse to retuse, 6.5–8.5 mm long, 2.5–3.5 mm wide; ovary trigonous-lanceolate, *ca.* 6.5 mm long, 3.5 mm across (wings excluded), pale green to pinkish, 3-winged, wings unequal, crescent-shaped, apically obtuse, longer wing 5–6.5 mm wide, shorter wings *ca.* 2.5 mm wide; locules 2, placentae 2 per locule; styles 3, fused at base, yellow, *ca.* 3 mm long, apically split and C-shaped; stigmas in a spiral band and papillose all around. **Fruit** pendent on a fine stalk 7–10 mm long, *ca.* 7 mm long, 2 mm across (wings excluded), wings crescent-shaped, unequal, 3–6 mm tall.

Leaf anatomy

Adaxial surface glabrous or with sparse, minute glandular hairs (Figure 3A). Cross section ca. 610 μm thick; epidermis single-layered on both surfaces, ca. 23–35 μm thick; hypodermis 2-layered on both surfaces, first-layer 70–85 μm thick, second-layer 180–240 μm thick; palisade tissue 1-cell layered, cells funnel-shaped, ca. 35 μm long; spongy tissue ca. 100 μm long, 4–6 cell-layered (Figure 3D). Abaxial surface with sparse, minute glandular hairs and multiseriate trichomes ca. 1–1.2 mm long (Figure 3B). Stomata complexes clustered (Figure 3B, C).

Chromosome cytology

Somatic chromosomes at mitotic metaphase of *Begonia moneta* were determined to be $2n = 30$ (Figure 4). Among the 30 chromosomes, two are comparatively longer (1.8–2.0 μm long) than the rest; the remaining 28 chromosomes gradually varied from 0.9 to 1.5 μm long. Several longer chromosomes were metacentric or submetacentric, however, centromere positions of shorter chromosomes could not be determined. Satellites were not observed.

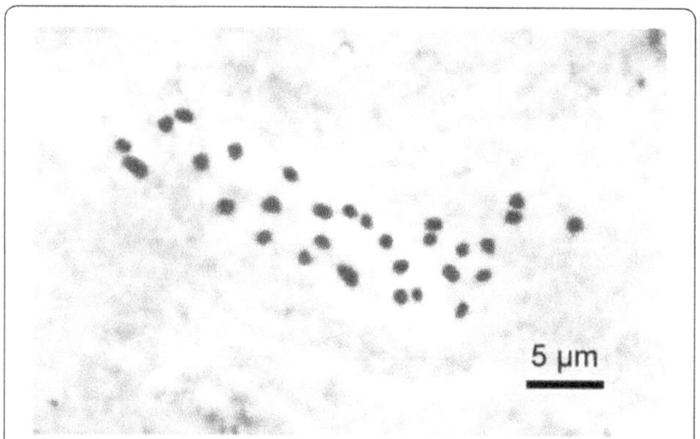

Figure 4 Somatic chromosomes at metaphase of Begonia moneta ($2n = 30$; *Peng 22344*, HAST).

In *Begonia* sect. *Baryandra*, chromosome numbers were documented for *B. gueritziana* ($2n = 28$), *B. blancii* ($2n = 30$) and *B. suborbiculata* ($2n = 30$) (Hughes et al. 2011). Our unpublished data also showed that $2n = 28$ and 30 are major chromosome numbers in this section.

Distribution and ecology

MALAYSIA. Borneo. Sabah. Kota Marudu District (Figure 5). Endemic in Kinabalu, growing in an ultramafic area at Serinsim substation, elev. ca. 300–350 m.

Etymology

The specific epithet refers to the thick, rounded leaves that are reminiscent of coins.

Additional specimen examined

MALAYSIA. Borneo, Sabah. Kota Marudu District: Kinabalu Park–Serinsim substation, Bat Cave Trail, elev. ca.350 m. 13 November 2009, *C.-I Peng 22344*, with *K. F. Chung, W. C. Leong & Rimi Repin* (HAST), same loc., *Yabainus et al. SP 10520* (SNP), *Masrina et al. SP 10538* (SNP), *Masrina et.al SP 10535* (SNP), *Antony van der Ent et. al., SNP 34512* (SNP); Serinsim substation, Mt. Nombuyukong Trail, Plot 1, *Antony van der Ent et .al., SNP 34591* (SNP).

Notes

Begonia moneta (sect. *Baryandra*) is similar to *B. gueritziana* Gibbs, a widespread species of the same section in Borneo, differing in the peltate (vs. basifixed) leaves and the smaller flower parts. Also, their chromosome numbers are different (*B. moneta*, $2n = 30$; *B. gueritziana*, $2n = 28$). The peltate and succulent foliage of *B. moneta* is also reminiscent of *B. burttii* Kiew & S. Julia and *B. payung* S. Julia & Kiew, both of sect. *Reichenheimia*, from Sarawak. *Begonia moneta* is distinct from the two species in having branched (vs. entire) placental lamellae. Additionally, *B. moneta* differs from *B. burttii* in having 4 (vs. 5) tepals in pistillate flowers and markedly unequal (vs. equal) fruit wings. *Begonia moneta* differs from *B. payung* in the smaller leaves and conspicuously winged (vs. wingless) capsules. Clustered stomata complexes were observed in *B. moneta* (Figure 3C), which are likely the way in which the species copes with periodical drought or fluctuating environment (Hughes et al. 2011; Rubite et al. 2014).

2. *Begonia peridoticola* Rimi, C.-I Peng & C. W. Lin, sp. nov. **-TYPE:** MALAYSIA. Borneo. Sabah. Kota Marudu District, Kinabalu Park – Serinsim substation, Bat Cave, elev. ca. 370 m. Erect herb to 60 cm or more; leaves green

Figure 5 Distribution map of Begonia moneta and B. peridoticola (Solid circle) in Sabah.

above, red beneath; rhizomes lacking. Plants sterile when collected on 13 November 2009. Type specimens with flowers/fruits pressed from plants brought back from the field and cultivated in the experimental greenhouse. *Ching-I Peng 22343-A*, with *Kuo-Fang Chung, Wai-Chao Leong & Rimi Repin* (holotype, SNP; isotypes, A, E, HAST, KEP, SAN).

橄欖岩秋海棠 Figures 6 and 7.

Plant monoecious, epipetric, perennial, cane-like. **Stem** crimson to olive green, 30–50 (–70) cm tall, 3–5 mm thick, glabrous, internodes 1.2–4.5 cm. **Stipules** persistent, pale green to reddish, broadly ovate-triangular, *ca.* 1.5 cm long, 1 cm wide, herbaceous, ridge keeled, margin entire, apex aristate, arista *ca.* 0.2 cm. **Leaves** numerous, simple, asymmetric, lanceolate-ovate, basifixed, asymmetric with a well-developed basal lobe on one side giving a cordate

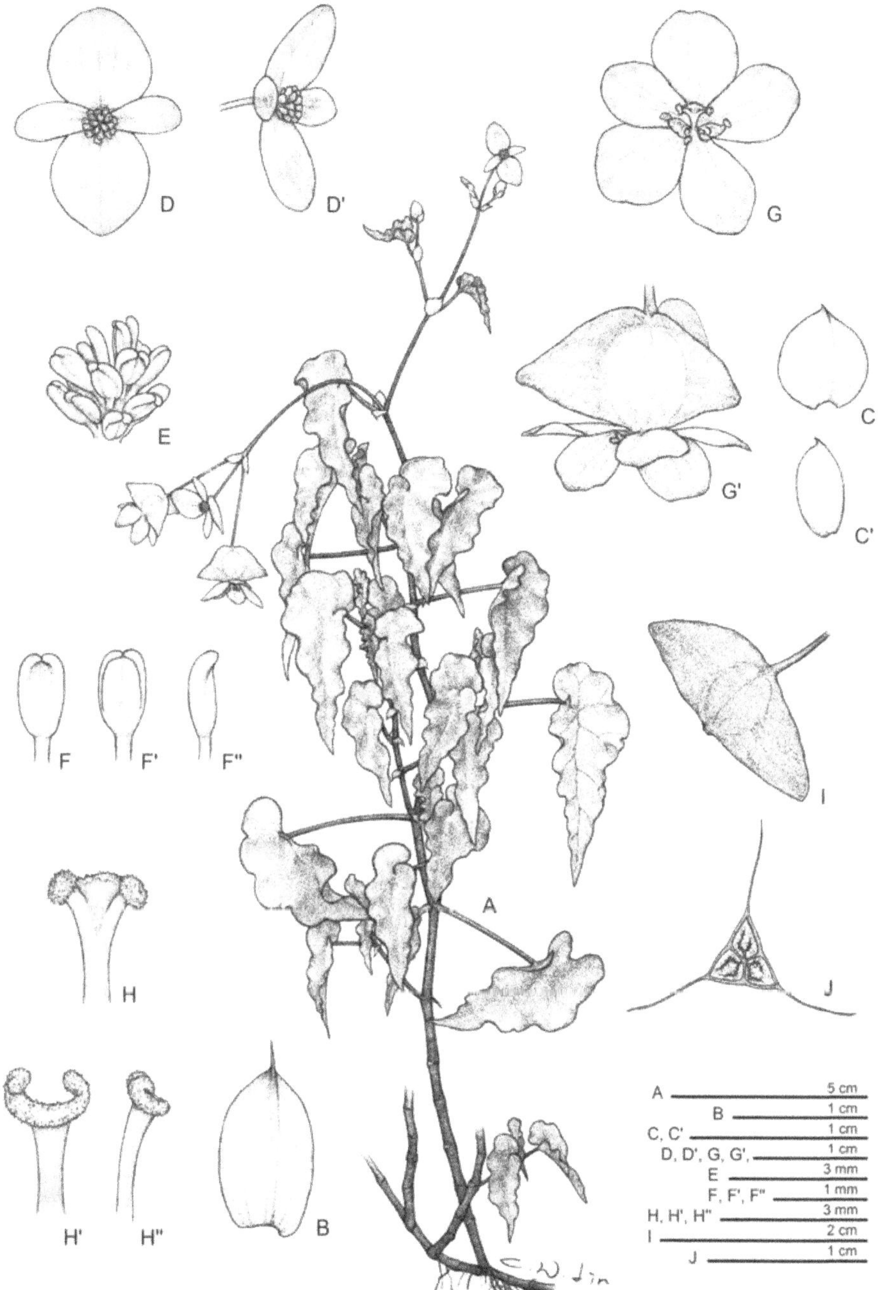

Figure 6 Begonia peridoticola Rimi, C.-I Peng & C. W. Lin. **A**. Habit; **B**. Stipule; **C, C'**. Lowermost bract in the inflorescence, face and side views; **D', D"**. Staminate flower; **E**, Androecium; **F, F', F"**. Anther, dorsal, ventral and side views; **G, G'**. Pistillate flower; **H, H', H"**. Style and stigmatic band, dorsal, ventral and side views; **I**. Fruit; **J**. Cross sections of immature fruit. All from *Peng 22343* (HAST).

appearance, margin subentire, strongly undulate, 3.2–7.5 cm long (basal lobes included), 1.7–3.4 cm wide, broad side to 1.9 cm wide, apex acuminate, base markedly unequal, basal lobes cordate, 0.8–1.7 cm long, adaxially malachite green to olive green, margin reddish, glossy, succulent, veins inconspicuous; abaxially crimson to reddish pale-green, glabrous, venation conspicuous, dark red, slightly prominent, primary veins distinguishable, 2.4–5.8 cm long, veins pinnate along primary veins, with 2–3 secondary veins on each side, branching dichotomously or nearly so, tertiary veins weakly percurrent. **Petiole** crimson to olive green, terete, 0.15–0.25 cm thick,1.8–2.6 cm long in upper leaves, to 7.1 cm in lower leaves, glabrous. **Bracts** deciduous, pale green, at first node of inflorescence broadly ovate to

Figure 7 (See legend on next page.)

(See figure on previous page.)
Figure 7 Begonia peridoticola Rimi, C.-I Peng & C. W. Lin. **A**. Habit and habitat; **B**. Habit, dorsal and ventral views; **C**. Stipule; **D**. Portion of leaf, adaxial surface; **E**. Portion leaf, abaxial surface; **F**. Lowermost cyme; **G**. Upper cymes with staminate flowers; **H**. Staminate flower; **I**. Pistillate flower; **J**. Fruit, side view; **K**. Cross section of an immature fruit. All from *Peng 22343* (HAST).

orbicular, *ca.* 6 mm long, 5.5 mm wide, margin entire, apex shortly aristate; at summit of inflorescence 3.8–5 mm long, 3.5–4.8 mm wide. **Inflorescence** bisexual, protogynous, axillary in upper axils, reddish, glabrous; lowermost cyme pendant, to 6 cm long, comprising 2 pistillate flowers with or without a staminate flower in-between; second from the lowermost cyme usually with 1 pistillate and 1 staminate flowers; upper cymes staminate, ascending, 3- or 4-flowered. **Staminate flower:** pedicel to 2.2 cm long, glabrous, tepals 4, white to pink, margin entire, apex rounded, glabrous, outer 2 broadly ovate, 7.5–10 mm long, 6.5–8 mm wide, those at lowermost cyme larger, 8–12 mm long, 7.5–10 mm wide, inner 2 narrowly ovate to narrowly obovate, 6–7.5 mm long, 2.8–3.2 mm wide , those at lowermost cyme 8–12 mm long, 7.5–8.5 mm wide; androecium symmetric, stamens *ca.* 25, anthers yellow, oblong-obovoid, slightly compressed, apex retuse, *ca.* 1 mm long, filaments *ca.* 0.3 mm long, slightly fused at base. **Pistillate flower:** pedicels 19–23 mm long, glabrous; tepals 5, white to pink, margin entire, glabrous, outer 2 broadly ovate to orbicular, 6.5–8.5 mm long, 5.8–8 mm wide, inner 3 broadly obovate, base cuneate, apex obtuse or rounded, 6–9 mm long, 3.5–6.5 mm wide, ovary trigonous-ellipsoid, 8–9 mm long, 6.5–7 mm across (wings excluded), 3-locular, placentation axile, placentae bifid; 3-winged, wings subequal, subtriangular, 9–11 mm long, 8–9.5 mm wide; styles 3, golden yellow, bifid, *ca.* 2.5 mm long, apically split and C-shaped; stigmas in a spiral band and papillose all around. **Fruit** pendent on stalk 5–10 mm long, capsule body 7–9 mm long, 5 mm across, glabrous, wings 3, subequal, subtriangular, margin subentire, 8–11 mm wide.

Leaf anatomy
Adaxial surface glabrous (Figure 3E). Cross section ca. 580 μm thick; upper epidermis single-layered, ca. 150 μm thick; lower epeidermis 80–130 μm; palisade tissue 1-cell layered, cells funnel-shaped, ca. 25–55 μm long; spongy tissue ca. 250 μm long, 5–6 cell-layered (Figure 3H). Abaxial surface with sparse, minute glandular hairs. Stomata complexes single, helicocytic (Figure 3F, G).

Distribution and ecology
MALAYSIA. Borneo. Sabah. Kota Marudu District. Endemic in Kinabalu Park, elev. ca. 300–400 m, growing on the crevices of peridotite breccia rock face in an ultramafic area.

Etymology
The specific epithet is derived from the peridotite breccia rock face at the Bat Cave cliffs near Serinsim substation from where it was collected.

Additional specimen examined
MALAYSIA. Borneo, Sabah, Kota Marudu District: Kinabalu Park – Serinsim substation, Bat Cave, ca. 370 m elevation. Erect herb, to 60 cm or more; plant sterile when collected, leaves green above, red beneath, rhizomes lacking. 13 Nov 2009, *C.-I Peng 22343* (HAST); same loc., *Dolois et al. SP 05965* (SNP), *Masrina et al. SP 10536* (SNP), *Antony van der Ent et al. SNP 31263* (SNP), *Antony van der Ent et al. SNP 31260* (SNP).

Notes
Begonia peridoticola resembles *B. keithii* Kiew (Kiew 1998) and *B. punchak* Kiew & S. Julia (Kiew & Julia 2007) in the cane-like stem and succulent leaves. *Begonia keithii* is endemic to limestone in the Semporna Forest Reserve in southeastern Sabah. The new species is readily distinguishable from *B. keithii* in the undulate, lanceolate-ovate (vs. narrowly lanceolate) leaves and cordate (vs. much prolonged) basal lobes. The new species is also similar to *B. punchak* from limestone areas in Kuching, Sarawak, differing in the persistent (vs. deciduous) stipules; yellow, spiral (vs. crimson, widely U-shaped) styles; and much larger capsular wings (8–11 mm vs. 2–3 mm wide).

Conclusion
A careful study of the literature, herbarium specimens and living plants, both in the wild and in cultivation, supports the recognition of the two new species. Detailed descriptions, line drawings, color plates, chromosome data, foliar SEM observations and comparisons with phenetically similar species are provided to aid in identification.

Competing interests
The authors declare that they have no competing interests.

Authors' contributions
RR organized the field trip and provided ecological information; RR, WCL and CIP collected the new species; WCL and KFC assisted the field work; CWL prepared line drawings; YK carried out the cytological study; CIP, CWL and RR prepared the manuscript. All authors read and approved the final manuscript.

Acknowledgments
We thank Sabah Parks for providing technical support and facilitating the procedure of collecting permits. This study was supported in part by grants from Academia Sinica to Ching-I Peng (HAST).

Author details
[1]Herbarium (HAST), Biodiversity Research Center, Academia Sinica, Nangang, Taipei 115, Taiwan. [2]Division of Botanical Garden, Taiwan Forestry Research Institute, No. 53, Nan-Hai Road, Taipei 100, Taiwan. [3]Sabah Parks, 88806 Kota Kinabalu, Sabah, Malaysia. [4]School of Forestry and Resource Conservation, National Taiwan University, Taipei 106, Taiwan.

References

Beaman JH, Anderson CR, Beaman RS (2001) The Plants of Mount Kinabalu 4. Dicotyledon Families Acanthaceae to Lythraceae. Natural History Publications (Borneo) Shn. Bhd.; Kew: Royal Botanic Gardens, Kota Kinabalu

Chung K-F, Leong W-C, Rubite RR, Rimi R, Kiew R, Liu Y, Peng C-I (2014) Phylogenetic analyses of Begonia sect. Coelocentrum and allied limestone species of China shed light on the evolution of Sino-Vietnamese karst flora. Bot Stud 55:e1, doi:10.1186/1999-3110-55-1 1209

Ding B, Nakamura K, Kono Y, Ho M-J, Peng C-I (2014) *Begonia jinyunensis* (Begoniaceae, section *Platycentrum*), a new palmately compound leaved species from Chongqing. China Bot Stud 55:e62, doi:10.1186/s40529-014-0062-6

Hughes M, Rubite RR, Kono Y, Peng C-I (2011) *Begonia blancii* (sect. *Diploclinium*, Begoniaceae), a new species endemic to the Philippine island of Palawan. Bot Stud 52:203–209

Kiew R (1998) Niche partitioning in limestone begonias in Sabah, Borneo, including two new species. Gard Bull Singapore 50(2):161–169

Kiew R (2001) The limestone begonias of Sabah, Borneo: Flagship species for conservation. Gard Bull Singapore 53(1–2):241–286

Kiew R, Julia S (2007) *Begonia* (Begoniaceae) from limestone hills in the Kuching Division, Sarawak, Borneo, including nine new species. Gard Bull Singapore 58(2):199–232

Kiew R, Tan JH (2004) *Begonia sabahensis* Kiew & J. H. Tan (Begoniaceae), a new yellow-flowered *Begonia* from Borneo. Gard Bull Singapore 56(1–2):73–77

Levan A, Fredga K, Sandberg AA (1964) Nomenclature for centromeric position on chromosomes. Hereditas 52:201–220

Lin C-W, Peng C-I (2014) *Begonia natunaensis* (sect. *Reichenheimia*, Begoniaceae), a new species from Natuna Island, Indonesia. Taiwania 59(4):368–373

Lin C-W, Chung S-W, Peng C-I (2014a) Three new species of *Begonia* (sect. *Petermannia*, Begoniaceae) from Sarawak, Borneo. Phytotaxa 191(1):129–140. http://dx.doi.org/10.11646/phytotaxa.191.1.8

Lin C-W, Chung S-W, Peng C-I (2014b) *Begonia hosensis* (sect. *Reichenheimia*, Begoniaceae), a new species from Sarawak, Malaysia. Taiwania 59(4): 326–330

Lin C-W, Chung S-W, Peng C-I (2015) *Begonia baik* and *B. padawanensis* spp. nov. (Begoniaceae) from sandstone areas in Sarawak, Malaysia. Nord. J Bot. doi:10.1111/njb.00641.

Nakamura K, Rubite R, Kono Y, Callado JR, Peng C-I (2013) *Begonia tandangii* (Begoniaceae, section *Baryandra*), a new species from Luzon Island, the Philippines. Phytotaxa 145(1):27–37, http://dx.doi.org/10.11646/phytotaxa.145.1.3

Peng C-I, Yang H-A, Kono Y, Chung K-F, Huang Y-S, Wu W-H, Liu Y (2013) Novelties in *Begonia* sect. *Coelocentrum*: *B. longgangensis* and *B. ferox* from limestone areas in Guangxi, China. Bot Stud 54:e44, doi:10.1186/1999-3110-54-44

Peng C-I, Wang H, Kono Y, Yang H-A (2014a) *Begonia wui-senioris* (sect. *Platycentrum*, Begoniaceae), a new species from Myanmar. Bot Stud 55:e13, doi:10.1186/1999-3110-55-13

Peng C-I, Jin X-H, Ku S-M, Kono Y, Huang H-Y, Yang H-A (2014b) *Begonia wuzhishanensis* (sect. *Diploclinium*, Begoniaceae), a new species from Hainan Island, China. Bot Stud 55:e24, doi:10.1186/1999-3110-55-24

Peng C-I, Ku S-M, Yang H-A, Leong W-C, Liu Y, Nguyen T-H, Kono Y, Chung K-F (2014c) Two new species of *Begonia* sect. *Coelocentrum, B. guixiensis* and *B. longa*, from Sino-Vietnamese limestone karsts. Bot. Stud. 55:e52. doi:10.1186/s40529-014-0052-8

Reza A, Kiew R (1998) *Begonia lazat* (Begoniaceae), a new culinary begonia from Borneo. Gard Bull Singapore 50(1):43–48

Rubite RR, Hughes M, Alejandro GJD, Peng C-I (2013) Recircumscription of *Begonia* sect. *Baryandra* (Begoniaceae): evidence from molecular data. Bot Stud 54:e38, doi:10.1186/1999-3110-54-38

Rubite RR, Callado JRC, Kono Y, Yang H-A (2014) *Begonia chingipengii* (sect. *Baryandra*, Begoniaceae), a new species from Luzon Island, Philippines. Phytotaxa 164(3):175–182

Somaweer R (2007) Kinabalu Mountain Range, University of Peradeniya, Sri Lanka, available from: http://phylodiversity.net/. (accessed: 12 Feb. 2014)

The endophytic fungi of *Salvia miltiorrhiza* Bge.f. alba are a potential source of natural antioxidants

Yan-Ling Li[1†], Xiao-Ming Xin[2†], Zheng-Yao Chang[1], Ren-Jiu Shi[1*], Zeng-Min Miao[1], Jing Ding[3] and Gang-Ping Hao[1*]

Abstract

Background: *Salvia miltiorrhiza* Bge. f. alba is a traditional Chinese herbal drug with special pharmacological effect on thromboangiitis obliterans. However, the nature source of *S.miltiorrhiza* Bge.f.alba is now in short supply because of the over-collection of the wild plant. To better utilize this resource, the diversity and antioxidant activity of endophytic fungi isolated from *S. miltiorrhiza* Bge. f. alba were investigated.

Results: A total of 14 endophytic fungi were isolated from different parts of *S. miltiorrhiza* Bge.f.alba. Based on morphological and molecular identification, the endophytic fungi isolated were classified into four genera (*Alternaria* sp., *Fusarium* sp., *Schizophyllum* sp. and *Trametes* sp.). These fungal extracts were prepared using ethanol and evaluated for their phytochemical compounds and antioxidant activity. *Alternaria alternata* SaF-2 and *Fusarium proliferatum* SaR-2 are of particular interest because they yielded all of nine phytochemicals including saponins, phenol, flavonoids, cardiac glycosides, steroids, tannins, alkaloids, anthroquinone and terpenoids. *F. proliferatum* SaR-2 and *A. alternata* SaF-2 also exhibited stronger antioxidant activities by FRAP and DPPH method, having the higher levels of phenol and flavonoid than those of plant root. The total amount of phenol and flavonoid quantified were of 21.75, 20.53 gallic acid equivalent per gram and 8.27 and 7.36 μg/mg of quercetin equivalent respectively. These two endophytic fungi (SaR-2 and SaF-2) were found to have comparable scavenging abilities on both FRAP (1682.21 and 1659.05 μmol/mg, respectively) and DPPH-free radicals (90.14% and 83.25%, respectively, at 0.1 mg/mL). This is the first report about isolation of endophytic fungi from *S. miltiorrhiza* Bge.f.alba and their antioxidant activities.

Conclusions: These results indicate that the endophytic fungi associated with *S. miltiorrhiza* Bge.f. alba can be a potential source of novel natural antioxidants.

Keywords: *Salvia miltiorrhiza* Bge.f.alba; Endophytic fungi; Identification; Phytochemicals; Antioxidant activity

Background

Salvia miltiorrhiza Bunge is a well-known medicinal plant, and its root, called "dan shen" in Chinese, is a traditional Chinese herbal drug used for the treatment of various kinds of diseases, especially for cardiovascular and cerebrovascular diseases (Zhou et al. 2012). *S. miltiorrhiza* Bge.f.alba is a white flowered varietas of *S. miltiorrhiza* Bunge and was present only in Shandong province of China. Studies showed that it had special pharmacological effect on thromboangiitis obliterans (Hao et al. 2009). The main bioactive constituents in root of two kinds of *S. miltiorrhiza* include lipid-soluble diterpenes (dihydrotanshione I, tanshinone I, cryptotanshinone and tanshinone IIA), and water-soluble phenolic compounds (salvianolic, rosmarinic, salvianic acid and protocatechuic) (Chen and Chen 1999; Liu et al. 2007). Previous research on phenolic compounds showed that there were important biological activities such as antioxidant and antithrombotic effects (Lam et al. 2007; Li 1997; Yan et al. 2006; Zhou et al. 2005). Compared with *S. miltiorrhiza* Bunge, *S. miltiorrhiza* Bge.f.alba has higher phenolic acids contents and higher pharmaceutical values with potential application in pharmaceutical industry (Hao et al. 2009; Hao et al. 2012). However, the nature source of *S.miltiorrhiza* Bge.f.alba is now in short supply because of the over-collection of the wild plant. Therefore, it is important to find a substitutable approach to produce the active compounds similar with the host plant to meet the medical demand.

* Correspondence: rjshi@tsmc.edu.cn; haogangping@163.com
[†]Equal contributors
[1]College of Life Sciences, Taishan Medical University, Tai'an 271016, China
Full list of author information is available at the end of the article

Plant endophytic fungi are symbiotic fungi that inhabit the interior of the healthy tissues of the host plants without causing apparent symptoms of disease (Saikkonen et al. 1998). They have been found in every plant species examined, and it is estimated to be around over one million endophytic fungi colonizing in plants. Endophytic fungi residing within these plants are able to produce bioactive compounds such as paclitaxel, podophyllotoxin and camptothecine (Aly et al. 2010), which were also produced by their respective host plants. This is advantageous for us to develop an alternative way for efficiently producing these valuable and scarce bioactive constituents. Thus plant endophytic fungi have been considered to be a novel and promising resource of natural bioactive compouds with extensive application in agriculture, industry and medicine field (Schulz et al. 2002; Strobel et al. 2004; Verma et al. 2009). Many valuable bioactive products from endophytes have been recognized as promising sources of antimicrobial, antioxidant, and anticancer substances (Jayanthi et al., 2011; Lin et al. 2007; Sadananda et al. 2011). These bioactive products could be classified as saponins, phenol, flavonoids, cardiac glycosides, steroids, tannins, alkaloids, anthroquinone and terpenoids (Tan and Zou 2001; Zhang et al. 2006).

Increasing evidence showed that reactive oxygen species (ROS) could cause oxidative damage of lipids, proteins, DNA and RNA, eventually enhancing the risk for aging, cardiovascular disease, cancer, atherosclerosis, diabetes, Alzheimer's disease and other diseases (Finkel and Holbrook 2000; Lachance et al. 2001). Therefore, antioxidants are believed to be highly effective radical scavengers in the prevention of this ROS mediated diseases. To date, there is little report about isolation and antioxidant activities of endophytic fungi from *S. miltiorrhiza* Bge.f.alba. The present study, therefore, was carried out to better understand the phytochemicals and antioxidant potential of endophytic fungi from *S. miltiorrhiza* Bge.f.alba.

Methods

Plant materials

Nine healthy *Salvia miltiorrhiza* Bge. f. alba were randomly collected in July 2010 from *Salvia miltiorrhiza* cultivation zone at three sites in Taishan Medical University, Shandong Province, China. The current-year roots (4–6 cm × 2–3 cm, length × diameter), stems (8–12 cm × 1–2 cm, length × diameter), leaves and flowers were placed in sterile plastic bags and stored at 4°C until isolation procedure was started.

Isolation of fungal endophytes

Plant tissues were surface-sterilized using the procedure described by Fisher et al. (1993) with minor modification. The root, stem, leaf and flower segments were cleaned under running tap water. After air-drying, the cleaned stems and roots were cut into small pieces, and then all the tissues were surface sterilized by immersion in 75% ethanol for 1 min, 5% sodium hypochlorite solution for 3 min and then washed three times in sterile distilled water for 1 min each time. The surface-sterilized samples were cut into small pieces using a sterile blade and placed on sterile potato dextrose agar (PDA) at 28°C. The hyphal tip of endophytic fungus growing out from the plant tissue was cut by a sterile pasture pipette and transferred to PDA plate supplemented with ampicillin (150 mg/L) and streptomycin (100 mg/L) to inhibit the bacterial growth until the mycelium or colony appeared surrounding the segments. The efficiency of surface sterilization procedure was checked for each sterilized plant segment following the imprint method. In addition, to check the presence of surface associated fungi, plant samples without surface sterilization were cultured in the same condition as negative controls. Colonization frequency (CF) was calculated as the total number of plant tissue segments colonized by an endophyte, divided by the total number of incubated segments.

Endophytic fungi from the roots, stems, leaves and flowers of *Salvia miltiorrhiza* Bge. f. alba have been numbered and codified as SaR1- SaR7, SaS1, SaL1 and SaF1- SaF5. All the isolated endophytic fungi have been stored on PDA slants at 4°C and kept at College of Life Sciences, Taishan Medical University, Shandong Province, China.

Identification of fungal endophytes

The identification procedure of endophytic fungi was based on morphology and molecular methods. The morphological characters included culture characteristics and the morphology of conidia. The molecular method was carried out to characterize some non-sporulating group using the ribosomal internal transcribed spacer (ITS) sequence. Fungal genomic DNA was extracted from fresh mycelia using an SDS extraction protocol described by Plaza et al. (2004). The primers ITS1 (5'-TCCGTAGGT GAACCTGCGG-3') and ITS4 (5'-TCCTCCGCTTATT GATATGC-3') were used to amplify the ITS region. The PCR products were subsequently purified and sequenced in two directions on an ABI 3700 automated sequencer. The resulting sequences were subjected to BLAST searches of the NCBI GenBank database to determine the identity of the fungi.

Fermentation and preparation of crude extracts

Each isolated endophytic fungus strain was inoculated in potato dextrose liquid medium with 150 rpm shaking at 25°C for 5–7 days. Mycelia and broths were separated by filteration through two layers of cheesecloth. Mycelia were thoroughly washed with sterile distilled water, air-dried in an oven at 60°C, and ground into fine powder.

The dried mycelia sample (2 g) was extracted with 50 mL of 95% ethanol for three times. Culture filtrates was treated by rotary evaporation under vacuum and then extracted thrice with a threefold of the volume of 95%(v/v) ethanol. The obtained extracts were filtered by Whatman No.1 filter paper and then concentrated under vacuum at 45°C to yield the crude extracts. As a control, 2 g of plant root power was extracted with 50 mL of 95% ethanol and stored at 4°C till further process.

Phytochemical screening

Preliminary phytochemical analysis of the crude extracts of fungi and plant root was carried out for the presence of the following metabolites such as saponins, phenol, flavonoids, cardiac glycosides, steroids, tannins, alkaloids, anthroquinone and terpenoids according to standard methods (Devi et al. 2012; Edeoga et al. 2005; Maobe et al. 2013).

Determination of total phenolic content

Total phenolic contents from endophytic fungi and plant root were respectively measured by Folin-Ciocalteu's colorimetric method (Taga et al. 1984) with some modification. Briefly, 100 mg of different ethanol extracts was added to 5 mL of 0.3% HCl in methanol/deionised water (60:40, v/v) respectively. The resulting mixture (100 μL) was added to 2 mL of 2% aqueous sodium carbonate. Then the mixture was incubated for 2 min at room temperature. 100 μL of 50% Folin-Ciocalteu's reagent was added to treated mixture and incubated for 30 min at room temperature, and absorbance was measured at 750 nm with the spectrophotometer against blank. The total phenol content was calculated on the basis of the standard curve of gallic acid. Phenol contents were expressed as mg of gallic acid equivalents (GAEs) per g of extract.

Determination of total flavonoid content

Total flavonoid content was estimated by a colorimetric method reported by Barros et al. (2007). The extract (250 μL) was mixed with distilled water (1.25 mL) and sodium nitrite solution (5%, 75 μL). After 5 min incubation at room temperature, aluminum chloride solution (10%, 150 μL) was added. After 6 min, sodium hydroxide (1 M, 500 μL) and distilled water (275 μL) were added to the mixture. The solution was mixed well and incubated at 25°C for 30 min. Absorbance was measured at 510 nm against blank. The content of flavonoid was calculated on the basis of the standard curve of quercetin and the results were expressed as mg of quercetin equivalents per g of extract.

Determination of antioxidant activity

Antioxidant activity of examined extracts was measured using ferric ion reducing antioxidant power (FRAP) and 1,1-diphenyl-2-picryl-hydrazyl (DPPH) assay.

FRAP assay

FRAP reagents was freshly prepared by mixing 25 mL acetate buffer (300 mM, pH 3.6), 2.5 mL 2,4,6-tris (2-pyridyl)-S-triazine (TPTZ) solution (10 mM TPTZ in 40 mM/L HCl) and 2.5 mL $FeCl_3$ (20 mM) water solution. Each sample (150 μL) (0.5 mg/mL) dissolved in methanol was added to 4.5 mL of freshly prepared FRAP reagent and stirred. After 5 min, absorbance was measured at 593 nm, using FRAP working solution as blank (Szőllősi and Szőllősi Varga 2002; Tomic et al. 2009). A calibration curve of ferrous sulfate (100–1000 μmol/L) was used and results were expressed in μmol Fe^{2+}/mg dry weight extract. The relative activity of the samples was compared with the standards ascorbic acid and butylated hydroxytoluene (BHT).

DPPH radical assay

The free radical scavenging activities of different extracts were carried out by using the 1,1-diphenyl-2-picryl-hydrazyl (DPPH) assay (Braca et al. 2001). Crude extract (0.1 mL) was mixed with 3 mL of a 0.004% methanol solution of DPPH. The mixture were reacted in the dark condition for 30 min and the absorbance was determined at 517 nm. The percentage inhibition activity was calculated using the following equation:

DPPH scavenging effect (The free radical scavenging) = $[(Abs_{control} - Abs_{sample})/Abs_{control}] \times 100$, Where $Abs_{control}$ is the absorbance of the control reaction, Abs_{sample} is the absorbance of the extract/standard.

Statistical analysis

Analysis of variance was performed by ANOVA. Means between treatment groups were compared for significance by using Duncan new multiple-range test. P values < 0.05 were considered to be significant and P values < 0.01 to be very significant. All experiments were performed in triplicate (n = 3) and results were reported as means ± standard deviation (SD).

Results

Isolation and identification of endophytic fungi

A total of 216 tissue segments (including 54 roots, 54 stems, 54 leaves and 54 flowers) collected from 9 individuals of *S. miltiorrhiza* Bge.f.alba at the three sites were processed, and 14 endophytic fungi isolates were recovered. Of these, the majority (n = 7, CF 12.96%) was recorded from roots, followed by flowers (n = 5, 9.26%), stems (n = 1, 1.85%) and leaves (n = 1, 1.85%), respectively (Table 1). 4 isolates mainly belonging to *Alternaria* sp. and Sterile mycelia were isolated from Site 1, 6 strains (*Alternaria* sp., *Fusarium* sp., and Sterile mycelia) were isolated from Site 2 and 4 strains (*Alternaria* sp., and Sterile mycelia) were isolated from Site 3. Among which, *Alternaria* sp. were the dominant species isolated

Table 1 Genera and number of endophytic fungi recovered from root, stem, leaf and flower of *S. miltiorrhiza* Bge. f. alba

Species	Number of isolates and Colonization frequency (%)			
	Sample tissues			
	Root	Stem	Leaf	Flower
	N (%)	N (%)	N (%)	N (%)
Alternaria sp.	0 (0)	0 (0)	1 (1.85)	5 (9.26)
Fusarium sp.	2 (3.70)	0 (0)	0 (0)	0 (0)
Sterile mycelia	5 (9.26)	1 (1.85)	0 (0)	0 (0)
Total	7 (12.96)	1 (1.85)	1 (1.85)	5 (9.26)

from leaves and flowers, Sterile mycelia occurred in roots and stems, and *Fusarium* sp. were only present in roots. Six isolates of non-sporulating Sterile mycelia were grouped into two morphotypes (morphotypes SaR3, SaR4, SaR5, SaS1, and morphotypes SaR6, SaR7), which were both *Basidiomycota* as indicated by promoting sporulation. Moreover, the selected two non-sporulating fungi (SaR-3, SaR-6), *Alternaria* sp. (SaF-2), and *Fusarium* sp. (SaR-2) were further identified by the internal transcribed spacer (ITS) rRNA gene sequence analysis. The accession numbers were provided by the Genbank. The result of identification showed that two non-sporulating fungi SaR-3 (JQ409156) and SaR-6 (JQ409166) were closely related to *Schizophyllum commune* (EU530002) and *Trametes hirsuta* (EF546240) (Table 2), while SaF-2 (JQ409154) and SaR-2 (JQ409155) were closely related to *Alternaria alternata* (FJ228163) and *Fusarium proliferatum* (EF546240) (Table 2).

Preliminary phytochemical screening

Preliminary phytochemical screening of fungal ethanolic extracts showed the presence of saponins, flavonoids, cardiac glycosides, terpenoids, steroids, tannins, phenol, anthroquinone and alkaloids. The result of qualitative analysis of the phytochemicals were summarised in Table 3. *A. alternata* SaF-2 and *F. proliferatum* SaR-2 showed presence of all of nine phytochemicals above, while *S. commune* SaR-3, *T. hirsuta* SaR-6 and plant extracts also had all the phytochemical except cardiac glycosides and anthraquinones. Cardiac glycosides was found to be present in the extracts of SaL-1, SaF-1, SaF-2, SaR-1, SaR-2, SaR-5, SaR-6 and SaS-1, while anthroquinone was found to be present in the extracts of SaF-2, SaF-3, SaR-2, SaR-5, SaR-6 and SaR-7. The presence of terpenoids and phenol was observed in all of the endophytes except *Alternaria* sp. SaF-5 and *Schizophyllum* sp. SaS-1, while flavonoids were found to be present in all the extracts except *Alternaria* sp. SaF-1 and *Trametes* sp. SaR-7. The results of phytochemical analysis indicated that the endophytic fungi from *S. miltiorrhiza* Bge.f.alba could produce the same or similar phytochemicals as those originally from the host plant, and new bioactive compounds that were not present in host plant.

Table 2 Identification of endophytic fungi based on sequence of ITS DNA

Isolate	Accession number	Size of ITS amplicon (bp)	Closest match in GenBank	Percentage identity
SaF-2	JQ409154	532	*Alternaria alternata* (FJ228163)	100
SaR-2	JQ409155	518	*Fusarium proliferatum* (EF546240)	99
SaR-3	JQ409156	598	*Schizophyllum commune* (EU530002)	99
SaR-6	JQ409166	624	*Trametes hirsuta* (EF546240)	99

Determination of total phenol content and total flavonoid content

The total phenolic content and total flavonoid content of four endophytic strains (*A. alternata*, *F. proliferatum*, *S. commune* and *T. hirsuta*) and host plant were expressed in terms of mg gallic acid equivalent (GAE)/g of extract and mg quercetin equivalents/g of extract respectively. The higher total phenolic content was found in the extract of *F. proliferatum* SaR-2 (21.75), followed by *A. alternata* SaF-2 extract (20.53) and Plant root extract (19.17)

Table 3 Phytochemical analysis for the ethanol extracts of different endophytic fungi and plant root

Genera	Isolates	A	B	C	D	E	F	G	H	I
Alternaria sp.	SaL1	-	+	+	+	+	+	+	-	-
	SaF1	+	-	+	+		+	+	-	-
	SaF2	+	+	+	+	+	+	+	+	+
	SaF3	-	+	-	+	+	-	+	+	-
	SaF4	-	+	-	+	-	-	+	-	-
	SaF5	+	+	-	-	-	+	-	-	+
Fusarium sp.	SaR1	-	+	+	+	-	+	+	-	-
	SaR2	+	+	+	+	+	+	+	+	+
Schizophyllum sp.	SaR3	+	+	-	+	+	+	+	-	+
	SaR4	-	+	-	+	-	+	+	-	-
	SaR5	+	+	+	+		+	+	+	-
	SaS1	-	+	+	-	+	-	-	-	+
Trametes sp.	SaR6	+	+	+	+	+	-	+	+	-
	SaR7	-	-	-	+		+	+	+	-
Plant root		+	+	-	+	+	+	+	-	+

+: Presence, –: absence; Sa: *Salvia miltiorrhiza* Bge.f.alba; SaR, SaS, SaL and SaF: isolate assignment code from root, stem, leaf, and flower tissues respectively; A: Saponins, B: Flavonoids, C: Cardiac glycosides, D: Terpenoids, E: Steroids, F: Tannins, G: Phenol, H: Anthroquinone, I: Alkaloids. Data is three replicates of each sample.

(Table 4). Similarly, the total flavonoid content was found to be higher in *F. proliferatum* SaR-6 (8.27), followed by *A. alternata* SaF-2 (7.36) and plant root (6.98) (Table 4). The more phenol and flavonoid content in the endophytic fungi than host plant may have contributed dramatically to their antioxidant activities.

Antioxidant capacity analysis

The antioxidant capacity of the ethanol extracts of four selected endophytes and plant extract were assessed for their ability to reduce TPRZ-Fe (III) complex to TPTZ-Fe (II) (Table 4). The reducing ability of the ethanol extracts was in the range of 512.25 to 1682.21 μmol Fe (II)/mg. The FRAP values for the extracts of *F. proliferatum* SaR-2 and *A. alternata* SaF-2 were significantly higher than those of ascorbic acid and BHT, while the FRAP values for the extracts of *S. commune* SaR-3 and *T. hirsuta* SaR-6 were significantly lower than that of ascorbic acid but higher than that of BHT.

The antioxidant activities of the extracts of endophytic fungi and plant root, as compared with the standards ascorbic acid and BHT, were determined by the capability to scavenge DPPH free radicals. Figure 1 showed the dose response curve of DPPH radical scavenging activity of all samples. It was observed that the ethanol extracts of *F. proliferatum* SaR-2 and *A. alternata* SaF-2 had higher radical scavenging activity than that of plant root. At a concentration of 0.1 mg/mL, the scavenging activity of the plant root extract reached 80.23%, while the scavenging activities of *F. proliferatum* SaR-2 and *A. alternata* SaF-2 were 90.14% and 83.25% respectively. Even though the DPPH radical scavenging abilities of the extract of *F. proliferatum* SaR-2 (90.14%) was lower than those of the standards ascorbic acid (95.43%) and BHT (94.14%) at 0.1 mg/mL, it still reached at high concentration. Thus the study showed that *F. proliferatum* SaR-2 could be a promising resource of natural antioxidants.

Discussion

During this study, an assessment was performed for the diversity of endophytic fungi from *Salvia miltiorrhiza* Bge. f.alba, which are only distributed in Shandong province of China and have been used for the treatment of various cardiovascular diseases (Zhou et al. 2012). A total of fourteen fungal endophytes were isolated and identified by the morphological and molecular method. Morphological investigations have resulted in the identification of two fungal species: *Alternaria* and *Fusarium* species. Six isolates belonging to non-sporulating fungi were identified on the basis of internal transcribed spacer (ITS) rRNA gene sequence analysis, which was consistent with the other reports from different hosts (Liu et al. 2007). The present study is the first report that several fungal endophytes are associated with *S. miltiorrhiza* Bge.f.alba.

It is interesting to record the predominance of *Alternaria* sp. from this host, as it was also dominant in many other reports from different host plants (Khan et al. 2010; Kumar et al. 2011; Liu et al. 2007). Besides *Alternaria* sp., other genera like *Fusarium* sp. were also reported frequently from a number of other plant species (Liu et al. 2007; Wang et al. 2008). Therefore, it appears that *Alternaria* and *Fusarium* species are able to associate endophytically with a wide range of host plants.

Phytochemical analysis was carried out to assess the diversity of chemical compounds produced by the endophytic fungi from *S. miltiorrhiza* Bge.f.alba. The results showed that there were significant amount of secondary metabolites including saponins, flavonoids, terpenoids, steroids, tannins, phenol, and alkaloids in the ethanolic extracts of the endophytic fungi, similar to or with more activity than those in the host plant root extracts. This result was consistent with the earlier reports (Govindappa et al. 2011; Sadananda et al. 2011). However, cardiac glycosides and anthraquinones were not present in the plant extracts but in the extracts of *A. alternata* SaF-2, *F. proliferatum* SaR-2, *Schizophyllum sp.* SaR-5 and *T. hirsuta* SaR-6. This result indicated that the fungal endophytes associated with *S. miltiorrhiza* Bge.f.alba had the capability of producing the same or similar phytochemicals as those present in the host plant, and new bioactive compounds that were not present in host plant. It is interesting to find that *Fusarium* sp. and *Alternaria* sp. could produce bioactive

Table 4 Total antioxidant activity and total phenolic content of ethanol extracts of four endophytic fungi and plant root

Samples	Phenol (mg/g)	Flavonoid (mg/g)	FRAP (μmol/L)
Alternaria alternata SaF-2	20.53 ± 0.08	7.36 ± 0.09	1659.05 ± 0.06
Fusarium proliferatum SaR-2	21.75 ± 0.11	8.27 ± 0.12	1682.21 ± 0.05
Schizophyllum commune SaR-3	12.96 ± 0.18	5.62 ± 0.15	586.65 ± 0.13
Trametes hirsuta SaR-6	11.21 ± 0.25	4.56 ± 0.08	512.25 ± 0.15
Plant root extract	19.17 ± 0.09	6.98 ± 0.13	1347.54 ± 0.11
Ascorbic acid			1655.25 ± 0.07
BHT			66.43 ± 0.16

Data represent means of three replicates.

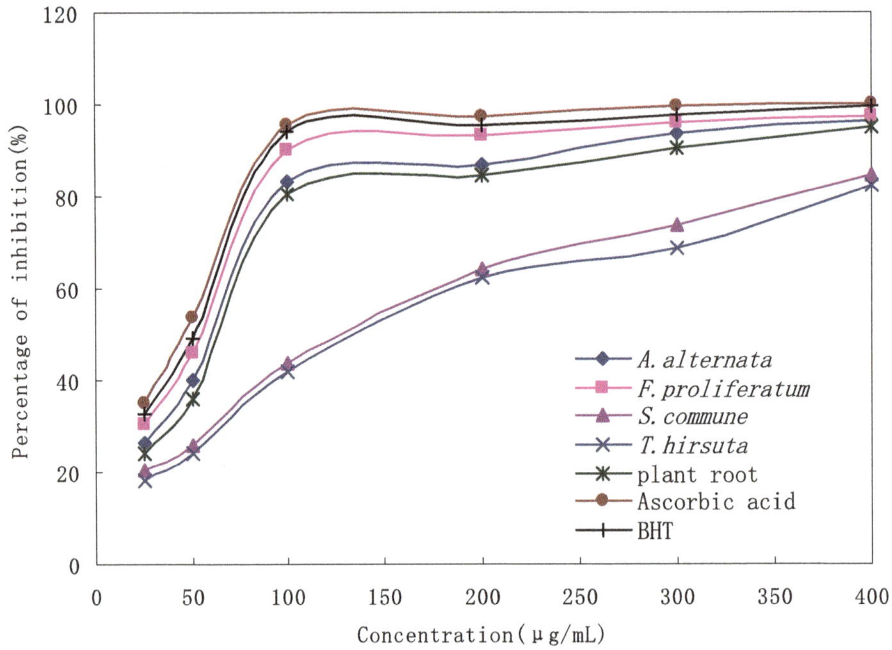

Figure 1 DPPH radical scavenging activity of the ethanol extracts of four selected endophytic fungi, plant root and the positive controls (ascorbic acid and BHT). All the experiments were repeated for three times.

compounds originally from their host plants. Similarly, *Fusarium* sp. and *Alternaria* sp. were found to produce active constituents such as paclitaxel, podophyllotoxin, camptothecine, which were also produced by their host plants (Zhao et al. 2011).

The presence of phytochemicals within fungal endophytes can be promising sources for medicinal or agrochemical use. The phytochemicals including phenols and flavonoids in fungal endophytes may be responsible for the antioxidant property (Hamilton et al. 2012). According to recent study, a highly positive correlation between total phenol content, total flavonoid content and antioxidant activity seems to be the trend in many endophytic fungi. In our study, strong antioxidant activities were present in the ethanol extracts of *F. proliferatum* SaR-2 and *A. alternata* SaF-2, comparable with the standards ascorbic acid and BHT. High phenolic and flavonoid content found in the ethanol extracts of *Fusarium* sp. and *Alternaria* sp. imply the contribution of these compounds to antioxidant activities, which was consistent with early study (Govindappa et al. 2011; Murthy et al. 2011; Sadananda et al. 2011). These findings indicate that endophytic fungi from *S. miltiorrhiza* Bge.f.alba may be effective as a promising potential for the development of novel antioxidant drugs. However, further studies are recommended to purify and characterize the structure of the biologically active constituents.

Conclusions

The present study demonstrated that fungal endophytes, *Alternaria alternata* SaF-2 and *Fusarium proliferatum* SaR-2 from *Salvia miltiorrhiza* Bge.f.alba yielded medically important phytochemical compounds. The antioxidant potential may be directly linked to the phenolic and flavonoid compounds present in the ethanol extracts of *A. alternata* SaF-2 and *F. proliferatum* SaR-2. To our knowledge, this is the first report that fungal endophytes associated with *S. miltiorrhiza* Bge.f.alba have been found to possess antioxidant potential.

Abbreviations

BHT: Butylated hydroxytoluene; CF: Colonization frequency; DPPH: 1,1-diphenyl-2-picryl-hydrazyl; FRAP: Ferric ion reducing antioxidant power; GAEs: Gallic acid equivalents; ITS: Internal transcribed spacer; mg/g: Milligram/gram; PDA: Potato dextrose agar; ROS: Reactive oxygen species; Sa: *Salvia miltiorrhiza* Bge.f.alba; SaF: SaL, SaR and SaS, Flower, Leaf, Root and Stem tissues of *Salvia miltiorrhiza* Bge.f.alba; TPC: Total Phenolic Content; TPTZ: 2,4,6-tris (2-pyridyl)-S-triazine.

Competing interests

The authors declare that they have no competing interests.

Authors' contributions

YL took charge of experimental design, isolation and identification of endophytic fungi, phytochemical screening, interpretation of results, and write of the manuscript. XM performed the determination of antioxidant activity and interpretation of results. GP and RJ participated in the experiment design, ZY and JD analyzed data. ZM and GP involved in the correction of the paper. All the authors read and approved the final manuscript.

Acknowledgments

This work was supported by National Natural Science Foundation of China (81403036, 81173489), Shandong Provincial Medical Science and Technology Development Project (2013WS0321), Shandong Provincial Natural Science Foundation (ZR2014HM048, ZR2010CM063), Tai'an Science and Technology Development Project (20132093) and Shandong Provincial Science and Technology Project of Population and Family Planning (2013–17).

Author details

[1]College of Life Sciences, Taishan Medical University, Tai'an 271016, China. [2]College of Pharmacy, Taishan Medical University, Tai'an 271016, China. [3]College of Chemical Engineering, Taishan Medical University, Tai'an 271016, China.

References

Aly AH, Debbab A, Kjer J, Proksch P (2010) Fungal endophytes from higher plants: a prolific source of phytochemicals and other bioactive natural products. Fungal Divers 41:1–16

Barros L, Calhelha RC, Vaz JA, Ferreira ICFR, Baptista P, Estevinho LM (2007) Antimicrobial activity and bioactive compounds of Portuguese wild edible mushrooms methanolic extracts. Eur Food Res Technol 225:151–156

Braca A, Tommasi ND, Bari LD, Pizza C, Politi M, Morelli I (2001) Natural antioxidants from plant material in phenolic compounds in food and their effects on health. J Nat Prod 64:892–895

Chen H, Chen F (1999) Effects of methyl jasmonate and salicylic acid on cell growth and cryptotanshinone formation in Ti transformed *Salvia miltiorrhiza* cell suspension cultures. Biotechnol Lett 21:803–807

Devi NN, Prabakaran JJ, Wahab F (2012) Phytochemical analysis and enzyme analysis of endophytic fungi from *Centella asiatica*. Asian Pacific Journal of Tropical Biomedicine 2:1280–1284

Edeoga HO, Okwu DE, Mbaebie BO (2005) Phytochemical constituents of some Nigerian medicinal plants. Afr J Biotechnol 4:685–688

Finkel T, Holbrook NJ (2000) Oxidants, oxidative stress and biology of aging. Nature 408:239–247

Fisher PJ, Petrini O, Sutton BC (1993) A comparative study of fungal endophytes in leaves, xylem and bark of *Eucalyptus* in Australia and England. Sydowia 45:338–345

Jayanthi G, Kamalraj S, Karthikeyan K, Muthumary J (2011) Antimicrobial and antioxidant activity of the endophytic fungus *Phomopsis* sp. GJJM07 isolated from *Mesua ferrea*. Int J Curr Sci 1:85–90

Govindappa M, Bharath N, Shruthi HB, Santoyo G (2011) *In vitro* antioxidant activity and phytochemical screening of endophytic extracts of *Crotalaria pallida*. Free Rad Antiox 1:79–86

Hamilton CE, Gundel PE, Helander M, Saikkonen K (2012) Endophytic mediation of reactive oxygen species and antioxidant activity in plants: a review. Fungal Divers 54:1–10

Hao GP, Shi RJ, Wang JM, Qi B (2009) Construction of full-length cDNA library of white flower *Salvia miltiorrhiza* Bge f.alba root and partial EST sequence analysis. Afr J Biotechnol 8:2027–2035

Hao GP, Ji HW, Li YL, Shi RJ, Wang JM, Feng L, Huang LQ (2012) Exogenous ABA and polyamines enhanced salvianolic acids contents in hairy root cultures of *Salvia miltiorrhiza* Bge. f.alba. POJ 5:446–452

Khan R, Shahaz S, Choudhary MI, Khan SA, Ahmad A (2010) Communities of endophytic fungi in medicinal plant *Withania somnifera*. Pak J Bot 42:1281–1287

Kumar S, Kaushik N, Edrada-Ebel R, Ebel R, Proksch P (2011) Isolation, characterization, and bioactivity of endophytic fungi of *Tylophora indica*. World J Microbiol Biotechnol 27:571–577

Lachance PA, Nakat Z, Jeong WS (2001) Antioxidants: an integrative approach. Nutrition 17:835–838

Lam FF, Yeung JH, Chan KM, Or PM (2007) Relaxant effects of danshen aqueous extract and its constituent danshensu on rat coronary artery are mediated by inhibition of calcium channels. Vasc Pharmacol 46:271–277

Li LN (1997) Water soluble active components of *Salvia miltiorrhiza* and related plants. J Chin Pharm Sci 6:57–64

Lin X, Lu CH, Huang YJ, Zheng ZH, Su WJ, Shen YM (2007) Endophytic fungi from a pharmaceutical plant, *Camptotheca acuminata*: isolation, identification and bioactivity. World J Microbiol Biotechnol 23:1037–1040

Liu AH, Guo H, Ye M, Lin YH, Sun JH, Xu M, Guo DA (2007) Detection, characterization and identification of phenolic acids in Danshen using high-performance liquid chromatography with diode array detection and electrospray ionization mass spectrometry. J Chromatogr A 1161:170–182

Maobe MAG, Gatebe E, Gitu L, Rotich H (2013) Preliminary phytochemical screening of eight selected medicinal herbs used for the treatment of diabetes, malaria and pneumonia in Kisii Region, Southwest Kenya. Eur J Appl Sci 5:01–06

Murthy NK, Pushpalatha KC, Joshi CG (2011) Antioxidant activity and phytochemical analysis of endophytic fungi isolated from *Lobelia nicotianifolia*. J Chem Pharm Res 3:218–225

Plaza GA, Upchurch R, Brigmon RL, Whitman WB, Ulfig K (2004) Rapid DNA extraction for screening soil filamentous fungi using PCR amplification. Pol J Environ Stud 13:315–318

Sadananda TS, Nirupama R, Chaithra K, Govindappa M, Chandrappa CP, Vinay Raghavendra B (2011) Antimicrobial and antioxidant activities of endophytes from *Tabebuia argentea* and identification of anticancer agent (lapachol). J Med Plants Res 5:3643–3652

Saikkonen K, Faeth SH, Helander M, Sullivan TJ (1998) Fungal endophytes: a continuum of interactions with host plants. Annu Rev Ecol Syst 29:319–343

Schulz B, Boyle C, Draeger S, Römmert A, Krohn K (2002) Endophytic fungi:a source of novel biologically active secondary metabolites. Mycol Res 106:996–1004

Strobel G, Daisy B, Castillo U, Harper J (2004) Natural products from endophytic microorganisms. J Nat Prod 67:257–268

Szőllősi R, Szőllősi Varga I (2002) Total antioxidant power in some species of Labiatae (Adaptation of FRAP method). Acta Biologica Szegediensis 46:125–127

Taga MS, Miller EE, Pratt DE (1984) Chia seeds as a source of natural lipid antioxidants. J Am Oil Chem Soc 61:928–993

Tan RX, Zou WX (2001) Endophytes:a rich source of functional metabolites. Nat Prod Rep 18:448–459

Tomic A, Petrovic S, Pavlovic M, Trajkovski B, Milenkovic M, Vucicevic D, Niketic M (2009) Antimicrobial and antioxidant properties of methanol extracts of two *Athamanta turbish* subspecies. Pharm Biol 47:314–319

Verma VC, Kharmar RN, Strobel GA (2009) Chemical and functional diversity of natural products from plant associated endophytic fungi. Nat Prod Commun 4:1511–1532

Wang YT, LO HS, Wang PH (2008) Endophytic fungi from *Taxus mairei* in Taiwan: first report of *Colletotrichum gloeosporioides* as an endophyte of *Taxus mairei*. Bot Stud 49:39–43

Yan Q, Shi M, Ng J, Wu JY (2006) Elicitor-induced rosmarinic acid accumulation and secondary metabolism enzyme activities in *Salvia miltiorrhiza* hairy roots. Plant Sci 170:853–858

Zhang HW, Song YC, Tan RX (2006) Biology and chemistry of endophytes. Nat Prod Rep 23:753–771

Zhao J, Shan T, Mou Y, Zhou L (2011) Plant-derived bioactive compounds produced by endophytic fungi. Mini Rev Med Chem 11:159–168

Zhou L, Zuo Z, Chow MS (2005) Danshen: an overview of its chemistry, pharmacology, pharmacokinetics and clinical use. J Clin Pharmacol 45:1345–1359

Zhou X, Chan K, Yeung JHK (2012) Herb–drug interactions with Danshen (*Salvia miltiorrhiza*): a review on the role of cytochrome P450 enzymes. Drug Metabol Drug Interact 27:9–18

26

Lipid accumulation in response to nitrogen limitation and variation of temperature in *Nannochloropsis salina*

Eman M Fakhry and Dahlia M El Maghraby*

Abstract

Background: This batch study deals with the relation between lipid as well as triglyceride contents in *Nannochloropsis salina* and variation in culture conditions such as nitrogen concentration and temperature.

Results: The tested parameters caused reduction in growth expressed as cell count, optical density and dry weight, as well strongly involved in lipids and triglycerides accumulation and significantly affected the lipid productivity. At the beginning of the work, the concentration of nitrogen in the medium was reduced to three quarter, half and quarter of the original f2 medium while the temperature kept constant. After that, the optimal nitrogen concentration (quarter of the original media) giving high lipid yield was tested with different temperature degrees from 15 to 35°C with five degree intervals. Although the growth was insignificantly influenced, a considerable increase in lipid and triglyceride (56.1 and 15.1% of dry weight respectively) was observed when the concentration of nitrogen in the medium was reduced to the quarter. Moreover, 59.3% lipid and 17.1% triglyceride on the basis of dry weight were obtained by the combination of 25% nitrogen concentration and 30°C. Simple regressions recommended that the interaction effect of nitrogen limitation and temperature on lipid and triglyceride accumulation was not as fundamental as for nitrogen limitation stress.

Conclusion: The degree of nitrogen availability in the combination of temperature effect has been identified as the critical determinant for the maximal production of lipid in *N. salina*. Nevertheless, major advances in this field can be considered by studying more stresses techniques and genetic strategies.

Keywords: Lipids and triglycerides accumulation; Nitrogen depletion; Temperature; *Nannochloropsis salina*

Background

Lipid-producing microalgae have the potential to permit lipid accumulation without demanding for arable land. Under favorable growth conditions, algal biomasses are successfully produced with relatively low lipid contents. Optimization of biomass and lipid in microalgae has been studied in batch, closed and open outdoor systems (Zou and Richmond 2000; Sandnes et al. 2005; Grobbelaar 2007). In those culture systems, the environmental parameters and nutrient supply were found to be vital factors for growth improvement and induction of lipid in microalgae. Accordingly high lipid productivity which is a main requirement for commercial production of microalgal oil-derived biodiesel was achieved (Sharma et al. 2012). However, increase in lipid productivity (mass of lipids per unit of area per unit of time) can be achieved by either enhancement the rate of biomass accumulation, or increase the biomass with high lipid percent on the basis of dry weight. To attain this goal, different approaches including nutrient starvation, bioprocess optimization and genetic engineering can be used (Hu et al. 2008). However, the maximum yield will be obtained by the algal response to the surroundings.

Under environmental stress conditions, many algae modify their lipid biosynthetic pathways to produce and accumulate neutral lipids, mainly in the form of triacylglycerol. However, the biosynthesis pathway of triacylglycerol may play active role in the stress response. On the other hand, it offers carbon and energy storage

* Correspondence: dahliamag@hotmail.com
Department of Botany and Microbiology, Faculty of Science, Alexandria University, Alexandria 21511, Egypt

function in the cell that allows microalgae to tolerate unfavorable environmental conditions (Hu et al. 2008). In general, algal biomass and triglycerides compete for photosynthetic assimilate and a renovating of physiological pathways is required to stimulate lipid biosynthesis (Sharma et al. 2012). The need to improve lipid production can potentially be addressed by survey microalgae with adequate lipid content. Microalgae are hopeful lipid-producers as they are adapted to grow over a wide range of environmental conditions with short doubling times (Mata et al. 2010), higher growth rate and photosynthetic efficiencies than conventional crops (Chisti 2007; Wu et al. 2013). Furthermore, they possess various and infrequent patterns of cellular lipids and have the ability to alter lipid metabolism in response to environmental changes by several factors, of which nutrients (Chen et al. 2011; Feng et al. 2011; Mairet et al. 2011), temperature (Li et al. 2011; Fuentes-Grünewald et al. 2012) and irradiance (Hu et al. 2008).

Numerous studies highlight the key role of nitrogen and temperature in algal lipid accumulation and the conditions applied in these studies differ extensively. Nitrogen- deficient conditions can affect the microalgal growth and change the metabolic pathway to the accumulation of storage lipids (Lombardi and Wangersky 1991). Actually, the increase in lipid content may be species and strain specific (El-Baky et al. 2004; Pal et al. 2011; Olofsson et al. 2014). *Dunaliella*, *Chlorella* and *Nannochloropsis* species are known to respond to nitrogen starvation by increasing lipid production (Lombardi and Wangersky 1995; Guevara et al. 2005; Converti et al. 2009; Dong et al. 2013). Also, it is evident that nitrogen is a limiting factor for the growth of many living species (Hansen et al. 2000; Li et al. 2010; Park et al. 2012). Temperature is one of the principal factors in culture conditions that affect the growth and lipid production in microalgae (Tzovenik et al. 2003; Roleda et al. 2013, Rukminasari 2013). The response of microalgal lipid content to high and low growth temperatures differs from species to species (Renaud et al. 2002; Wu et al. 2013).

Genus *Nannochloropsis* is a microalga belonging to Eustigmatophyceae. The genus is broadly well appreciated in aquaculture (Roncarati et al. 2004; Bentley et al. 2008) due to its comparatively high growth rate, resistance to mixing and contamination together with high nutritional values and high lipid content (Rodolfi et al. 2003; Olofsson et al. 2012).

Current study aimed to obtain experimental data that clarify how *N. salina* will respond to stress conditions for maximum production of lipids as biodiesel feedstock and as component for valuable foodstuff and health products. Algal biomass and corresponding total lipid and triglyceride contents were estimated under nitrogen depletion in growth medium with variation in the temperature degrees during cultivation.

Methods

Organism and growth condition

N. salina was obtained from the Culture Collection of the Med Algae laboratory, Faculty of Science, Alexandria University, Egypt. All the glassware and media were always sterilized prior to cultivation. The cultures were grown in 1 L Erlenmeyer flasks with 600 mL f/2 medium (Guillard 1975) using the atmospheric CO_2 as carbon source and kept under controlled environmental conditions of approximately light intensity (150 μmol m^{-2} s^{-1}), light/dark cycle (12:12 h) and temperature (25 ± 2°C). The cultures were shaken twice daily to avoid sticking. This was referred to as control culture.

Experimental design

To investigate the effects of nitrogen deficiency on biomass and lipid yield, *N. salina* was batch cultured autotrophically and axenically in N-deficient conditions using the original source of nitrogen in f/2 medium (sodium nitrate). N-deficiency was achieved by cultivation of the microalga in conical flasks containing 1 L. of culture medium with normal nitrogen concentration and three additional cultivations were run under N-deficient conditions (25, 50 and 75% of the original levels used in the control medium). Subsequently, recultivation of cells in further experiment was provided to evaluate the effect of temperature under the chosen N concentration. The temperature values tested were from 15 to 35°C with five degree intervals. The media were buffered with sodium bicarbonate and HCl to pH 7.5. Each treatment consisted of triplicate flasks and continuous aeration was provided.

Growth measurements

The cell density of *N. salina* in culture system was estimated daily by counting under light microscope using the haemocytometer counting chamber. The biomass of the cultures was determined for every 24 h by measuring the optical density at wavelength of 680 nm (Huang et al. 2002) via Perkin-Elmer spectrum RXIFT-IR System. Each sample was measured twice and the mean value was calculated. To estimate the dry weight, microalgal biomass was harvested by centrifugation at 2000 g for 15 min. The cell pellets were washed twice with distilled water. The collected pellets were oven dried at 60°C till constant weight. The weights of the dry mass were evaluated in relation to the relative biomass of cultures.

Extraction of total cellular lipids and triglycerides

Lipid content was estimated at zero time and after 12 days (exponential phase) of cultivation. The algal lipid was extracted following the protocol of Bligh and Dyer (1959). The cells were harvested by centrifugation at 2000 g for 15 min. The pellet was subjected to wet weight estimation and then dried in oven at 60°C till constant weight. Algal sample was extracted with chloroform: methanol mixture (2:1 v/v) and kept for 24 hours at 25°C. The mixture was vigorously agitated in vortex for few minutes. The homogenate was centrifuged at 4000 rpm for 15 min. The lower layer was separated and the procedure was again repeated with the pellet. The phase containing dissolved lipids were transferred into a separatory funnel and shaken for 5 min. The lipid fractions were separated in a clean pre-weighed vial (first wt) and the solvent was evaporated using rotary evaporator. The weight of the vial was again recorded (second wt). Lipid content was calculated by subtracting first wt from second wt. Triglycerides were extracted using n-hexane, purified by separation on thin layer chromatography and quantified by comparing to a standard curve generated from known amounts of TAG standard. The weight of both lipid content and triglyceride was determined and calculated in correlation to the dry algal biomass. Lipid productivity was calculated as follows:

$$P_{Lipid}\left(g\ L^{-1}d^{-1}\right) = \frac{C_f \times DCW_f - C_i \times DCW_i}{T}$$

Where: P_{Lipid}: lipid productivity, C_f: final lipid content after 12 days of cultivation, DCW_f: final biomass of the microalgae in the lipid producing phase after 12 days of cultivation, C_i: initial lipid content at zero time, DCW_i: initial biomass in the lipid producing phase at zero time, T: the cultivation time.

Statistical analysis

All experiments were repeated three times independently, and data were recorded as the mean. Statistical analyses were performed using simple linear regressions (R) (version 2.12.0, R Foundation for Statistical Computing, Vienna, Austria) to estimate the relationship between the dependent variables (total lipid and triglyceride contents) and the independent variables (nitrogen concentrations (%) and temperatures). The effects of the treatments were tested by one-way analysis of variance (ANOVA). Means were compared between the treatments using the LSD (least significant difference) test at the 0.05 probability level.

Results

Effect of nitrogen limitation

Growth measurements of *N. salina* grown under normal condition and different nitrogen concentrations were reported in Table 1. The micro-alga shows maximum cell count ($\times 10^6$/ml), optical density (OD_{680}) and dry weight (g L^{-1}) on the twelfth day. The highest cell number obtained was under control condition (9.9 × 10^6/ml culture) with 0.77 optical density, 0.61 g L^{-1} dry weight and 0.228 g L^{-1} day^{-1} lipid productivity. Media with 75% and 50% nitrogen concentration yielded cell number of 9.1 and 8.2 × 10^6/ml culture giving 0.57 and 0.53 g L^{-1} dry weight and 0.237 and 0.298 g L^{-1} day^{-1} lipid productivity respectively. The lowest cell number, optical density and dry weight obtained were in medium with 25% nitrogen concentration (7.9 × 10^6/ml culture, 0.63 and 0.48 g L^{-1} respectively) with highest lipid productivity of 0.370 g L^{-1} day^{-1}.

Although *N. salina* maintained the ability to grow under nitrogen depletion with a reduced rate in cell number and dry weight comparing to that grown in control (Table 1), nitrogen depletion induced an increase in the lipid content which is correlated with the accumulation of triglycerides with highest percentages after 12 days of culturing (Figures 1 and 2). The maximum lipid and triglyceride contents (56.1and 15.1% of dry weight respectively) were gained at 25% nitrogen concentration followed by 50% nitrogen concentration (45.2 and 12.5 of dry weight respectively), while the minimum contents (36.1 and 10.7% of dry weight respectively) were obtained at 75% nitrogen concentration. With regard to the control, the lipid and triglyceride contents were 34.6 and 9.7% of dry weight respectively. Simple regression analysis (Figure 3) showed significant relationships between lipid and triglyceride contents and nitrogen concentrations giving 91% and 94% (simple regression: $y_{lipids} = -0.300x + 61.6$, $R^2 = 0.919$, and $y_{triglycerides} = -0.074x + 16.51$, $R^2 = 0.947$) of the variation, respectively. Consequently, lipids and triglycerides tend to accumulate more in the cells under condition of nitrogen limitation.

Effect of variation in temperature

To study the combined effect of nitrogen limitation and variation in temperature on batch growth of *N. salina*, cells grown in 25% nitrogen concentration were re-cultivated under different temperatures (15, 20, 25, 30 and 35°C) up to the stationary phase. The maximum algal growth was achieved on the 10th day at 30°C. Later, the micro-algal growth was gradually dropped and the pale green color of cells was noticed. By comparing the results of growth measurements to the control (Table 1), decrease in cell number, optical density and dry weight of *N. salina* accompanied by increase in the lipid productivity were marked. Number of cells were ranged from 6.9 × 10^6 to 8.8 × 10^6 /ml culture with optical density of 0.53 and 0.72 and dry weight of 0.42 and 0.53 respectively. However, the maximum (0.431 g L^{-1} day^{-1}) and minimum (0.311 g L^{-1} day^{-1}) lipid productivity were

Table 1 Growth measurements and lipid productivity of *Nannochloropsis salina* operated at different nitrogen concentrations and temperatures

	Cell count ($\times 10^6$/ml)	Optical density (OD_{680})	Dry weight of cells ($g\ L^{-1}$)	Lipid productivity ($g\ L^{-1}\ day^{-1}$)
Control	9.93 ± 0.25	0.77 ± 0.04	0.61 ± 0.03	0.23 ± 0.0
75	9.07 ± 0.45[a*]	0.71 ± 0.03[a*]	0.57 ± 0.06	0.24 ± 0.0[a*]
50	8.23 ± 0.45[a*b*]	0.65 ± 0.02[a*b*]	0.53 ± 0.05[a*]	0.30 ± 0.0[a*b*]
25	7.86 ± 0.45[a*b*]	0.63 ± 0.02[a*b*]	0.48 ± 0.02[a*b*]	0.37 ± 0.0[a*b*c*]
Temperature°C				
15	6.87 ± 0.55	0.53 ± 0.03	0.42 ± 0.05	0.31 ± 0.0
20	7.57 ± 0.35[a]	0.59 ± 0.02[a]	0.45 ± 0.04	0.35 ± 0.01[a]
25	8.23 ± 0.25[ab]	0.66 ± 0.01[ab]	0.50 ± 0.04[a]	0.37 ± 0.01[ab]
30	8.67 ± 0.06[ab]	0.71 ± 0.03[abc]	0.53 ± 0.02[ab]	0.43 ± 0.0[abc]
35	8.57 ± 0.25[ab]	0.72 ± 0.02[abc]	0.51 ± 0.02[a]	0.41 ± 0.01[abcd]

Mean ± SE (n = 3) followed by different letters indicate significant difference at P ≤ 0.05, according to one-way analysis of variance (ANOVA) and the LSD (least significant difference) test (a*: significant with control at ≤0.05, b*: significant with 75% N at ≤0.05, c*: significant with 50% N at ≤0.05, a: significant with 15°C at ≤0.05, b: significant with 20°C at ≤0.05, c: significant with 25°C at ≤0.05, d: significant with 30°C at ≤0.05).

acquired at 30°C and 15°C respectively after 10 days of culturing.

Lipid and triglyceride contents on the second, sixth and tenth days of culturing were demonstrated in Figures 1 and 2. Reduction in nitrogen concentration up to 25% accompanied by raising the temperature from 15 to 35°C resulted in an increase in both lipid and triglyceride contents with maximum values at 30°C. Considerable increase in lipid and triglyceride contents was observed since the biomass decreased with respect to the control. The lipid and triglyceride contents were 59.3 and 17.1% of dry weight respectively yielded from a biomass of 0.53 g L^{-1} at 10th day under 25% N-limitation and 30°C, whereas under control condition, they were found to be 34.6 and 9.7% of dry weight respectively with biomass of 0.61 g L^{-1} at 12th day. Simple regression analysis (Figure 4) explained significant relationships between lipid and triglyceride contents and temperature with 74% and 87% (simple regression, $y_{lipid} = 0.482x + 43.11$, $R^2 = 0.743$, $y_{triglyceride} = 0.252x + 8.866$, $R^2 = 0.875$).

Discussion

The optimization of culture condition for lipid production was affected by many interrelated factors. Significant differences in growth and cellular components of microalgal cells have been observed depending on how the alga will act in response to variations in culture conditions such as limitation or starvation of essential nutrients (Pinto et al. 2003; Ip and Chen 2005). Nitrogen is the most critical macronutrient affecting growth and lipid metabolism in algae, since it is an essential constituent of functional processes and cell structure (Sharma et al. 2012). A general trend of lipid accumulation, mostly triglycerides, in response to nitrogen

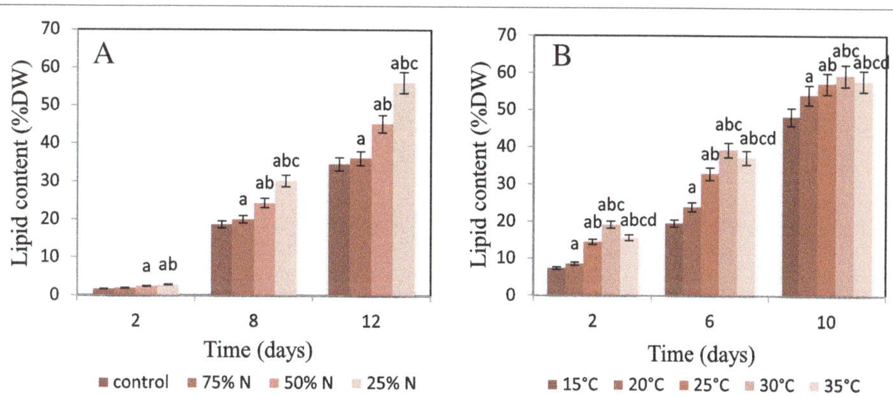

Figure 1 Cellular lipid content of *Nannochloropsis salina*. The lipid contents were monitored **(A)** after 2, 8 and 12 days of incubation under optimal conditions and in response to three depleted nitrogen concentrations **(B)** after 2, 6 and 10 days of incubation in response to five different temperature degrees. The error bars indicate mean ± SE (n = 3). The effects of the treatments were tested by one-way analysis of variance (ANOVA). Means were compared between the treatments using the LSD (least significant difference) test at the 0.05 probability level.

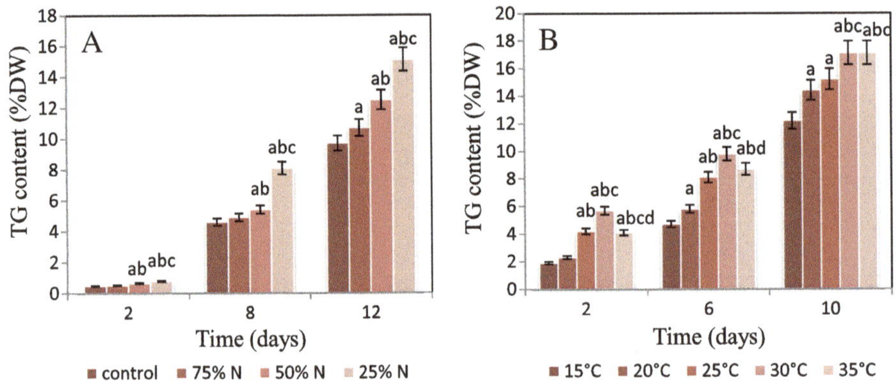

Figure 2 Triglyceride content of *Nannochloropsis salina*. The triglyceride contents were measured **(A)** after 2, 8 and 12 days of cultivation under optimal conditions and three depleted nitrogen concentrations **(B)** after 2, 6 and 10 days of cultivation under five different temperature degrees. The error bars indicate mean ± SE (n = 3). The effects of the treatments were tested by one-way analysis of variance (ANOVA). Means were compared between the treatments using the LSD (least significant difference) test at the 0.05 probability level.

deficiency has been observed in several species of different microalgae (Hsieh and Wu 2009; Yeh and Chang 2011; Sun et al. 2014). Furthermore, temperature is one of the key factors that affects the growth, lipids and the types of fatty acids produced by microalgae (Renaud et al. 2002; Converti et al. 2009; Taoka et al. 2009). In the present investigation, *N. salina* was characterized in terms of their responses to reduction in nitrogen concentration and different cultivation temperatures. Cell counting, optical density and biomass dry weight are effortless methods easy to use for *N. salina* growth evaluation. The relationship between cell number and optical density is dependent on culture conditions, for instance culture media and cell age. In fact, the size and weight of each cell is not always the same. Results indicated that cell counts, optical density and biomass dry weight of the tested alga decreased when grown under reduced nitrogen concentrations simultaneously with enhancement in lipid productivity, total lipid and triglyceride contents. This result is in accordance with Yeesang and Cheirsilp (2011) who reported the loss of biomass when green alga, *Botrycoccus sp.* was exposed to nitrogen deficient conditions. Mandal and Mallick (2009) and Gouveia and Oliveira (2009) have also reported decreased growth pattern in *Scenedesmus obliquus*, under nitrogen deficient conditions as well as for *Chlorella pyrenoidosa* (Nigam et al. 2011).

In this study, there is an inverse relationship between lipid content and nitrogen concentration. *N. salina* was able to stay alive under the whole tested nitrogen concentrations and induces increase in cellular lipid content. This is attributable to the unusual pattern of lipids together with the ability to adapt lipid metabolism efficiently in response to changes in cultivation conditions (Guschina and Harwood 2006). Culture aging or senescence also affects lipid and fatty acid content and composition. The total lipid content of cells increased with age in the green alga *Chlorococcum macrostigma* (Collins and Kalnins 1969), and the diatom *Thalassiosira fluviatillis* (Conover 1975) and *Coscinodiscus eccentricus*

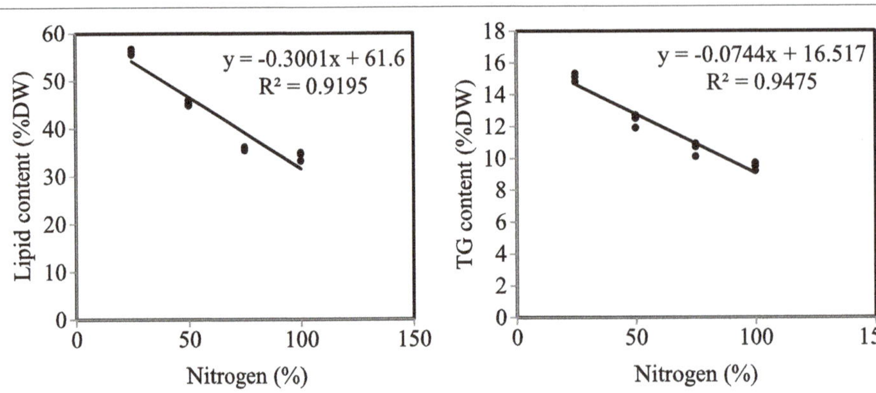

Figure 3 Statistical data relating to nitrogen limitation. The correlation between lipid and triglyceride contents as dependent variables and nitrogen concentrations as independent variable was performed using simple linear regression analysis.

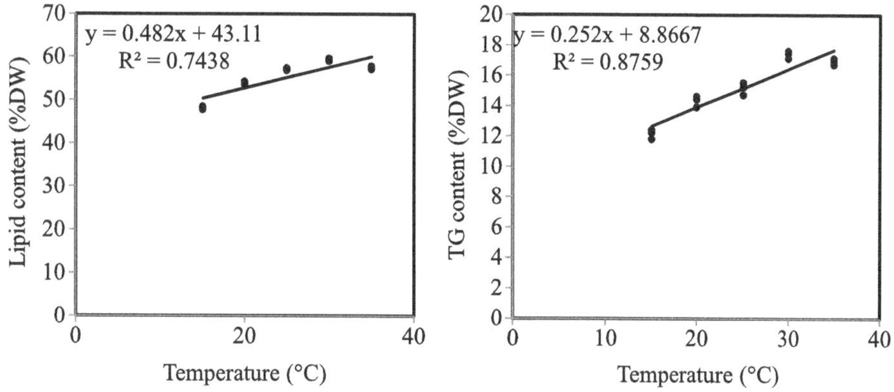

Figure 4 Statistical data relating to variation in temperature. The correlation between lipid and triglyceride contents as dependent variables and cultivation temperature degrees as independent variable was carried out via simple linear regression analysis.

(Pugh 1971). However, this increase in total lipids in *N. salina* cells was mainly of neutral lipids, mostly triglycerides and the maximum increase was noticed for the duration of stationary phase. This is in agreement with Bigogno et al. (2002) who reported that triglycerides increased in the green alga *Parietochloris incise* from 43% in the logarithmic phase to 77% in the stationary phase. However, this increase in triglycerides may due to the alteration in lipid metabolism from membrane lipid synthesis to storage of neutral lipids. However, biosynthesis and conversion of some existing membrane polar lipids into triglycerols can contribute to the enhancement in triglyceride (Xiao et al. 2013). Therefore, triglycerides may account for as much as 80% of the total lipid content in the cell (Tornabene et al. 1983; Suen et al. 1987; Tonon et al. 2002; Hu et al. 2008).

Presented here, data of lipid productivity was contrary associated with tested nitrogen concentrations indicating that mechanisms associated with nitrogen metabolisms might be involved in lipid biosynthesis (Wagenen et al. 2012) and accordingly enhancement of lipid productivity. Cultivation of *N. salina* under reduced nitrogen concentrations generates the necessity for nitrogen and encourages the accumulation of preserved lipids and increasing in lipid productivity. However, lipid accumulation may be lower in higher nitrogen concentrations and consequently resulting in a lesser amount of lipid productivity. Comparable results showed that transferring algal biomass from a nitrogen-sufficient phase to a nitrogen-deficient phase induces lipid production and lipid productivity (Su et al. 2011).

In this work, the combined effect of nitrogen-limited conditions (25% of the original concentration in the medium) and different temperature degrees was studied. The microalga exhibits reduction in growth and accretion in lipid productivity when compared to control. Moreover, low temperatures had a minimal effect while high temperatures had a maximal effect on lipids and triglycerides improvement. The ability of *N. salina* to grow under the entire range of the studied temperature degrees and 25% nitrogen concentration illustrates that the combined effect of these two studied stressors create an additional response to accumulate cellular lipids and triglycerides and accordingly increase in lipid productivity. Illman et al. (2000) and Liu et al. (2008) have revealed that the quantity and quality of lipids within the cell may well differ in relation to changes in growth conditions such as temperature or light intensity. Actually, temperature affects the physiological processes by changing the rate of chemical reactions and the stability of cellular components (Sandnes et al. 2005; Wagenen et al. 2012). However, lipid production and lipid productivity is a strain-specific function of physiological responses to many factors such as cultivation temperature (Griffiths and Harrison 2009; Wagenen et al. 2012). The lipid content in the chrysophytan *Ochromonas danica* (Aaronson 1973) and the eustigmatophyte *N. salina* (Boussiba et al. 1987) increases with increasing temperature. In contrast, no significant change in the lipid content was observed in *Chlorella sorokiniana* grown at various temperatures (Patterson 1970). In this work, simple regression analysis is used to assess the relative impact of nitrogen concentrations and temperatures as variables on lipid and triglyceride production in *N. salina*. Referring to statistical results, reduction of nitrogen concentration to 75% of the original medium significantly explained the increase in lipid and triglyceride produced, while less significant was observed when testing the combined effect between 25% nitrogen concentration and temperature of 30°C. This may be attributed to the lipid metabolism, particularly the biosynthetic pathways of lipid and triglycerides. Synthesis and accumulation of lipid correlated with significant changes in triglycerides in the cell occur when algae are cultivated under stress conditions. This accumulation forced by chemical or physical stimuli, either individually or in combination (Hu et al. 2008). Based on this study, nitrogen limitation needs to be more considered rather than combination with temperature when monitoring lipids and triglycerides production in *N. salina*.

Conclusion

Of particular interest was the influence of nitrogen depletion and variation of temperature on lipids as well as triglycerides production in *N. salina*. As nitrogen concentration in the medium was reduced by 75% and at temperature of 30°C, an increase in lipid and triglyceride contents with high lipid productivity was noticed. Based on dry weight, algal biomass can serve as lipid-rich feedstock with increased lipid content (59.3%) compared to control (34.6%). Simple regression provides significant differences in percentages between variables. For application at commercial scale, future work should be directed to additional practice to improve algal lipid yield per unit time.

Competing interests

The authors declare that they have no competing interests.

Authors' contributions

Both authors read and approved the final manuscript.

Acknowledgements

The authors acknowledge the technical support provided by MED-ALGAE project, funded by the European Union ENPI program (grant number I-B/202/099).

References

Aaronson S (1973) Effect of incubation temperature on the macromolecular and lipid content of the phytoflagellate *Ochromonas danic*. J Phyco 9:111–113

Bentley CD, Carroll PM, Watanabe WO, Riedel AM (2008) Intensive rotifer production in a pilot-scale continuous culture recirculating system using nonviable microalgae and an ammonia neutralizer. J World Aquacult Soc 39:625–635

Bigogno C, Khozin-Goldberg I, Boussiba S, Vonshak A, Cohen Z (2002) Lipid and fatty acid composition of the green oleaginous alga *Parietochloris incisa*, the richest plant source of arachidonic acid. Phytochemistry 60:497–503

Bligh EG, Dyer WJ (1959) A rapid method of total lipid extraction and purification. Can J Biochem Physiol 37(8):911–917

Boussiba S, Vonshak A, Cohen Z, Avissar Y, Richmond A (1987) Lipid and biomass production by the halotolerant microalga *Nanochloropsis salina*. Biomass 12:37–47

Chen M, Tang H, Ma H, Holland TC, Ng KYS, Salley SO (2011) Effects of nutrients on growth and lipid accumulation in the green algae *Dunaliella tertiolecta*. Bioresour Technol 102:1649–1655

Chisti Y (2007) Biodiesel from microalgae. Biotechnol Adv 25:294–306

Collins RP, Kalnins K (1969) The fatty acids of *Cryptomonas ovata* var palustris. Phyton 26:47–50

Conover SAM (1975) Partitioning of nitrogen and carbon in cultures of the marine diatom *Thalassiosira fluviatillis* supplied with nitrate, ammonium, or urea. Mar Biol 32:231–246

Converti A, Casazza AA, Ortiz EY, Perego PD, Borghi M (2009) Effect of temperature and nitrogen concentration on the growth and lipid content of *Nannochloropsis oculata* and *Chlorella vulgaris* for biodiesel production. Chem Eng Process 48:1146–1151

Dong H-P, Williams E, Wang D-Z, Xie Z-X, Hsia R-C, Jenck A, Halden R, Li J, Chen F, Place AR (2013) Responses of *Nannochloropsis oceanica* IMET1 to long-term nitrogen starvation and recovery. Plant Physiol 162:1110–1126

El-Baky HHA, El-Baz FK, El-Baroty GS (2004) Production of lipids rich in omega 3 fatty acids from the halotolerant alga *Dunaliella salina*. Biotechnology 3:102–108

Feng D, Chen Z, Xue S, Zhang W (2011) Increased lipid production of the marine oleaginous microalgae *Isochrysis zhangjiangensis* (Chrysophyta) by nitrogen supplement. Bioresour Technol 102:6710–6716

Fuentes-Grünewald C, Garcés E, Alacid E, Sampedro N, Rossi S, Camp J (2012) Improvement of lipid production in the marine strains *Alexandrium minutum* and *Heterosigma akashiwo* by utilizing abiotic parameters. J Ind Microbiol Biotechnol 39:207–216

Gouveia L, Oliveira AC (2009) Microalgae as a raw material for biofuels production. J Ind Microbiol Biotechnol 36:269–274

Griffiths MJ, Harrison STL (2009) Lipid productivity as a key characteristic for choosing algal species for biodiesel production. J Appl Phycol 21:493–507

Grobbelaar JU (2007) Photosynthetic characteristics of *Spirulina platensis* grown in commercial-scale open outdoor raceway ponds: what do the organisms tell us? J Appl Phycol 19:591–598

Guevara M, Lodeiros C, Gómez O, Lemus N, Núñez P, Romero L, Vásquez A, Rosales N (2005) Carotenogénesis de cinco cepas del alga *Dunaliella sp.* (Chlorophyceae) aisladas de lagunas hipersalinas de Venezuela. Revista De Biologia Tropical 53(3–4):331–337

Guillard RRL (1975) Culture of marine phytoplankton for feeding marine invertebrates. In: Smith WL, Chanley MH (eds) Culture of marine invertebrate animals. Plenum Press, New York, USA, pp 26–60

Guschina IA, Harwood JL (2006) Lipids and lipid metabolism in eukaryotic algae. Prog Lipid Res 45:160–186

Hansen JW, Pedersen A-GU, Berntsen J, Rønbøg IS, Hansen LS, Lomstein BA (2000) Photosynthesis, respiration, and nitrogen uptake by different compartments of a Zostera marina community. Aquat Bot 66:281–295

Hsieh CH, Wu WT (2009) Cultivation of microalgae for oil production with a cultivation strategy of urea limitation. Bioresour Technol 100:3921–3926

Hu Q, Sommerfeld M, Jarvis E, Ghirardi M, Posewitz M, Seibert M, Darzins A (2008) Microalgal triacylglycerols as feedstocks for biofuel production: perspectives and advances. Plant J 54:621–639

Huang XH, Li CL, Liu CW, Wang ZD, Chen JJ (2002) Studies on the N and P nutrient demand in *Nannochloropsis oculata*. Mar Sci (Chinese) 26:13–17

Illman AM, Scragg AH, Shales SW (2000) Increase in *Chlorella* strains calorific values when grown in low nitrogen medium. Enzyme Microb Technol 27:631–635

Ip PF, Chen F (2005) Employment of reactive oxygen species to enhance astaxanthin formation in *Chlorella zofingiensis* in heterotrophic culture. Process Biochem 40:3491–3496

Li X, Hu H-Y, Gan K, Sun Y-X (2010) Effects of different nitrogen and phosphorus concentrations on the growth, nutrient uptake, and lipid accumulation of a freshwater microalga *Scenedesmus sp.* Bioresour Technol 101:5494–5500

Li X, Hu H-Y, Zhang Y-P (2011) Growth and lipid accumulation properties of a freshwater microalga *Scenedesmus sp.* under different cultivation temperature. Bioresour Technol 102:3098–3102

Liu ZY, Wang GC, Zhou BC (2008) Effect of iron on growth and lipid accumulation in *Chlorella vulgaris*. Bioresour Technol 99:4717–4722

Lombardi AT, Wangersky PJ (1991) Influence of phosphorus and slicon on lipid class production by the marine diatom *Chaetoceros gracilis* grown in turbidostat cage cultures. Mar Ecol Prog Ser 77:39–47

Lombardi AT, Wangersky PJ (1995) Particulate lipid class composition of 3 marine phytoplankters *chaetoceros-gracilis*, *isochrysis-galbana* (Tahiti) and *Dunaliella tertiolecta* grown in batch culture. Hydrobiologia 306:1–6

Mairet F, Bernard O, Masci P, Lacour T, Sciandra A (2011) Modelling neutral lipid production by the microalga *Isochrysis aff. galbana* under nitrogen limitation. Bioresour Technol 102:142–149

Mandal S, Mallick N (2009) Microalga *Scenedesmus obliquus* as a potential source for biodiesel production. Appl Microbiol Biotechnol 84:281–291

Mata TM, Martins AA, Caetano NS (2010) Microalgae for biodiesel production and other applications: a review. Renew Sustain Energ Rev 14:217–232

Nigam S, Rai MP, Sharma R (2011) Effect of nitrogen on growth and lipid content of *Chlorella pyrenoidosa*. Am J Biochem Biotechnol 7(3):124–129

Olofsson M, Lamela T, Nilsson E, Bergé J-P, del Pino V, Uronen P, Legrand C (2012) Seasonal variation of lipids and fatty acids of the microalgae *Nannochloropsis oculata* grown in outdoor large-scale photobioreactors. Energies 5:1577–1592

Olofsson M, Lamela T, Nilsson E, Bergé J-P, del Pino V, Uronen P, Legrand C (2014) Combined effects of nitrogen concentration and seasonal changes on the production of lipids in *Nannochloropsis oculata*. Mar Drugs 12:1891–1910

Pal D, Khozin-Goldberg I, Cohen Z, Boussiba S (2011) The effect of light, salinity, and nitrogen availability on lipid production by *Nannochloropsis sp.* Appl Microbiol Biotechnol 90:1429–1441

Park S-J, Choi Y-E, Kim EJ, Park W-K, Kim CW, Yang J-W (2012) Serial optimization of biomass production using microalga *Nannochloris oculata* and corresponding lipid biosynthesis. Bioprocess Biosyst Eng 35:3–9

Patterson G (1970) Effect of temperature on fatty acid composition of *Chlorella sorokiniana*. Lipids 5:597–600

Pinto E, Sigaud-Kutner T, Leitao M, Okamoto O (2003) Heavy metal induced oxidative stress in algae. J Phycol 39:1008–1018

Pugh PR (1971) Changes in the fatty acid composition of *Coscinodiscus eccentricus* with culture-age and salinity. Mar Biol 11:118–124

Renaud SM, Thinh LV, Lambrinidis G, Parry DL (2002) Effect of temperature on growth, chemical composition and fatty acid composition of tropical Australian microalgae grown in batch cultures. Aquaculture 211:195–214

Rodolfi L, Zittelli GC, Barsanti L, Rosati G, Tredici MR (2003) Growth medium recycling in *Nannochloropsis sp.* mass cultivation. Biomol Eng 20:243–248

Roleda MY, Stephen PS, Raymond JGL, John GD, Elanor MB, Michele SS (2013) Effects of temperature and nutrient regimes on biomass and lipid production by six oleaginous microalgae in batch culture employing a two-phase cultivation strategy. Bioresour Technol 129:439–449

Roncarati A, Meluzzi A, Acciarri S, Tallarico N, Melotti P (2004) Fatty acid composition of different microalgae strains (Nannochloropsis sp., Nannochloropsis oculata (Droop) Hibberd, Nannochloris atomus Butcher and Isochrysis sp.) according to the culture phase and the carbon dioxide concentration. J World Aquacult Soc 35:401–411

Rukminasari N (2013) Effect of temperature and nutrient limitation on the growth and lipid content of three selected microalgae (*Dunaliella tertiolecta*, *Nannochloropsis* sp. and *Scenedesmus* sp.) for biodiesel production. Int J Mar Sci 3(17):135–144

Sandnes JM, Kallqvist T, Wenner D, Gislerod HR (2005) Combined influence of light and temperature on growth rates of *Nannochloropsis oceanica*: linking cellular responses to large-scale biomass production. J Appl Phycol 17:515–525

Sharma KK, Schuhmann H, Schenk PM (2012) High lipid induction in microalgae for biodiesel production. Energies 5:1532–1553

Su C-H, Chien L-J, Gomes J, Lin Y-S, Yu Y-K, Liou J-S, Syu R-J (2011) Factors affecting lipid accumulation by *Nannochloropsis oculata* in a two-stage cultivation process. J Appl Phycol 23:903–908

Suen Y, Hubbard JS, Holzer G, Tornabene TG (1987) Total lipid production of the green alga *Nannochloropsis sp.* QII under different nitrogen regimes. J Phycol 23:289–297

Sun X, Cao Y, Xu H, Liu Y, Sun J, Qiao D, Cao Y (2014) Effect of nitrogen-starvation, light intensity and iron on triacylglyceride/carbohydrate production and fatty acid profile of *Neochloris oleoabundans* HK-129 by a two-stage process. Bioresour Technol 155:204–212

Taoka Y, Nagano N, Okita Y, Izumida H, Sugimoto S, Hayashi M (2009) Influences of culture temperature on the growth, lipid content and fatty acid composition of *Aurantiochytrium sp.* strain mh0186. Mar Biotechnol 11:368–374

Tonon T, Larson TR, Graham IA (2002) Long chain polyunsaturated fatty acid production and partitioning to triacylglycerols in four microalgae. Phytochemistry 61:15–24

Tornabene TG, Holzer G, Lien S, Burris N (1983) Lipid composition of the nitrogen starved green alga *Neochloris oleabundans*. Enzyme Microbial Technol 5:435–440

Tzovenik I, De Pauw N, Sorgeloos P (2003) Optimization of T-ISO biomass production rich in essential fatty acids I. Effect of different light regimes on growth and biomass production. Aquaculture 216:203–222

Wagenen JV, Miller TW, Hobbs S, Hook P, Crowe B, Huesemann M (2012) Effects of light and temperature on fatty acid production in *Nannochloropsis Salina*. Energies 5:731–740

Wu LF, Chen PC, Lee CM (2013) The effects of nitrogen sources and temperature on cell growth and lipid accumulation of microalgae. Int Biodeterior Biodegradation 85:506–510

Xiao Y, Zhang J, Cui J, Feng Y, Cui Q (2013) Metabolic profiles of *Nannochloropsis oceanica* IMET1 under nitrogen-deficiency stress. Bioresour Technol 130:731–738

Yeesang C, Cheirsilp B (2011) Effect of nitrogen, salt, and iron content in the growth medium and light intensity on lipid production by microalgae isolated from freshwater sources in Thailand. Bioresour Technol 102:3034–3040

Yeh KL, Chang JS (2011) Nitrogen starvation strategies and photobioreactor design for enhancing lipid production of a newly isolated microalga *Chlorella vulgaris* esp-31: implications for biofuels. Biotechnol J 6:1358–1366

Zou N, Richmond A (2000) Light-path length and population density in photoacclimation of *Nannochloropsis sp.* (Eustigmatophyceae). J Appl Phycol 12:349–354

Effects of yam dioscorin interventions on improvements of the metabolic syndrome in high-fat diet-induced obese rats

Shen-Liang Shih[1†], Yin-Shiou Lin[2†], Shyr-Yi Lin[3,4*] and Wen-Chi Hou[2,5,6*]

Abstract

Background: The metabolic syndrome (MS) is termed a cluster of multiple metabolic risk criteria which is positively correlated with cardiovascular disease and type 2 diabetes mellitus (DM). Yam dioscorins have been reported to exhibit biological activities, however, little is known their preventive effects on the MS. Therefore, a high-fat (HF) diet was used to induce Wistar rat obesity and then yam dioscorin (50 mg/kg, dio50) was intervened daily concurrent HF diet (HF diet + dio50) for five weeks to check the changes of weights of body and tissues, blood pressures, and impaired glucose tolerances. The in vitro peptic hydrolysates of dioscorin with molecular mass between 3 kDa and 10 kDa and less than 3 kDa were used to determine dipeptidyl peptidase IV (DPP IV) inhibitory activities which DPP IV inhibitor has been reported to prevent and treat type 2 DM.

Results: There were no significant difference in body weights, feed intakes, feed conversion, and weights of adipose tissues of obese rats in groups of HF and (HF diet + dio50). However, the systolic blood pressures in obese rats of 2-, 3- and 4-week dioscorin interventions were showed significantly lower ($P < 0.05$) compared to the HF group. The dioscorin intervention (HF+ dio50) was showed significantly different ($P < 0.05$) and improved the impaired glucose tolerances compared to HF group in obese rats by the oral glucose tolerance tests. It was also found that the fraction with different molecular mass of dioscorin peptic hydrolysates (5 mg/ml) showed inhibitory activities against DPP IV using sitagliptin phosphate as positive controls.

Conclusions: Yam dioscorins exhibit improved MS activities in obese rats which the related mechanisms may need further investigations.

Keywords: Dioscorin; Dipeptidyl peptidase IV (DPP IV); Metabolic syndrome (MS); Peptic hydrolysates; Systolic blood pressure; Yam

Background

The metabolic syndrome (MS) is termed a cluster of five multiple metabolic risk criteria, including (1) abdominal obesity (central obesity), (2) hypertriglyceridemia [fasting triglycerides (TGs) ≥ 150 mg/dl], (3) low serum high-density lipoprotein cholesterol (HDL ≤ 40 mg/dl), (4) elevated blood pressure (systolic ≥ 130 mmHg or diastolic ≥ 85 mmHg), and (5) high fasting blood glucose (≥110 mg/dl), and an increasing obesity and sedentary lifestyles relate positively the rising MS worldwide prevalence (Day 2007; Alberti et al. 2009). A general diagnosis with three out of above-mentioned five criteria can recognize as the MS (Day 2007). Insulin resistance and impaired glucose tolerance are also two important metabolic risk criteria which may involve in factors for the MS general diagnosis (Day 2007). The health risks of MS associated with obesity vary among individuals, but consistently include type 2 diabetes mellitus (type 2 DM), hypertension, coronary heart disease and cancer (Zimmet 1982; Prentice 2006). The abdominal obesity may play the central role in MS which is the situation of exceeding visceral fat deposited in peritoneal cavity, and will initiate inflammation and dyslipidemia, increase the blood pressure

* Correspondence: sylin@tmu.edu.tw; wchou@tmu.edu.tw
†Equal contributors
[3]Department of Primary Care Medicine, Taipei Medical University Hospital, Taipei, Taiwan
[2]Graduate Institute of Pharmacognosy, Taipei Medical University, Taipei, Taiwan
Full list of author information is available at the end of the article

and decrease insulin sensitivity accompanied with abnormal blood glucose. Obesity is blamed as a major contributing factor in over 0.3 million deaths per year in the America and related economic costs over US 100 billion per year (Daniels 2006; Rodgers et al. 2012). In fact, obesity associated with diabetes are considered not only a clinical problem but also a public health issue in many countries, and 80% of overweight people are also diagnosed with type 2 DM which are referred as "the twin epidemics" (Smyth and Heron 2005).

Several animal models are established and suitable for one or more metabolic risk criteria in MS studies (Panchal and Brown 2011), such as genetic rodent models, including *db/db* mice, *ob/ob* mice, Zucker diabetic fatty rats, and Otsuka Long-Evans Tokushima fatty (OLETF) rats etc., however, such genetic mutations in established rodent animals are few reported in humans; on the other hand, the diet-induced rodent models may mimic closely to MS symptoms in humans, including fructose-induced, sucrose-induced, and high-fat (HF) diet induced rodent animals (Panchal and Brown 2011). The HF diets are reported to induce overweight, obesity, dyslipidemia, insulin resistance, and high blood pressures in rodent animals (Agulia and Mandarim-de-Lacerda 2003; Woods et al. 2003; Kobayasi et al. 2010), and are widely used for MS study (Frigolet et al. 2013).

The dipeptidyl peptidase IV (DPP IV) is a serine-type proteinase (EC 3.4.14.5) which metabolizes peptide hormones, such as glucagon-like peptide-1 (GLP-1), an insulinotropic peptide hormone can stimulate glucose-dependent insulin secretions (Mentlein 2005; Drucker 2006; Idris and Donnelly 2007). DPP IV inhibitors that control the glycermia by modulating the GLP-1 levels are currently developed for type 2 DM treatments (Smyth and Heron 2005; Idris and Donnelly 2007). DPP IV cleaves the Pro or Ala at the second position of the active *N*-terminal GLP-1(7–37) or GLP-1(7–36) amide which results in functional inactive GLP-1(9–37) or GLP-1(9–36) amide (Drucker 2006; Idris and Donnelly 2007). Therefore, researchers attempted to isolate potential DPP IV inhibitory peptides from protein hydrolysates, such as pepsin-pancreatin hydrolysates of sodium caseinate, skim milk powders and milk protein concentrates (Lacroix and Li-Chan 2012), Umamizyme G hydrolysates of defatted rice bran (Hatanaka et al. 2012), and pepsin-treated whey protein (Lacroix and Li-Chan 2013).

Yams are recognized as herbal plants since the tuber dried slices are widely used as Chinese herbal medicines, and the fresh tuber has also been a staple food in West Africa, Southern Asia, and the Caribbean. The tubers of yam storage protein, dioscorin, account for about 90% of the extractable water-soluble proteins from different species as estimated by the immune staining method (Hou et al. 2000). The yam dioscorin and/or its peptic hydrolysates have been reviewed for biological activities in vitro and/or in vivo (Lu et al. 2012), among which the antihypertensive activities (Hsu et al. 2002; Lin et al. 2006; Liu et al. 2009a,b; Lin et al. 2014) and antioxidant activities (Hou et al. 2001; Liu et al. 2006; Han et al. 2013, 2014a,b,c) may involve in MS studies. Therefore, a HF diet which the fat composition provide 60% of total calories is used to induce obese rats and then yam dioscorin is intervened to observe the changes of body weights, blood pressures, and glucose tolerances. It is also to test the pepsin hydrolysates of yam dioscorin to evaluate DPP IV inhibitory activities.

Methods

Materials

DPP IV (lyophilized powder, ≥ 10 units/mg protein, D-7052), glucose, Gly-Pro *p*-nitroanilide, pepsin (3460 units/mg solid, P-6887), and sitagliptin phosphate were purchased from Sigma Chemical Co. (St. Louis, MO). The HF diet for obesity induction (fat composition provide 60% of total calories, D12942) was purchased from Research Diets, Inc. (NJ, USA). The standard mouse/rat chow (fat composition provide 12.137% of total calories, Prolab® RMH2500, 5P14 Diet; PMI Nutrition International, MO, USA).

HF diets-induced obese rats for MS studies

Male 10-week-old Wistar rats (N = 24) were purchased from National Laboratory Animal Center (Taipei, Taiwan) and housed in wire-bottomed stainless steel cages in a temperature- and humidity-controlled room (at 22°C) with a 12-h light/dark cycle which had free access to the feeds and water. All animal experimental procedures were reviewed and approved by the Institutional Animal Care and Use Committee, Taipei Medical University (LAC-100-0038). After acclimation for one week, rats were randomly divided into three groups (N = 8 for each group), including a blank group for standard mouse/rat chow and two HF diet-induced obese groups (one control as HF group and one dioscorin-intervened group) for 70-days. In the dioscorin intervened group, yam dioscorin (50 mg/kg, dio50) was intervened daily concurrent HF diets (HF + dio50) from day-36 to day-70. The rat weights and feed intakes were recorded during experiments. The feed conversion is defined as a ratio of amounts of feed intakes (g) divided by rat weight gains (g) during experimental periods. Rats were sacrificed, organs (heart, lung, liver, kidney, and spleen) and adipose tissues (retroperitoneal fat, mesenteric fat, visceral fat, and testicle fat) were collected and weighted for comparisons.

Changes of blood pressures in the dioscorin intervention of HF diet-induced obese rats

The systolic blood pressure (SBP) and diastolic blood pressure (DBP) of rats were measured at the end of each week after dioscorin intervention for 1-, 2-, 3-, and 4-weeks by using an indirect tail-cuff blood pressure meter (BP-98A, Softron Co. Ltd. Tokyo, Japan). Distilled water (0.5 ml) was administered to the rats in the normal diet group (the blank) and HF group (the control) instead of dioscorin solutions for comparisons.

Impaired glucose tolerances in the dioscorin interventions of HF diet-induced obese rats by oral glucose tolerance tests

Impaired glucose tolerance of HF diet-induced obese rats were measured by oral glucose tolerance tests (OGTT) following the previous report (Ito et al. 2001; Andrikopoulos et al. 2008) with modifications. Rats of each group at the ends of the 10th-week were fasted for 16 hours and glucose was administered by oral gavage at 1 g/kg body weight. Blood (0.1 ml) was obtained from the tail vein of rats at 0, 5, 30, 60, 90, and 120 min after the glucose loads. Plasma glucose were determined by using RANDOX glucose kit (Randox Laboratories-US, Ltd. USA). The assay principle is based on the hydrogen peroxide generation catalyzed by glucose oxidase and further reacted with phenol and 4-aminophenazone to produce reddish violet quinoneimine dye with absorbance at 505 nm.

DPP IV inhibitory activities of peptic hydrolysates of yam dioscorin

The procedure for peptic hydrolysates of yam dioscorin was following the previous reports (Hsu et al. 2002; Liu et al. 2006). The hydrolytic ratio (wt/wt) of yam dioscorin (g) to pepsin (g) was set at 5/1 in 0.1 M KCl buffer (pH 2.0) with stirring at 4°C for 3 days. After hydrolysis, 1.0 M Tris–HCl buffer (pH 8.3) was added to pH 7.5 to stop hydrolysis. The molecular cut-off centricon or centriprep device (YM-10, 10 kD; YM-3, 3 kD; Millipore Co., USA) was used to separate peptic hydrolysates of yam dioscorin into three portions as followings: higher than 10 kD (peptide > 10 kDa), between 3 to 10 kD (3 kD < peptide < 10 kD), and less than 3 kDa (peptide < 3kD). The fractions of between 3 to 10 kD (3 kD < peptide < 10 kD) and less than 3 kDa (peptide < 3kD) were further analyzed by reverse-phase C18 Spherical HPLC column (10 mm × 250 mm). The mobile phase was mixed in stepwise gradients with solvent A (deionized water containing 0.1% trifluoroacetic acid) and solvent B (100% acetonitrile containing 0.1% trifluoroacetic acid) as followings, 0 to 5 min, 100% to 79% solvent A and 0 to 21% solvent B; 21 min, 75% solvent A and 25% solvent B; 22 min to 27 min, 60% solvent A and 40% solvent B; 40 min, 0% solvent A and 100% solvent B. Flow rate was 3 ml/min. The detector was set at 220 nm. The latter two fractions were lyophilized for DPP IV inhibitory assays. The DPP IV inhibitory activity was assayed according the previous report (Lacroix and Li-Chan 2013) with modifications. The DPP IV enzyme powder (≥10 units/mg protein) was re-dissolved in 1 ml of 100 mM Tris buffer (pH 8.0) as a stock enzyme solution and a 50-fold dilution before uses as a working enzyme solution. The substrate, Gly-Pro *p*-nitroanilide, was prepared as 4 mM stock solutions. The sitagliptin phosphate, the DPP IV inhibitor (Drucker et al. 2007) as the positive control, was prepared as 1 μM stock solutions. Each 50 μl of working DPP IV enzyme solution and peptic dioscorin fractions was mixed at 37°C for 10 min, then 50 μl of substrate was added and adjusted to 200 μl by 100 mM Tris buffer (pH 8.0). The sitagliptin phosphate was used instead of tested dioscorin fraction for the control. The 100 mM Tris buffer (pH 8.0) was used instead of tested dioscorin fraction for the blank. The absorbance at 405 nm was measured at 60 min by using an ELISA reader (TECAN Sunrise microplate reader; Männedorf, Switzerland). The DPP IV inhibition (%) was calculated as $(A405_{blank}\text{-}A405_{sample\ or\ control}/A405_{blank}) \times 100\%$.

Statistical analyses

Data were expressed as mean ± SEM in the rat weight changes and impaired glucose tolerances by OGTT methods, others were expressed as mean ± SD. Multiple group comparisons under the same treated time were performed using one-way analysis of variance (ANOVA) followed by the *post hoc* Tukey's test, and values that have not been indicated with the same alphabet were significantly different ($P < 0.05$). Statistical analysis was performed using the GraphPad Prism 5.0 software (San Diego, CA, USA).

Results

Effects of dioscorin interventions on body weights of HF diet-induced obese rats

At the day 36, the average weight of rats in the group of normal diet, HF diet, and (HF diet + dio50) was 389.65 ± 22.79 (g), 462.25 ± 11.06 (g), and 459.32 ± 38.94 (g), respectively. At the day 68, the average weight of rats in the group of normal diet, HF diet, and (HF diet + dio50) was 427.17 ± 18.20 (g), 514.90 ± 13.44 (g), and 509.76 ± 28.54 (g), respectively. Rats fed with the standard chow (the normal diet) showed lighter average body weights and significant difference ($P < 0.05$) compared to those fed with HF diets without or with dioscorin interventions during 70-day experiments (Figure 1A). While, the average weight of rats in two groups of HF-induced obese rats, namely HF diet group or with yam dioscorin interventions (HF diet + dio50) group showed no significant differences ($P > 0.05$) at the beginning (day 36) and the ends (day 68) of experiments (Figure 1A). It meant that yam dioscorin interventions at 50 mg/kg for 5 weeks showed no weight reduction activity against obese-induced rats at the present protocol designs. Figure 1B showed the feed

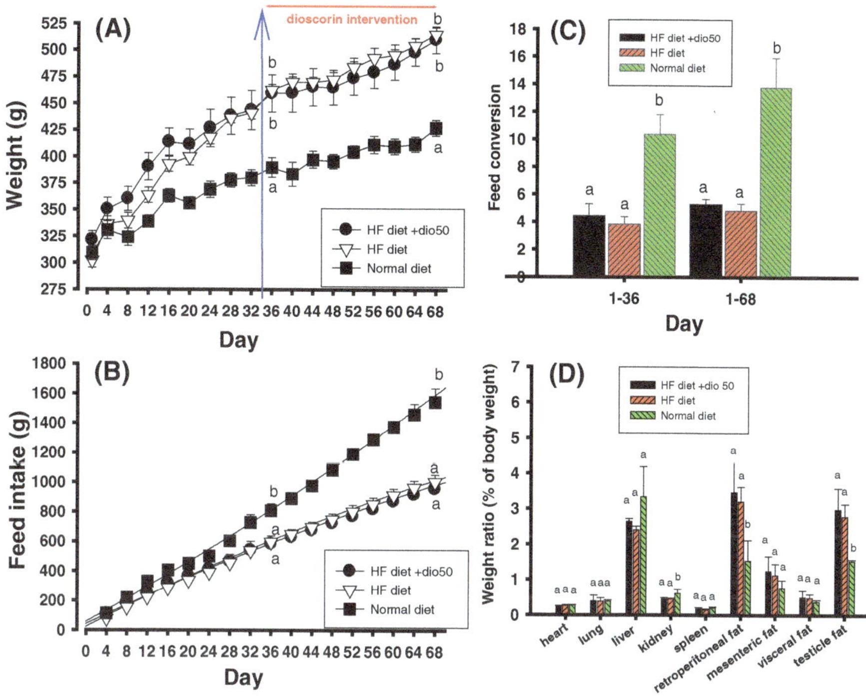

Figure 1 Effects of dioscorin interventions (50 mg/kg) on (A) body weights, (B) feed intakes, (C) feed conversion (feed intakes/weight gains), and(D) weights of organ and fat tissues of high fat-diet induced obese rats. Arrow indicates the dioscorin intervention from day 36 to day 70. Data were expressed as mean ± SEM in the rat weight changes and others were expressed as mean ± SD. Multiple group comparisons under the same treated time were performed using one-way analysis of variance followed by the *post hoc* Tukey's test, and values that have not been indicated with the same alphabet were significantly different ($P < 0.05$). HF, a high-fat diet; (HF diet + dio50), yam dioscorin (50 mg/kg, dio50) was intervened daily concurrent HF diet.

intakes during experiments. The feed intakes in three rat groups were showed linearly increases, while, rats fed with the standard chow (the normal diet) showed higher feed intakes and significant difference ($P < 0.05$) compared to those fed with HF diets without or with dioscorin interventions during 70-day experiments (Figure 1B). Figure 1C showed the feed conversion rate of three rat groups during obesity-induced and the whole experimental stages which was calculated from feed intakes (data in the Figure 1B) divided by weight gained (data in the Figure 1A). Rats fed with the standard chow (the normal diet) showed higher feed conversion rate and significant difference ($P < 0.05$) compared to those fed with HF diets without or with dioscorin interventions during obesity-induced and the whole experimental stages (Figure 1C). It meant that more of standard chow (the normal diet) were needed compared to those of HF diet to gain the same weight. Figure 1D showed the organ weights of three rat groups. Generally, rats in the normal diet group showed significant difference ($P < 0.05$) and lower weights of adipose tissues (such as retroperitoneal fat and testicle fat) than those in the HF diet groups. Other organs except from kidney showed no significant difference among three groups ($P > 0.05$).

Effects of dioscorin interventions on blood pressures of HF diet-induced obese rats

Figure 2 showed the changes of SBP (Figure 2A) and DBP (Figure 2B) in three rat groups after obesity-induced stages (day 36 to day 70). From the results of Figure 2A, the rats in the normal diet group showed lower SBP and significant differences ($P < 0.05$) compared to those in the HF diet group at 6-, 7-, 8-, and 9-week. While, dioscorin intervention group (HF diet + dio50) showed lower SBP and significant differences ($P < 0.05$) compared to those in the HF diet group at 7-, 8-, and 9-week (corresponding to dioscorin intervention for 2 weeks, 3 weeks, and 4 weeks, respectively) and comparable to those in the normal diet group. From the results of Figure 2B, the rats in the normal diet group showed lower DBP and significant differences ($P < 0.05$) compared to those in the HF diet group at 6-, 7-, 8-, and 9-week. Dioscorin intervention group (HF diet + dio50) showed lowered SBP, however, did not exhibit significant differences ($P > 0.05$) compared to those in the HF diet group at 7-, 8-, and 9-week. These results showed that the dioscorin intervention could improve blood pressure in obese rats, especial for SBP.

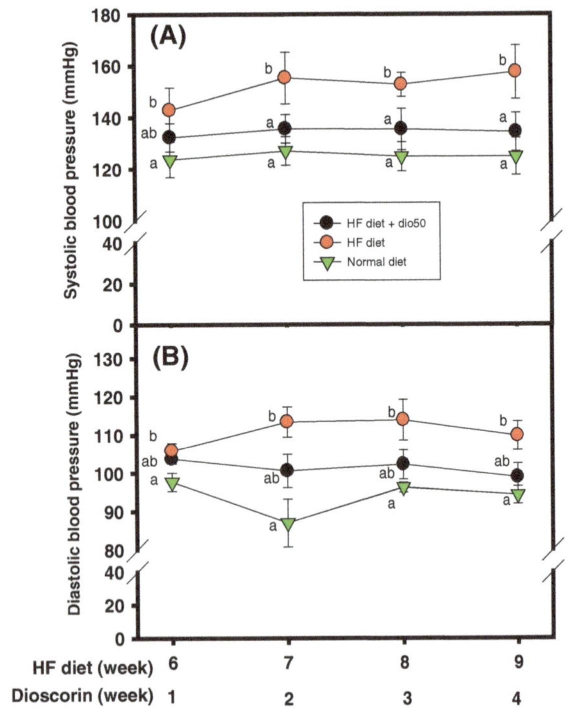

Figure 2 Effects of dioscorin interventions (50 mg/kg) on (A) systolic blood pressure and (B) diastolic blood pressures of high fat-diet induced obese rats. Blood pressure was measured at the end of each week after dioscorin intervention for 1-, 2-, 3-, and 4-weeks by using an indirect tail-cuff blood pressure meter (BP-98A, Softron Co. Ltd. Tokyo, Japan). Data were expressed as mean ± SD. Multiple group comparisons under the same treated time were performed using one-way analysis of variance followed by the *post hoc* Tukey's test, and values that have not been indicated with the same alphabet were significantly different ($P < 0.05$). HF, a high-fat diet; (HF diet + dio50), yam dioscorin (50 mg/kg, dio50) was intervened daily concurrent HF diet.

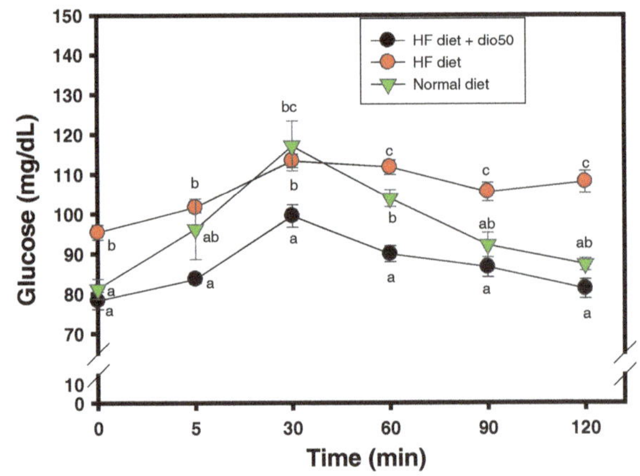

Figure 3 Effects of dioscorin interventions (50 mg/kg) on impaired glucose intolerances of high fat-diet induced obese rats by oral glucose tolerance tests. Data were expressed as mean ± SEM. Multiple group comparisons under the same treated time were performed using one-way analysis of variance followed by the *post hoc* Tukey's test, and values that have not been indicated with the same alphabet were significantly different ($P < 0.05$). HF, a high-fat diet; (HF diet + dio50), yam dioscorin (50 mg/kg, dio50) was intervened daily concurrent HF diet.

Effects of dioscorin interventions on impaired glucose tolerances of HF diet-induced obese rats

After fasting overnight, the impaired glucose tolerance of rats in three groups were measured by OGTT (Figure 3). Plasma glucose concentrations (mg/dL) of the normal diet group in the 0, 5, 30, 60, 90, and 120 min were 81.19 ± 6.09, 96.22 ± 18.67, 117.13 ± 15.22, 103.90 ± 5.22, 92.14 ± 7.57, and 87.24 ± 3.77, respectively; plasma glucose concentrations (mg/dL) of HF diet group were 95.40 ± 5.66, 101.67 ± 4.07, 113.39 ± 5.04, 111.75 ± 5.51, 105.48 ± 7.02, and 108.05 ± 8.34, respectively; plasma glucose concentrations (mg/dL) of (HF diet + dio50) group were 78.31 ± 6.66, 83.66 ± 3.06, 99.50 ± 8.42, 89.92 ± 6.08, 86.59 ± 7.52, and 81.14 ± 7.21, respectively. At the beginning, rats in the normal diet group and (HF diet + dio50) group showed lower plasma glucose concentrations and significant differences ($P < 0.05$) compared to those in the HF diet group. Later, rats in the (HF diet + dio50) group kept the lowest plasma glucose concentrations and significant differences ($P < 0.05$) among three rat groups at each 5-min, 30-min, 60-min, 90-min, and 120-min time intervals and comparable to rats in the normal diet group at 90-min and 120-min time intervals. The plasma glucose concentrations of rats in the normal diet groups and (HF diet + dio50) group were then back to the baseline at 120 min, however, the plasma glucose concentrations of rats in the HF diet group were still kept at higher levels (108.05 ± 8.34 mg/dL). From the OGTT data, it meant that the dioscorin intervention could improve impaired glucose tolerances of HF diet-induced obese rats.

DPP IV inhibitory activities of peptic hydrolysates of yam dioscorin

The reverse phase HPLC chromatograms of peptic hydrolysates of yam dioscorin with molecular mass between 3 kD and 10 kD (3 kD < peptide < 10 kD, Figure 4A) and less than 3 kD (peptide < 3kD, Figure 4B) were shown at Figure 4. These two hydrolytic fractions were used to analyze the DPP IV inhibitory activities compared to the positive control of sitagliptin phosphate (Figure 4C). The sitagliptin phosphate showed dose-dependent inhibitions against DPP IV. Under 5 mg/ml doses, the peptide fractions of 3 kD < peptide < 10 kD and peptide < 3kD, respectively, showed 80.82 ± 2.68 (%) and 54.42 ± 3.51 (%) DPP IV inhibitory activities which might be equivalent to sitagliptin phosphate of 250.86 nM and 61.79 nM, respectively.

Discussion

The present animal experiments showed that the dioscorin interventions at dose of 50 mg/kg daily for five

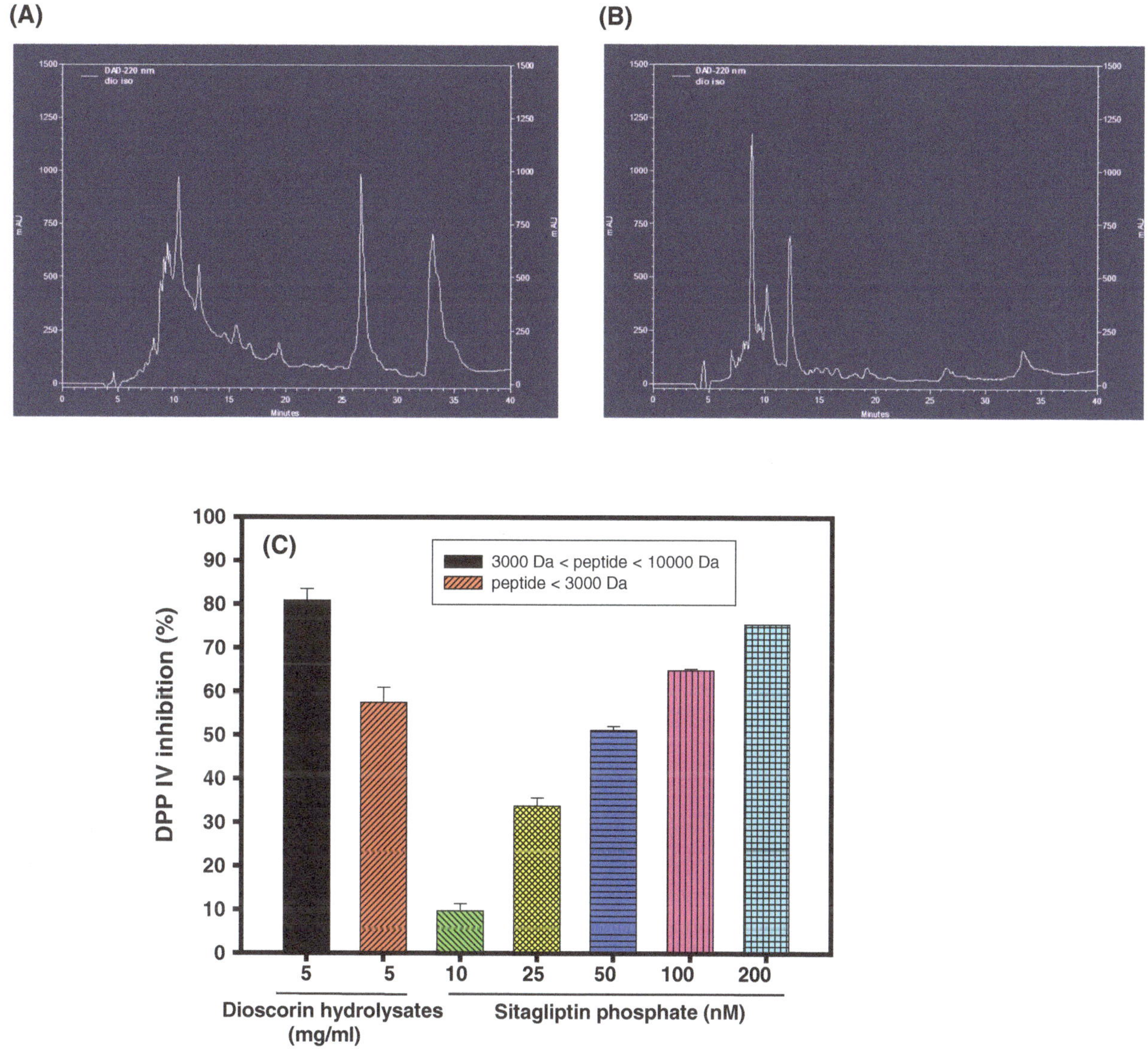

Figure 4 DPP IV inhibitory activities of peptic hydrolysates of yam dioscorin. (A) HPLC chromatograms of yam dioscorin peptic hydrolysates with molecular mass between 3 kD and 10 kD, **(B)** HPLC chromatograms of yam dioscorin peptic hydrolysates with molecular mass less than 3 kD, and **(C)** DPP IV inhibitory activities of two fractions (5 mg/ml) of dioscorin peptic hydrolysates from molecular cut-off membrane devices, and sitagliptin phosphate was acted as positive controls.

weeks could improve SBP and impaired glucose tolerances, but had no effects on reductions of body weight and fat tissues, of 10-week HF diet-induced obese rats which the improved metabolic risks were involved in MS criteria. The dioscorin and its peptic hydrolysates at doses of 40 mg/kg were showed to have antihypertensive activities against spontaneously hypertensive rats (SHR) (Lin et al. 2006); the dioscorin at doses of 20 mg/kg or 80 mg/kg showed to lower oxidative stress of BALB/c mice induced by subcutaneous galactose injections (Han et al. 2014b). The obesity may play the central role in MS which the renin-angiotensin system in dysfunctional adipocytes will initiate inflammation and dyslipidemia, increase blood pressure and decrease insulin sensitivity accompanied with abnormal blood glucose (Frigolet et al. 2013). Therefore, anti-obesity or prevention against obesity may reduce directly the risks of MS. There were several reports concerning anti-obesity activity from natural resources in rodent models induced by HF diets, such as α-lipoic acid (Kim et al. 2004), cryptotanshinone from *Salvia militorrhiza* (Kim et al. 2007), rutin and *o*-coumaric acid (Hsu et al. 2009), honokiol and magnolol (Kim et al. 2013), pectin pentaoligosaccharide (Li et al. 2013), high taurine and glycine contents of scallop protein (Tastesen et al. 2014). The above-mentioned models for HF diet-induced obese studies were generally applied either by HF diet pretreatment for a period of time and then tested sample interventions or co-treatment of HF-diet and tested samples in the same time. The vasorelaxing peptides of Arg-Phe and Ile-His-Arg-Phe derived from rice glutelin protein showed to lower food intakes in rodent models (Kagebayashi et al. 2012; Kontani et al. 2014) which maybe have anti-obesity activity. At present, it is not sure that less amounts of dioscorin used with the similar MS preventive activities, or higher amounts of dioscorin used or co-treatments of HF-diet and dioscorin used with anti-obesity activities. Reagan-Shaw et al. (2007) suggested to use body surface area normalization for dose translation from animal to human studies. It was calculated that the human equivalent dose was 8.11 mg/kg of human body weight from dioscorin intervention of 50 mg/kg of rat body weight in the present experiment. An adult of 60 kg weigh might have to consume about 490 mg dioscorin/day to achieve benefits of improving MS and need further investigations.

The yam dioscorin and its peptic hydrolysates or synthesized peptides derived from dioscorin had been reported to lower blood pressures using SHR as animal models (Lin et al. 2006; Liu et al. 2009a,b; Lin et al. 2014). From the present results of Figure 2, dioscorin intervention at least for two weeks showed to lower SBP of HF diet-induced obese rats and comparable to the normal diet fed ones. The dysfunctional adipocytes in obesity will increase circulating renin-angiotensin systems, and the generated angiotensin II from angiotensin I by angiotensin converting enzyme (ACE) hydrolysis resulted in the higher blood pressures (Frigolet et al. 2013). It was reported that obesity may elevate systemic oxidative stress from overloaded nutrients of HF diets and advanced glycation endproduct generations (Kahn et al. 2006). Glycation is the non-enzymatic modification of proteins through the reduction of sugars and their metabolized intermediates, such as glyoxal or methylglyoxal, and leads to the irreversible formation of advanced glycation end products (Kikuchi et al. 2003). The hypertensive rat might increase oxidative stress and methylglyoxal amounts in vascular smooth muscle cells (Wu and Juurlink 2002). The peptic hydrolysates of dioscorin showed ACE inhibitory activities (Hsu et al. 2002) and the synthesized peptides derived from pepsin hydrolysates *in silico* vasorelaxing activities (Lin et al. 2014). The synthesized peptides derived from pepsin hydrolysates *in silico* exhibited antioxidant and antiglycation activities in vitro and in vivo (Han et al. 2013; Han et al. 2014a,c). It was proposed that the dioscorin intervention showed to lower SBP of HF diet-induced obese rats might be from ACE inhibitory and vasorelaxing activities, and in part from antioxidant and/or antiglycation activities of active peptides after dioscorin ingestions and need further investigations.

From the results of Figure 3, it was found that the impaired glucose tolerances in HF diet-induced obese rats were improved after dioscorin interventions by OGTT methods. At the beginning and the end of OGTT test, the obese rats exhibited significantly higher glucose levels compared to rats in normal diet group and (HF + dio50) group. From the results of Figure 4, it was found that yam dioscorin peptic hydrolysates with molecular mass 3 kD < peptide < 10 kD and less than 3 kD at doses of 5 mg/ml showed DPP IV inhibitory activities in vitro which might be equivalent to effects of 250.86 nM and 61.79 nM of sitagliptin phosphate, respectively. Moreover, the undigested dioscorin was also used to evaluate DPP IV inhibitory activity, under 5 mg/ml concentration, which showed 40.7% DPP IV inhibition (data not show). The native dioscorin showed less anti-DPP IV activity compared to the 3 kD < peptide < 10 kD and less than 3 kD of dioscorin peptic hydrolysates. The results of DPP IV activity of the native and digested dioscorin is in agreement with the general observation that short-chain biologically active peptides can be released and absorbed in the small intestine after oral administration (Phelan et al. 2009) which may correlate with the improved OGTT. The sitagliptin phosphate, DPP IV inhibitor approved by US FDA in 2006, can prolong GLP-1 biological activities to stimulate glucose-dependent insulin secretions and for type 2 DM treatment in vivo (Idris and Donnelly 2007; Drucker et al. 2007). The DPP IV inhibitor (valine-pyrrolidide) showed the improved glucose tolerance and insulin secretion in HF diet-fed C57BL/6 J mice (Ahrén et al. 2000). The mice fed a HF diet (58% fat) together with DPP IV inhibitor (NVP DPP728) in the drinking water for 8 weeks showed the improved glucose tolerance and increased circulating levels of insulin and GLP-1 compared to HF diet only (Reimer et al. 2002). It was suggested that the improved glucose tolerance in HF diet-induced obese rats after dioscorin interventions might be in part from DPP IV inhibitory activities of active peptides from dioscorin after being ingested and need further investigations.

Conclusions

In conclusion, yam dioscorin interventions exhibit the improved MS activities in obese rats and peptic hydrolysates of yam dioscorin in vitro exhibit DPP IV inhibitory activities which the related mechanisms may need further investigations.

Competing interests

The authors declare that they have no competing interests.

Authors' contributions

SLS, SYL, and WCH participated the discussion and concepts of experimental designs, MS writing and revision; YSL performed experiments. All authors read and approved the final manuscript.

Acknowledgements
The authors would like to express thanks to Yuan's General Hospital, Kaohsiung, Taiwan (103YGH-TMU-02-3) and Ministry of Science and Technology, Republic of China (NSC 102-2313-B-038 -004 -MY3) for financial supports.

Author details
[1]Yuan's General Hospital, Department of Breast Surgery and Cancer Center, Kaohsiung, Taiwan. [2]Graduate Institute of Pharmacognosy, Taipei Medical University, Taipei, Taiwan. [3]Department of Primary Care Medicine, Taipei Medical University Hospital, Taipei, Taiwan. [4]Department of General Medicine, School of Medicine, Taipei Medical University, Taipei, Taiwan. [5]Traditional Herbal Medicine Research Center, Taipei Medical University Hospital, Taipei, Taiwan. [6]Program for the Clinical Drug Discovery from Botanical Herbs, Taipei Medical University, Taipei, Taiwan.

References

Agulia MB, Mandarim-de-Lacerda CA (2003) Heart and blood pressure adaptations in Wistar rats fed with different high-fat diets for 18 months. Nutrition 19:347–352

Ahrén B, Holst JJ, Mårtensson H, Balkan B (2000) Improved glucose tolerance and insulin secretion by inhibition of dipeptidyl peptidase IV in mice. Eur J Pharmacol 404:239–245

Alberti KGMM, Eckel RH, Grundy SM, Zimmet PZ, Cleeman JI, Donato KA, Fruchart J-C, James WPT, Loria CM, Smith SC Jr (2009) Harmonizing the metabolic syndrome: a joint interim statement of the international diabetes federation task force on epidemiology and prevention; national heart, lung, and blood institute; american heart association; world heart federation; international atherosclerosis society; and International association for the study of obesity. Circulation 120:1640–1645

Andrikopoulos A, Blair AR, Deluca N, Fam BC, Proietto J (2008) Evaluating the glucose tolerance test in mice. Am J Physiol Endocrinol Metab 295:E1323–E1332

Daniels J (2006) Obesity: America's epidemic. Am J Nurs 106:40–49

Day C (2007) Metabolic syndrome, or what you will: definitions and epidemiology. Diabetes Vasc Dis Res 4:32–38

Drucker DJ (2006) The biology of incretin hormones. Cell Metab 3:153–165

Drucker D, Easley C, Kirkpatrick P (2007) Sitagliptin. Nat Rev Drug Discover 6:109–110

Frigolet ME, Nimbe Torres N, Armando R, Tovar AR (2013) The renin–angiotensin system in dipose tissue and its metabolic consequences during obesity. J Nutr Biochem 24:2003–2015

Han CH, Liu JC, Fang SU, Hou WC (2013) Antioxidant activities of the synthesized thiol-contained peptides derived from computer-aided pepsin hydrolysis of yam tuber storage protein, dioscorin. Food Chem 138:923–930

Han CH, Lin YS, Lin SY, Hou WC (2014a) Antioxidant and antiglycation activities of the synthesised dipeptide, Asn-Trp, derived from computer-aided simulation of yam dioscorin hydrolysis and its analogue, Gln-Trp. Food Chem 147:195–202

Han CH, Lin YF, Lin YS, Lee TL, Huang WJ, Lin SY, Hou WC (2014b) Effects of yam tuber protein, dioscorin, on attenuating oxidative status and learning dysfunction in D-galactose-induced BALB/c mice. Food Chem Toxicol 65:356–363

Han CH, Lin YS, Lee TL, Liang HJ, Hou WC (2014c) Asn-Trp dipeptides improve the oxidative stress and learning dysfunctions in $_D$-galactose-induced BALB/c mice. Food Funct 5:2228–2236

Hatanaka T, Inoue Y, Arima J, Kumagai Y, Usuki H, Kawakami K, Kimura M, Mukaihara T (2012) Production of dipeptidyl peptidase IV inhibitory peptides from defatted rice bran. Food Chem 134:797–802

Hou WC, Chen HJ, Lin YH (2000) Dioscorins from different *Dioscorea* species all exhibit both carbonic anhydrase and trypsin inhibitor activities. Bot Bull Acad Sin 41:191–196

Hou WC, Lee MH, Chen HJ, Liang WL, Han CH, Liu YW, Lin YH (2001) Antioxidant activities of dioscorin, the storage protein of yam (*Dioscorea batatas* Decne) tuber. J Agric Food Chem 49:4956–4960

Hsu FL, Lin YH, Lee MH, Lin CL, Hou WC (2002) Both dioscorin, the tuber storage protein of yam (*Dioscorea alata* cv. Tainong No. 1), and its peptic hydrolysates exhibited angiotensin converting enzyme inhibitory activities. J Agric Food Chem 50:6109–6113

Hsu CL, Wu CH, Huang SL, Yen GC (2009) Phenolic compounds rutin and o-coumaric acid ameliorate obesity induced by high-fat diet in rats. J Agric Food Chem 57:425–431

Idris I, Donnelly R (2007) Dipeptidyl peptidase-IV inhibitors: a major new class of oral antidiabetic drug. Diabetes Obes Metab 9:153–165

Ito M, Kondo Y, Nakatani A, Hayashi K, Naruse A (2001) Characterization of low dose streptozotocin-induced progressive diabetes in mice. Envir Texicol Pharmacol 9:71–78

Kagebayashi T, Kontani N, Yamada Y, Mizushige T, Arai T, Kino K, Ohinata K (2012) Novel CCK-dependent vasorelaxing dipeptide, Arg-Phe, decreases blood pressure and food intake in rodents. Mol Nutr Food Res 56:1456–1463

Kahn SE, Hull RL, Utzschneider KM (2006) Mechanisms linking obesity to insulin resistance and type 2 diabetes. Nature 444:840–846

Kikuchi S, Shinpo K, Takeuchi M, Yamagishi S, Makita Z, Sasaki N, Tashiro K (2003) Glycation—a sweet tempter for neuronal death. Brain Res Rev 41:306–323

Kim MS, Park JY, Namkoong C, Jang PG, Ryu JW, Song HS, Yun JY, Namgoong IS, Ha J, Park IS, Lee IK, Viollet B, Youn JH, Lee HK, Lee KU (2004) Anti-obesity effects of α-lipoic acid mediated by suppression of hypothalamic AMP-activated protein kinase. Nat Med 10:727–733

Kim EJ, Jung SN, Son KH, Kim SR, Ha TY, Park MG, Jo IG, Park JG, Choe W, Kim SS, Ha J (2007) Antidiabetes and antiobesity effect of cryptotanshinone via activation of AMP-activated protein kinase. Mol Pharmacol 72:62–72

Kim YJ, Choi MS, Cha BY, Woo JT, Park YB, Kim SR, Jung UJ (2013) Long-term supplementation of honokiol and magnolol ameliorates body fat accumulation, insulin resistance, and adipose inflammation in high-fat fed mice. Mol Nut food Res 57:1988–1998

Kobayasi R, Akamine EH, Davel AP, Rodrigues MAM, Carvalho CRO, Rossoni LV (2010) Oxidative stress and inflammatory mediators contribute to endothelial dysfunction in high-fat diet-induced obesity in mice. J Hypertens 28:2111–2119

Kontani N, Omae R, Kagebayashi T, Kaneko K, Yamada Y, Mizushige T, Kanamoto R, Ohinata K (2014) Characterization of Ile-His-Arg-Phe, a novel rice-derived vasorelaxing peptide with hypotensive and anorexigenic activities. Mol Nutr Food Res 58:359–364

Lacroix IME, Li-Chan ECY (2012) Dipeptidyl peptidase-IV inhibitory activity of dairy protein hydrolysates. Int Dairy J 25:97–102

Lacroix IME, Li-Chan ECY (2013) Inhibition of dipeptidyl peptidase (DPP)-IV and α-glucosidase activities by pepsin-treated whey proteins. J Agric Food Chem 61:7500–7506

Li T, Zhu R, Dong Y, Liu Y, Li S, Chen G (2013) Effects of pectin pentaoligosaccharide from hawthorn (*Crataegus pinnatifida* Bunge. var. Major) on the activity and mRNA levels of enzymes involved in fatty acid oxidation in the liver of mice fed a high-fat diet. J Agric Food Chem 61:7599–7605

Lin CL, Lin SY, Lin YH, Hou WC (2006) Effects of tuber storage protein of yam (*Dioscorea alata* cv. Tainong No. 1) and its peptic hydrolyzates on spontaneously hypertensive rats. J Sci Food Agric 86:1489–1494

Lin YS, Lu YL, Wang GJ, Liang HJ, Hou WC (2014) Vasorelaxing and antihypertensive activities of synthesized peptides derived from computer-aided simulation of pepsin hydrolysis of yam dioscorin. Bot Stud 55:49, doi:10.1186/s40529-014-0049-3

Liu YH, Liang HJ, Cheng HC, Liu YW, Hou WC (2006) Comparisons of in vitro antioxidant activities of storage proteins in tuber of two *Dioscorea* species. Bot Stud 47:231–237

Liu DZ, Liang HJ, Han CH, Lin SY, Chen CT, Fan M, Hou WC (2009a) Feeding trial of instant food containing lyophilised yam powder in hypertensive subjects. J Sci Food Agric 89:138–143

Liu YH, Lin YS, Liu DZ, Han CH, Chen CT, Fan M, Hou WC (2009b) Effects of different types of yam (*Dioscorea alata*) products on the blood pressure of spontaneously hypertensive rats. Biosci Biotechnol Biochem 73:1371–1376

Lu YL, Chia CY, Liu YW, Hou WC (2012) Biological activities and applications of dioscorins, the major tuber storage proteins of yam. J Tradit Complement Med 2:41–46

Mentlein R (2005) Therapeutic assessment of glucagon-like peptide-1 agonists compared with dipeptidyl peptidase IV inhibitors as potential antidiabetic drugs. Expert Opin Investig Drugs 14:57–64

Panchal SK, Brown L (2011) Rodent models for metabolic syndrome research. J Biomed Biotech 2011:351982, doi: 10.1155/2011/351982

Phelan M, Aherne A, FitzGerald RJ, O'Brien NM (2009) Casein-derived bioactive peptides: biological effects, industrial uses, safety aspects and regulatory status. Int Dairy J 19:43–654

Prentice AM (2006) The emerging epidemic of obesity in developing countries. Int J Epidemiol 35:93–99

Reagan-Shaw S, Nihal M, Ahmad N (2007) Dose translation from animal to human studies revisited. FASEB J 22:659–661

Reimer MK, Holst JJ, Ahrén B (2002) Long-term inhibition of dipeptidyl peptidase IV improves glucose tolerance and preserves islet function in mice. Eur J Endocrinol 146:717–727

Rodgers RJ, Tschöp MH, Wilding JPH (2012) Anti-obesity drugs: past, present and future. Dis Model Mech 5:621–626
Smyth S, Heron A (2005) Diabetes and obesity: the twin epidemics. Nat Med 12:75–80
Tastesen HS, Keenan AH, Madsen L, Kristiansen K, Liaset B (2014) Scallop protein with endogenous high taurine and glycine content prevents high-fat, high-sucrose-induced obesity and improves plasma lipid profile in male C57BL/6 J mice. Amino Acids 46:1659–1671
Woods SC, Seeley RJ, Rushing PA, D'Alessio D, Tso P (2003) A controlled high-fat diet induces an obese syndrome in rats. J Nutr 133:1081–1087
Wu L, Juurlink BHJ (2002) Increased methylglyoxal and oxidative stress in hypertensive rat vascular smooth muscle cells. Hypertension 39:809–814
Zimmet P (1982) Type 2 (non-insulin-dependent) diabetes – an epidemiological overview. Diabetologia 22:399–411

Study on the effect of magnetic field treatment of newly isolated *Paenibacillus* sp.

Jie Li[1], Yanli Yi[1*], Xilei Cheng[2], Dageng Zhang[1] and Muhammad Irfan[3]

Abstract

Background: Symbiotic nitrogen fixation in plants occurs in roots with the help of some bacteria which help in soil nitrogen fertility management. Isolation of significant environment friendly bacteria for nitrogen fixation is very important to enhance yield in plants.

Results: In this study effect of different magnetic field intensity and treatment time was studied on the morphology, physiology and nitrogen fixing capacity of newly isolated *Paenibaccilus* sp. from brown soil. The bacterium was identified by 16S rDNA sequence having highest similarity (99%) with *Paenibacillus* sp as revealed by BLAST. Different magnetic intensities such as 100mT, 300mT and 500mT were applied with processing time of 0, 5, 10, 20 and 30 minutes. Of all these treatment 300mT with processing time of 10 minutes was found to be most suitable treatment. Results revealed that magnetic treatment improve the growth rate with shorter generation time leading to increased enzyme activities (catalase, peroxidase and superoxide dismutase) and nitrogen fixing efficiencies. High magnetic field intensity (500mT) caused ruptured cell morphology and decreased enzyme activities which lead to less nitrogen fixation.

Conclusion: It is concluded that appropriate magnetic field intensity and treatment time play a vital role in the growth of soil bacteria which increases the nitrogen fixing ability which affects the yield of plant. These results were very helpful in future breading programs to enhance the yield of soybean.

Keyword: Magnetic treatment; *Paenibaccilus* sp; Catalase; Peroxidase; Superoxide dismutase

Background

Paenibacillus genus of bacteria was first included in Bacillus genus and then reclassified to a separate genus in 1993 (Ash et al. 1993). These bacteria found in variety of environments like soil, water, forage, rhizosphere, insect larvae, vegetable matter and in clinical samples (McSpadden Gardener 2004, (Montes et al. 2004; Ouyang et al. 2008; Lal & Tabacchioni 2009)). These bacteria are of prime importance in agriculture for nitrogen fixation and industrial importance due to production of antibiotics and enzymes (Mavingui & Heulin 1994; Von der Weid et al. 2003). These bacteria produce plant growth hormones, suppress phytopathogens and solubilize organic phosphate (Mavingui & Heulin 1994; Lebuhn et al. 1997; Pires & Seldin 1997).

Nitrogen is very essential nutrient for the growth of plants. So, these bacteria fix nitrogen from the air and provide this nitrogen to plants in the form of ammonium ions or other nitrogenous compounds essential for growth. From this symbiotic association, plant provides some organic compounds synthesized from photosynthesis (Sawada et al. 2003). These bacteria not only fix the nitrogen but also enrich the soil fertility, increase plant production, and improve the quality, degrade organic pollutants and production of vitamin B series compounds (Sierra et al. 1999; Agus et al. 2000). The nitrogen deficiency was recovered by these rhizobia (Fisher & Long 1992). In this process, plant produced some reactive oxygen species including the hydrogen peroxide and hydroxyl radicals and superoxide anion by defence reaction (Lamb & Dixon 1997; Santos et al. 2001). So it was necessary to study the rhizobia catalase, peroxidase and superoxide dismutase active changes.

* Correspondence: yiyanli@126.com
[1]Key Laboratory of Preservation of Northeast Cultivated Land, Ministry of Agriculture, National Engineering Laboratory for Efficient Utilization of Soil and Fertilizer, College of Land and Enviroment, Shenyang Agricultural University, 120 dongling road, Shenyang 110866, Liaoning, China
Full list of author information is available at the end of the article

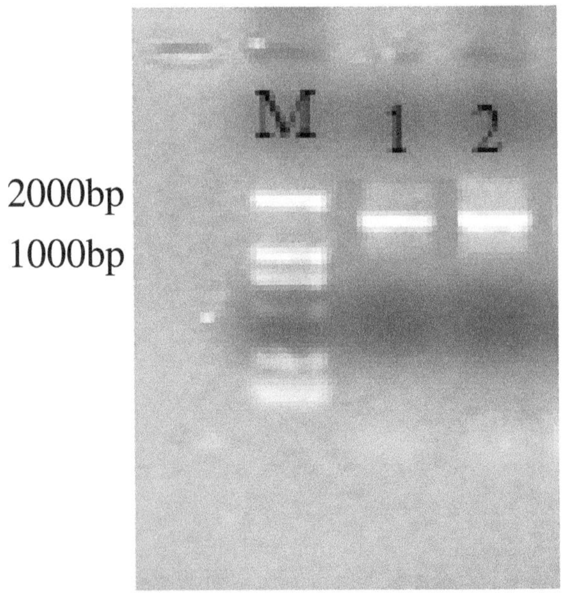

Figure 1 Amplified product of 16S rDNA from newly isolated bacterial species.

A lot of research showed that the magnetic treatments have certain stimulative effect on crop production and development and, it also affect the genetic quality of seeds ((Zhu et al. 1996; Liu et al. 1996; Yan et al. 1997; He et al. 1999; Mao et al. 2002); Jia et al. 2002; (Liu et al. 2003)). Enzyme as protein with catalytic activity has an important role in the life process, and as a catalyst it was increasingly being attention (Cheng et al. 2007). Magnetic field on the influence of the enzyme activity has been reported (He et al. 1998; Li et al. 2007; Hua et al. 2008), and this area now attracts more and more people's attention, but most of these studies focused on animals, plants and very little research on bacteria. So this study was aimed to check the effect of magnetic field on soybean rhizobia isolated from brown soil and their enzyme activities (peroxidase, catalase and superoxide dismutase) under the influence various intensity of magnetic treatment.

Methods

Materials

The Brown soil samples were collected from Shenyang Agriculture University, Shenyang Liaoning P.R. China. The samples were kept in sterile plastic bags and transferred aseptically to the lab.

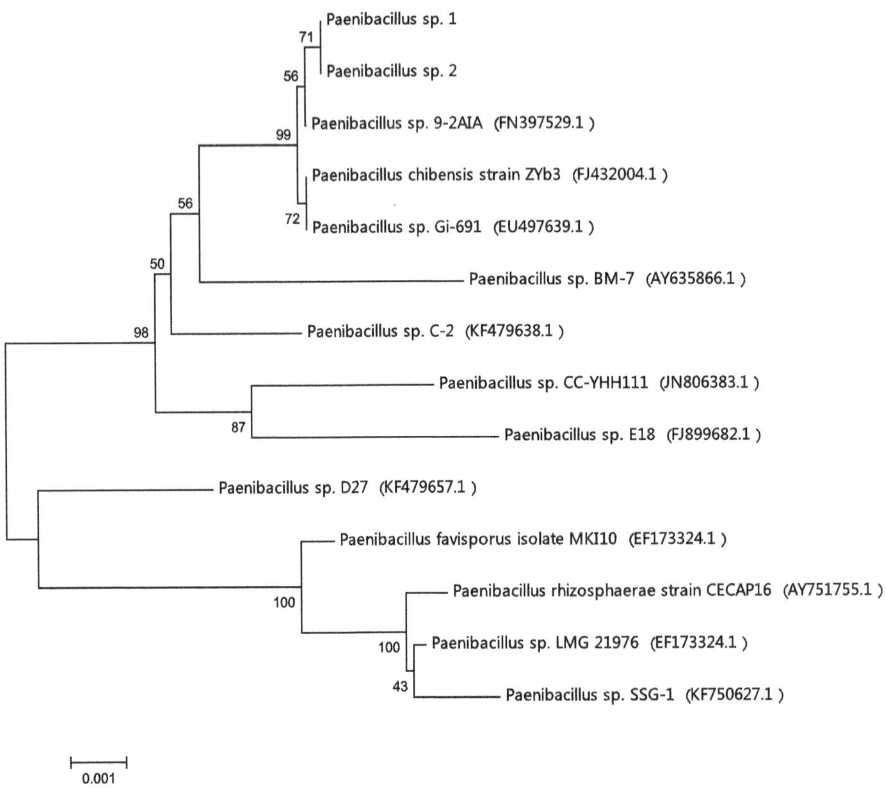

Figure 2 Neighbor-joining phylogenetic analysis of 16S rDNA sequences (1442 bp) of newly isolated *Paenibacillus* sp. compared with the sequence of standard strains. The bar represents a 0.002% of nucleotide variance.

Table 1 Effect of magnetic field treated soil on soybean nodular and nitrogen content

Treatment		Bacterial dry weight		Effective number of root nodule		Total nitrogen	
Magnetic field (mT)	Time (min)	Weight (g dry wt pl^{-1})	Percentage change (%)	Amount (No./plant)	Percentage change (%)	Content (mg/g)	Percentage change (%)
Control (CK)		0.06^h	-	7 ± 1.23	-	3.23 ± 0.12	-
100	5	0.20^c	233	24 ± 3.47	243	3.46 ± 0.13	7.12
	10	0.19^d	217	18 ± 1.76	157.	3.51 ± 0.12	8.67
	20	0.17^e	183	17 ± 1.84	143	3.35 ± 0.11	3.72
	30	0.15^f	150	16 ± 1.34	129	3.25 ± 0.12	0.62
300	5	0.27^a	350	37 ± 1.63	429	4.33 ± 0.17	34.06
	10	0.26^b	333	28 ± 2.95	300	4.55 ± 0.15	40.87
	20	0.15^f	150	24 ± 2.58	243	3.58 ± 0.13	10.84
	30	0.09^g	50	13 ± 1.72	86	3.51 ± 0.16	8.67
500	5	0.04^i	−33	7 ± 1.25	0	3.34 ± 0.13	3.41
	10	0.06^h	0	7 ± 1.13	0	3.25 ± 0.12	0.62
	20	0.05^i	−17	6 ± 1.13	−14	3.17 ± 0.11	−1.86
	30	0.05^i	−17	7 ± 1.12	0	3.14 ± 0.13	−2.79

The different letters show significant difference ($P < 0.05$).

Isolation of *Paenibacillus*

The *Paenibacillus* sp. were isolated using standard procedures, and were purified by repeatedly streaking the bacteria on yeast extract-mannitol agar (YMA) medium (Vincent 1970) and stored at 4°C.

Molecular identification of *Paenibacillus*

Genomic DNA of the newly isolated bacterial strain was extracted by method as described by Ausubel et al. (1994). The DNA was amplified using universal primers 27 F:5′ -GAGAGTTTGATCCTGGCTCAG-3′ and 1492R:5′ -GGYTACCTTGTTACGACTT-3′. PCR reactions were performed in 50 l volume containing 1 μL template DNA, 4 μL $MgCl_2$ (25 mmol/L), 5 μL 10× PCR buffer (Mg^{2+}free), 4 μL dNTP(10 mmol/L), 1 μL of each primer (10 μmol/L), 0.5 μL of TaqDNA polymerase (5u/μL) and 33.5 μL ddH_2O. PCR amplification conditions as follows: Initial denaturation at 94°C for 5 min followed by 30 cycles of denaturation at 94°C for 30 s, annealing at 55°C for 30 s and extension 72°C for 1 min, final extension at 72°C for 10 min. Amplification products were separated by 1·0% agarose gel electrophoresis and visualized under UV light after staining with ethidium bromide. The amplified 16S rRNA gene was sequenced using ABI 3730xl DNA Analyzer (Applied Biosystems, USA). The sequences were identified based on similarity using the Basic Local Alignment Search Tool (BLAST) program National Centre for Biotechnology Information (NCBI) online standard (http://www.ncbi.nlm.nih.gov/).

Table 2 The magnetic treatment of soybean purification number and generation of rhizobium time influence

Treatment		*Paenibacillus* sp. I		*Paenibacillus* sp. II	
Magnetic field (mT)	Time (min)	Number	Generation of *Paenibacillus* sp. (h)	Number	Generation of *Paenibacillus* sp. (h)
CK		42	10.0	40	3.9
100	5	65	8.6	75	3.0
	10	101	8.2	68	3.1
	20	79	8.4	62	3.3
	30	55	8.6	64	3.5
300	5	85	8.0	122	3.1
	10	127	8.0	83	2.8
	20	95	8.1	77	3.1
	30	66	8.1	81	3.1
500	5	94	9.6	115	3.8
	10	102	9.6	84	3.9
	20	86	9.8	82	3.9
	30	42	9.7	78	4.0

Magnetic treatment of soil

The soil was treated by magnetic field in 100 mT, 300 mT and 500 mT for 5 min, 10 min, 20 min and 30 min respectively. Soybean was planted in the treated soil samples using phosphate and potash fertilizers (75 mg $kg^{-1}P_2O_5$;75 mg kg^{-1} K_2O). After harvestation the plants and soil was used to determine the soybean nodulation and nitrogen fixation capacities.

Magnetic treatment of *Paenibacillus* sp.

The *Paenibacillus* sp. was inoculated in 100 mL of YMA medium, incubated at 28°C for 36 h with agitation speed 200 rpm. The cell growth was measured by taking OD

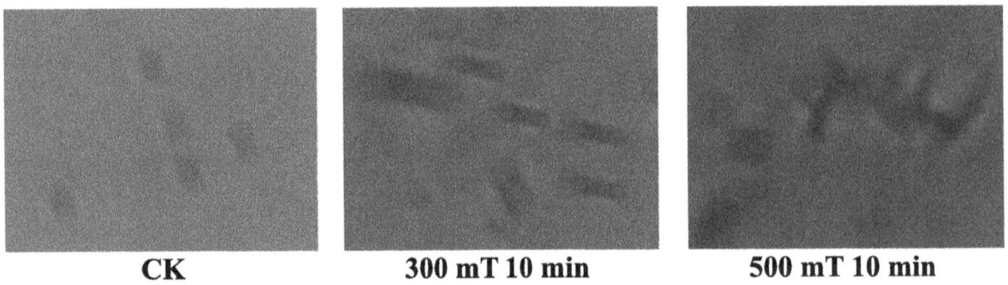

Figure 3 Effects of magnetic fields on the morphology of *Paenibacillus* sp.

at 520 nm. After the cell growth, 25 mL of *Paenibacillus* sp. cell suspension was taken in a test tube and treated it with different magnetic fields like 100, 300 and 500 mT with different time period such as 0, 5, 10, 20 and 30 min. Each experiment was conducted in triplicates and *Paenibacillus* sp. without magnetic treatment was taken as control.

Enzyme assay

The *Paenibacillus* sp. broth was centrifuge at 5000 × g, 4°C for 10 min. After centrifugation the supernatant was discarded and the pellet was suspended in 50 mmol L^{-1} phosphate buffer (pH 7.0) and then subjected to sonication. The homogenate solution was centrifuged for 10 min at 10000 × g, 4°C. After centrifugation, the supernatant was used for determination of peroxidase (POD), superoxidase dismutase (SOD) and catalase (CAT) activities. Catalase activity was assay of hydrogen peroxide based on the formation of its stable complex with ammonium molkbdate and the OD was measured at 405 nm (Fang et al. 2004). One unit of catalase activity was defined as the decomposition of 1 μ mol of hydrogen peroxide per minute under standard assay conditions. Peroxidase activity was determined by hydrogen peroxide-dependent oxidation of guaiacol. Samples were mixed with guaiacol solution (20 mmol/L guaiacol in 0.1 mol/L phosphate buffer (pH 6.8) and 0.03% (v/w) hydrogen peroxide) (Bergmeger et al. 1983). Increase in absorbance at 470 nm was recorded using UV-visible spectrophotometer. One unit of POD activity was defined as the change in absorbance of 0.01 per minute at room temperature. Total SOD activity was assayed by the inhibition of the photochemical reduction of pyrogallol (PAPG) by following the photo reduction of nitroblue tetrazolium (Cai et al. 2006). One unit of SOD activity was defined as amount of enzyme producing a 50% suppression of PAPG reduction. All the Enzyme specific activity is expressed as U/ml.

Total nitrogen determination

Total plant nitrogen (N) concentration was analysed with Kjeldahl determination and colorimetric method as described by Baethgen and Alley (Baethgen & Alley 1989). Nitrogen fixed was calculated as the total plant nitrogen content at harvest, minus the total nitrogen content at the start of the treatments.

Statistical analysis

The data obtained after experimentation was statistically evaluated using ANOVA at significance level of $p < 0.05$ by using computer based programme SPSS.

Results and discussion

Molecular identification of *Paenibacillus* sp.

The newly isolated bacterial strain was identified by molecular techniques using 16S rDNA sequencing. Product of 1442 bp was obtained after PCR amplification (Figure 1). The sequencing result was compared by BLAST for homology analysis. The isolated strain had 99% similarity with *Paenibacillus* sp. 9-2AIA (FN397529.1), *Paenibacillus* sp. Gi-691 (EU497639.1), *Paenibacillus chibensis strain ZYb3* (FJ432004.1), 98% similarity with *Paenibacillus sp. C-2* (KF479638.1), *Paenibacillus* sp. CC-YHH111 (JN806383.1), *Paenibacillus* sp. BM-7 (AY6-

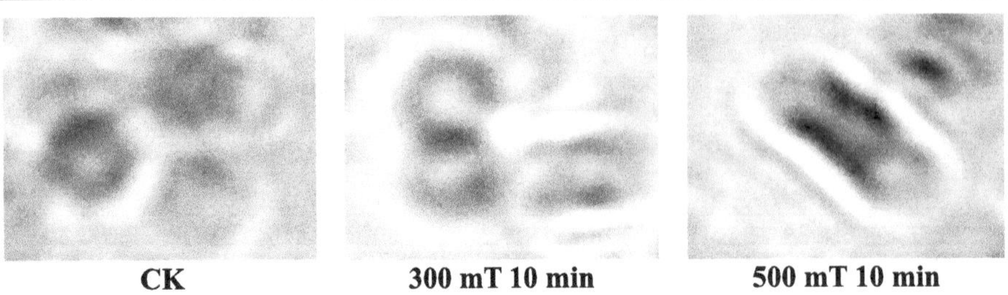

Figure 4 Effect of magnetic fields on the morphology of *Paenibacillus* sp.II.

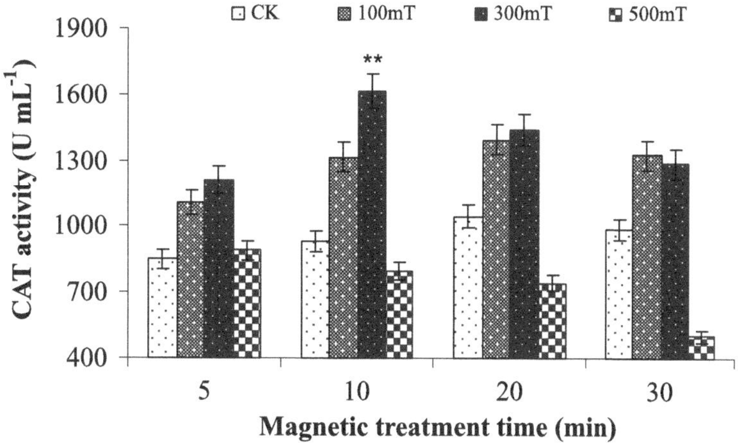

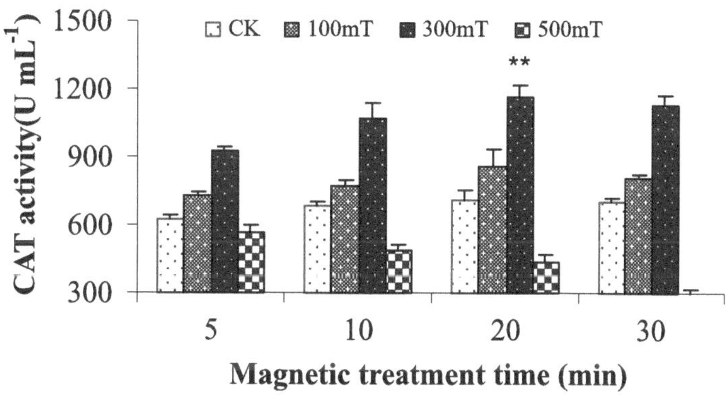

Figure 5 Effect of magnetic fields on CAT activity of *Paenibacillus* sp.I (upper) and *Paenibacillus* sp.II (lower). Error bars indicate SD among triplicates. Stars represent significance level at $P > 0.05$.

35866.1), 97% similarity with *Paenibacillus* sp. E18 (FJ899682.1), *Paenibacillus* sp. D27 (KF479657.1), *Paenibacillus favisporus* isolate MKI10 (EF173324.1) and 96% similarity with *Paenibacillus rhizosphaerae* strain CECAP16 (AY751755.1) and *Paenibacillus* sp. SSG-1 (KF750627.1) as shown in phylogenetic tree (Figure 2).

Effect of magnetic field treated soil soybean nodular and nitrogen fixation

The soybean was planted in the brown soil treated with magnetic field, and the plants were harvested to measure the nodular situation and nitrogen content. The soybean was planted in the absence of nitrogen fertilizer and the difference in nitrogen content before and after treatment was used to compare the treatment effects. Thus we can use plant nitrogen to respond to the difference in the amount of nitrogen-fixing performance. Results shown in the Table 1 described that the soybean plant root dry weight, effective number of *Paenibacillus* sp. and total nitrogen content which treated by 100 mT, 300 mT magnetic field in plant were significantly higher as compared to control (CK). Results of this study was in agreement with Jing et al. (Jing et al. 1992) who reported that appropriate intensity of magnetic field increased 4–5 time nitrogen fixing efficiency of the *Bradyrhizobium japonicum* 005. High treatment (500 mT) of magnetic field resulted decreased number of effective rhizobia thus leading to decreased total nitrogen content. This low nitrogen content leads to less growth because nitrogen is the main nutrition factor that influences the growth in plants (Marschner 1995; Barker & Bryson 2007). Nitrate is inorganic nitrogen which is dominant in agricultural soils and present in three or more times higher than natural soils (Hagedorn et al. 2001; Owen & Jones 2001).

Effect of magnetic treatment on generation time of *Paenibacillus* sp.

In order to check the effect of magnetic field on *Paenibacillus* sp. morphology and physiology, all the effects were studied before and after treatment using various magnetic intensities and treatment time. After the magnetic field treatment, *Paenibacillus* sp. I and *Paenibacillus* sp. II quantity and generation time were changed. On magnetic treatment, the number of *Paenibacillus* sp. was increased as compared to control (CK). This effect

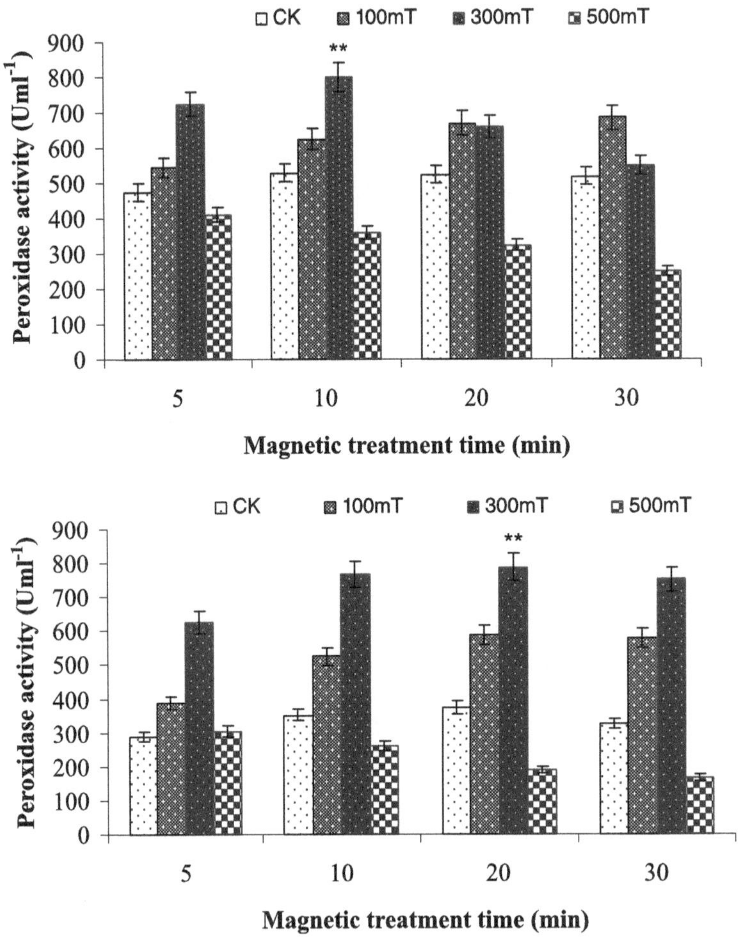

Figure 6 Effect of magnetic field on Peroxidase activity of *Paenibacillus* sp.I (upper) and *Paenibacillus* sp.II (lower). Error bars indicate SD among triplicates. Stars represent significance level at $P > 0.05$.

was changed with respect to magnetic intensity and time duration. Results (Table 2) showed that magnetic intensity of 300 mT had strong effect on the population of *Paenibacillus* sp. The growth of *Paenibacillus* sp. was influenced by magnetic intensity and magnetic treatment time. By increasing the treatment time, number of *Paenibacillus* sp. was increased and further increase in treatment time resulted decline in *Paenibacillus* sp. growth. Of all the treatment time, 10 minutes of magnetic treatment gave better yield. The generation time of 100 mT, 300 mT magnetic field intensity treatments were shortened as compared to control (CK) while 500 mT magnetic field intensity treatments has no significant effect on the generation time as compared to control. These results indicated that magnetic field treatment significantly enhance the *Paenibacillus* sp. population in shorter time. Cheng and Yi (Cheng & Yi 2009) reported that magnetic field intensity of 300 mT had a significant positive effect on the generation time of slow-growing rhizobium (USDA110) and fast-growing rhizobium (USDA191).

Effect of magnetic treatment on morphology of the *Paenibacillus* sp.

The effect of magnetic field on morphology of *Paenibacillus* sp. was also studied as shown in the Figures 3 and 4. Results showed that morphology of *Paenibacillus* sp. was affected by magnetic treatment. Only 10 minutes of magnetic treatment time with intensity of 300 mT and 500 mT was studied. In 300 mT treatment the cell becomes thin and longer while in case of 500 mT treatment the cell membrane was ruptured and the cytoplasmic sap was released leading to the death of the *Paenibacillus* sp. These results revealed that suitable processing time with specific intensity can promote the growth of the *Paenibacillus* sp. but high and long time magnetic processing inhibited the growth thus leading to the death of *Paenibacillus* sp. Our findings were similar as reported by Fadel et al. (Fadel et al. 2003). Various reports suggested that exposure of microorganisms to magnetic field caused changes in morphology and growth (Mohamed et al. 1997; Gaafar et al. 2006).

Effect of magnetic field treatment on enzyme activity of *Paenibacillus* sp.

The influence of magnetic field was also checked on the catalase, peroxidase and superoxide dismutase activity of *Paenibacillus* sp. I and II. The catalases activity of *Paenibacillus* sp. I and II were enhanced by 300 mT intensity as shown in Figure 5. *Paenibacillus* sp. I showed more catalase activity as compared to *Paenibacillus* sp. II. The effective intensity and magnetic treatment time for *Paenibacillus* sp. I and II were 300 mT intensity for 10 and 20 minutes respectively. Both 100 mT and 300 mT increased the catalase activity while 500 mT decreased catalase activity as compared to control. The highest growth rate of *Paenibacillus* sp. I and II was 77% and 95% respectively. These results suggested that magnetic treatment time and intensity had strong influence on metabolic activity of *Paenibacillus* sp.

Figure 6 showed the influence of different magnetic field on peroxidase activity of *Paenibacillus* sp. The 100 and 300 mT magnetic field treatment have growth trend to the peroxidase activity of *Paenibacillus* sp. The highest growth rate of *Paenibacillus* sp. I and II was 68% and 203% in 10 min and 30 min magnetic treatment respectively. Both *Paenibacillus* sp. showed maximum peroxidase activity at 10 min of treatment time with intensity of 300 mT. This activity behavior is almost similar to that of catalase activity.

Figure 7 depicted the effect of different magnetic field on superoxide dismutase activity from *Paenibacillus* sp. Superoxide dismutase activity was found maximum at 20 min of magnetic treatment with intensity of 300 mT. This enzyme activity also showed the same trend as the previous ones (catalase and peroxidase activity) but here increased treatment time resulted increased superoxide dismutase activity. In all the three enzyme activities (catalase, peroxidase and superoxide dismutase) magnetic field intensity of 500 mT inhibited the enzyme activity because this magnetic intensity ruptured the cell shape which leads to decline in enzyme activities. The growth rate at 300 mT magnetic treatment was 340% and 153% for *Paenibacillus* sp. I and II respectively. So this study was in good agreement with Liu et al. (Liu et al. 1996) who reported that appropriate magnetic field intensity enhanced the activities of hydrogen peroxidases, invertases, amylases and phosphatases in the three tested soils. Another study also revealed that magnetic field enhances the catalase and superoxide dismutase activity isolated from the roots of soybean (Celik et al. 2009).

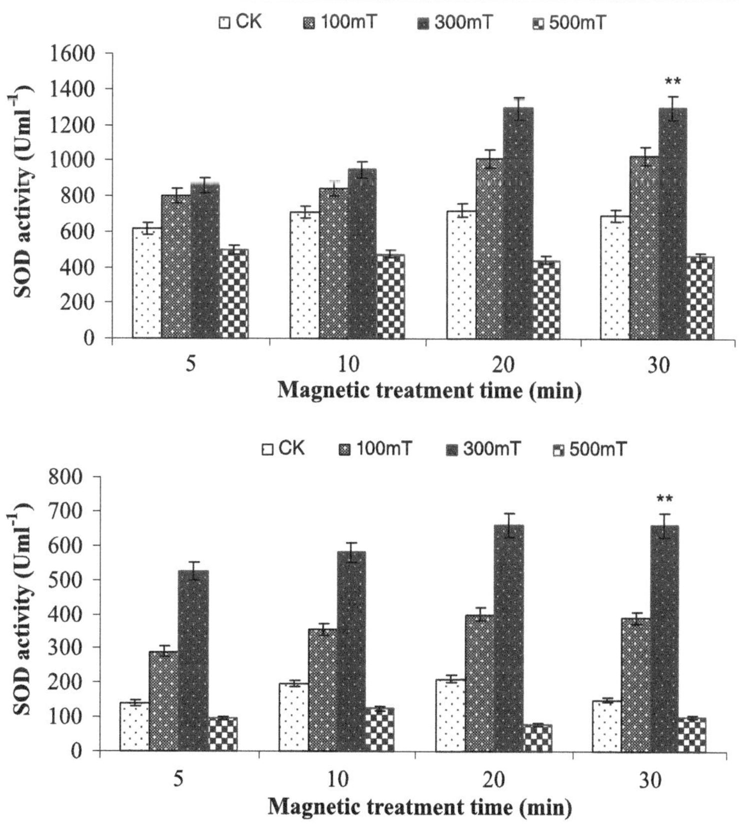

Figure 7 Effect of magnetic fields on Superoxide dismutase activity (SOD) of *Paenibacillus* sp.I (upper) and *Paenibacillus* sp.II (lower). Error bars indicate SD among triplicates. Stars represent significance level at $P > 0.05$.

Conclusion

In conclusion the magnetic treatment significantly enhances the bacterial population with shorter generation time. This increased population of *Paenibacillus* sp. would increase the nitrogen fixing efficiency thus leading to greater yield. The enzyme activities were also increased under the influence of magnetic treatment. Increased magnetic field intensity and longer magnetic processing time resulted ruptured bacterial cell which leads to cell death, thus reduction in nitrogen fixation efficiency. To achieve the better yield, appropriate magnetic field intensity and magnetic processing time is very important for this whole process.

Competing interest

The authors declare that they have no competing interests.

Authors' contribution

YY designed the study project, JL, XC performed experiments and collected data, DZ performed data analysis, MI performed bioinformatics analysis and draft manuscript. All authors read and approved the final manuscript.

Acknowledgment

This study was supported by the National Natural Science Foundation of China (Grant No. 40771111) and the Shenyang agricultural university youth fund (Grant No. 20070136).

Author details

[1]Key Laboratory of Preservation of Northeast Cultivated Land, Ministry of Agriculture, National Engineering Laboratory for Efficient Utilization of Soil and Fertilizer, College of Land and Enviroment, Shenyang Agricultural University, 120 dongling road, Shenyang 110866, Liaoning, China. [2]Liaoning Academy of Environmental Sciences, Shenyang, China. [3]Bioscience and Biotechnology College, Shenyang Agricultural University, 120 dongling road, Shenyang 110866, China.

References

Agus JE, Steven DG, Brian ND (2000) Isolation and characterization of 2,3-dichloro-Ipropanol-edgrading. Appl Environ Microbiol 66(7):2882–2887

Ash C, Priest FG, Collins MD (1993) Molecular identification of rRNA group 3 *Bacilli* (Ash, Farrow, Wallbanks and Collins) using a PCR probe test. Proposal for the creation of a new genus *Paenibacillus*. Antonie Van Leeuwenhoek 64:253–260

Ausubel M, Brent R, Kingston RE, Moore DD, Smit JA, Seidman JC, Struhl KS (1994) Current protocols in molecular biology, section 2.4. John Wiley and Sons, New York

Baethgen WE, Alley MM (1989) A manual colorimetric procedure for measuring ammonium nitrogen in soil and plant. Commun Soil Sci Plant Anal 20 (9&10):961–969

Barker AV, Bryson GM (2007) Nitrogen. In: Barker AV, Pilbeam DJ (eds) Handbook of Plant nutrition. CRC Press, Boca Raton, pp 21–50

Bergmeger H, Bergmeyer J, Grabl M (1983) Methods of enzymatic Analysis, vol 3, 3rd edn. Verlag Chemie Press, Weinheim

Cai Y, Lai Z, Shen J (2006) Detection of superoxide dismutase activities in different organsim *Phyllanthus emblica* L. Chin J Trop Crops 27(4):29–33

Celik Ö, Büyükuslu N, Atak C, Rzakoulieva A (2009) Effects of magnetic field on activity of superoxide dismutase and catalase in *Glycine max* (L.) Merr. Roots. Pol J Environ Stud 18(2):175–182

Cheng X, Yi Y (2009) Effect of Magnetic Treatment on Amount and Generation Time of Slow-Growing Rhizobium (USDA110) and Fast-growing Rhizobium (USDA191). Southwest Chin J Agri Sci 22(5):1400–1403

Cheng X, Yi Y, Du A (2007) Effects of Magnetic treatment on the Azotobacter contents in brown Earth. Chin J Soil Sci 38(5):1025–1027

Fadel MA, Wael SM, Mostafa RM (2003) Effect of 50 Hz, 0.2 mT magnetic fields on RBC properties and heart functions of albino rats. Bioelectromagnetics 24:535–545

Fang F, Li Y, Du GC (2004) Themo-alkali stable catalase *Themoascus aurantiacus* and its potential use in textile bleaching process. Chin J Biotechnol 20(3):423–428

Fisher RF, Long SR (1992) Rhizobium-plant signal exchange. Nature 357(6380):655–660

Gaafar EA, Hanafy MS, Tohamy EY, Ibrahim MH (2006) Stimulation and control of *E. coli* by using an extremely low frequency magnetic field. Rom J Biophys 16(4):283–296

Hagedorn F, Bucher JB, Schleppi P (2001) Contrasting dynamics of dissolved inorganic and organic nitrogen in soil and surface waters of forested catchments with Gleysols. Geoderma 100:173–192

He H, Zhu Y, Zhong K (1998) Effect of magnetic field on celhlase activity and conformation. J Jishou Univ 19(1):42–46

He H, Zhu Y, Fan Q (1999) The Effect of Magnetic Field on *Escherichia Coli* and Glutamic Acid Decarbolxylase. J Jishou Univ 20(3):26–29

Hua H, Shen Y, Wu W (2008) Effects of magnetic field on seed quality, POD and SOD of Pinus massoniana. J Nanjing For Univ 32(3):39–42

Jia Y, Ma Y, Wang Z (2002) The enzymatic activity of tomato seeds with magnetic field treatment. Biotechnology 10(2):14–17

Jing Y, Zhang B, Wang Y, Lin X (1992) Effect of magnetic field on symbiotic nitrogen fixation of soybean nodules. Acta Botanica Sinica 34(5):364–368

Lal S, Tabacchioni S (2009) Ecology and biotechnological potential of *Paenibacillus* polymyxa: a minireview. Indian J Microbiol 49:2–10

Lamb C, Dixon RA (1997) The oxidative burst in plant disease resistance. Annu Rev Plant Physiol Plant Mol Biol 48:251–275

Lebuhn M, Heulin T, Hartmann A (1997) Production of auxin and other indolic and phenolic compounds by *Paenibacillus polymyxa* strains isolated from different proximity to plant roots. FEMS Micr Eco 22:325–334

Li J, Jiao Y, Yi Y (2007) Effects of magnetic field on catalase and peroxidase activities in brown earth. J Shenyang Agric Univ 38(1):70–74

Liu XY, Yi YL, Xia LH (1996) Effect of magnetic field on enzyme activities in main soils of Northeast China. Pedosphere 6(4):341–348

Liu YH, Xu LH, Tang X (2003) The Effect of Magnetic Field Treatment on Artificially Aging Pepper Seeds. J Shandong Inst Build Mat 17(3):286–288

Mao N, Huang Y, Zhang Z (2002) Study on Biological Effect of Magnetic Resonance and Magnetized Water on *Agricus bisporus* Strain 176. J Fujian Teachers Univ (Nat Sci) 18(3):61–65

Marschner H (1995) Mineral Nutrition of Higher Plants. Academic Press Limited, San Diego, London

Mavingui P, Heulin T (1994) In vitro chitinase antifungal activity of a soil, rhizosphere and rhizoplane populations of *Bacillus polymyxa*. Soil Biol Biochem 26:801–803

McSpadden Gardener BB (2004) Ecology of *Bacillus* and *Paenibacillus* spp. in Agricultural Systems. Phytopathology 94:1252–1258

Mohamed AA, Ali FM, Gaafar EA, Magda HR (1997) Effects of magnetic field on the biophysical, biochemical properties and biological activity of *Salmonella typhi.*, Master thesis submitted for Biophysics department, Faculty of science, Cairo University, Egypt.

Montes MJ, Mercade E, Bozal N, Guinea J (2004) *Paenibacillus antarcticus* sp. nov., a novel psychrotolerant organism from the Antarctic environment. Int J Syst Evol Microbiol 54:1521–1526

Ouyang J, Pei Z, Lutwick L, Dalal S, Yang L, Cassai N, Sandhu K, Hanna B, Wieczorek RL, Bluth M, Pincus MR (2008) Case report: *Paenibacillus thiaminolyticus*: a new cause of human infection, inducing bacteremia in a patient on hemodialysis. Ann Clin Lab Sci 38:393–400

Owen AG, Jones DL (2001) Competition for amino acids between wheat roots and rhizosphere microorganisms and the role of amino acids in plant N acquisition. Soil Biol Biochem 33:651–657

Pires MN, Seldin L (1997) Evaluation of Biology system for identification of strains of *Paenibacillus azotofixans*. Ant Van Leeu 71:195–200

Santos R, Herouart D, Sigaud S, Touati D, Puppo A (2001) Oxidative burst in alfalfa-Sinorhisobium meltloti symbiotic interaction. Mol Plant Micobe Interact 14(1):86–89

Sawada H, Kuykendall LD, Young JM (2003) Changing concepts in the systematics of bacterial nitrogen-fixing legume symbionts. J Gen Appl Microbiol 49(3):155–79

Sierra S, Rodelas B, Martinez-Toledo MV, Pozo C, González-López J (1999) Production of B-group vitamins by two Rhizobium stains in chemically defined media. J Appl Microbiol 86:851–858

Vincent JM (1970) A Manual for the Practical Study of Root-Nodule Bacteria. Blackwell Scientific, Oxford, IBP Handbook, 15
Von der Weid I, Alviano DS, Santos ALS, Soares RMA, Alviano CS, Seldin L (2003) Antimicrobial activity of *Paenibacillus peoriae* against a broad spectrum of phytopathogenic bacteria and fungi. J Appl Microbiol 95:1143–1151
Yan L, Zhu Y, He S, Cao Z (1997) Effect of static magnetic field on activity of immobilized a-amylase. Chinese Sci Bull 42(2):127–130
Zhu Y, Zhong K, He S (1996) Efect of Magnetic Field on Activity of Immobilized LDH. J Hunan Univ 23(5):57–61

Philodendron luisae (Araceae), a new species from Rio de Janeiro State, Brazil

Luana Silva Braucks Calazans[1*], Nerivaldo Gomes Antas[2] and Cassia Mônica Sakuragui[1]

Abstract

Background: *Philodendron* is the second largest genus of Araceae, being highly diverse in the Atlantic Forest biome, with nearly one third of the Brazilian species occurring in Southern Brazil, particularly in Rio de Janeiro state. During a local inventory in Silva Jardim municipality, we found a peculiar population of *Philodendron* growing in lowland rainforest.

Results: After morphological analysis and comparisons with similar species, the population proved to be a new undescribed species of subgenus *Philodendron* section *Macrobelium*.

Conclusions: The new species, named *Philodendron luisae*, is here described, illustrated and compared to morphologically close species.

Keywords: Aroids; Taxonomy; Conservation; Atlantic forest; Silva Jardim

Background

Philodendron Schott is the second largest genus of Araceae, with ca. 480 exclusively Neotropical species (Boyce and Croat 2014). The genus is highly diverse in tropical rainforests, such as the Atlantic Forest of coastal Brazil. Despite the elevated levels of deforestation in the Atlantic Forest (Fundação SOS Mata Atlântica – Instituto Nacional de Pesquisas Espaciais 2014), this biome accounts for ca. 12% of the genus diversity in its overall range and 40% of the genus diversity in Brazil (Sakuragui et al. 2014). These numbers are being constantly increased by the recognition of new species, especially from Southeastern Brazil (e.g. Buturi et al. 2014, Calazans and Sakuragui 2013, Coelho 2010, Gonçalves 2011).

In Rio de Janeiro state, Southern Brazil, the genus is represented by an impressive number of species, despite the reduced geographical range and high levels of habitat fragmentation (Sakuragui et al. 2011). Of the 168 Brazilian species, 30 occur in the Rio de Janeiro state (Sakuragui et al. 2014). Although efforts to catalogue the regional flora dates back to the eighteen century, many localities remain poorly known or even uncollected, especially outside the metropolitan area of the Rio de Janeiro City, justifying intensive sampling efforts.

Here we describe *Philodendron luisae sp. nov.*, an only recently collected and recognized species from Rio de Janeiro State's lowlands.

* Correspondence: luanasbcalazans@gmail.com
[1]Instituto de Biologia, Departamento de Botânica, Universidade Federal do Rio de Janeiro, CCS, Av. Carlos Chagas Filho, 373 - Sala A1-088 - Bloco A, Ilha do Fundão, Rio de Janeiro, RJ CEP 21941-902, Brazil
Full list of author information is available at the end of the article

Methods

During a flora inventory conducted in the Atlantic Forest remnants in Silva Jardim municipality, Rio de Janeiro State, we found an indeterminable *Philodendron* species. It was recognized as belonging to subgenus *Philodendron* section *Macrobelium* and analyzed through the two more comprehensive and updated keys of the section (Croat 1997, Sakuragui et al. 2005). The species was also compared with the type specimens and descriptions of morphologically similar species. The descriptive terminology follows Stearn (2004) and Mayo (1991). Morphological analyzes of fresh and dry materials were performed with the aid of a stereoscopic microscope. The extent of occurrence and area of occupancy were calculated using the GeoCAT tool (Bachman et al. 2011).

Results and discussion

Philodendron luisae Calazans, *sp. nov.* (Figures 1 and 2).

Type:—BRAZIL. Rio de Janeiro: Silva Jardim, road RJ-126, Sítio Além do Horizonte, 22° 32′ 47.8″ S, 42°

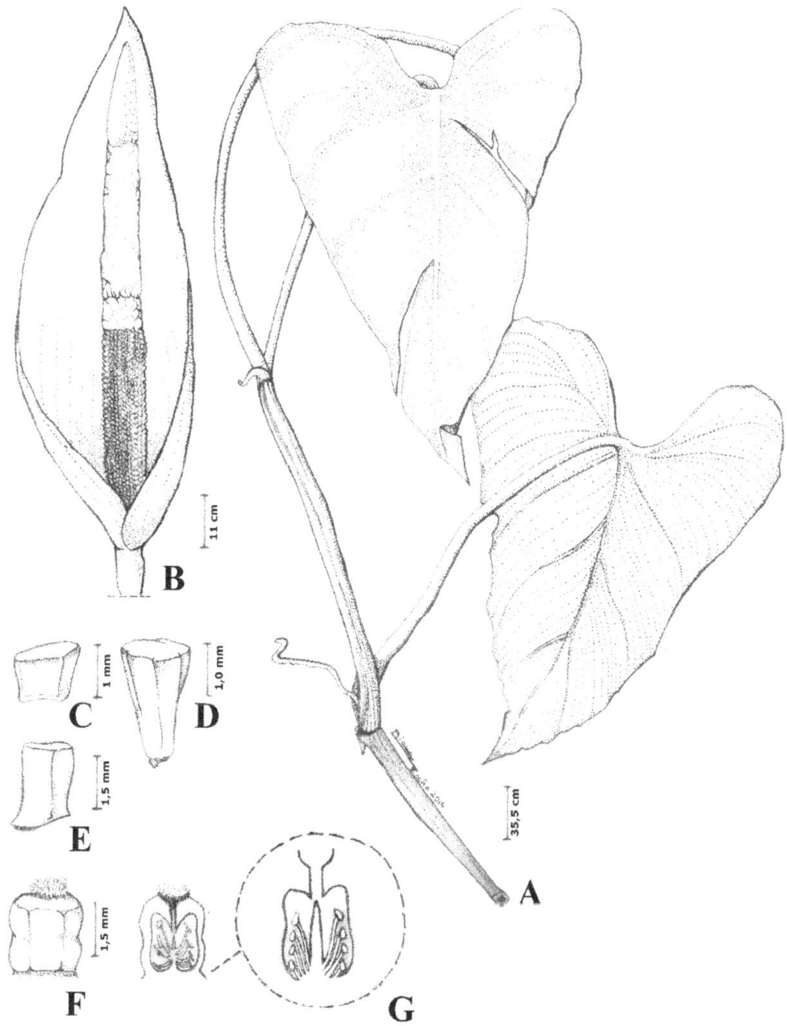

Figure 1 ***Philodendron luisae.*** **A**. Habit, x ½. **B**. Inflorescence, x 1. **C.** Apical staminode. **D**. Stamen. **E**. Itermediate staminode. **F**. Gynoecium **G**. Longitudinal cut of gynoecium and the basal placentation (detail). **A** from *Antas et al. 188* (RB); **B-G** from *Antas 181* (RB).

27′ 54.1″ W, 07 November 2013, *N.G. Antas 181* (holotype, RB; isotypes, NY, K).

Herb hemi-epiphytic. ***Internodes*** 4.7–11(–15) cm long, usually shorter in flowering shoots, 3–4-angular, keeled, greenish becoming light brown, drying often cracked with rhytidome-like layers; **intravaginal squamules** up to 3 per node, inconspicuous, deciduous, becoming dark. ***Prophyll*** 6.8–10 × 0.1–1.5 cm, triangular, deciduous, slightly keeled, smooth, yellowish becoming cream, drying brown. ***Petiole*** 11–17 × 0.4–0.6 cm, adaxially flattened, abaxially rounded, glossy green, slightly striated, drying dark brown; **leaf blade** 16–20 × 9–11.3 cm, triangular to cordate-sagittate, smooth, glossy green, abaxially paler, drying membranous, striated, olive-green, strongly discolorous, margin entire, apex acuminate, acumen 1–2 cm long, sometimes curved, base cordate; **anterior division** 12–15.5 cm long, midrib impressed on both faces, drying dark brown, **primary lateral veins** 3–4 pairs, arising from midrib at 70°, 40° and 25–35° angle respectively from the base to the apex, arcuate to margin, impressed on both faces, drying discrete adaxially, dark brown abaxially, **secondary veins** indistinct, parallel to primary veins, numerous, drying evident on both faces, prominent adaxially; **posterior divisions** 3.5–5 cm long, cordate, primary acroscopic veins 2(–3), basal denudation absent. ***Inflorescence*** solitary; **peduncle** 1.52–2.5 cm long, cylindrical; **spathe** 10–11 cm long, ovate, acuminate, acumen ca. 1 cm long, constriction not evident, externally green becoming cream towards the apex, striated, internally cream, reddish at the base, resin canals internally visible; **stipe** absent; **spadix** 8–9 cm long, slender; apical sterile zone 1.3–2.1 cm long, yellowish; fertile male zone 3–3.6 cm long, yellowish; intermediate sterile zone 0.6–0.8 cm long, cream; female zone 3–4.3 cm long, light green; apical staminodes ca. 1 mm long, prismatic; stamens ca.

Figure 2 Habitat and morphology of *Philodendron luisae*. **A**. Habitat in a fragment of Atlantic Ombrophilous Dense Submontane Forest in Silva Jardim municipality. **B**. An hemi-epiphytic individual growing on *Dicksonia* sp. **C**. Young leaves. **D**. Flowering shoot with an inflorescence in pre-anthesis and a leaf ripped. **E**. Detail of the flattened petiole. **F**. Detail of the angular internodes. **G**. Detail of the inflorescence manually opened to show the interior.

1 mm long, prismatic; intermediate staminodes ca. 1.5 mm long, prismatic; gynoecium 1.5–2.0 mm long, ovary ca. 1.5 mm long, barrel-shaped, (6)–7–8–locular, 3–4–ovulate, placentation basal, stylar region ca. 0.75 mm long, as wide as the ovary, stigmatic region ca. 0.75 mm long. ***Berries*** unknown. ***Seeds*** unknown.

Phenology

Collected in flower in November.

Etymology

The species is named in memory of the Biology undergraduate student Luisa Pinho Sartori, who inspired the conservation and educational initiatives promoted by Sartori family.

Distribution and ecology

Only known from three records in reduced Atlantic Ombrophilous Dense Submontane Forest fragments in Silva Jardim municipality, a rural zone of Rio de Janeiro state (Figure 3). The species can be found in small patches of submontane and seasonally flooded forests in areas of regeneration, growing mainly in primary *Tabebuia* formations associated to the phorophytes *Tabebuia cassinoides* (Lam.) DC. and *Dicksonia* sp.. This is indicative of the species' tolerance to open habitats, which may be interesting

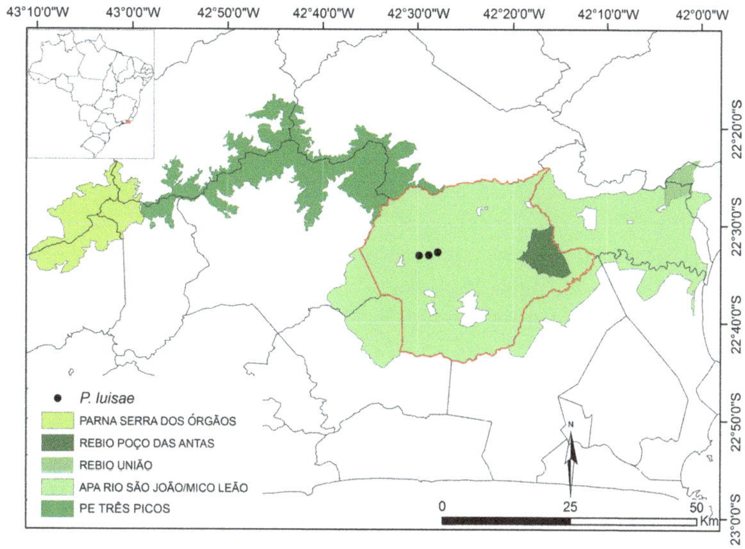

Figure 3 Distribution map of *Philodendron luisae*. The red spot show the focus area in the Rio de Janeiro state. Green areas represent the ecological corridor Mosaico Central Fluminense composed of different conservation unities. Silva Jardim municipality is highlighted by the red line. APA – Área de Proteção Ambiental; PARNA – Parque Nacional; PE – Parque Estadual; REBIO – Reserva Biológica.

in reforestation projects. The species is also frequently found growing together with *P. nadruzianum* Sakur.

Conservation

Our data so far indicate the species as *critically endangered* (CR) - B1ab (ii, iii, iv) - following the categories and criteria of International Union for Conservation of Nature (2012), with extent of occurrence estimated in 32.2 km^2. In this category are inserted species facing a high risk of extinction in the wild, with extent of occurrence <100 km^2 and very fragmented, known from a single location, with continuing decline inferred for the number of individuals and habitat quality.

All records of *P. luisae* up to now are strongly related to stream margins, an area designated as permanently protected by Brazilian federal law in order to preserve the biodiversity and environmental resources in strategic areas. Additionally, the species is only known from particular properties within the Área de Proteção Ambiental da Bacia do Rio São João/ Mico-Leão-Dourado, a conservation unity of sustainable use (Ministério do Meio Ambiente – Instituto Chico Mendes de Conservação da

Table 1 Comparison between *P. luisae* and closely related species

Species	Internodes dimension (cm) and shape	Extrafloral nectaries	Leaf dimension (cm)	Primary lateral veins	Inflorescences per sympodium	Apical sterile zone	Spathe color outside, constriction and opening at anthesis	Resin canals inside the spathe
P. fragile	1–5, terete	absent	15–37 × 9–24	4–5	1(–2)	absent	greenish becoming cream towards the apex, slightly constricted, moderately opened	not visible
P. luisae	4.7–11(–15), markedly angular	absent	16–20 × 9–11.3	3–4	1	present	green becoming cream towards the apex, not constricted, moderately opened	visible
P. millerianum	2.4–8.5, terete	absent	20.2–21.7 × 4–11.1	4	1	present	completely white, strongly constricted, slightly opened	not visible
P. simonianum	2–4, terete	absent	36–42 × 14–20	3–4	3–4	present	completely white, not constricted, reflexed	not visible
P. tenuispadix	3–4, terete	present	51–59 × 24–27	6–8	1–3	present	completely green, slightly constricted, almost completely opened, but not reflexed	not visible

Information from *P. millerianum*, *P. simonianum* and *P. tenuispadix* according to the original publications (Coelho and Sakuragui 2007, Sakuragui 2001 and Gonçalves 2002, respectively); from *P. fragile* according to Sakuragui et al. (2005).

Biodiversidade 2008). The region suffered historical fragmentation due to selective extraction of timber and farming practices (Carvalho et al. 2006, Guedes-Bruni et al. 2006), however, the forest remnants are very representative of the Atlantic Ombrophilous Dense Submontane Forest, being of high priority for conservation (Carvalho et al. 2006). These remnants compose the ecological corridor Mosaico Central Fluminense along with the conservation units Parque Estadual dos Três Picos, Parque Nacional da Serra dos Órgãos, Reserva Biológica Poço das Antas and Reserva Biológica União (Instituto Estadual do Ambiente 2013) (Figure 3). Probably, populations of *P. luisae* may be found in the lowlands of this ecological corridor, assured within protected areas.

The association between *P. luisae* and the phorophyte *T. cassinoides* is interesting from the conservational viewpoint since this tree is currently threatened and listed in the Red List of the Brazilian Flora (Lohmann et al. 2013). The tree has suffered an intensive selective extraction due to its high quality timber, used mainly for the manufacturing of shoes (Lohmann et al. 2013). This activity probably caused impact on the natural populations of *P. luisae* in the *Tabebuia* formation, contributing to the fragmented distribution currently known.

Paratype

Brazil, Rio de Janeiro: Silva Jardim, road RJ-126, Fazenda Novo Horizonte, 22° 33′ 08.0″ S, 42° 29′ 49.6″ W, 13 November 2013, *N.G. Antas et al. 188* (MBML, RB, SPF).

Features and affinities

Philodendron luisae can be promptly recognized by its small and fragile leaves (often ripped when adult) and 3–4-angular, light brown colored stem. The overall leaf shape and inflorescence with apical sterile zone makes this species close to *P. simonianum* Sakur. and *P. tenuispadix* E.G.Gonç., but it differs by its longer internodes, smaller leaf blade, number of primary lateral veins, number of inflorescences per sympodium, absence of extrafloral nectaries and spathe features (Table 1). When the leaves are still young, *P. luisae* also resembles *P. fragile* Nadruz & Mayo and *P. millerianum* Nadruz & Sakur. by its triangular and fragile leaves without well developed posterior divisions, but the species show a number of differences between them (Table 1).

Moreover, among these related species, *P. luisae* is the unique known to grown in swamp forest, a vegetation formation frequently more open and exposed to seasonality than the humid and shaded forests where the most of *Philodendron* species occurs.

Conclusions

Philodendron luisae is a new species easily recognizable and well supported and represents the 31st species record for Rio de Janeiro State. The species, only known from one locality, exemplifies the importance and urgency of local and regional floras to a broaden knowledge of the Brazilian biodiversity.

Competing interests

The authors declare that they have no competing interests.

Authors' contributions

LSBC, NGA and CMS collected and discovered the new species; LSBC carried out the morphological analyses; NGA provided the floristic information; all authors prepared, read and approved the final manuscript.

Acknowledgements

We are grateful to Fábio Sartori for encouraging field expeditions and flora inventories in the Atlantic Forest remnants in his properties, to Diana Rocha Monteiro for the illustration, to Marco Octávio Pellegrini and to Rodrigo Theófilo Valadares for the assistance with the figures and for suggestions to an earlier version of the manuscript. CMS is a CNPq research fellow.

Author details

[1]Instituto de Biologia, Departamento de Botânica, Universidade Federal do Rio de Janeiro, CCS, Av. Carlos Chagas Filho, 373 - Sala A1-088 - Bloco A, Ilha do Fundão, Rio de Janeiro, RJ CEP 21941-902, Brazil. [2]Fundação Luisa Sartori, Sartori Árvores Nativas e Reflorestamento, Fazenda Além do Horizonte Rodovia RJ-126, km 28, s/n – Gaviões, Silva Jardim, Rio de Janeiro, RJ, Brazil.

References

Bachman S, Moat J, Hill AW, Torre J, Scott B (2011) Supporting Red List threat assessments with GeoCAT: geospatial conservation assessment tool. ZooKeys 150:117–126

Boyce PC, Croat TB (2014) The Überlist of Araceae: totals for published and estimated number of species in aroid genera. Available via http://www.aroid.org/genera/140601uberlist.pdf. Accessed 14 July 2014

Buturi CV, Temponi LG, Sakuragui CM (2014) A new species of *Philodendron* (Araceae) in Paraná, Brazil. Phytotaxa 174(3):144–148

Calazans LSB, Sakuragui CM (2013) A new species of *Philodendron* (Araceae) and a key to Brazilian Atlantic Forest species of *P.* subgenus *Pteromischum*. Phytotaxa 94(2):49–55

Carvalho FA, Nascimento MT, Braga JMA (2006) Composição e riqueza florística do componente arbóreo da Floresta Atlântica submontana na região de Imbaú, Município de Silva Jardim, RJ. Acta Bot Bras 20(3):727–740

Coelho MAN (2010) Espécies novas de *Anthurium* e *Philodendron* (Araceae) do sudeste brasileiro. Bol Mus Biol Mello Leitão (N Ser) 28:21–40

Coelho MAN, Sakuragui CM (2007) A new species of *Philodendron* Schott (Araceae) from Brazil. Kew Bull 62(4):629–631

Croat TB (1997) A revision of *Philodendron* subgenus *Philodendron* (Araceae) for Mexico and Central America. Ann Miss Bot Gard 84(3):311–704

Fundação SOS Mata Atlântica – Instituto Nacional de Pesquisas Espaciais (2014) Atlas dos remanescentes florestais de Mata Atlântica período 2012–2013. ArcPlan, São Paulo

Gonçalves EG (2002) New aroid taxa from Brazil. Aroideana 25:16–35

Gonçalves EG (2011) *Philodendron bernardopazii* and *P. ricardoi* spp. nov. (Araceae) from Espírito Santo State, Brazil. Nordic J Bot 29:385–390

Guedes-Bruni RR, Silva Neto SJ, Morim MP, Mantovani W (2006) Composição florística e estrutura de trecho de Floresta Ombrófila Densa Atlântica aluvial na Reserva Biológica de Poço das Antas, Silva Jardim, Rio de Janeiro, Brasil. Rodriguésia 57(3):413–428

Instituto Estadual do Ambiente (2013) Parque Estadual dos Três Picos: plano de manejo. INEA, Rio de Janeiro

International Union for Conservation of Nature (2012) IUCN Red List Categories and Criteria: Version 3.1, 2nd edn. IUCN, Gland and Cambridge

Lohmann LG, Sfair JC, Monteiro NP, Santos Filho LAF (2013) In: Martinelli G, Moraes MA (orgs.) Livro Vermelho da Flora do Brasil, 1st edn. Andrea Jakobsson, Rio de Janeiro

Mayo SJ (1991) A revision of *Philodendron* subgenus *Meconostigma* (Araceae). Kew Bull 46(4):601–681

Ministério do Meio Ambiente – Instituto Chico Mendes de Conservação da Biodiversidade (2008) Plano de manejo da Área de Proteção Ambiental da Bacia do Rio São João/ Mico-Leão-Dourado. MMA, Rio de Janeiro

Sakuragui CM (2001) Two new species of *Philodendron* (Araceae) from Brazil. Novon 11(1):102–104

Sakuragui CM, Mayo SJ, Zappi DC (2005) Taxonomic revision of Brazilian species of *Philodendron* section *Macrobelium*. Kew Bull 60(4):465–513

Sakuragui CM, Calazans LSB, Morais EB, Coelho MAN, Pellegrini MOO (2011) Diversity and conservation of *Philodendron* Schott (Araceae) in Atlantic Forest of Rio de Janeiro State, Brazil. Feddes Repert 122(7–8):472–496

Sakuragui CM, Calazans LSB, Soares ML (2014) *Philodendron* In: Lista de Espécies da Flora do Brasil. Jardim Botânico do Rio de Janeiro. Available via http://www.reflora.jbrj.gov.br/jabot/floradobrasil/FB5015. Accessed 14 July 2014

Stearn WT (2004) Botanical Latin. Timber Press, Portland

30

Nostoc sp. extract induces oxidative stress-mediated root cell destruction in *Mimosa pigra* L.

Siriphan Sukkhaeng, Nuttha Sanevas and Srisom Suwanwong*

Abstract

Background: *Mimosa pigra* is an invasive weed in some regions of South East Asia and Australia. Our previous study has revealed that a cyanobacterium, *Nostoc* sp., extract can inhibit root growth in *M. pigra* seedlings. In this study, some physiological processes involve oxidative stress-mediated cell death and root ultrastructure were investigated to clarify the mechanisms of root growth suppression and bioherbicidal potential of the extract.

Results: *Nostoc* sp. extract enhanced overproduction of reactive oxygen species (ROS) at 24 h, the intensity of red fluorescence increased at 72 h, and caused a slightly increased H_2O_2 consistent with the activation of scavenging enzymes (catalase, ascorbic acid peroxidase, glutathione reductase, and peroxidases). This suggests that oxidative stress occurred in the presence of the extract which was supported by increased cell death and lipid peroxidation at 24 h. Reduction of malondialdehyde content and an increase in cell death at 72 h indicated oxidative damage and cellular leakage. Ultrastructural changes were determined at 72 h by scanning electron micrographs which confirmed the damage of epidermal and root cap cells and the disaggregation and destruction of root tip cells. Transmission electron micrographs showed the dissolution of the middle lamella, deposition of some substances in vacuoles, and abnormal mitochondria (swollen mitochondria and indistinct cristae).

Conclusions: *Nostoc* sp. extract enhance oxidative stress by ROS production resulting in lipid peroxidation and massive cell death despite the activation of antioxidative enzymes. Understanding mechanism of action of *Nostoc* sp. extract will provide information for application of the extract to use as natural herbicide for control of *M. pigra*.

Keywords: *Nostoc* sp; Oxidative stress; Lipid peroxidation; Cell death; Root ultrastructure; *Mimosa pigra*

Background

Cyanobacteria are known to produce various kinds of secondary metabolites that can affect many biochemical processes within cells and can influence the growth of surrounding organisms (Leflaive and Ten-Hage 2007). Natural products from cyanobacteria exhibit various biological inhibitory effects which are cytotoxic, antibacterial, antifungal, and antialgal (Etchegaray et al. 2004; Kreitlow et al. 1999; Kulik 1995). Cyanobacteria have been recognized as options for novel bioactive natural products and use as bio-control agents and herbicides due to their inhibitory activities against some aquatic and terrestrial plants, especially weeds (Duke et al. 2002; Berry et al. 2008).

Nostoc sp., a filamentous cyanobacterium (Nostocales), has been reported to produce some chemical substances affecting on various organisms. For example, an alkaloid nostocarboline had an inhibitory activity on *Microcystis aeruginosa* (Blom et al. 2006). Nostocyclamide, a cyclic peptide, had antialgal activity and growth inhibitory activities against cyanobacteria and Chlorophyceae (Kobayashi and Kajiyama 1998). The effect on higher plants was reported by Hirata et al. (2003) that nostocine A produced by *Nostoc spongiaeforme* TISTR 8169 exhibited an inhibitory activity on root growth of barnyard grass (*Echinochloa crus-galli* (L.) P. Beauv.). It is possible that *Nostoc* species has weed suppressing potential. Not much is known about the mechanism of *Nostoc* extract.

Understanding the mechanism of natural plant compounds is important for research in natural herbicides. Under unfavorable conditions, such as high or low temperature, water deficit, salinity, and some chemical substances, normal metabolisms of plants are disturbed and toxic molecules are produced (Mano 2002). One of these toxic molecules is ROS which can abstract

* Correspondence: fscisss@ku.ac.th
Department of Botany, Faculty of Science, Kasetsart University, Bangkok 10900, Thailand

hydrogen atoms from polyunsaturated fatty acids (the composition of biological membranes), initiate lipid peroxidation, and cause cell death. During this processes, plants suffer from oxidative stress and malondialdehyde (MDA) is produced as a by-product (Halliwell and Chirico 1993; Gutteridge 1995). Because of ROS toxicity, plant cells have mechanisms to reduce these toxic compounds by ROS scavenging via antioxidative enzyme systems (Mittler 2002). Root ultrastructural changes have been used to study the damage of structures and organelles in root tips exposed to some allelochemicals. These include the decreases in the number of ribosome, dictyosome, mitochondria, endoplasmic reticulum, and metabolic products, mitochondrial swelling and loss of cristae, dissolution of the middle lamella, and the irregular-shaped cells (Burgos et al. 2004; Jiang and Liu 2010; Yang et al. 2011).

In our previous study, crude extract from *Nostoc* sp. was effective on shoot and root growth suppression of *M. pigra* and had an adverse effect on mitotic cell division (Sukkhaeng et al. 2014), but the inhibition of cell division might not be the only major mechanism of root growth suppression. The aims of this study were to explore the effect of *Nostoc* sp. extract on ROS-mediated oxidative stress (superoxide anion radical ($O_2{\bullet}^-$) and lipid peroxidation), ROS metabolism in terms of the alterations of antioxidative enzymes, and oxidative damage (cell death and root ultrastructural changes).

Methods

Cyanobacteria culture and extraction

A cyanobacterium, *Nostoc* sp. (BotKU C3004, voucher specimen is deposited at herbarium of the Department of Botany, Faculty of Science, Kasetsart University) was collected from rice straw in Bangkok, Thailand and isolated by the streak plate method. *Nostoc* cells were cultured in BG-11 liquid medium (Rippka et al. 1979), pH 6.5 under daylight fluorescent lamps (300 μmol m^{-2} s^{-1}) at ambient temperature (32 ± 3°C). The cells in the exponential phase of growth (15 days after culture) were harvested by means of filtration and subsequently dried at 50°C for 72 h. The dried cells were ground to powder and extracted with 80% methanol for 24 h at ambient temperature. The solution was filtered and the supernatant was evaporated in a rotary evaporator to obtain a crude brown gum. The residue was extracted twice and the supernatant was combined with the first one.

Bioassay

The gummy substance was dissolved in water containing 0.1% dimethyl sulfoxide (DMSO) at 0, 0.1, 0.3 and 0.5%. Test solutions of various concentrations (1.5 ml) were added to petri dishes (5 cm in diameter) containing filter paper. Six seeds of *M. pigra* were placed on the filter paper and kept in the dark at ambient temperature for 24 and 72 h. The water containing 0.1% DMSO was used as a control.

Determination of antioxidative enzyme activities

Three hundred milligrams of frozen roots were crushed to a fine powder in a mortar under liquid nitrogen and homogenize with 3 ml of 25 mM potassium phosphate buffer (pH 7.8) containing 0.4 mM EDTA-4H, 1 mM ascorbic acid, and 2% PVPP. The homogenate was centrifuged at 12,000 rpm for 20 min at 4°C and the supernatant was used as an enzyme extract for the assays of catalase (CAT; EC 1.11.1.6), ascorbic acid peroxidase (APX; EC 1.11.1.11), glutathione reductase (GR; EC 1.6.4.2), and peroxidases (PODs; EC 1.11.1.7) activities.

CAT activity was assayed by determination of the initial rate of hydrogen peroxide disappearance according to Aebi (1983). Activity was measured spectrophotometrically at 240 nm (E = 0.0394 mM^{-1} cm^{-1}). GR activity was determined by the method of Halliwell and Foyer (1978). Activity was determined at 340 nm by measuring the decrease of NADPH absorbance (E = 6.1 mM^{-1} cm^{-1}). APX activity was assayed by the method of Nakano and Asada (1987). Activity was determined at 290 nm by measuring the decrease of ascorbic acid (E = 2.8 mM^{-1} cm^{-1}). PODs activity was determined by the method of Putter (1974). PODs activity was measured spectrophotometrically at 436 nm by the formation of guaiacol dehydrogenation products (E = 25.5 μM^{-1} cm^{-1}).

Histochemical detection of ROS

ROS production was estimated using a superoxide anion radical-specific oxidation-sensitive fluorescence dye, dihydroethidium (DHE) (Sunohara and Matsumoto 2008) with a minor modification. Treated roots were stained with 10 μM DHE in 100 μM $CaCl_2$, pH 4.75, by gently shaking for 2 h at 25°C in the dark. The roots were then soaked in 100 μM $CaCl_2$ for 5 min and observed using a fluorescence microscope (Nikon E600 with a B-2A filter combination, excitation 450–490 nm, emission ≥ 520 nm).

H_2O_2 determination

Hydrogen peroxide content was determined according to Velikova et al. (2000). This method is based on potassium iodide (KI) oxidation by H_2O_2 to give iodide ion. Iodide reacts with iodine to form triiodide (I_3^-) which show absorption at 285 nm (Junglee et al. 2014). For every sample, distilled water was used instead of KI for tissue coloration background. H_2O_2 standard solutions were prepared for calibration curve.

Determination of root cell death

Treated roots were stained with Evans blue solution (0.025% (w/v) in 100 μM $CaCl_2$, pH 5.6) for 30 min. Stained roots were washed 3 times with 100 μM $CaCl_2$ (pH 5.6), until the dye no longer eluted from the roots (Yamamoto et al. 2001). Ten root tips were excised (3 mm) and allowed to soak in 200 μl of *N,N*-dimethylformamide for 24 h. The absorbance of released Evans blue was measured at 600 nm.

Determination of lipid peroxidation

The thiobarbituric acid test, which determines MDA as an end product of lipid peroxidation (Velikova et al. 2000), was used. The amount of thiobarbituric acid-reactive substances, red pigments, was calculated from the extinction coefficient 155 mM^{-1} cm^{-1}.

Root ultrastructural studies

For scanning electron microscopy, excised root tips of 5 mm (0 and 0.5% at 72 h treated roots) were prefixed in 2.5% glutaraldehyde in a 0.1 M sodium phosphate buffer (pH 7.2), postfixed in 2% osmium tetroxide for 2 h., dehydrated in an acetone series, and dried in critical point dryer (Emitech; K850). After sputter-coating with gold by using an ion coater (Eiko Engineer; IB-2), they were examined with a scanning electron microscope (SEM) (Jeol; JSM5600LV) which operated at 10 kV. For transmission electron microscopy, 3 mm root tip segments were fixed overnight in 2.5% glutaraldehyde in a 0.1 M sodium phosphate buffer (pH 7.2). They were postfixed for 2 h in 2% osmium tetroxide in water. The roots were embedded in a Spurr's resin and polymerized for 7 h at 80°C. Next, 60 nm cross-sliced sections, probably located in the middle part between the root tip and the quiescent center in the root cap, were made on an ultramicrotome (Leica; UCT) and stained with 5% (w/v) uranyl acetate and 2% (w/v) lead citrate. After fixing, the specimens were examined with a transmission electron microscope (TEM) (Jeol; JEM1220) which operated at 80 kV.

Statistical analysis

All statistical analyses were performed as analysis of variance (ANOVA) and means were compared using Duncan's multiple range test (DMRT) at $p \leq 0.05$.

Results and discussion

Responses of antioxidative enzyme activities to *Nostoc* sp. extract were determined as an indicator of oxidative stress. Antioxidative enzymes are involved in the detoxification and balance of reactive oxygen species (ROS) when oxidative stress occurs (Mittler 2002). After 24 h of exposure, CAT, APX, GR, and PODs activities significantly increased with increasing extract concentration (Figure 1a-d). CAT and PODs are enzymes which convert H_2O_2 to water and protect the cells from the damaging effects of H_2O_2 – a precursor of hydroxyl radical ($HO^{\bullet}$) which is a high oxidative power molecule. APX and GR are enzymes which are involved in the ascorbate-glutathione cycle. CAT and the ascorbate-

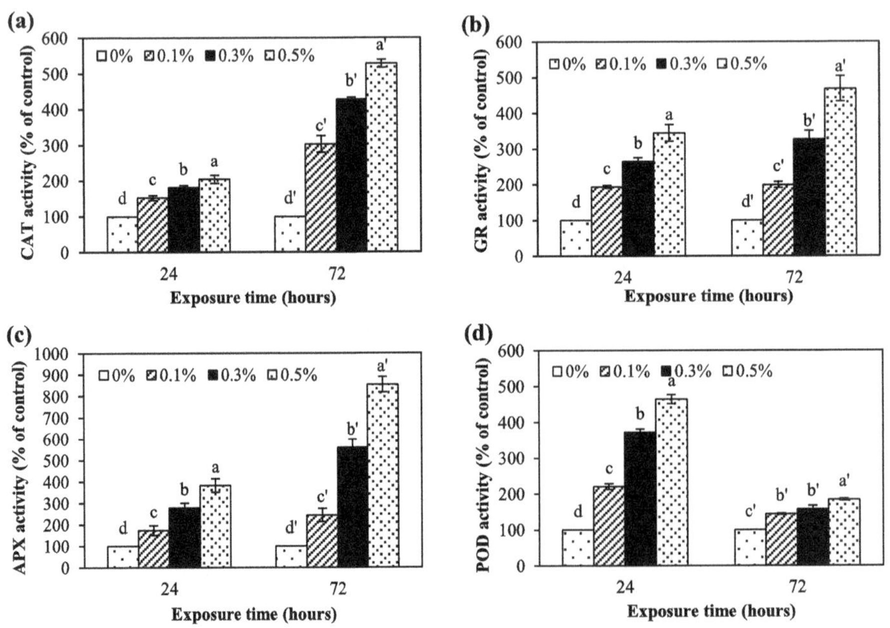

Figure 1 Antioxidative enzyme activities in *M. pigra* roots after exposure to the extract at different concentrations for 24 and 72 h. CAT **(a)**, GR **(b)**, APX **(c)**, and PODs **(d)** Data are mean values of three replications and SD are indicated by vertical bars. The different letters above the column indicate significant differences ($p \leq 0.05$) by DMRT.

glutathione cycle are important in H_2O_2 scavenging. CAT has a high reaction rate, but a low affinity for H_2O_2, so its function is only removing the bulk of H_2O_2. APX has a higher affinity for H_2O_2 so it can scavenge small amounts of H_2O_2 in more specific locations (Dat et al. 2000). For PODs activity that highly increased at 24 h after treatment but drastically reduced at 72 h indicating a sensitivity of PODs at the initial stress condition (24 h) but not after prolonged stress (72 h). This was similar to the response of guaiacol peroxidases to cadmium that was decreased with high concentration but not with low concentration (Cho and Seo 2005). Other enzymes involved in H_2O_2 scavenging (APX and CAT) seemed to be principal than PODs at 72 h after treatment. Several studies have been reported that there is an increase in antioxidative enzyme activities upon allelochemical stress; for example, Pflugmacher et al. (2006) reported that enhanced activities of SOD, CAT, PODs, and GR in alfalfa (*Medicago sativa* L.) in response to oxidative stress caused by the exposure to cyanobacterial metabolites. Essential oils from redstem wormwood (*Artemisia scoparia* Waldst & Kit.) induced oxidative stress in wheat (*Triticum aestivum* L.) roots and elevated antioxidative enzyme activities (Kaur et al. 2012). All antioxidative enzyme activities kept increasing within 72 h, except for PODs (Figure 1a-d). These results suggest that *Nostoc* sp. extract might cause oxidative stress resulting in a significant activation of antioxidative enzyme activities.

Effect of *Nostoc* sp. extract on ROS (superoxide anion radicals ($O_2\bullet^-$) and H_2O_2) production were detected. Dihydroethidium has been widely used to evaluate ROS production, mainly of $O_2\bullet$. It can freely permeate cell membranes and react with $O_2\bullet^-$ to form a red fluorescent product (Gomes et al. 2005). Our results showed a slightly red fluorescence at 0.3 and 0.5% of the extract concentrations after exposure for 24 h. After 72 h, roots were stained stronger and produced a brighter fluorescence at a concentration 0.1-0.5% than in the control (Figure 2). This indicated that the extract induced ROS production (presumably $O_2\bullet^-$), but the exact mechanism is not known. There were reports that ROS can be generated as by-products of biotransformation reactions of toxins (Lee and Farrell 2001; Pflugmacher 2004; Lushchak 2012). Superoxide anion radicals can be produced from mitochondria when respiration is suppressed, NADPH oxidase on plasma membranes, peroxisome, and endoplasmic reticulum and serves as the precursor of other harmful ROS (Sharma et al. 2012).

H_2O_2 content in treated roots was not drastically different from the control at 24 and 72 h after treatment but the trends of the H_2O_2 content were increased at both times of exposure (Figure 3). H_2O_2 is one of ROS molecule which can be converted from $O_2\bullet^-$ and these two molecules were used to indicate oxidative stress level in many researches (Bai et al. 2009; Lara-nuñez et al. 2006). The results showed that the treated roots induced the higher production of $O_2\bullet^-$ and H_2O_2 than in the control confirming the potential of *Nostoc* sp. extract in oxidative stress induction. However, ROS also correlated with antioxidative enzyme activities. The results of antioxidative enzyme activities except for PODs showed the increase at 24 and 72 h indicating the occurrence of oxidative stress. The cells have to induce the antioxidative enzyme activities to balance ROS level resulting in slight difference of H_2O_2 content between treatment and control. This was supported by the study of Singh et al. (2007) who found that arsenic induced oxidative stress in mung bean (*Phaseolus aureus* Roxb.) which was corresponded to the increase of antioxidative enzyme activities, but not to H_2O_2 accumulation.

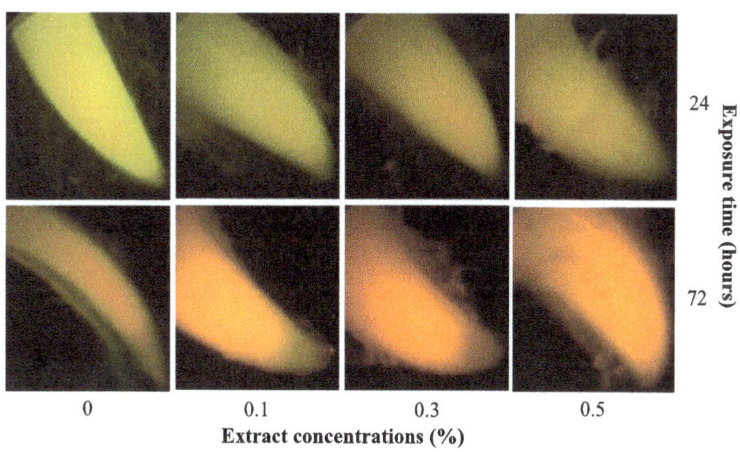

Figure 2 ROS production in *M. pigra* roots after exposure to the extract at different concentrations for 24 and 72 h (upper and lower rows). Red fluorescence shows ROS production (presumably $O_2\bullet^-$).

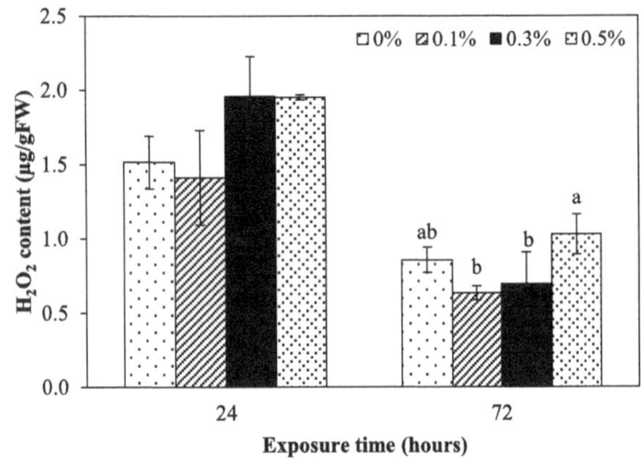

Figure 3 Hydrogen peroxide content in *M. pigra* roots after exposure to the extract at different concentrations for 24 and 72 h. Data are mean values of three replications and SD are indicated by vertical bars. The different letters above the column indicate significant differences ($p \leq 0.05$) by DMRT.

Cell death occurred after roots were exposed to the extract for 24 and 72 h. This was indicated by relative Evans blue uptake (Figure 4). At 24 h, the uptake of Evans blue was not different from the control with roots exposed to the extract at a concentration of 0.1%, but the uptake gradually increased at 0.3 to 0.5%. Evans blue uptake of treatment with 0.3 and 0.5% dramatically increased about 7 and 11 times more than that of the control at 72 h, respectively (Figure 4). Evans blue is a non-permeating stain which can pass through a damaged plasma membrane of a dead cell and is generally used to indicate the degradation of the plasma membrane and cell death (Peterson et al. 2008). Our results indicated that the extract caused the loss of membrane integrity and strongly induced cell death at 72 h. This result was supported by Volk and Mundt (2007) who studied the cytotoxicity of exometabolites from *Nostoc insulare* culture, resulting in human amniotic epithelial cell death. There are some reports about the effect of cyanobacterial toxins on plant cell death. For example, microcystin-RR from *Microcystis aeruginosa* decreased cell viability of tobacco (*Nicotiana tabacum* L. cv. BY-2) cell line after treatment for 144 h (Yin et al. 2005). Sanevas et al. (2007) reported that crude extract from *Hapalosiphon* sp. caused cell death in onion (*Allium cepa* L. cv. Raputa II) and wheat (*Triticum aestivum* L. cv. Norin 61) roots. In general an increase in the number of cell death when the concentration and time were increased correlated with activation of ROS production, suggesting ROS caused oxidative stress and cell death despite increased antioxidative enzyme activities. This may be due to an imbalance that the production of ROS is over antioxidant defenses.

Effect of *Nostoc* sp. extract on lipid peroxidation was investigated by determining malondialdehyde (MDA) content, which is a by-product of the peroxidation of membrane lipids. MDA content at 24 h significantly increased with higher concentrations. There was no significant difference between MDA content at various concentrations and the control at 72 h after exposure (Figure 5). The results showed that elevated lipid peroxidation, indicator of oxidative damage, is an important feature of the toxicity of the extract to *M. pigra* roots. These results agree with other cyanobacterial extract that have involved MDA content and lipid peroxidation. It has been reported that microcystins from cyanobacterial bloom material induced oxidative damage, such as lipid peroxidation, in alfalfa (*Medicago sativa* L.) seedlings (Pflugmacher et al. 2006). Microcystin-LR can

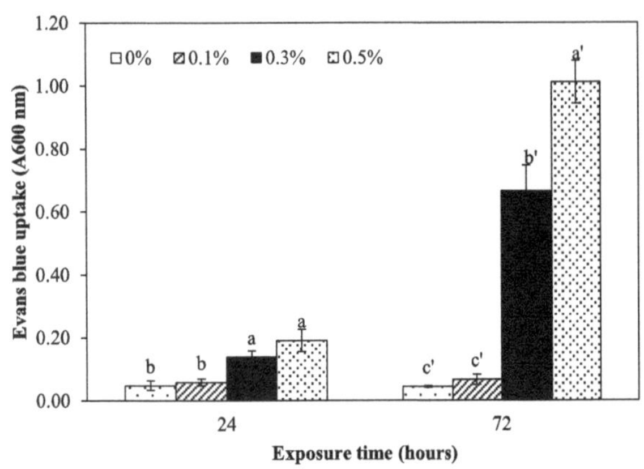

Figure 4 Uptake of Evans blue in *M. pigra* root tips after exposure to the extract at different concentrations for 24 and 72 h. Data are mean values of three replications and SD are indicated by vertical bars. The different letters above the column indicate significant differences ($p \leq 0.05$) by DMRT.

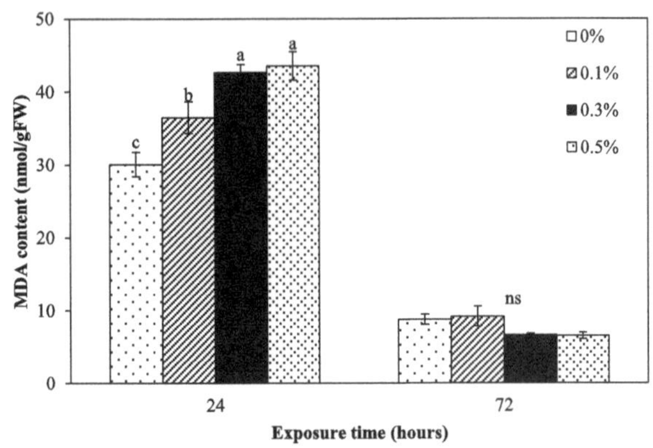

Figure 5 MDA content of *M. pigra* roots exposed to different concentrations of the extract for 24 and 72 h. Data are mean values of three replications and SD are indicated by vertical bars. The different letters above the column indicate significant differences ($p \leq 0.05$) at each time by DMRT. ns = no significant difference.

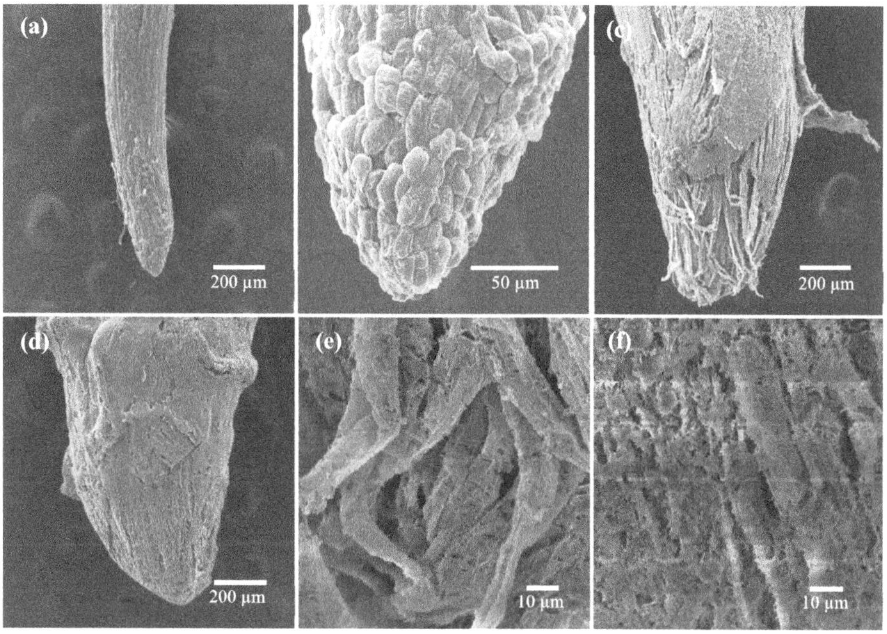

Figure 6 SEM micrographs of *M. pigra* root tips treated with an extract from *Nostoc* sp. for 72 h. (a) control (magnification = 80x) **(b)** healthy root cap from the control (500x) **(c, d)** roots grown in 0.5% of the extract (80x) and **(e, f)** root tip cell damage in the treatment with 0.5% of the extract (1,000x).

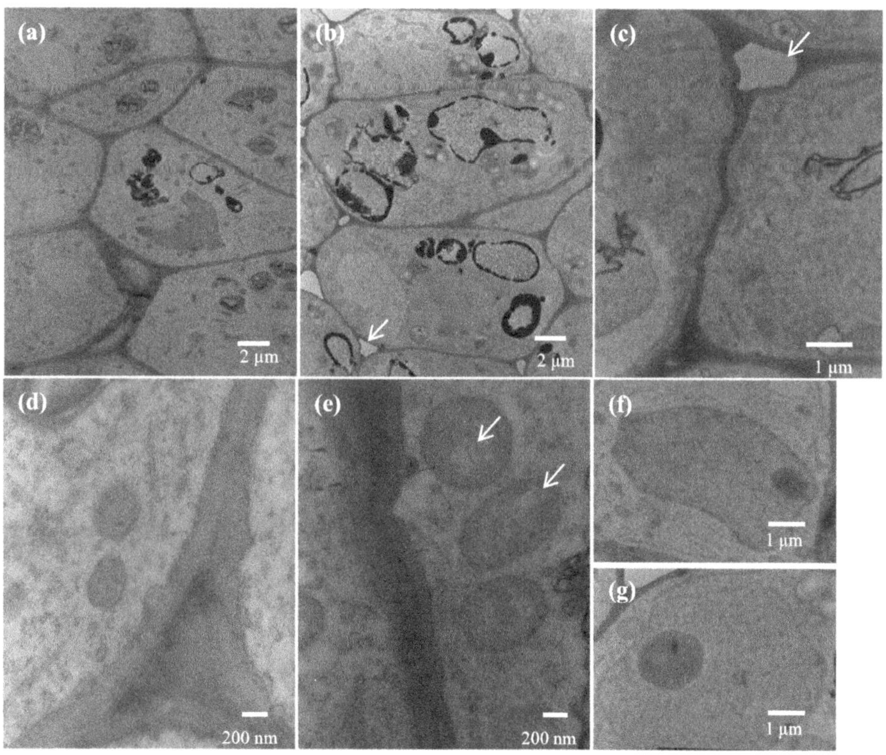

Figure 7 TEM micrographs of *M. pigra* root tips treated with an extract from *Nostoc* sp. for 72 h. (a) control (magnification = 2,500x) **(b)** treated root (2,500x) showing some substances stored vacuoles and the dissolution of the middle lamella (white arrow) **(c)** treated root (8,000x) showing the dissolution of the middle lamella (white arrow) **(d)** mitochondria of the control (20,000x) **(e)** swollen mitochondria in treated root (20,000x) showing indistinct cristae (white arrows) **(f, g)** normal nucleus of control and treated root (5,000x).

increase MDA content in *Vallisneria natans* (Lour.) Hara (Jiang et al. 2011). Koodkaew et al. (2012a, b) showed that ambiguine D isonitrile and hapalocyclamide from the cyanobacterium, *Hapalosiphon* sp., increased MDA content and caused cell death in lettuce (*Lactuca sativa* L.) seedlings. Normally, MDA levels can be correlated with ROS levels (Yin et al. 2005; Koodkaew et al. 2012a, b). The extract might induce ROS production in the cells and cause oxidative damage such as peroxidation of lipids, causing initial cell death at 24 h. The lack of a difference of MDA content at 72 h might be the result of extremely damaged cell membrane leading to metabolite leakage under longer of exposure because it correlated with the severe cell death at 72 h after treatment (Figure 4).

Changes of root ultrastructure after *Nostoc* sp. extract exposure were investigated by using electron microscope. After 72 h of exposure, SEM micrographs showed completely changed root tip cells/structures between the control and the treated roots. The control roots were narrower and more elongated than those of the treated roots (Figure 6). The root tip cells and epidermal cells of the control roots were intact while the epidermis peeled off and the root cap destructed in the treated roots (Figure 6a-b, c-d). These ultrastructural changes confirmed that cell damage was caused by the extract. It has been reported that some allelochemicals such as two sesquiterpene-derivatives from *Ageratina adenophora* (Asteraceae) caused irregular shapes of root tip cells with a greater number of separated cells and the collapse of the root cap cells (Yang et al. 2011). Burgos et al. (2004) reported that allelochemical from rye (*Secale cereal* L.) reduced the differentiation of root cap cells of cucumber (*Cucumis sativus* L.) seedlings. The extract had severe effects on root cap and epidermal cells. It might be caused by the susceptibility of root cap and epidermal cells to the environment. The extract could easily penetrate the cell walls and enter the cells.

Disaggregation and destruction of root tip cells (Figure 6e-f) indicate that the extract affected the cell structure by disintegrating the middle lamella as shown in a TEM micrograph (Figure 7b-c). Beers (1997) reported that the dissolution of middle lamella and cell wall occur when plant cells undergo cell death. Roots response to environmental stress by the increase of lignin production resulting cell wall lignification and stunted roots (Siegel 1993). Disaggregation of root tip cells might be caused by the continued expansion of the stele and a rigidification of the roots. TEM micrographs also showed swollen mitochondria and indistinct cristae in treated roots (Figure 7e). Mitochondria are one of the major sources of ROS in roots (Mano 2002; Gill and Tuteja 2010) which can cause oxidative damage in cells. Jones (2000) reported that swollen organelles and broken plasma membranes are stages of cell death when metabolic homeostasis fails and mitochondria contribute to apoptotic cell death. Continuous damage of mitochondria becomes irreversible and ensures cell death (Van Loo et al. 2002). Diaz-Tielas et al. (2012) showed that alteration of mitochondrial membranes, swollen mitochondria, and the induction of irreversible cell death were due to the phytotoxicities of chalcone on *Arabidopsis thaliana* (L.) Heynh. seedlings.

Conclusions

Based on our results, we conclude that a *Nostoc* sp. extract enhanced oxidative stress by ROS production resulting in lipid peroxidation of cell membranes and cell death despite the activation of antioxidative enzymes. This was confirmed by destruction of root tip cells and aberrant mitochondria. These results are consistent with our previous study that the extract inhibited root growth of *M. pigra* by inducing oxidative stress and cell death. To clarify the mechanism of action of *Nostoc* sp. extract provides benefit information for application to use as natural herbicide and control *M. pigra* – invasive weed in some parts of the world. The extract might be developed to use as bioherbicide for weed control. The effects of this extract on other weed species, field experiment, and purification of bioactive compounds need to be investigated.

Abbreviations

APX: Ascorbic acid peroxidase; CAT: Catalase; DHE: Dihydroethidium; GR: Glutathione reductase; H_2O_2: Hydrogen peroxide; $HO^{\bullet}$: Hydroxyl radical; MDA: Malondialdehyde; NADPH: Nicotinamide adenine dinucleotide phosphate; $O_2{\bullet}^-$: Superoxide anion radical; PODs: Peroxidases; ROS: Reactive oxygen species; SEM: Scanning electron microscope; TEM: Transmission electron microscope.

Competing interests

The authors declare that they have no competing interests.

Authors' contributions

SSRP and SNT are responsible for cyanobacteria culture. SSS and SNT supplied all laboratory and chemical substances used in this study. SSRP, SNT, and SSS designed the experiment. SSRP determined, analysed the data, and also drafted the manuscript. All authors read and approved the final manuscript.

Acknowledgements

We are thankful to the Thai Science Scholar Project, Bilateral Research Cooperation (BRC) from Faculty of Sciences, and Kasetsart University Research and Development Institute (KURDI) for financial support. We also thank Scientific Equipment and Research Division, Kasetsart University for using their electron microscope.

References

Aebi H (1983) Catalase. In: Bergmeyer HU (ed) Methods in enzymatic analysis, vol 3. Academic Press, New York, pp 276–286

Bai R, Ma F, Liang D, Zhao X (2009) Phthalic acid induces oxidative stress and alters the activity of some antioxidant enzymes in roots of *Malus prunifolia*. J Chem Ecol 35:488–494

Beers EP (1997) Programmed cell death during plant growth and development. Cell Death Differ 4:649–661

Berry JP, Gantar M, Perez MH, Berry G, Noriega FG (2008) Cyanobacterial toxins as allelochemicals with potential applications as algaecides, herbicides and insecticides. Mar Drugs 6:117–146

Blom JF, Brutsch T, Barbaras D, Bethuel Y, Locher HH, Hubschwerlen C, Gademann K (2006) Potent algicides based on the cyanobacterial alkaloid nostocarboline. Org Lett 8(4):737–740

Burgos NR, Talbert RE, Kim KS, Kuk YI (2004) Growth inhibition and root ultrastructure of cucumber seedlings exposed to allelochemicals from rye (*Secale cereal*). J Chem Ecol 30(3):671–689

Cho U, Seo N (2005) Oxidative stress in *Arabidopsis thaliana* exposed to cadmium is due to hydrogen peroxide accumulation. Plant Sci 168:113–120

Dat J, Vandenabeele S, Vranová E, Montagu MV, Inzé D, Breusegem FV (2000) Daul action of the active oxygen species during plant stress responses. Cell Mol Life Sci 57:779–795

Díaz-Tielas C, Graña E, Sotelo T, Reigosa MJ, Sánchez-Moreiras AM (2012) The natural compound trans-chalcone induces programmed cell death in *Arabidopsis thaliana* roots. Plant Cell Environ 35:1500–1517

Duke SO, Rimando AM, Schrader KK, Aliotta G, Oliva A, Romagni JG (2002) Chemicals from nature for weed management. Weed Sci 50:138–151

Etchegaray A, Rabello E, Dieckmann R, Moon DH, Fiore MF, Döhren H, Tsai SM, Neilan BA (2004) Algicide production by the filamentous cyanobacterium *Fischerella* sp. CENA 19. J Appl Phycol 16:237–243

Gill SS, Tuteja N (2010) Reactive oxygen species and antioxidant machinery in abiotic stress tolerance in crop plants. Plant Physiol Biochem 48:909–930

Gomes A, Fernandes E, Lima JLFC (2005) Fluorescence probes used for detection of reactive oxygen species. J Biochem Biophys Methods 65:45–80

Gutteridge JMC (1995) Lipid peroxidation and antioxidants as biomarkers of tissue damage. Clin Chem 41(12):1819–1828

Halliwell B, Chirico S (1993) Lipid peroxidation: its mechanism, measurement, and significance. Am J Clin Nutr 57:715–725

Halliwell B, Foyer CH (1978) Properties and physiological function of a glutathione reductase purified from spinach leaves by affinity chromatography. Planta 139:9–17

Hirata K, Yoshitomi S, Dwi S, Iwabe O, Mahakhant A, Plochai J, Miyamoto K (2003) Bioactivities of Nostocine a produced by a freshwater cyanobacterium *Nostoc spongiaeforme* TISTR 8169. J Biosci Bioeng 95(5):512–517

Jiang W, Liu D (2010) Pb-induced cellular defense system in the root meristematic cells of *Allium sativum* L. BMC Plant Biol 10:40

Jiang J, Gu X, Song R, Wang X, Yang L (2011) Microcystin-LR induced oxidative stress and ultrastructural alterations in mesophyll cells of submerged macrophyte *Vallisneria natans* (Lour.) Hara. J Hazard Mater 190:188–196

Jones A (2000) Does the plant mitochondrion integrate cellular stress and regulate programmed cell death? Trends Plant Sci 5(5):225–230

Junglee S, Urban L, Sallanon H, Lopez-Lauri F (2014) Optimized assay for hydrogen peroxide determination in plant tissue using potassium iodide. Am J Anal Chem 5:730–736

Kaur S, Singh HP, Batish DR, Kohli RK (2012) *Artemisia scoparia* essential oil inhibited root growth involves reactive oxygen species (ROS)-mediated disruption of oxidative metabolism: *in vivo* ROS detection and alterations in antioxidant enzymes. Biochem Syst Ecol 44:390–399

Kobayashi A, Kajiyama S (1998) Secondary metabolites of cyanobacteria *Nostoc* sp. Chin J Oceanol Limnol 16:109–117

Koodkaew I, Sunohara Y, Matsuyama S, Matsumoto H (2012a) Isolation of ambiguine D isonitrile from *Hapalosiphon* sp. and characterization of its phytotoxic activity. Plant Growth Regul 68:141–150

Koodkaew I, Sunohara Y, Matsuyama S, Matsumoto H (2012b) Phytotoxic action mechanism of hapalocyclamide in lettuce seedlings. Plant Physiol Biochem 58:23–28

Kreitlow S, Mundt S, Lindequist U (1999) Cyanobacteria-a potential source of new biologically active substances. J Biotechnol 70(1–3):61–63

Kulik MM (1995) The potential for using cyanobacteria (blue-green algae) and algae in the biological control of plant pathogenic bacteria and fungi. Eur J Plant Pathol 101(6):585–599

Lara-Nuñez A, Romero-Romero T, Ventura JL, Blancas V, Anaya AL, Cruz-Ortega R (2006) Allelochemical stress causes inhibition of growth and oxidative damage in *Lycopersicon esculentum* Mill. Plant Cell Environ 29:2009–2016

Lee AU, Farrell GC (2001) Mechanism of azathioprine-induced injury to hepatocytes: roles of glutathione depletion and mitochondrial injury. J Hepatol 35:756–764

Leflaive J, Ten-Hage L (2007) Algal and cyanobacterial secondary metabolites in freshwaters: a comparison of allelopathic compounds and toxins. Freshwater Biol 52:199–214

Lushchak VI (2012) Glutathione homeostasis and functions: potential targets for medical interventions. J Amino Acids 2012:1–26

Mano J (2002) Early events in environmental stresses in plants–induction mechanisms of oxidative stress. In: Inzé D, Montagu MV (eds) Oxidative stress in plant. Taylor & Francis, London

Mittler R (2002) Oxidative stress, antioxidant and stress tolerance. Plant Sci 7(9):405–410

Nakano Y, Asada K (1987) Purification of ascorbate peroxidase in spinach chloroplasts; its inactivation in ascorbate-depleted medium and reactivation by monodehydroascorbate radical. Plant Cell Physiol 28:131–140

Peterson RL, Peterson CA, Melville LH (2008) Technique plant anatomy through creative laboratory exercises. National Research council, Canada

Pflugmacher S (2004) Promotion of oxidative stress in the aquatic macrophyte *Ceratophyllum demersum* during biotransformation of the cyanobacterial toxin microcystin-LR. Aquat Toxicol 70:169–178

Pflugmacher S, Jung K, Lundvall L, Neumann S, Peuthert A (2006) Effects of cyanobacterial toxins and cyanobacterial cell-free crude extract on germination of alfalfa (*Medicago sativa*) and induction of oxidative stress. Environ Toxicol Chem 25(9):2381–2387

Putter J (1974) Peroxidases. In: Bergmeyer HU (ed) Methods of enzymatic analysis, vol 2. Academic Press, New York, pp 685–690

Rippka R, Deruelles J, Waterbury JB, Herdman M, Stanier RY (1979) Genetic assignments, strain histories and properties of pure cultures of cyanobacteria. J Gen Microbiol 111:1–61

Sanevas N, Sunohara Y, Matsumoto H (2007) Characterization of reactive oxygen species-involved oxidative damage in *Hapalosiphon* species crude extract-treated wheat and onion roots. Weed Biol Manag 7:172–177

Sharma P, Jha AB, Dubey RS, Pessarakli M (2012) Reactive oxygen species, oxidative damage, and antioxidative defense mechanism in plants under stressful conditions. J Bot 2012. doi:10.1155/2012/217037

Siegel BZ (1993) Plant peroxidases-an organismic perspective. Plant Growth Regul 12:303–312

Singh HP, Batish DR, Kohli RK, Arora K (2007) Arsenic-induced root growth inhibition in mung bean (*Phaseolus aureus* Roxb.) is due to oxidative stress resulting from enhanced lipid peroxidation. Plant Growth Regul 53:65–73

Sukkhaeng S, Sanevas N, Suwanwong S (2014) Inhibition of seedling growth in giant mimosa and reduction of mitotic activity in onion root tips caused by cyanobacterial extract. Chiang Mai J Sci 41(5.1):1150–1156

Sunohara Y, Matsumoto H (2008) Quinclorac-induced cell death is accompanied by generation of reactive oxygen species in maize root tissue. Phytochemistry 69:2312–2319

Van Loo G, Saelens X, Van Gurp M, Mac Farlane M, Martin SJ, Vandenabeele P (2002) The role of mitochondrial factors in apoptosis: a Russian roulette with more than one bullet. Cell Death Differ 9:1031–1042

Velikova V, Yordanov I, Edreva A (2000) Oxidative stress and some antioxidant systems in acid rain-treated bean plants protective role of exogenous polyamines. Plant Sci 151:59–66

Volk R, Mundt S (2007) Cytotoxic and non-cytotoxic exometabolites of the cyanobacterium *Nostoc insulare*. J Appl Phycol 19:55–62

Yamamoto Y, Kobayashi Y, Matsumoto H (2001) Lipid peroxidation is an early symptom triggered by aluminum, but not the primary cause of elongation inhibition in pea roots. Plant Physiol 125:199–208

Yang G, Wan F, Guo J, Liu W (2011) Cellular and ultrastructural changes in the seedling roots of upland rice (*Oryza sativa*) under the stress of two allelochemicals from *Ageratina adenophora*. Weed Biol Manag 11:152–159

Yin L, Huang J, Huang W, Li D, Wang G, Liu Y (2005) Microcystin-RR-induced accumulation of reactive oxygen species and alteration of antioxidant systems in tobacco BY-2 cells. Toxicon 46:507–512

Permissions

All chapters in this book were first published in BS, by Springer; hereby published with permission under the Creative Commons Attribution License or equivalent. Every chapter published in this book has been scrutinized by our experts. Their significance has been extensively debated. The topics covered herein carry significant findings which will fuel the growth of the discipline. They may even be implemented as practical applications or may be referred to as a beginning point for another development.

The contributors of this book come from diverse backgrounds, making this book a truly international effort. This book will bring forth new frontiers with its revolutionizing research information and detailed analysis of the nascent developments around the world.

We would like to thank all the contributing authors for lending their expertise to make the book truly unique. They have played a crucial role in the development of this book. Without their invaluable contributions this book wouldn't have been possible. They have made vital efforts to compile up to date information on the varied aspects of this subject to make this book a valuable addition to the collection of many professionals and students.

This book was conceptualized with the vision of imparting up-to-date information and advanced data in this field. To ensure the same, a matchless editorial board was set up. Every individual on the board went through rigorous rounds of assessment to prove their worth. After which they invested a large part of their time researching and compiling the most relevant data for our readers.

The editorial board has been involved in producing this book since its inception. They have spent rigorous hours researching and exploring the diverse topics which have resulted in the successful publishing of this book. They have passed on their knowledge of decades through this book. To expedite this challenging task, the publisher supported the team at every step. A small team of assistant editors was also appointed to further simplify the editing procedure and attain best results for the readers.

Apart from the editorial board, the designing team has also invested a significant amount of their time in understanding the subject and creating the most relevant covers. They scrutinized every image to scout for the most suitable representation of the subject and create an appropriate cover for the book.

The publishing team has been an ardent support to the editorial, designing and production team. Their endless efforts to recruit the best for this project, has resulted in the accomplishment of this book. They are a veteran in the field of academics and their pool of knowledge is as vast as their experience in printing. Their expertise and guidance has proved useful at every step. Their uncompromising quality standards have made this book an exceptional effort. Their encouragement from time to time has been an inspiration for everyone.

The publisher and the editorial board hope that this book will prove to be a valuable piece of knowledge for researchers, students, practitioners and scholars across the globe.

List of Contributors

Pablo Bolaños-Villegas
Fabio Baudrit Agricultural Experimental Station, University of Costa Rica, La Garita de Alajuela, P.O. Box 183-4050, Alajuela, Costa Rica

Cian-Ling Guo and Guang-Yuh Jauh
Institute of Plant and Microbial Biology, Academia Sinica, 128 Sec. 2, Academia Rd, Nankang,
Taipei 11529, Taiwan

José Roberto Medina-Medrano, Norma Almaraz-Abarca, M. Socorro González-Elizondo, José Natividad Uribe-Soto, Laura Silvia González-Valdez and Yolanda Herrera-Arrieta
Centro Interdisciplinario de Investigación para el Desarrollo Integral Regional, Unidad Durango, Instituto Politécnico Nacional, Sigma 119, Fraccionamiento 20 de Noviembre II, 34220 Durango, Dgo., Mexico

Aylin Güney, Danielle Kerr, Ayça Sökücü, Reiner Zimmermann and Manfred Küppers
University of Hohenheim, Institute of Botany, Garbenstrasse 30, 70599 Stuttgart, Germany

Rosario Rivera Rubite, Llogene B De Layola and Arthur Gregory N Virata
Department of Biology, College of Arts and Sciences, University of the Philippines Manila, Padre Faura, Manila 1000, Philippines

Mark Hughes
Royal Botanic Garden Edinburgh, 20a Inverleith Row, Edinburgh EH3 5LR, UK

Patrick Blanc
CNRS, 3 rue Michel-Ange, 75794 Paris, France

Kuo-Fang Chung
School of Forestry and Resource Conservation, National Taiwan University, Taipei 106, Taiwan

Hsun-An Yang, Yoshiko Kono and Ching-I Peng
Biodiversity Research Center, Academia Sinica, Taipei 115, Taiwan

Grecebio J D Alejandro
College of Science and Research Centre for the Natural and Applied Sciences, University of Santo Tomas, España, Manila 1015, Philippines

Hsin-Hung Lin and Hsiao-Feng Lo
Department of Horticulture and Landscape Architecture, National Taiwan University, Taipei, Taiwan

Kuan-Hung Lin and Su-Ching Chen
Department of Horticulture and Biotechnology, Chinese Culture University, Taipei, Taiwan

Yu-Hsing Shen
Institute of plant and microbial biology, Academia Sinica, Taipei, Taiwan

Yuh-Hwa Liu
Division of Gastroenterology, Shin Kong Wu Ho-Su Memorial Hospital, Taipei, Taiwan

Shyr-Yi Lin and Yuh-Hwa Liu
Department of General Medicine, Taipei Medical University, Taipei, Taiwan

Yin-Shiou Lin, Kuan-Ling Lin, Chao-Hsiang Chen and Wen-Chi Hou
Graduate Institute of Pharmacognosy, Taipei Medical University, Taipei, Taiwan

Yeh-Lin Lu
Department of Pharmacy, Taipei Medical University, Taipei, Taiwan

Chao-Hsiang Chen and Mei-Yin Chien
Ko Da Pharmaceutical Co. Ltd, Tao-Yuan, Taiwan

Huey-Fang Shang
Department of Microbiology and Immunology, Taipei Medical University, Taipei, Taiwan

Haiying Cui, Xuejing Zhang, Hui Zhou, Chengting Zhao and Lin Lin
School of Food & Biological Engineering, Jiangsu University, Zhenjiang 212013, P. R. China

Jie-Cai Zhao, Jun Luo, Chun-Ping Yang and Guo-Xing Cao
Department of Forestry, Sichuan Agricultural University, Yaan 625014, China

Guo-Xing Cao
Xishuangbanna Tropical Botanical Garden, Chinese Academy of Sciences, Mengla 666303, China

Yu Pan and Heng-Yu Huang
College of Traditional Chinese Medicine, Yunnan University of Traditional

Chinese Medicine, Kunming, China

Yu Pan, Ji Zhang, Yan-Li Zhao and Yuan-Zhong Wang
Institute of Medicinal Plants, Yunnan Academy of Agricultural Sciences, Kunming, China

Jayanta Kumar Patra, Gitishree Das and Kwang-Hyun Baek
School of Biotechnology, Yeungnam University, Gyeongsan, Gyeongbuk 712-749, Republic of Korea

Xue Li, Wei Lu, Guyue Hu, Xiao Chan Wang, Yu Zhang, Guo Xiang Sun and Zhichao Fang
College of Engineering, Nanjing Agricultural University, Nanjing, Jiangsu, People's Republic of China

Xiao Chan Wang, Yu Zhang and Guo Xiang Sun
Jiangsu Province Engineering Lab for Modern Intelligent Facilities of Agriculture Technology & Equipment, Nanjing, Jiangsu, People's Republic of China

Yun-Ting Tseng and Kuo-Chih Lin
Department of Life Science, National Dong Hwa University, Hualien County 974, Taiwan

Antony van der Ent
Centre for Mined Land Rehabilitation, Sustainable Minerals Institute, The University of Queensland, St. Lucia, QLD 4072, Australia
Laboratoire Sols et Environnement, Université de Lorraine-INRA, UMR 1120, 54000 Nancy, France

K. M. Wong
Singapore Botanic Gardens, 1 Cluny Road, Singapore 259569, Singapore

Yun-Hong Tan
Center for Integrative Conservation, Xishuangbanna Tropical Botanical Garden, Chinese Academy of Sciences, Menglun, Mengla 666303, Yunnan, China

Yan-Nan Liu and Sheng-Xiang Yu
State Key Laboratory of Systematic and Evolutionary Botany, Institute of Botany, Chinese Academy of Sciences, Beijing 100093, People's Republic of China

Hong Jiang
Yunnan Academy of Forestry/Yunnan Laboratory for Conservation of Rare, Endangered and Endemic Forest Plants, State Forestry Administration, Kunming 650201, China

Xin-Xin Zhu
Shanghai Chenshan Plant Science Research Center, Chinese Academy of Sciences, Shanghai Chenshan Botanical Garden, Shanghai 201602, China

Wei Zhang
Key Laboratory of Economic Plants and Biotechnology, Kunming Institute of Botany, Chinese Academy of Sciences, Kunming 650201, Yunnan, China

Ya-Lun Huang and Wen-Yuan Kao
Institute of Ecology and Evolutionary Biology, National Taiwan University, 1, Roosevelt Rd., Sec. 4, Taipei 106, Taiwan

Wen-Yuan Kao
Department of Life Science, National Taiwan University, 1, Roosevelt Rd., Sec. 4, Taipei 106, Taiwan

Chih-Wei Tsai and Chih-Yu Chiu
Biodiversity Research Center, Academia Sinica, Taipei 11529, Taiwan

Guanglong Tian
Environmental Monitoring and Research Division, Monitoring and Research Department, Metropolitan Water Reclamation District of Greater Chicago (MWRD), Lue-Hing R&D Laboratory, 6001 W. Pershing Road, Cicero, IL 60804, USA

Ai Kobayashi and Hisashi Kato-Noguchi
Department of Applied Biological Science, Faculty of Agriculture, Kagawa University, Miki, Kagawa 761-0795, Japan

Chyi-Rong Chiou
School of Forestry and Resource Conservation, National Taiwan University, No. 1, Sec. 4, Roosevelt Rd., Taipei 10617, Taiwan (R.O.C.)

Tung-Yu Hsieh
Shanghai Institutes for Biological Sciences, Chinese Academy of Sciences, 320 Yue Yang Rd.,
Shanghai 200031, China
Shanghai Chenshan Plant Science Research Center, Chinese Academy of Sciences, 3888 Chenhua Road, Songjiang, Shanghai 201602, China
Shanghai Key Laboratory of Plant Functional Genomics and Resources, Shanghai Chenshan Botanical Garden, 3888 Chenhua Road, Songjiang, Shanghai 201602, China

Chang-Chi Chien
College of Business, Chung Yuan Christian University, 200, Chung Pei Rd., Chung Li 32023, Taiwan (R.O.C.)

Hsuan Chen, Mei-Chu Chung, Yuan-Ching Tsai, Fu-Jin Wei and Yue-Ie C. Hsing
Institute of Plant and Microbial Biology, Academia Sinica, Taipei 115, Taiwan

Jaw-Shu Hsieh and Hsuan Chen
Department of Agronomy, National Taiwan University, Taipei 106, Taiwan

Chao-Lin Kuo
Department of Chinese Pharmaceutical Sciences and Chinese Medicine Resources, China Medical University, Taichung, Taiwan

Dinesh-Chandra Agrawal, Hung-Chi Chang, Ya-Ting Chiu, Chu-Peng Huang, Shih-Hung Huang and Hsin-Sheng Tsay
Department of Applied Chemistry, Chaoyang University of Technology, Taichung, Taiwan

Hung-Chi Chang
Department of Golden-Ager Industry Management, Chaoyang University of Technology, Taichung, Taiwan

Yi-Lin Chen
Department of Biotechnology and Animal Science, National Ilan University, Yilan City, Taiwan

Hsin-Sheng Tsay
Department of Agronomy, National Chung Hsing University, Taichung, Taiwan

Kenta Watanabe
Okinawa National College of Technology, 905 Henoko, Nago, Okinawa 905-2192, Japan

T Y Aleck Yang and Tai-Liang Huang
Department of Botany, National Museum of Natural Science, Kuan Chien Rd, Taichung 404, Taiwan

T Y Aleck Yang
Department of Life Science, National Chung Hsin University, Taichung 402, Taiwan

Chihiro Nishihara
774-1-302, Umusa, Nago, Okinawa 905-0006, Japan

Koh Nakamura
Botanic Garden, Field Science Center for Northern Biosphere, Hokkaido University, North 3, West 8, Chuo-ku, Sapporo 060-0003, Japan

Ching-I Peng
Biodiversity Research Center, Academia Sinica, Taipei, Nangang 115, Taiwan

Takashi Sugawara
Makino Herbarium, Graduate School of Science, Tokyo Metropolitan University, 1-1 Minami-Ohsawa, Hachioji, Tokyo 192-0397, Japan

Ching-I Peng, Hsun-An Yang and Yoshiko Kono
Herbarium (HAST), Biodiversity Researach Center, Academia Sinica, Taipei, Nangang 115, Taiwan

Che-Wei Lin
Division of Botanical Garden, Taiwan Forestry Research Institute, No. 53, Nan–Hai Road, Taipei 10066, Taiwan

Hieu Quang Nguyen
Center for Plant Conservation (CPC), Vietnam Union of Science and Technology Associations, No. 25/32, Lane 191, Lac Long Quan Road, Nghia DoCau Giay District, Hanoi, Vietnam

Hsin-Yi Liu and Ting-Ting Jong
Department of Chemistry, National Chung Hsing University, 250 Kuo-kuang Road, Taichung 402, Taiwan

Hsin-Yi Peng and Su-Tze Chou
Department of Food and Nutrition, Providence University, 200, Sec. 7, Taiwan Boulevard, Shalu Dist., Taichung 43301, Taiwan

Shih-Lan Hsu
Department of Education and Research, Taichung Veterans General Hospital, 1650, Sec. 4, Taiwan Boulevard, Taichung City 40705, Taiwan

Ching-I Peng, Yoshiko Kono and Wai-Chao Leong
Herbarium (HAST), Biodiversity Research Center, Academia Sinica, Nangang, Taipei 115, Taiwan

Che-Wei Lin
Division of Botanical Garden, Taiwan Forestry Research Institute, No. 53, Nan-Hai Road, Taipei 100, Taiwan

Rimi Repin
Sabah Parks, 88806 Kota Kinabalu, Sabah, Malaysia

Wai-Chao Leong and Kuo-Fang Chung
School of Forestry and Resource Conservation, National Taiwan University, Taipei 106, Taiwan

Yan-Ling Li, Zheng-Yao Chang, Ren-Jiu Shi, Zeng-Min Miao and Gang-Ping Hao
College of Life Sciences, Taishan Medical University, Tai'an 271016, China

Xiao-Ming Xin
College of Pharmacy, Taishan Medical University, Tai'an 271016, China

Jing Ding
College of Chemical Engineering, Taishan Medical University, Tai'an 271016, China

Eman M Fakhry and Dahlia M El Maghraby
Department of Botany and Microbiology, Faculty of Science, Alexandria University, Alexandria 21511, Egypt

Shen-Liang Shih
Yuan's General Hospital, Department of Breast Surgery and Cancer Center, Kaohsiung, Taiwan

Yin-Shiou Lin and Wen-Chi Hou
Graduate Institute of Pharmacognosy, Taipei Medical University, Taipei, Taiwan

Shyr-Yi Lin
Department of Primary Care Medicine, Taipei Medical University Hospital, Taipei, Taiwan

Department of General Medicine, School of Medicine, Taipei Medical University, Taipei, Taiwan

Wen-Chi Hou
Traditional Herbal Medicine Research Center, Taipei Medical University Hospital, Taipei, Taiwan
Program for the Clinical Drug Discovery from Botanical Herbs, Taipei Medical University, Taipei, Taiwan

Jie Li, Yanli Yi and Dageng Zhang
Key Laboratory of Preservation of Northeast Cultivated Land, Ministry of Agriculture, National Engineering Laboratory for Efficient Utilization of Soil and Fertilizer, College of Land and Enviroment, Shenyang Agricultural University, 120 dongling road, Shenyang 110866, Liaoning, China

Xilei Cheng
Liaoning Academy of Environmental Sciences, Shenyang, China

Muhammad Irfan
Bioscience and Biotechnology College, Shenyang Agricultural University, 120 dongling road,
Shenyang 110866, China

Luana Silva Braucks Calazans and Cassia Mônica Sakuragui
Instituto de Biologia, Departamento de Botânica, Universidade Federal do Rio de Janeiro, CCS, Av. Carlos Chagas Filho, 373 - Sala A1-088 - Bloco A, Ilha do Fundão, Rio de Janeiro, RJ CEP 21941-902, Brazil

Nerivaldo Gomes Antas
Fundação Luisa Sartori, Sartori Árvores Nativas e Reflorestamento, Fazenda Além do Horizonte
Rodovia RJ-126, km 28, s/n - Gaviões, Silva Jardim, Rio de Janeiro, RJ, Brazil

Siriphan Sukkhaeng, Nuttha Sanevas and Srisom Suwanwong
Department of Botany, Faculty of Science, Kasetsart University, Bangkok 10900, Thailand